HARCOURT BRACE JOVANOVICH COLLEGE OUTLINE SERIES

PRE-ALGEBRA

Alan Wise

Department of Mathematics, University of San Diego

Carol Wise

Books for Professionals
Harcourt Brace Jovanovich, Publishers
San Diego   New York   London

Requests for permission to make copies of any part of the work should be mailed to:
Permissions Department
Harcourt Brace Jovanovich, Publishers
8th Floor
Orlando, Florida 32887

Printed in the United States of America

Library of Congress Cataloging-in-Publication Data

Wise, Alan.
    Pre-algebra / Alan Wise. Carol Wise.
        p. cm. — (Books for professionals) (Harcourt Brace Jovanovich college outline series)
    Includes index.
    ISBN 0-15-601518-8 : $12.95 (est.)
    1. Mathematics.   I. Wise, Carol. 1940- . II. Title.
III. Series.   IV. Series: Harcourt Brace Jovanovich college outline series.
QA39.2.W57   1991
513 — dc20   91-12589
                                                       CIP

ISBN 0-15-601518-8

First edition

A B C D E

PREFACE

The purpose of this book is to present a complete course in pre-algebra in the clear, concise form of an outline. This outline provides an in-depth review of the principles of pre-algebra for independent study, and contains essential supplementary material for mastering those principles. Or, this outline can simply be used as a valuable, self-contained refresher course on the practical applications of pre-algebra.

Regular features at the end of each chapter are specially designed to supplement your textbook and course work in pre-algebra.

SOLVED PROBLEMS Each chapter of this outline contains a set of exercises and word problems and their step-by-step solutions. Undoubtedly the most valuable feature of this outline, these problems allow you to become proficient in the fundamental skills and applications of pre-algebra. Along with the sample midterm and final exams, they also give you ample exposure to the kinds of questions that you are likely to encounter on a typical exam. To make the most of these solved problems, try writing your own solutions first. Then compare your answers to the detailed solutions provided in the book. If you have trouble understanding a particular problem, or set of problems, use the example reference numbers to locate and review the appropriate instruction and parallel step-by-step examples that were developed earlier in the chapter.

SUPPLEMENTARY EXERCISES Each chapter of this outline concludes with a set of drill exercises and practical applications and their answers. The supplementary exercises are designed to help you master and retain all the newly discussed skills and concepts presented in the given chapter, and also contain example references.

Of course, there are other features of this outline that you will find very helpful, too. One is the format itself, which serves both as a clear guide to important ideas and as a convenient structure upon which to organize your knowledge. A second is the attention devoted to methodology and the practical applications of pre-algebra. Yet a third is the careful listing of learning objectives in each section denoted by the uppercase letters A, B, C, D, or E.

We wish to thank the people who made this text possible, including Susan McColl for painstakingly checking the solutions and answers to each exercise and word problem.

ALAN WISE

San Diego, California

CAROL WISE

CONTENTS

1 WHOLE NUMBERS

THIS CHAPTER IS ABOUT

☑ Using Place Value
☑ Adding Whole Numbers
☑ Subtracting Whole Numbers
☑ Multiplying Whole Numbers
☑ Dividing Whole Numbers
☑ Factoring Whole Numbers
☑ Solving Word Problems Involving Whole Numbers

1-1. Using Place Value

A. Identifying digits in whole numbers

The numbers you use to count things are called the **counting numbers** or **natural numbers.**

EXAMPLE 1-1 List all the counting numbers.

Solution The counting numbers can be listed as: 1, 2, 3, 4, 5, 6, 7, 8, 9, 10, 11, 12, . . .

or just: 1, 2, 3, . . .

Note: An **ellipsis** symbol (. . .) following the counting numbers indicates that the counting number *pattern* shown continues on forever.

When you add the number **zero** (0) to the list of counting numbers, you have the list of **whole numbers.** Every whole number is made up of one or more of ten **digits:** 0, 1, 2, 3, 4, 5, 6, 7, 8, and 9.

EXAMPLE 1-2 List all the whole numbers.

Solution The whole numbers can be listed as: 0, 1, 2, 3, 4, 5, 6, 7, 8, 9, 10, 11, 12, . . .

or just: 0, 1, 2, 3, . . .

EXAMPLE 1-3: List the digits that make up the whole number 36,580.

Solution The digits that make up 36,580 are 3, 6, 5, 8, and 0.

To help locate the correct digit given its **place-value name,** you can use a **place-value chart.**

hundred billions	ten billions	billions	hundred millions	ten millions	millions	hundred thousands	ten thousands	thousands	hundreds	tens	ones

← place-value names

EXAMPLE 1-4   Find the digit that is in the ten-thousands place in 750,408,391,062. Draw a line under this digit.

Solution   Use a place-value chart:

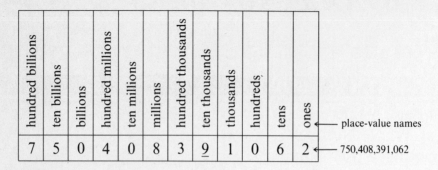

You can see that the 9 is in the ten-thousands place.

The commas in large numbers like 750,408,391,062 separate the number into **periods**—starting from the right, the periods are units, thousands, millions, billions, and so on.

Billions	Millions	Thousands	Units ← period names
7 5 0 ,	4 0 8 ,	3 9 1 ,	0 6 2

EXAMPLE 1-5   Write 750,408,391,062 using period names instead of commas.

Solution   750 billion 408 million 391 thousand 62

Note: All period names *except for units* must be written out. The whole number 62, for example, can be written as "62 units," or just "62."

To help identify the correct place-value name of a given digit, you can use a place-value chart.

EXAMPLE 1-6   Write the place-value name of the underlined digit in 246,137,958.

Solution

Billions			Millions			Thousands			Units ← period names		
hundred billions	ten billions	billions	hundred millions	ten millions	millions	hundred thousands	ten thousands	thousands	hundreds	tens	ones ← place-value names
			2	4	6	1	3	7	9	5	8

The place-value name of the underlined digit is *hundred millions*.

Once the place-value name of a given digit is identified, you can write the **value** of that digit.

EXAMPLE 1-7   What is the value of 5 in 2,035,819?

Solution   In 2,035,819, the value of 5 is 5 thousands, or 5000.

Note: In a four-digit number like 5000, you do not have to write the comma separating thousands and units. However, in all whole numbers with five digits or more, you must separate each period with a comma.

B. Writing whole numbers

To write a whole number when only **one-digit values** greater than zero are given, you must be careful to write a zero (0) each time a one-digit value is missing.

EXAMPLE 1-8   Write 6 thousands 8 tens 5 ones as a whole number.

Solution

$$
\begin{array}{ll}
\text{6 thousands} \quad \text{8 tens} \quad \text{5 ones} = & \text{6 thousands} \quad \overbrace{\text{no hundreds}}^{\text{missing value}} \quad \text{8 tens} \quad \text{5 ones} \\
= & \text{6 thousands} \quad \text{0 hundreds} \quad \text{8 tens} \quad \text{5 ones} \longleftarrow \text{one-digit values} \\
= & 6085 \longleftarrow \text{whole number}
\end{array}
$$

Note: "No hundreds" and "0 hundreds" mean the same thing.

The following **number-word names,** whole numbers, and one-digit values should be memorized.

Special Number-Word Names

one	two	three	four	five	six
1	2	3	4	5	6
1 one	2 ones	3 ones	4 ones	5 ones	6 ones
seven	eight	nine	ten	eleven	twelve
7	8	9	10	11	12
7 ones	8 ones	9 ones	1 ten	1 ten 1 one	1 ten 2 ones
thirteen	fourteen	fifteen	sixteen	seventeen	eighteen
13	14	15	16	17	18
1 ten 3 ones	1 ten 4 ones	1 ten 5 ones	1 ten 6 ones	1 ten 7 ones	1 ten 8 ones
nineteen	twenty	thirty	forty	· · ·	ninety
19	20	30	40	· · ·	90
1 ten 9 ones	2 tens	3 tens	4 tens	· · ·	9 tens
one hundred	two hundred	three hundred	four hundred	· · ·	nine hundred
100	200	300	400	· · ·	900
1 hundred	2 hundreds	3 hundreds	4 hundreds	· · ·	9 hundreds
one thousand	two thousand	three thousand	four thousand	· · ·	nine thousand
1000	2000	3000	4000	· · ·	9000
1 thousand	2 thousands	3 thousands	4 thousands	· · ·	9 thousands

To write a whole number when the number-word name is given, you may find it helpful to first write one-digit values.

EXAMPLE 1-9   Write eight thousand thirty as a whole number.

$$
\begin{array}{ll}
\textit{Solution} \quad \text{eight thousand thirty} = & \text{8 thousands} \quad \overbrace{\text{no hundreds}}^{\text{missing value}} \quad \text{3 tens} \quad \overbrace{\text{no ones}}^{\text{missing value}} \\
= & \text{8 thousands} \quad \text{0 hundreds} \quad \text{3 tens} \quad \text{0 ones} \longleftarrow \text{one-digit values} \\
= & 8030 \longleftarrow \text{whole number}
\end{array}
$$

Similarly, to write the number-word name when the whole number is given, you may find it helpful to first write one-digit values.

EXAMPLE 1-10   Write 6207 as a number-word name.

Solution   6207 = 6 thousands   2 hundreds   0 tens   7 ones ⟵—— one-digit values

= six thousand two hundred seven ⟵—— number-word name

Note: **A zero value** (like 0 tens) is never written as part of the number-word name.

When it takes two words to write a number-word name for whole numbers between 20 and 100, you must use a **hyphen** (-).

EXAMPLE 1-11   Write the number-word name for (**a**) 23, (**b**) 438, and (**c**) 5089.

Solution

(**a**) 23 = 2 tens   3 ones   (**b**) 438 = 4 hundreds   3 tens   8 ones

= twenty-three   = four hundred thirty-eight

zero value

(**c**) 5089 = 5 thousands   0 hundreds   8 tens   9 ones

= five thousand eighty-nine

Caution: Never use the word *and* when writing a whole-number word name.

EXAMPLE 1-12   Write the whole-number word name for 9007.

zero values

Solution   9007 = 9 thousands   0 hundreds   0 tens   7 ones

= nine thousand seven   (Do not write "nine thousand and seven.")

C. Comparing two whole numbers

To compare two whole numbers, you have to decide whether the two numbers are *equal* to each other or whether one number is *greater* or *less* than the other. You can use one of the following three symbols to show these comparisons:

= means *is equal to* (for example, 3 is equal to 3, or 3 = 3)

> means *is greater than* (for example, 3 is greater than 2, or 3 > 2)

< means *is less than* (for example, 2 is less than 3, or 2 < 3)

The symbol = is called an **equality symbol.** The symbols > and < are called **inequality symbols.** Another symbol, ≠ (*is not equal to*), is also an inequality symbol.

To compare two whole numbers, you may find it helpful to write one number under the other while aligning like place values.

EXAMPLE 1-13   Which is greater, 25,831,506 or 25,829,417?

Solution

Millions			Thousands			Units		
hundred millions	ten millions	millions	hundred thousands	ten thousands	thousands	hundreds	tens	ones
	2	5	8	3	1	5	0	6
	2	5	8	2	9	4	1	7

Write one number under the other while aligning like values.

same   └— different (3 ten thousands > 2 ten thousands)

25,831,506 is greater than 25,829,417—or, using the symbol, 25,831,506 > 25,829,417—because 3 ten thousands is greater than 2 ten thousands, or 3 ten thousands > 2 ten thousands.

D. Ordering whole numbers

To order two or more whole numbers, you list them from largest to smallest or from smallest to largest.

To order whole numbers from *largest to smallest*, you write the largest number first, then the next largest number, and so on until you have listed all the given numbers.

EXAMPLE 1-14   Order the following whole numbers from largest to smallest:

$$8516 \qquad 263 \qquad 10,519 \qquad 8520 \qquad 900$$

Solution

10,519 ←——— largest number

8 520

8 516

900

263 ←——— smallest number

To order whole numbers from *smallest to largest*, you just reverse the largest-to-smallest order.

EXAMPLE 1-15   Order the following whole numbers from smallest to largest:

$$7513 \qquad 2409 \qquad 2490 \qquad 5713 \qquad 7531 \qquad 2049$$

Solution

2049 ←——— smallest number

2409

2490

5713

7513

7531 ←——— largest number

E. Rounding whole numbers

When an exact answer is not needed for a problem, you may want to make the computation easier and **round** the given whole numbers before computing. The place to which you round the numbers will depend on how accurate your answer needs to be.

If the digit to the right of the digit to be rounded is 5 or more, then **round up** as follows:

(1) Increase the digit to be rounded by 1.

(2) Replace the digits to the right of the rounded digit with zeros.

EXAMPLE 1-16   Round 82,635 to the nearest thousand.

thousands

Solution $82,635 = 82,635$ Draw a line under the digit to be rounded.

$= 82,635$ Look at the next digit to the right. Is that digit 5 or more?
In this case, it is: 6 > 5

$\approx 83,000$ Round up: Increase 2 by 1 to get 3.
Replace 6, 3, and 5 with zeros.

Note: The symbol ≈ means *is approximately equal to* and is used whenever you round a number.

If the digit to the right of the digit to be rounded is less than 5, then **round down** as follows:

(1) Leave the digit to be rounded the same.

(2) Replace the digits to the right of the rounded digit with zeros.

EXAMPLE 1-17   Round 82,635 to the nearest ten thousand.

	ten thousands	

Solution   82,635 = 8̲2,635   Draw a line under the digit to be rounded.

= 8̲2,635   Look at the next digit to the right. Is that digit 5 or more? In this case, it is not: $2 < 5$

$\approx 80{,}000$ Round down: Leave 8 the same. Replace 2, 6, 3, and 5 with zeros.

1-2. Adding Whole Numbers

To join two or more amounts together, you **add.**

EXAMPLE 1-18   Write "☐☐ joined together with ☐☐☐" as addition.

Solution   "☐☐ joined together with ☐☐☐" equals $2 + 3$. ⟵ addition

The **addition problem** $2 + 3$ is read as "two **plus** three." The symbol " + " is called an **addition sign** or **addition symbol.**

A. Basic addition facts

The following example shows one way to solve an addition problem.

EXAMPLE 1-19   Add $2 + 3$.

Solution $2 + 3 = 5$ because ☐☐ joined together with ☐☐☐ equals ☐☐☐☐☐.

The **number sentence** $2 + 3 = 5$ is called a **basic addition fact.** To add whole numbers quickly and accurately, you must know the 100 basic addition facts from memory [see Table 1-1]. The basic addition fact $2 + 3 = 5$ is written in **horizontal form.** Basic addition facts can also be written in **vertical form.**

EXAMPLE 1-20   Write the number sentence $2 + 3 = 5$ in vertical form.

Solution

$$
\begin{array}{c} 2 \\ +3 \\ \hline 5 \end{array} \quad \text{or} \quad \begin{array}{c} 3 \\ +2 \\ \hline 5 \end{array}
$$

2 or 3 can be used as the top number.
The other number is used as the bottom number.
The answer (5) is written under the addition bar.

The numbers to be added in an addition problem are called the **addends,** and the answer to the addition problem is called the **sum.** So, in Example 1-20, 2 and 3 are the addends and 5 is the sum.

B. Adding, given horizontal form

If you're given whole numbers with two or more digits in horizontal form, you can add these numbers by first writing them in vertical form.

TABLE 1-1 BASIC ADDITION FACTS

0 +0 0	0 +1 1	0 +2 2	0 +3 3	0 +4 4	0 +5 5	0 +6 6	0 +7 7	0 +8 8	0 +9 9
1 +0 1	1 +1 2	1 +2 3	1 +3 4	1 +4 5	1 +5 6	1 +6 7	1 +7 8	1 +8 9	1 +9 10
2 +0 2	2 +1 3	2 +2 4	2 +3 5	2 +4 6	2 +5 7	2 +6 8	2 +7 9	2 +8 10	2 +9 11
3 +0 3	3 +1 4	3 +2 5	3 +3 6	3 +4 7	3 +5 8	3 +6 9	3 +7 10	3 +8 11	3 +9 12
4 +0 4	4 +1 5	4 +2 6	4 +3 7	4 +4 8	4 +5 9	4 +6 10	4 +7 11	4 +8 12	4 +9 13
5 +0 5	5 +1 6	5 +2 7	5 +3 8	5 +4 9	5 +5 10	5 +6 11	5 +7 12	5 +8 13	5 +9 14
6 +0 6	6 +1 7	6 +2 8	6 +3 9	6 +4 10	6 +5 11	6 +6 12	6 +7 13	6 +8 14	6 +9 15
7 +0 7	7 +1 8	7 +2 9	7 +3 10	7 +4 11	7 +5 12	7 +6 13	7 +7 14	7 +8 15	7 +9 16
8 +0 8	8 +1 9	8 +2 10	8 +3 11	8 +4 12	8 +5 13	8 +6 14	8 +7 15	8 +8 16	8 +9 17
9 +0 9	9 +1 10	9 +2 11	9 +3 12	9 +4 13	9 +5 14	9 +6 15	9 +7 16	9 +8 17	9 +9 18

EXAMPLE 1-21 Add 215 + 63.

Solution

(1) Write in vertical form. Line up like values.

(2) Add ones.
5 + 3 = 8 (ones)

(3) Add tens.
1 + 6 = 7 (tens)

(4) Add hundreds.
2 + 0 = 2 (hundreds)

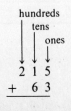

C. Adding with renaming

Values like 12 ones and 13 tens are called **two-digit values.** A two-digit value can always be renamed as two one-digit values.

EXAMPLE 1-22   Rename (**a**) 12 ones and (**b**) 13 tens as two one-digit values.
Solution

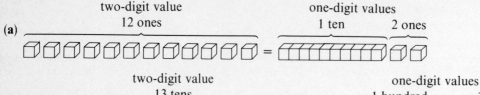

(a)

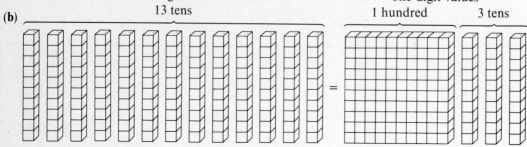

(b)

When adding, you must rename each two-digit value, unless it is the last step.

EXAMPLE 1-23   Add 689 + 743.

Solution

(1) Add ones.
9 + 3 = 12 (ones)
Rename ones:
12 ones = 1 ten   2 ones

(2) Add tens.
1 + 8 + 4 = 13 (tens)
Rename tens:
13 tens = 1 hundred   3 tens

(3) Add hundreds.
1 + 6 + 7 = 14 (hundreds)
Last step—Don't rename!

D. Adding more than two whole numbers

To add more than two whole numbers, add as you would with two whole numbers.

EXAMPLE 1-24   Add 98 + 7846 + 385 + 275.
Solution

(1) Add ones.
8 + 6 + 5 + 5 = 24

(2) Add tens.
2 + 9 + 4 + 8 + 7 = 30

(3) Add hundreds.
3 + 8 + 3 + 2 = 16

(4) Add thousands.
1 + 7 = 8

1-3. Subtracting Whole Numbers

Recall that to join two or more amounts together, you add. But what should you do when you want to do the opposite of add? That is, what should you do when you want to take one amount away from another amount?

To take one amount away from another amount, you **subtract.**

EXAMPLE 1-25   Write "☐☐☐☐☐ take away ☐☐☐" as subtraction.

Solution   "☐☐☐☐☐ take away ☐☐☐" equals 5 − 3. ⟵ subtraction

The **subtraction problem** 5 − 3 is read as "five **minus** three." The symbol " − " is called a **subtraction sign** or **subtraction symbol.**

A. Basic subtraction facts

The following example shows one way to solve a subtraction problem.

EXAMPLE 1-26   Subtract 5 − 3.

Solution   5 − 3 = 2 because

☐☐☐☐☐ take away ☐☐☐ equals ☐☐☒☒☒ or ☐☐

The number sentence 5 − 3 = 2 is called a **basic subtraction fact.** To subtract whole numbers quickly and accurately, you must know the 100 basic subtraction facts from memory [see Table 1-2]. The basic subtraction fact 5 − 3 = 2 is written in horizontal form. Basic subtraction facts can also be written in vertical form.

EXAMPLE 1-27   Write the number sentence 5 − 3 = 2 in vertical form.

Solution

$$
\begin{array}{r}
5 \\
-3 \\
\hline
2
\end{array}
$$

5 ⟵ The first number in 5 − 3 = 2 is the top number.
−3 ⟵ The second number in 5 − 3 = 2 is the bottom number.
2 ⟵ The answer (2) is written under the subtraction bar.

The number you subtract from in a subtraction problem is called the **minuend,** the number to be subtracted is called the **subtrahend,** and the answer is called the **difference.** So in Example 1-27, 5 is the minuend, 3 is the subtrahend, and 2 is the difference.

Subtraction is related to addition in the following way:

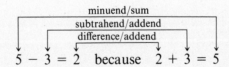

minuend/sum
subtrahend/addend
difference/addend
5 − 3 = 2   because   2 + 3 = 5

The difference plus the subtrahend equals the minuend; that is, when the difference and the subtrahend of a subtraction problem are addends, the sum is equal to the minuend. So the subtraction problem 5 − 3 = ? and the **missing addend problem** ? + 3 = 5 are really just two different ways of stating the same problem. The answer to both problems is 2.

The subtraction problem 5 − 3 = ? asks: If you have 5 and take 3 away, how many are left?

The missing addend problem ? + 3 = 5 asks: How many more are needed if you have 3 but want 5?

B. Subtracting, given horizontal form

To subtract two whole numbers with two or more digits given in horizontal form, you first write the numbers in vertical form.

TABLE 1-2 BASIC SUBTRACTION FACTS

0 −0 ‾0	1 −0 ‾1	2 −0 ‾2	3 −0 ‾3	4 −0 ‾4	5 −0 ‾5	6 −0 ‾6	7 −0 ‾7	8 −0 ‾8	9 −0 ‾9
1 −1 ‾0	2 −1 ‾1	3 −1 ‾2	4 −1 ‾3	5 −1 ‾4	6 −1 ‾5	7 −1 ‾6	8 −1 ‾7	9 −1 ‾8	10 − 1 ‾9
2 −2 ‾0	3 −2 ‾1	4 −2 ‾2	5 −2 ‾3	6 −2 ‾4	7 −2 ‾5	8 −2 ‾6	9 −2 ‾7	10 − 2 ‾8	11 − 2 ‾9
3 −3 ‾0	4 −3 ‾1	5 −3 ‾2	6 −3 ‾3	7 −3 ‾4	8 −3 ‾5	9 −3 ‾6	10 − 3 ‾7	11 − 3 ‾8	12 − 3 ‾9
4 −4 ‾0	5 −4 ‾1	6 −4 ‾2	7 −4 ‾3	8 −4 ‾4	9 −4 ‾5	10 − 4 ‾6	11 − 4 ‾7	12 − 4 ‾8	13 − 4 ‾9
5 −5 ‾0	6 −5 ‾1	7 −5 ‾2	8 −5 ‾3	9 −5 ‾4	10 − 5 ‾5	11 − 5 ‾6	12 − 5 ‾7	13 − 5 ‾8	14 − 5 ‾9
6 −6 ‾0	7 −6 ‾1	8 −6 ‾2	9 −6 ‾3	10 − 6 ‾4	11 − 6 ‾5	12 − 6 ‾6	13 − 6 ‾7	14 − 6 ‾8	15 − 6 ‾9
7 −7 ‾0	8 −7 ‾1	9 −7 ‾2	10 − 7 ‾3	11 − 7 ‾4	12 − 7 ‾5	13 − 7 ‾6	14 − 7 ‾7	15 − 7 ‾8	16 − 7 ‾9
8 −8 ‾0	9 −8 ‾1	10 − 8 ‾2	11 − 8 ‾3	12 − 8 ‾4	13 − 8 ‾5	14 − 8 ‾6	15 − 8 ‾7	16 − 8 ‾8	17 − 8 ‾9
9 −9 ‾0	10 − 9 ‾1	11 − 9 ‾2	12 − 9 ‾3	13 − 9 ‾4	14 − 9 ‾5	15 − 9 ‾6	16 − 9 ‾7	17 − 9 ‾8	18 − 9 ‾9

EXAMPLE 1-28   Subtract $897 - 35$.

Solution

(1) Write in vertical form. Line up like values.

(2) Subtract ones. $7 - 5 = 2$ (ones)

(3) Subtract tens. $9 - 3 = 6$ (tens)

(4) Subtract hundreds. $8 - 0 = 8$ (hundreds)

```
 hundreds
  | tens
  |  | ones
  ↓  ↓  ↓
  8  9  7              8 9 7              8 9 7              8 9 7
-    3  5            -   3 5            -   3 5            -   3 5
                          2                  6 2                8 6 2  ← difference
                         ↑                  ↑                  ↑
                       ones               tens              hundreds
```

To *check subtraction*, you can use addition.

EXAMPLE 1-29   Check $897 - 35 = 862.$ ←—— proposed difference

Solution

$$
\begin{array}{r}
8\ 6\ 2 \\
+\ \ 3\ 5 \\
\hline
8\ 9\ 7
\end{array}
$$ ←—— 862 checks

Think: $897 - 35 = 862$ if $862 + 35 = 897$

C. Subtracting with one renaming

The following example shows how to rename two one-digit values (like 3 tens 5 ones) to get more ones.

EXAMPLE 1-30   Rename 3 tens 5 ones to get more ones.

Solution   3 tens 5 ones = 2 tens   1 tens   5 ones

        = 2 tens   10 ones   5 ones

        = 2 tens   15 ones ←—— more ones

Think: 3 tens = 2 tens   1 ten

     1 ten = 10 ones

When the digits in the ones column cannot be subtracted, you must rename to get more ones.

EXAMPLE 1-31   Subtract $935 - 216.$

Solution

(1) Subtract ones? No.
5 − 6 does not have a whole-number answer. You need more than 5 ones, so you must rename.

(2) Rename to get more ones.
3 tens   5 ones = 2 tens   15 ones

(3) Subtract as before.
15 − 6 = 9 (ones)
2 − 1 = 1 (tens)
9 − 2 = 7 (hundreds)

$$
\begin{array}{r}
9\ 3\ 5 \\
-2\ 1\ 6 \\
\hline
\end{array}
$$

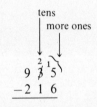

tens
| more ones

$$
\begin{array}{r}
9\ \overset{2}{\cancel{3}}\,\overset{1}{5} \\
-2\ 1\ 6 \\
\hline
\end{array}
$$

$$
\begin{array}{r}
9\ \overset{2}{\cancel{3}}\,\overset{1}{5} \\
-2\ 1\ 6 \\
\hline
7\ 1\ 9
\end{array}
$$

When the digits in any given place-value column cannot be subtracted, you rename to get more of that place value in the minuend (top number).

EXAMPLE 1-32   Subtract $837 - 692.$

Solution

(1) Subtract ones.
7 − 2 = 5 (ones)

(2) Subtract tens? No.
3 − 9 does not have a whole-number answer. You need more than 3 tens, so you must rename.

(3) Rename to get more tens.
8 hundreds   3 tens = 7 hundreds 13 tens
Subtract as before.

$$
\begin{array}{r}
8\ 3\ 7 \\
-6\ 9\ 2 \\
\hline
5
\end{array}
$$

tens

$$
\begin{array}{r}
8\ 3\ 7 \\
-6\ 9\ 2 \\
\hline
5
\end{array}
$$

hundreds
| more tens

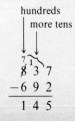

$$
\begin{array}{r}
\overset{7}{\cancel{8}}\,\overset{1}{3}\ 7 \\
-6\ 9\ 2 \\
\hline
1\ 4\ 5
\end{array}
$$

D. Subtracting with two or more renamings

To subtract with two or more renamings, you should write each renaming neatly to avoid errors.

EXAMPLE 1-33   Subtract 7516 − 3827.

Solution

(1) Subtract ones.   **(2)** Subtract tens.   **(3)** Subtract hundreds.   **(4)** Subtract thousands.
16 − 7 = 9   10 − 2 = 8   14 − 8 = 6   6 − 3 = 3

$$
\begin{array}{r}
{}^{0}\;{}^{1}\\
7\,5\,\cancel{1}\,6\\
-\,3\,8\,2\,7\\
\hline
9
\end{array}
\qquad
\begin{array}{r}
{}^{4}\;{}^{10}\;{}^{1}\\
7\,\cancel{5}\,\cancel{1}\,6\\
-\,3\,8\,2\,7\\
\hline
8\,9
\end{array}
\qquad
\begin{array}{r}
{}^{6}\;{}^{14}\;{}^{10}\;{}^{1}\\
\cancel{7}\,\cancel{5}\,\cancel{1}\,6\\
-\,3\,8\,2\,7\\
\hline
6\,8\,9
\end{array}
\qquad
\begin{array}{r}
{}^{6}\;{}^{14}\;{}^{10}\;{}^{1}\\
\cancel{7}\,\cancel{5}\,\cancel{1}\,6\\
-\,3\,8\,2\,7\\
\hline
3\,6\,8\,9
\end{array}
$$

E. Subtracting across zeros

When the minuend (the top number) contains a zero, you will usually have to rename across zero in order to subtract.

EXAMPLE 1-34   Subtract 506 − 238.

Solution

(1) Subtract ones? No.
6 − 8 does not have a
whole-number answer.
Rename 0 tens 6 ones
to get more ones? No.
You cannot take 1 away
from 0.

(2) Rename across zero.
50 tens 6 ones = 49 tens   16 ones

(3) Subtract as before.
16 − 8 = 8
9 − 3 = 6
4 − 2 = 2

$$
\begin{array}{r}
\text{No}\\
\downarrow\\
5\,0\,6\\
-\,2\,3\,8\\
\hline
\end{array}
\qquad
\begin{array}{r}
\text{tens}\quad\text{more ones}\\
\downarrow\quad\downarrow\\
4\;\;9\;\;{}^{1}\\
\cancel{5}\,\cancel{0}\,6\\
-\,2\,3\,8\\
\hline
\end{array}
\qquad
\begin{array}{r}
4\;\;9\;\;{}^{1}\\
\cancel{5}\,\cancel{0}\,6\\
-\,2\,3\,8\\
\hline
2\,6\,8
\end{array}
$$

To rename across two or more zeros, rename in much the same way as you did across one zero.

EXAMPLE 1-35   Subtract 8000 − 654.

Solution

(1) Rename across zeros.
800 tens 0 ones = 799 tens 10 ones

(2) Subtract as before.

(3) Check as before.

$$
\begin{array}{r}
7\;9\;9\;{}^{1}\\
\cancel{8}\,\cancel{0}\,\cancel{0}\,0\\
-\;\;\;6\,5\,4\\
\hline
\end{array}
\qquad
\begin{array}{r}
7\;9\;9\;{}^{1}\\
\cancel{8}\,\cancel{0}\,\cancel{0}\,0\\
-\;\;\;6\,5\,4\\
\hline
7\,3\,4\,6
\end{array}
\qquad
\begin{array}{r}
1\;1\;1\\
7\,3\,4\,6\\
+\;\;\;6\,5\,4\\
\hline
8\,0\,0\,0
\end{array}
\;\longleftarrow\; 7346\text{ checks}
$$

1-4. Multiplying Whole Numbers

Multiplication is a short way to write repeated addends.

EXAMPLE 1-36   Write the repeated addends 4 + 4 + 4 as a multiplication problem.

Solution   You can think of 4 + 4 + 4 as "3 equal amounts of 4," or 3 × 4. ⟵ multiplication

The **multiplication problem** 3 × 4 is read as "three **times** four." The symbol " × " is called a **multiplication sign** or **multiplication symbol.**

A. Basic multiplication facts

The multiplication problem 3 × 4 can be represented by **sets,** or **collections,** of objects.

EXAMPLE 1-37 Represent the multiplication problem 3 × 4 using sets.

Solution

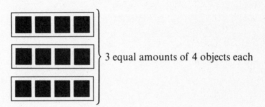

3 equal amounts of 4 objects each

The multiplication problem 3 × 4 can also be represented by a **rectangular array** of objects.

EXAMPLE 1-38 Represent the multiplication problem 3 × 4 using a rectangular array.

Solution

3 rows of 4 objects each

Note: In any case, 3 × 4 = 12 because: 4 + 4 + 4 = 8 + 4 = 12; that is,

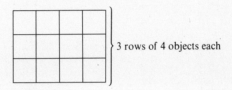

The number sentence 3 × 4 = 12 is called a **basic multiplication fact.** Basic multiplication can be written in horizontal form: 3 × 4 = 12. Or it can be written in vertical form:

$$\begin{array}{r} 4 \\ \times 3 \\ \hline 12 \end{array}$$

The number to be multiplied in a multiplication problem is called the **multiplicand,** the number to be multiplied by is called the **multiplier,** and the answer is called the **product.** So in 4 × 3 = 12, 4 is the multiplicand, 3 is the multiplier, and 12 is the product. The multiplicand and the multiplier are also called **factors.**

To multiply whole numbers quickly and accurately, you must know the 100 basic multiplication facts from memory [see Table 1-3].

There are several different ways to write the multiplication problem "3 times 4."

TABLE 1-3   BASIC MULTIPLICATION FACTS

0 ×0 ― 0	0 ×1 ― 0	0 ×2 ― 0	0 ×3 ― 0	0 ×4 ― 0	0 ×5 ― 0	0 ×6 ― 0	0 ×7 ― 0	0 ×8 ― 0	0 ×9 ― 0
1 ×0 ― 0	1 ×1 ― 1	1 ×2 ― 2	1 ×3 ― 3	1 ×4 ― 4	1 ×5 ― 5	1 ×6 ― 6	1 ×7 ― 7	1 ×8 ― 8	1 ×9 ― 9
2 ×0 ― 0	2 ×1 ― 2	2 ×2 ― 4	2 ×3 ― 6	2 ×4 ― 8	2 ×5 ― 10	2 ×6 ― 12	2 ×7 ― 14	2 ×8 ― 16	2 ×9 ― 18
3 ×0 ― 0	3 ×1 ― 3	3 ×2 ― 6	3 ×3 ― 9	3 ×4 ― 12	3 ×5 ― 15	3 ×6 ― 18	3 ×7 ― 21	3 ×8 ― 24	3 ×9 ― 27
4 ×0 ― 0	4 ×1 ― 4	4 ×2 ― 8	4 ×3 ― 12	4 ×4 ― 16	4 ×5 ― 20	4 ×6 ― 24	4 ×7 ― 28	4 ×8 ― 32	4 ×9 ― 36
5 ×0 ― 0	5 ×1 ― 5	5 ×2 ― 10	5 ×3 ― 15	5 ×4 ― 20	5 ×5 ― 25	5 ×6 ― 30	5 ×7 ― 35	5 ×8 ― 40	5 ×9 ― 45
6 ×0 ― 0	6 ×1 ― 6	6 ×2 ― 12	6 ×3 ― 18	6 ×4 ― 24	6 ×5 ― 30	6 ×6 ― 36	6 ×7 ― 42	6 ×8 ― 48	6 ×9 ― 54
7 ×0 ― 0	7 ×1 ― 7	7 ×2 ― 14	7 ×3 ― 21	7 ×4 ― 28	7 ×5 ― 35	7 ×6 ― 42	7 ×7 ― 49	7 ×8 ― 56	7 ×9 ― 63
8 ×0 ― 0	8 ×1 ― 8	8 ×2 ― 16	8 ×3 ― 24	8 ×4 ― 32	8 ×5 ― 40	8 ×6 ― 48	8 ×7 ― 56	8 ×8 ― 64	8 ×9 ― 72
9 ×0 ― 0	9 ×1 ― 9	9 ×2 ― 18	9 ×3 ― 27	9 ×4 ― 36	9 ×5 ― 45	9 ×6 ― 54	9 ×7 ― 63	9 ×8 ― 72	9 ×9 ― 81

Note: $0 \times 1 = 1 \times 0$, $1 \times 2 = 2 \times 1$, $2 \times 3 = 3 \times 2$, and so on. So it doesn't matter what order the factors are in.

EXAMPLE 1-39   Write "3 times 4" in five different ways.

Solution

$$3 \text{ times } 4 = 3 \times 4$$
$$= 3 \cdot 4$$
$$= 3(4) \quad \left.\right\} \text{all equivalent ways of writing "3 times 4"}$$
$$= (3)4$$
$$= (3)(4)$$

B. Multiplying by one digit

To multiply a whole number with two or more digits by a one-digit whole number, you multiply in vertical form. To multiply by one digit given vertical form, you first multiply ones, then tens, then hundreds, and so on.

EXAMPLE 1-40 Multiply $\begin{array}{r} 512 \\ \times\ 3 \end{array}$ ◄—— vertical form

Solution

(1) Multiply ones.
3 × 2 = 6 (ones)

(2) Multiply tens.
3 × 1 = 3 (tens)

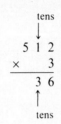

(3) Multiply hundreds.
3 × 5 = 15 (hundreds)

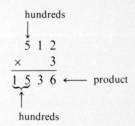

To multiply given horizontal form, you must first change to vertical form.

EXAMPLE 1-41   Multiply 6043 × 2. ◄—— horizontal form

Solution

(1) Change to vertical form.

$$\begin{array}{r} 6\ 0\ 4\ 3 \\ \times\qquad 2 \\ \hline \end{array}$$

(2) Multiply as before.

$$\begin{array}{r} 6\ 0\ 4\ 3 \\ \times\qquad 2 \\ \hline 1\ 2{,}0\ 8\ 6 \end{array}$$

When multiplying, you must rename two-digit values (like 35 ones, 18 tens, or 21 hundreds) unless it is the last step in the problem.

EXAMPLE 1-42   Rename the two-digit values for the multiplication problem 437 × 5.

Solution

(1) Multiply ones.
5 × 7 = 35 (ones)
Rename:
35 ones = 3 tens 5 ones

(2) Multiply tens.
5 × 3 = 15 (tens)
Add renaming:
15 + 3 = 18 (tens)
Rename:
18 tens = 1 hundred 8 tens

(3) Multiply hundreds.
5 × 4 = 20 (hundreds)
Add renaming:
20 + 1 = 21 (hundreds)
Last step—Don't rename!

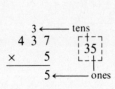

C. Multiplying by two or more digits

To multiply by two or more digits, you first multiply to get **partial products** and then add the partial products.

To multiply by two digits, you first multiply by ones and then by tens.

EXAMPLE 1-43   Multiply 625 × 43.

Solution

(1) Multiply by ones.
625 × 3 = 1875 (ones)

```
    1
  6 2 5
×   4 3 ←── ones
  ─────
  1 8 7 5
      ↑
     ones
```

(2) Multiply by tens.
625 × 4 = 2500 (tens)

```
    2
  1 1
  6 2 5
×   4 3 ── tens
  ─────
  1 8 7 5
  2 5 0 0
      ↑
     tens
```

(3) Add partial products.
1875 + 25,000 = 26,875

```
    2          2
  1 1            └── 2500 tens
  6 2 5
×   4 3
  ─────
  1 8 7 5← partial
+ 2 5 0 0 ← products
  ─────────
  2 6,8 7 5 ← product
```

Note: Make sure you always place the rightmost digit of each partial product directly below the digit you're multiplying by in the multiplier. For example, when you're multiplying by the digit in the tens place, the rightmost digit of that partial product should be in the tens place.

To multiply by more than two digits, you first find each partial product and then add the partial products.

EXAMPLE 1-44   Multiply 316 × 258.

Solution

(1) Find each partial product.
258 × 6 (ones) = 1548 (ones)
258 × 1 (ten) = 258 (tens)
258 × 3 (hundreds) = 774 (hundreds)

```
    1 2
    3 4
    2 5 8
×   3 1 6 ── hundreds
  ───────
  1 5 4 8
    2 5 8
  7 7 4
    ↑
  hundreds
```

(2) Add partial products.
1548 + 2580 + 77,400 = 81,528

258 tens ─┘   └── 774 hundreds

```
    1 2
    3 4
    2 5 8
×   3 1 6
  ───────
  1 5 4 8
    2 5 8
+ 7 7 4
  ─────────
  8 1,5 2 8
```

D. Shortcuts in multiplying

To save time and effort while multiplying, you can use the following shortcuts.

When a whole-number factor ends in one or more zeros, you can group and bring down those zeros before multiplying.

EXAMPLE 1-45   Multiply 500 × 29,000 using a shortcut.

Solution

(1) Group zeros together.

```
      zeros
  2 9 0 0 0
×   5 0 0
```

(2) Shortcut! Bring down zeros.

```
  2 9 0 0 0
×   5 0 0
  ─────────
  0 0 0 0 0
```

(3) Multiply as before.
5 × 29 = 145

```
    4
  2 9,0 0 0
×   5 0 0
  ─────────
1 4,5 0 0,0 0 0
```

When a whole-number factor has one or more zeros in the middle, you can write that factor on the bottom as the multiplier.

EXAMPLE 1-46 Multiply 658 × 7003 using a shortcut.

Solution

(1) Multiply by ones.
658 × 3 = 1974 (ones)

(2) Shortcut! Multiply by zeros.
658 × 0 = 0 (tens)
658 × 0 = 0 (hundreds)

(3) Multiply by thousands and add.
658 × 7 = 4606 (thousands)
1974 + 4,606,000 = 4,607,974

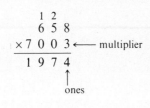

Writing the factor with zeros in the middle on the bottom as the multiplier is a shortcut because you have to write and add fewer partial products. Compare the shortcut method to the long method in the following examples.

$$\begin{array}{r} 658 \\ \times\,7\,0\,0\,3 \\ \hline 1\,9\,7\,4 \\ 4\,6\,0\,6\,0\,0 \\ \hline 4{,}6\,0\,7{,}9\,7\,4 \end{array} \text{two partial products}$$

$$\begin{array}{r} 7\,0\,0\,3 \\ \times\;\;\;6\,5\,8 \\ \hline 5\,6\,0\,2\,4 \\ 3\,5\,0\,1\,5 \\ 4\,2\,0\,1\,8 \\ \hline 4{,}6\,0\,7{,}9\,7\,4 \end{array} \text{three partial products}$$

1-5. Dividing Whole Numbers

Recall that to join two or more equal amounts together, you multiply. But what should you do when you want to do the opposite of multiply? That is, what should you do when you want to separate an amount into two or more equal amounts?

To separate an amount into two or more equal amounts, you **divide.**

EXAMPLE 1-47 Write "☐☐☐☐☐☐ separated into equal amounts of ☐☐☐" as division.

Solution "☐☐☐☐☐☐ separated into equal amounts of ☐☐☐" is 6 ÷ 3. (division)

The **division problem** 6 ÷ 3 is read as "six **divided by** three." The symbol " ÷ " is called a **division sign** or **division symbol.**

A. Basic division facts

The following example shows one way to solve a division problem.

EXAMPLE 1-48 6 ÷ 3 = 2 because

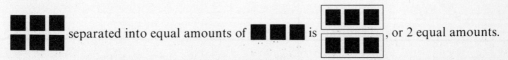

The number sentence 6 ÷ 3 = 2 is called a **basic division fact.** To divide quickly and accurately, you must know the 90 basic division facts from memory [see Table 1-4]. The basic division fact 6 ÷ 3 = 2 is written in horizontal form. Basic division facts can also be written in **division box form** (⟌).

TABLE 1-4   BASIC DIVISION FACTS

$\overset{0}{1\,\overline{)0}}$	$\overset{1}{1\,\overline{)1}}$	$\overset{2}{1\,\overline{)2}}$	$\overset{3}{1\,\overline{)3}}$	$\overset{4}{1\,\overline{)4}}$	$\overset{5}{1\,\overline{)5}}$	$\overset{6}{1\,\overline{)6}}$	$\overset{7}{1\,\overline{)7}}$	$\overset{8}{1\,\overline{)8}}$	$\overset{9}{1\,\overline{)9}}$
$\overset{0}{2\,\overline{)0}}$	$\overset{1}{2\,\overline{)2}}$	$\overset{2}{2\,\overline{)4}}$	$\overset{3}{2\,\overline{)6}}$	$\overset{4}{2\,\overline{)8}}$	$\overset{5}{2\,\overline{)10}}$	$\overset{6}{2\,\overline{)12}}$	$\overset{7}{2\,\overline{)14}}$	$\overset{8}{2\,\overline{)16}}$	$\overset{9}{2\,\overline{)18}}$
$\overset{0}{3\,\overline{)0}}$	$\overset{1}{3\,\overline{)3}}$	$\overset{2}{3\,\overline{)6}}$	$\overset{3}{3\,\overline{)9}}$	$\overset{4}{3\,\overline{)12}}$	$\overset{5}{3\,\overline{)15}}$	$\overset{6}{3\,\overline{)18}}$	$\overset{7}{3\,\overline{)21}}$	$\overset{8}{3\,\overline{)24}}$	$\overset{9}{3\,\overline{)27}}$
$\overset{0}{4\,\overline{)0}}$	$\overset{1}{4\,\overline{)4}}$	$\overset{2}{4\,\overline{)8}}$	$\overset{3}{4\,\overline{)12}}$	$\overset{4}{4\,\overline{)16}}$	$\overset{5}{4\,\overline{)20}}$	$\overset{6}{4\,\overline{)24}}$	$\overset{7}{4\,\overline{)28}}$	$\overset{8}{4\,\overline{)32}}$	$\overset{9}{4\,\overline{)36}}$
$\overset{0}{5\,\overline{)0}}$	$\overset{1}{5\,\overline{)5}}$	$\overset{2}{5\,\overline{)10}}$	$\overset{3}{5\,\overline{)15}}$	$\overset{4}{5\,\overline{)20}}$	$\overset{5}{5\,\overline{)25}}$	$\overset{6}{5\,\overline{)30}}$	$\overset{7}{5\,\overline{)35}}$	$\overset{8}{5\,\overline{)40}}$	$\overset{9}{5\,\overline{)45}}$
$\overset{0}{6\,\overline{)0}}$	$\overset{1}{6\,\overline{)6}}$	$\overset{2}{6\,\overline{)12}}$	$\overset{3}{6\,\overline{)18}}$	$\overset{4}{6\,\overline{)24}}$	$\overset{5}{6\,\overline{)30}}$	$\overset{6}{6\,\overline{)36}}$	$\overset{7}{6\,\overline{)42}}$	$\overset{8}{6\,\overline{)48}}$	$\overset{9}{6\,\overline{)54}}$
$\overset{0}{7\,\overline{)0}}$	$\overset{1}{7\,\overline{)7}}$	$\overset{2}{7\,\overline{)14}}$	$\overset{3}{7\,\overline{)21}}$	$\overset{4}{7\,\overline{)28}}$	$\overset{5}{7\,\overline{)35}}$	$\overset{6}{7\,\overline{)42}}$	$\overset{7}{7\,\overline{)49}}$	$\overset{8}{7\,\overline{)56}}$	$\overset{9}{7\,\overline{)63}}$
$\overset{0}{8\,\overline{)0}}$	$\overset{1}{8\,\overline{)8}}$	$\overset{2}{8\,\overline{)16}}$	$\overset{3}{8\,\overline{)24}}$	$\overset{4}{8\,\overline{)32}}$	$\overset{5}{8\,\overline{)40}}$	$\overset{6}{8\,\overline{)48}}$	$\overset{7}{8\,\overline{)56}}$	$\overset{8}{8\,\overline{)64}}$	$\overset{9}{8\,\overline{)72}}$
$\overset{0}{9\,\overline{)0}}$	$\overset{1}{9\,\overline{)9}}$	$\overset{2}{9\,\overline{)18}}$	$\overset{3}{9\,\overline{)27}}$	$\overset{4}{9\,\overline{)36}}$	$\overset{5}{9\,\overline{)45}}$	$\overset{6}{9\,\overline{)54}}$	$\overset{7}{9\,\overline{)63}}$	$\overset{8}{9\,\overline{)72}}$	$\overset{9}{9\,\overline{)81}}$

Note: Division by zero (0) is not defined.

EXAMPLE 1-49   Write $6 \div 3 = 2$ in division box form.

Solution

$$\overset{2}{3\,\overline{)6}} \longleftarrow \text{ the first number in } 6 \div 3 = 2 \text{ goes in the division box}$$

the second number in $6 \div 3 = 2$ goes in front of the division box

In a division problem, the number you are dividing by is the **divisor,** the number you're dividing is the **dividend,** and the answer is the **quotient.** So in $6 \div 3 = 2$, 3 is the divisor, 6 is the dividend, and 2 is the quotient.

There are several different ways to write the division problem "6 divided by 3."

EXAMPLE 1-50   Write "6 divided by 3" in six different ways.

Solution

$$
\begin{aligned}
6 \text{ divided by } 3 &= 6 \div 3 \\
&= 6/3 \\
&= \frac{6}{3} \\
&= 6 \cdot \tfrac{1}{3} \\
&= \tfrac{1}{3} \cdot 6 \\
&= 3\,\overline{)6}
\end{aligned}
\right\} \text{ all equivalent ways of writing "6 divided by 3"}
$$

Every basic division fact can be related to a basic multiplication fact, and vice versa.

The product of the quotient and the divisor is the dividend. That is, when the quotient and the divisor in a division problem are factors in a multiplication problem, the product is equal to the dividend.

So the division problem $6 \div 3 = ?$ and the related **missing factor problem** $? \times 3 = 6$ are really just two different ways of stating the same problem. The answer to both problems is 2.

EXAMPLE 1-51   Write the related basic multiplication fact for $6 \div 3 = 2$.

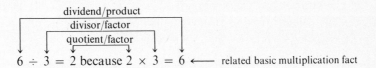

Solution $6 \div 3 = 2$ because $2 \times 3 = 6$ ←—— related basic multiplication fact

EXAMPLE 1-52   Write $6 \div 3 = ?$ and $? \times 3 = 6$ in words.

Solution $6 \div 3 = ?$ asks: If you separate 6 into equal amounts of 3 each, how many equal amounts are there in all?

　　　　$? \times 3 = 6$ asks: How many equal amounts of 3 each are needed to make 6 all together?

If a division problem is given, you can always write the related missing factor problem.

EXAMPLE 1-53   Write the related missing factor problem for $8 \div 2 = ?$.

Solution $8 \div 2 = ?$ means $? \times 2 = 8$ ←—— related missing factor problem

To help find a basic division fact, you can use the related missing factor problem and your multiplication facts.

EXAMPLE 1-54   Use a missing factor problem to solve the division problem $42 \div 7$.

Solution $42 \div 7 = ?$ means $? \times 7 = 42$ ←—— missing factor problem

　　　　　and **6** $\times 7 = 42$ ←—— basic multiplication fact

　　　　　so $42 \div 7 =$ **6** ←—— related basic division fact

B. Dividing by one digit

To divide by a one-digit whole number when there are two or more digits in the dividend, you always start the division process by trying to divide into the first digit in the dividend.

When a divisor does not divide a dividend exactly, there will always be a nonzero **remainder**, R.

EXAMPLE 1-55   Divide $35 \div 4$. ←—— horizontal form

Solution

(1) Write in division box form.

(2) Divide tens? No. There are no fours in 3.

(3) Divide ones? Yes. How many fours in 35 ones? Are there 7? Yes: $4 \times 7 = 28$ Are there 8? Yes: $4 \times 8 = 32$ Are there 9? No: $4 \times 9 = 36$ ⟵ too big There are **8** fours in 35.

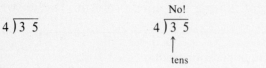

$$4 \overline{)3\ 5}$$

No!
$$4 \overline{)3\ 5}$$
↑
tens

$$\overset{8}{4 \overline{)3\ 5}}$$
↑
ones

(4) Multiply.
$4 \times 8 = 32$

(5) Subtract.
$35 - 32 = 3$

(6) Compare. Is the remainder 3 less than the divisor 4? Yes: $3 < 4$, so 8 is correct.

(7) Write the remainder.

$$\overset{8}{4 \overline{)3\ 5}} \\ \quad 3\ 2$$

$$\overset{8}{4 \overline{)\ \ 3\ 5}} \\ \quad -3\ 2 \\ \quad \overline{\quad 3}$$

$$\boxed{3<} \quad \overset{8 \longleftarrow \text{correct}}{4 \overline{)\ \ 3\ 5}} \\ \quad\quad -3\ 2 \\ \quad\quad \overline{\quad 3}$$

$$\overset{8\,\text{R3}}{4 \overline{)\ \ 3\ 5}} \\ \quad -3\ 2 \quad \text{remainder} \\ \quad \overline{\quad 3}$$

Note: In $35 \div 4 = 8$ R3, 8 is the quotient, 3 is the remainder, 8 R3 is the **division answer**, and R is the symbol for "remainder."

To check a division answer when the remainder is not zero, you multiply the proposed quotient by the divisor and then add the remainder.

EXAMPLE 1-56 Check $35 \div 4 = 8$ R3. ⟵ proposed division answer

Solution Multiply the proposed quotient (8) by the divisor (4) and then add the remainder (3) to see if you get the original dividend (35).

$$\begin{array}{r} 8 \longleftarrow \text{quotient} \\ \times 4 \longleftarrow \text{divisor} \\ \hline 3\ 2 \\ +\ \ 3 \longleftarrow \text{remainder} \\ \hline 3\ 5 \longleftarrow \text{8 R3 checks} \end{array}$$

Caution: For a division answer to be correct, the remainder must always be *less than* the divisor.

EXAMPLE 1-57 Find the error in the following step-by-step solution:

(1) Divide.   **(2)** Multiply.   **(3)** Subtract.   **(4)** Write remainder.   **(5)** Check.

$$\overset{6}{8 \overline{)5\ 6}}$$

$$\overset{6}{8 \overline{)5\ 6}} \\ \quad 4\ 8$$

$$\overset{6}{8 \overline{)\ \ 5\ 6}} \\ \quad -4\ 8 \\ \quad \overline{\quad 8}$$

$$\overset{6\,\text{R8}}{8 \overline{)\ \ 5\ 6}} \\ \quad -4\ 8 \\ \quad \overline{\quad 8}$$

$$\begin{array}{r} 6 \longleftarrow \text{quotient} \\ \times 8 \longleftarrow \text{divisor} \\ \hline 4\ 8 \\ +\ \ 8 \longleftarrow \text{remainder} \\ \hline 5\ 6 \longleftarrow \text{6 R8 checks?} \end{array}$$

Solution The error is in step 1, but becomes obvious in step 4:

$$\overset{6\,\text{R8}}{8 \overline{)5\ 6}} \longleftarrow \text{wrong (the remainder is not less than the divisor)} \qquad \overset{7}{8 \overline{)5\ 6}} \longleftarrow \text{correct quotient}$$

Note: This error would have been caught following the "Subtract" step if the "Compare" step had not been omitted. **NEVER omit the "Compare" step!**

EXAMPLE 1-58 Show how the "Compare" step will catch the error in Example 1-57.

Solution

(1) Divide. **(2)** Multiply. **(3)** Subtract. **(4)** Compare.

```
      6              6                6                           6  ←── wrong
  8) 5 6         8) 5 6          8)  5 6         8 =|   8) 5 6
                   4 8              − 4 8
                                      8
```

Think: 8 (the remainder) is not less than 8 (the divisor) which means 6 is not correct—it has to be larger. So you increase the quotient 6 by one to 7 and start the division process over again.

The correct solution to 56 ÷ 8 is found as follows:

(1) Divide, multiply, and subtract. **(2)** Compare. **(3)** Write the remainder. **(4)** Check.

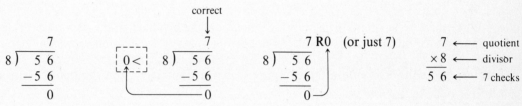

```
                              correct
                                ↓
      7                         7                 7 R0  (or just 7)         7  ←── quotient
  8)  5 6         0<|  8)  5 6            8)  5 6                          × 8  ←── divisor
   − 5 6                  − 5 6              − 5 6                          5 6  ←── 7 checks
      0                     0                   0
```

Note 1: When the remainder is zero, it's not necessary to write the remainder as part of the division answer: 7 R0 = 7

Note 2: To check a division answer when the remainder is zero, multiply the proposed quotient by the divisor to see if you get the original dividend [see step 4 in the previous example].

Note 3: When there is one digit in the divisor, the six basic division steps are

(1) Divide   **(4)** Compare
(2) Multiply   **(5)** Write the remainder (if not zero)
(3) Subtract   **(6)** Check

To divide when there are two or more digits in the quotient, you must **bring down** one or more digits from the dividend.

EXAMPLE 1-59 Divide 184 ÷ 5.

Solution

(1) Divide hundreds? No. There are no fives in 1.

(2) Divide tens? Yes.

```
      3        means   about 3
  5) 15              5) 18
```

(3) Multiply, subtract, and compare.

```
   No!
  5) 1 8 4
     ↑
   hundreds
```

```
       tens
        ↓
        3
  5) 1 8 4
        ↑
      tens
```

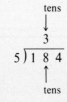

```
                correct
                  ↓
                  3
  3<|  5)  1 8 4
          − 1 5
              3
              ↑
            tens
```

(4) Bring down:
3 tens 4 ones = 34 ones

(5) Divide ones? Yes.

$$\begin{array}{r} 6 \\ 5\overline{)30} \end{array} \text{ means } \begin{array}{r} \text{about } 6 \\ 5\overline{)34} \end{array}$$

(6) Multiply, subtract, compare, and write the remainder.

$$\begin{array}{r} 3 \\ 5\overline{)184} \\ -15\downarrow \\ \hline 34 \\ \uparrow \\ \text{ones} \end{array}$$

$$\begin{array}{r} \text{ones} \\ \downarrow \\ 36 \\ 5\overline{)184} \\ -15 \\ \hline 34 \end{array} \quad \textit{Think: } 5\overline{)34}$$

$$\begin{array}{r} \text{correct} \\ \downarrow \\ 36\ \text{R4} \\ 4< \quad 5\overline{)184} \\ -15 \\ \hline 34 \\ -30 \\ \hline 4 \end{array}$$

(7) Check.

$$\begin{array}{r} 3 \\ 36 \quad \longleftarrow \text{quotient} \\ \times\quad 5 \quad \longleftarrow \text{divisor} \\ \hline 180 \\ +\quad 4 \quad \longleftarrow \text{remainder} \\ \hline 184 \quad \longleftarrow \text{36 R4 checks} \end{array}$$

Note: When there are no more digits to bring down from the dividend and the remainder is less than the divisor, you write the remainder (if not zero) as part of the division answer and then check.

When you cannot divide after bringing down a digit from the dividend, write a zero above that digit in the quotient.

EXAMPLE 1-60   Divide $617 \div 3$.

Solution

(1) Divide hundreds.

$$\begin{array}{r} 2 \\ 0< \quad 3\overline{)617} \\ -6 \\ \hline 0 \end{array}$$

(2) Bring down tens.

$$\begin{array}{r} 2 \\ 3\overline{)617} \\ -6 \\ \hline \cancel{0}\,1 \\ \uparrow \end{array}$$

When zero is the first digit of a whole number, you can cross it out.

(3) Divide tens? No.
There are no threes in 1.

$$\begin{array}{r} 20 \quad \longleftarrow \text{write a zero in the} \\ 3\overline{)617} \quad \text{quotient to show you} \\ -6 \quad \text{cannot divide} \\ \hline \cancel{0}\,1 \end{array}$$

Think: $3\overline{)1}$

(4) Bring down ones.

$$\begin{array}{r} 20 \\ 3\overline{)617} \\ -6 \\ \hline \cancel{0}\,17 \end{array}$$

(5) Divide ones.

$$\begin{array}{r} 205\ \text{R2} \\ 2< \quad 3\overline{)617} \\ -6 \\ \hline \cancel{0}\,17 \quad \textit{Think: } 3\overline{)17} \\ -15 \\ \hline 2 \end{array}$$

(6) Check.

$$\begin{array}{r} 205 \quad \longleftarrow \text{quotient} \\ \times\quad 3 \quad \longleftarrow \text{divisor} \\ \hline 615 \\ +\quad 2 \quad \longleftarrow \text{remainder} \\ \hline 617 \quad \longleftarrow \text{205 R2 checks} \end{array}$$

C. Dividing by two digits

To divide by a two-digit divisor, you can use the first digit of the divisor to **find an estimate** for each digit of the quotient. To find an estimate for a one-digit quotient, you can use basic division facts.

EXAMPLE 1-61 Divide 260 ÷ 31.

Solution

(1) Find an estimate.

$$\frac{8}{3\overline{)24}} \quad \text{means} \quad \frac{\text{about } 8}{3\overline{)26}}$$

$$\text{means} \quad \frac{\text{about } 8}{31\overline{)260}}$$

(2) Multiply (8 × 31 = 248), subtract (260 − 248 = 12), compare (12 < 31), and write the remainder (R = 12).

(3) Check.

$$
\begin{array}{r}
3\ 1 \longleftarrow \text{divisor} \\
\times\quad 8 \longleftarrow \text{quotient} \\
\hline
2\ 4\ 8 \\
+\quad 1\ 2 \longleftarrow \text{remainder} \\
\hline
2\ 6\ 0 \longleftarrow \text{8 R12 checks}
\end{array}
$$

estimate
$$
\begin{array}{r}
8 \\
31\overline{)2\ 6\ 0}
\end{array}
$$

correct
$$
\boxed{12<} \quad
\begin{array}{r}
8\ \text{R12} \\
31\overline{)\ 2\ 6\ 0} \\
-2\ 4\ 8 \longleftarrow \text{8 × 31} \\
\hline
1\ 2
\end{array}
$$

Note 1: Another way to state "$\frac{\text{about } 8}{3\overline{)26}}$ means $\frac{\text{about } 8}{31\overline{)260}}$" is "There are about 8 threes in 26, which means there should be about 8 thirty-ones in 260."

Note 2: When there are two or more digits in the divisor, the six basic division steps are

(1) Find an estimate

(2) Multiply

(3) Subtract

(4) Compare

(5) Write the remainder (if not zero)

(6) Check

Caution: When the first digit of the divisor is used to find an estimate, you may get an estimate that is too large. If an estimate is too large, you will not be able to subtract. When this happens, you must decrease the estimate until you can subtract.

EXAMPLE 1-62 Divide 365 ÷ 46.

Solution

(1) Find an estimate.

$$\frac{9}{4\overline{)36}} \quad \text{means} \quad \frac{\text{about } 9}{46\overline{)365}}$$

too large
$$
\begin{array}{r}
9 \\
46\overline{)3\ 6\ 5} \\
4\ 1\ 4 \longleftarrow \text{9 × 46}
\end{array}
$$
cannot subtract

(2) Decrease the estimate.

9 − 1 = 8 (new estimate)

too large also
$$
\begin{array}{r}
8 \\
46\overline{)3\ 6\ 5} \\
3\ 6\ 8 \longleftarrow \text{8 × 46}
\end{array}
$$
cannot subtract

(3) Decrease the estimate again.

8 − 1 = 7 (new estimate)

correct
$$
\boxed{43<} \quad
\begin{array}{r}
7\ \text{R43} \\
46\overline{)\ 3\ 6\ 5} \\
-3\ 2\ 2 \longleftarrow \text{7 × 46} \\
\hline
4\ 3
\end{array}
$$

(4) Check.

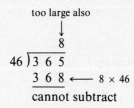

$$
\begin{array}{r}
4 \\
4\ 6 \longleftarrow \text{divisor} \\
\times\quad 7 \longleftarrow \text{quotient} \\
\hline
3\ 2\ 2 \\
+\quad 4\ 3 \longleftarrow \text{remainder} \\
\hline
3\ 6\ 5 \longleftarrow \text{7 R43 checks}
\end{array}
$$

To divide by two digits when there are two or more digits in the quotient, you can use the first digit of the divisor to find an estimate for each digit in the quotient.

EXAMPLE 1-63   Divide $8932 \div 29$.

Solution

(1) Find an estimate.

$$2\overline{)8} \quad \text{means} \quad \overset{\text{about }4}{29\overline{)89}}$$

too large
$$\overset{4}{}$$
$$29\overline{)\,8932}$$
$$\underline{1\,1\,6} \longleftarrow 4 \times 29$$
cannot subtract

(2) Decrease the estimate.
$4 - 1 = 3$ (new estimate)

correct
$$\overset{3}{}$$
$$29\overline{)\,8932}$$
$$\underline{-8\,7} \longleftarrow 3 \times 29$$
$$2$$

(3) Bring down and find an estimate. There are no twenty-nines in 23.

write zero
$$3\,0$$
$$29\overline{)\,8932}$$
$$\underline{-8\,7\downarrow}$$
$$2\,3 \qquad \textit{Think: } 29\overline{)23}$$

(4) Bring down and find an estimate.

$$2\overline{)2} \;\text{means}\; \overset{\text{about }11}{2\overline{)23}} \;\text{means}\; \overset{\text{about }9}{29\overline{)232}}$$

Note: 9 is the closest *single-digit* estimate to 11 you can make.

too large
$$3\,0\,9$$
$$29\overline{)\,8932}$$
$$\underline{-8\,7}\downarrow$$
$$2\,3\,2$$
$$\underline{2\,6\,1} \longleftarrow 9 \times 29$$
cannot subtract

(5) Decrease the estimate.
$9 - 1 = 8$ (new estimate)

correct
$$3\,0\,8$$
$$29\overline{)\,8932}$$
$$\underline{-8\,7}$$
$$2\,3\,2$$
$$\underline{-2\,3\,2} \longleftarrow 8 \times 29$$
$$0$$

(6) Check.

$$3\,0\,8 \longleftarrow \text{quotient}$$
$$\underline{\times\;\;2\,9} \longleftarrow \text{divisor}$$
$$2\,7\,7\,2$$
$$\underline{6\,1\,6}$$
$$8\,9\,3\,2 \longleftarrow 308 \text{ checks}$$

Note: In step 4, the estimate for $2\overline{)23}$ is the two-digit number 11. When an estimate has two digits (greater than 9), you first decrease the estimate to the one-digit number 9 and then see if you can subtract.

D. Dividing by three or more digits

To divide by three or more digits, you can use the first digit of the divisor to find an estimate for each digit in the quotient.

EXAMPLE 1-64   Divide $9500 \div 253$.

(1) Find an estimate.

$$2\overline{)8} \;\text{means}\; \overset{\text{about }4}{2\overline{)9}} \;\text{means}\; \overset{\text{about }4}{253\overline{)950}}$$

too large
$$\overset{4}{}$$
$$253\overline{)\,9500}$$
$$\underline{1\,0\,1\,2} \longleftarrow 4 \times 253$$
cannot subtract

(2) Decrease the estimate.
$4 - 1 = 3$ (new estimate)

correct
$$\overset{3}{}$$
$$191 < \quad 253\overline{)\,9500}$$
$$\underline{-7\,5\,9} \longleftarrow 3 \times 253$$
$$1\,9\,1$$

(3) Bring down and find an estimate.

$$\overset{9}{2\overline{)18}} \text{ means } \overset{\text{about } 9}{2\overline{)19}} \text{ means } \overset{\text{about } 9}{253\overline{)1910}}$$

too large
↓
$$\begin{array}{r} 3\ 9 \\ 253\overline{)\ 9\ 5\ 0\ 0} \\ -7\ 5\ 9 \\ \hline 1\ 9\ 1\ 0 \\ 2\cdot2\ 7\ 7 \longleftarrow 9 \times 253 \\ \hline \text{cannot subtract} \end{array}$$

(4) Decrease the estimate until you can subtract.

correct
↓
$$\begin{array}{r} 3\ 7 \text{ R}139 \\ 253\overline{)\ 9\ 5\ 0\ 0} \\ -7\ 5\ 9 \\ \hline 1\ 9\ 1\ 0 \\ -1\ 7\ 7\ 1 \\ \hline 1\ 3\ 9 \end{array}$$

$139 <$

(5) Check.

$$\begin{array}{r} 2\ 5\ 3 \longleftarrow \text{divisor} \\ \times\ \ \ 3\ 7 \longleftarrow \text{quotient} \\ \hline 1\ 7\ 7\ 1 \\ +7\ 5\ 9 \\ \hline 9\ 3\ 6\ 1 \\ +\ \ \ 1\ 3\ 9 \longleftarrow \text{remainder} \\ \hline 9\ 5\ 0\ 0 \longleftarrow 37 \text{ R}139 \text{ checks} \end{array}$$

1-6. Factoring Whole Numbers

Recall: The multiplier and multiplicand in a multiplication problem are called *factors* of the product. For example, in $3 \times 4 = 12$, 3 and 4 are factors of the product 12.

A given whole number is a factor of a second whole number if the given whole number divides the second whole number evenly

EXAMPLE 1-65 (a) Is 2 a factor of 12? (b) Is 5?

Solution

(a) 2 is a factor of 12 because 2 divides 12 evenly: $12 \div 2 = 6$ or $6 \times 2 = 12$

(b) 5 is not a factor of 12 because 5 does not divide 12 evenly: $12 \div 5 = 2 \text{ R}2$

To find all the whole-number factors of another whole number, you can use *division*.

EXAMPLE 1-66 Find all the whole-number factors of 12.

Solution

$12 \div 1 = 12$ means 1 is a factor of 12.　$12 \div 7 = 1 \text{ R}5$ means 7 is not a factor of 12.

$12 \div 2 = 6$ means 2 is a factor of 12.　$12 \div 8 = 1 \text{ R}4$ means 8 is not a factor of 12.

$12 \div 3 = 4$ means 3 is a factor of 12.　$12 \div 9 = 1 \text{ R}3$ means 9 is not a factor of 12.

$12 \div 4 = 3$ means 4 is a factor of 12.　$12 \div 10 = 1 \text{ R}2$ means 10 is not a factor of 12.

$12 \div 5 = 2 \text{ R}2$ means 5 is not a factor of 12.　$12 \div 11 = 1 \text{ R}1$ means 11 is not a factor of 12.

$12 \div 6 = 2$ means 6 is a factor of 12.　$12 \div 12 = 1$ means 12 is a factor of 12.

So the factors of 12 are 1, 2, 3, 4, 6, and 12.

If a whole number greater than zero has exactly two different factors—itself and 1—the number is called a **prime number.**

EXAMPLE 1-67   List all the prime numbers from 0 to 10.

Solution

0 is not a prime number because 0 has more than two different factors.
($0 \div 1 = 0, 0 \div 2 = 0, 0 \div 3 = 0, \ldots$ means every nonzero whole number is a factor of 0.)

1 is not a prime number because 1 has only one factor.
($1 \div 1 = 1$ means the only factor of 1 is 1.)

2 is a prime number because 2 has exactly two different factors.
($2 \div 1 = 2$ and $2 \div 2 = 1$ means the factors of 2 are 1 and 2.)

3 is a prime number because 3 has exactly two different factors.
($3 \div 1 = 3$ and $3 \div 3 = 1$ means the factors of 3 are 1 and 3.)

4 is not a prime number because 4 has more than two different factors.
($4 \div 1 = 4, 4 \div 2 = 2$, and $4 \div 4 = 1$ means the factors of 4 are 1, 2, and 4.)

5 is a prime number because 5 has exactly two different factors.
($5 \div 1 = 5$ and $5 \div 5 = 1$ means the factors of 5 are 1 and 5.)

6 is not a prime number because 6 has more than two different factors.
($6 \div 1 = 6, 6 \div 2 = 3, 6 \div 3 = 2$, and $6 \div 6 = 1$ means the factors of 6 are 1, 2, 3, and 6).

7 is a prime number because 7 has exactly two different factors.
($7 \div 1 = 7$, and $7 \div 7 = 1$ means the factors of 7 are 1 and 7.)

8 is not a prime number because 8 has more than two different factors.
($8 \div 1 = 8, 8 \div 2 = 4, 8 \div 4 = 2$, and $8 \div 8 = 1$ means the factors of 8 are 1, 2, 4, and 8.)

9 is not a prime number because 9 has more than two different factors.
($9 \div 1 = 9, 9 \div 3 = 3$, and $9 \div 9 = 1$ means the factors of 9 are 1, 3, and 9.)

10 is not a prime number because 10 has more than two different factors.
($10 \div 1 = 10, 10 \div 2 = 5, 10 \div 5 = 2$, and $10 \div 10 = 1$ means the factors of 10 are 1, 2, 5, and 10.)

So the prime numbers from 0 to 10 are 2, 3, 5, and 7.

A whole number is **even** if it ends in 0, 2, 4, 6 or 8.
A whole number is **odd** if it ends in 1, 3, 5, 7, or 9.

EXAMPLE 1-68   Identify 0, 5, 12, 29, 136, and 407 as odd or even.

Solution   0 is even because it ends in 0.

5 is odd because it ends in 5.

12 is even because it ends in 2.

29 is odd because it ends in 9.

136 is even because it ends in 6.

407 is odd because it ends in 7.

The only even prime number is 2. All of the prime numbers greater than 2 are odd because every even number greater than 2 has at least three different factors: 1, 2, and the even number itself.

Caution: Not all odd numbers greater than 1 are prime numbers.

EXAMPLE 1-69   Find an odd number greater than 1 that is not prime.

Solution   9 is an odd number because it ends in 9. But 9 is not a prime number because 9 has more than two different factors (1, 3, 9).

Note 1: Other odd numbers that are not prime are 1, 9, 15, 21, 25, 27, 33, 35, ...

Note 2: Every whole number is either even or odd.

A factor of a given whole number that is also a prime number is called a **prime factor.**

EXAMPLE 1-70   Find all the prime factors of 12.

Solution   The factors of 12 are 1, 2, 3, 4, and 6 [see Example 1-66]. The only prime factors of 12 are 2 and 3.

A whole number greater than zero with more than two different factors is called a **composite number.**

EXAMPLE 1-71   List all the composite numbers between 0 and 10.

Solution   1 is not a composite number because 1 has only one factor.

2 is not a composite number because 2 has exactly two different factors.

3 is not a composite number because 3 has exactly two different factors.

4 is a composite number because 4 has more than two different factors.

5 is not a composite number because 5 has exactly two different factors.

6 is a composite number because 6 has more than two different factors.

7 is not a composite number because 7 has exactly two different factors.

8 is a composite number because 8 has more than two different factors.

9 is a composite number because 9 has more than two different factors.

The composite numbers between 0 and 10 are 4, 6, 8, and 9.

Note 1: The whole numbers 0 and 1 are neither prime nor composite numbers.

Note 2: No prime number is a composite number because prime numbers have exactly two different factors and composite numbers always have more than two different factors.

Note 3: Every whole number is either 0, 1, prime, or composite.

To **factor** a whole number, you write it as a product of primes and/or composite numbers.

EXAMPLE 1-72   Factor 12 in as many different ways as possible.

Solution $\left.\begin{array}{l} 12 = 1 \times 12 \\ 12 = 2 \times 6 \\ 12 = 3 \times 4 \\ 12 = 2 \times 2 \times 3 \end{array}\right\}$ 12 can be factored in exactly four different ways.

Note: The factorizations 1×12 and 12×1 are not considered to be different factorizations, but **equivalent factorizations.**

A product of factors that are all prime numbers is called a **product of primes.**

EXAMPLE 1-73   Identify 12 as a product of primes, choosing from the factorizations from Example 1-72.

Solution   12 is not factored as a product of primes as 1×12 because 1 and 12 are not prime.

12 is not factored as a product of primes as 2×6 because 6 is not prime.

12 is not factored as a product of primes as 3×4 because 4 is not prime.

12 is factored as a product of primes as $2 \times 2 \times 3$ because 2 and 3 are both prime numbers.

Note: The factorizations $2 \times 2 \times 3$, $2 \times 3 \times 2$, and $3 \times 2 \times 2$ are all considered equivalent factorizations because each contains two factors of 2 and one factor of 3.

The following statement is one of the most important facts about the arithmetic of whole numbers.

The Fundamental Rule of Arithmetic

Every composite number can be factored as a product of primes in exactly one way (except for equivalent factorizations).

To factor a composite number as a product of primes, you can use division.

EXAMPLE 1-74 Use division to factor 12 as a product of primes.

Solution

prime divisor
↓

$12 \div 2 = 6$ ⟵ composite quotient (Continue the division process.)

$6 \div 2 = 3$ ⟵ prime quotient (*Stop!*)

$12 = 2 \times 2 \times 3$ ⟵ product of primes

Note 1: When the quotient is a composite number, continue the division process by dividing that quotient by another prime number.

Note 2: When the quotient is a prime number, stop the division process and write each prime divisor and the prime quotient as the product of primes.

To factor a whole number as a product of primes using division, it is usually easier to use a shortcut division method. Turn the division box upside down; then divide by a prime number and write the quotient below the division box, like this:

$$2\underline{|12} \longrightarrow 2\underline{|12} \longrightarrow 2\underline{|12} \longrightarrow 2\underline{|12} \longrightarrow 2\underline{|12}$$
$$6 \qquad \underline{|6} \qquad 2\underline{|6} \qquad 2\underline{|6}$$
$$3 \longleftarrow \text{prime quotient } (Stop!)$$

$$12 = 2 \times 2 \times 3 \longleftarrow \text{product of primes}$$

EXAMPLE 1-75 Factor (**a**) 15, (**b**) 18, and (**c**) 56 as products of primes.

Solution

(**a**) $3\underline{|15}$
             5

$15 = 3 \times 5$

(**b**) $2\underline{|18}$
          $3\underline{|9}$
             3

$18 = 2 \times 3 \times 3$

(**c**) $2\underline{|56}$
          $2\underline{|28}$
          $2\underline{|14}$
             7

$56 = 2 \times 2 \times 2 \times 7$

To factor a composite number as a product of primes, you can also use a **factor tree.**

EXAMPLE 1-76 Factor 12 as a product of primes using a factor tree.

Solution

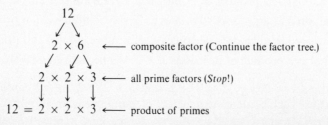

12

2×6 ⟵ composite factor (Continue the factor tree.)

$2 \times 2 \times 3$ ⟵ all prime factors (*Stop!*)

$12 = 2 \times 2 \times 3$ ⟵ product of primes

Note 1: When a factor in the bottom row of a factor tree is a composite number, continue the factor tree by factoring the composite number.

Note 2: When all the factors in the bottom row of a factor tree are prime numbers, stop the factor tree and write each prime factor in the bottom row as the product of primes.

Note 3: When a factor tree is continued with one or more prime numbers in the bottom row, bring down each prime number to the next row.

Factor trees for composite numbers always have two or more rows.

EXAMPLE 1-77   Factor (**a**) 15, (**b**) 18, and (**c**) 56 using a factor tree.

Solution

(**a**)

$$15 = 3 \times 5$$

(**b**)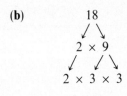

$$18 = 2 \times 3 \times 3$$

(**c**)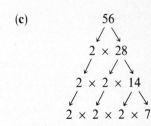

$$56 = 2 \times 2 \times 2 \times 7$$

1-7. Solving Word Problems Involving Whole Numbers

To solve the word problems in this section, you must add, subtract, multiply, and/or divide.

A. Choosing the correct operation ($+, -, \times, \div$)

To join two or more amounts together, you *add*.

To take one amount away from another amount, you *subtract*.

To join two or more equal amounts together, you *multiply*.

To separate an amount into two or more equal amounts, you *divide*.

EXAMPLE 1-78   Choose the correct operation to solve these word problems.

(**a**) Emily ran 15 km in the first race. Then she ran 8 km in the second race. How far did Emily run all together?

(**b**) Peter worked 15 hours this week. Last week he worked 8 hours. How many more hours did Peter work this week than last week?

Solution

(**a**) *Understand:*   The question asks you to join amounts of 15 and 8 together.

Decide: To join amounts together, you *add*.

(**b**) *Understand:*   The question asks you to take the amount of 8 away from 15.

Decide: To take an amount away, you *subtract*.

EXAMPLE 1-79   Choose the correct operation to solve these word problems.

(**a**) There are 6 balls in one can. There are 3 balls in another can. How many balls are there in all?

(**b**) There are 6 cans. There are 3 balls in each can. How many balls are there in all?

Solution

(**a**) *Understand:*   The question asks you to join amounts of 6 and 3 together.

Decide: To join amounts together, you *add*.

(**b**) *Understand:*   The question asks you to join 6 equal amounts of 3 together.

Decide: To join equal amounts together, you *multiply*.

EXAMPLE 1-80   Choose the correct operation to solve these word problems.

(a) Evelyn needs 40 woodscrews. Each package contains 5 woodscrews. How many packages should Evelyn buy?

(b) Myrle needs 40 woodscrews. Each woodscrew costs 5 cents. How much will the woodscrews cost Myrle?

Solution

(a) *Understand:*   The question asks you to separate 40 into equal amounts of 5.

   Decide: To separate into equal amounts, you *divide*.

(b) *Understand:*   The question asks you to join 40 equal amounts of 5 together.

   Decide: To join amounts together, you *multiply*.

B. Solving word problems

To solve *addition word problems,* you find

(a) how much two or more amounts are all together;

(b) the sum of one amount plus another amount.

To solve *subtraction words problems,* you find

(a) how much is left;

(b) how much more is needed;

(c) the difference between two amounts.

To solve *multiplication word problems,* you find

(a) how much two or more equal amounts are all together;

(b) the product of one amount times another amount.

To solve *division word problems,* you find

(a) how many equal amounts there are in all;

(b) how many are in each equal amount;

(c) how many times larger one amount is than another amount.

EXAMPLE 1-81   Solve the following word problem using addition or subtraction:

Carol had $20. She spent $12. How much money did Carol have left?

Solution

(1) *Identify:*   Carol had ($20). She spent ($12).   Circle each fact.
   How much money did Carol have left?   Underline the question.

(2) *Understand:*   The question asks you to find how much is left when 12 is taken away from 20.

(3) *Decide:*   To take 12 away from 20, you subtract.

(4) *Compute:* $20 - 12 = 8$

(5) *Interpret:*   The answer 8 means that after starting with $20 and then spending $12, Carol was left with **$8.** ⟵ solution

(6) *Check:*   Is $8 + $12 equal to the original $20? *Yes:* $8 + 12 = 20$

EXAMPLE 1-82   Solve the following word problem using addition, subtraction, or multiplication:

The high school track is 1320 feet (ft) around. A one-mile race is 4 times around the track. How many feet are in one mile?

Solution

(1) *Identify:*   The high-school track is (1320 feet (ft)) around.   Circle each fact.
   A one-mile race is (4 times) around the track.
   How many feet are in one mile?   Underline the question.

(2) *Understand:*   The question asks you to find 4 times 1320 ft.

(3) *Decide:*   To find the product of one amount times another amount, you multiply.

(4) *Compute:* $4 \times 1320 = 5280$

(5) *Interpret:*   5280 means that 4 times around the 1320-ft track is **5280 ft.** ⟵ solution

EXAMPLE 1-83   Solving the following word problem using addition, subtraction, multiplication, and/or division:

A restaurant needs at least 1000 eggs for the week. How many whole dozen eggs should be ordered if there are 12 eggs per dozen?

Solution

(1) *Identify*   A restaurant needs at least (1000 eggs)   Circle each fact.
for the week. How many whole dozen   Underline the question.
eggs should be ordered if there are
(12 eggs) per dozen?

(2) *Understand:*   The question asks you to find how many equal amounts of 12 there are in 1000, and then add 1 equal amount of 12 if there are any remaining.

(3) *Decide:*   To find how many equal amounts there are in all, you divide.
To find if there is anything left over, you look for a nonzero remainder.

(4) *Compute:*

$$
\begin{array}{r}
8\ 3 \text{ R4} \longleftarrow \text{nonzero remainder} \\
12\ )\overline{1\ 0\ 0\ 0} \\
-\ 9\ 6 \\
\hline
4\ 0 \\
-3\ 6 \\
\hline
4
\end{array}
$$

(5) *Interpret:*   83 R4 means that to receive at least 1000 eggs while ordering only whole dozens, the restaurant needs to order $(83 + 1)$ dozen, or **84 dozen.** ⟵ solution

(6) *Check:*   Is 84 dozen eggs at least 1000 eggs? Yes: 84 dozen $= 84 \times 12 = 1008$ (eggs)

SOLVED PROBLEMS

PROBLEM 1-1   Draw a line under the digit in each whole number that has the given place value:

(a) billions: 258,391,648,072

(b) tens: 169,418,720,000

(c) hundred thousands: 648,002,316,205

(d) ten millions: 715,800,321,296

(e) hundred billions: 403,489,267,185

(f) thousands: 350,813,765,048

(g) hundreds: 915,048,267,132

(h) millions: 512,346,992,817

(i) ten billions: 832,148,569,237

(j) ones: 209,136,004,218

(k) ten thousands: 108,655,431,789

(l) hundred millions: 621,856,413,729

Solution   Recall that to locate the correct digit given its place-value name, you can use a place-value chart [see Example 1-4]:

(a) 25<u>8</u>,391,648,072   **(b)** 169,418,720,0<u>0</u>0   **(c)** 648,0<u>0</u>2,316,205   **(d)** 715,<u>8</u>00,321,296

(e) <u>4</u>03,489,267,185   **(f)** 350,813,76<u>5</u>,048   **(g)** 915,048,267,<u>1</u>32   **(h)** 512,34<u>6</u>,992,817

(i) 8<u>3</u>2,148,569,237   **(j)** 209,136,004,21<u>8</u>   **(k)** 108,655,4<u>3</u>1,789   **(l)** 621,<u>8</u>56,413,729

PROBLEM 1-2   Write each of the following whole numbers using period names instead of commas:

(a) 56,250   (b) 25,092   (c) 37,000   (d) 8,356,003

(e) 75,025,000   (f) 170,000,000   (g) 5,210,034,002   (h) 900,000,000,000

Solution   Recall that the given period names in descending order are billions, millions, thousands, and units [see Example 1-5]:

(a) 56 thousand 250   (b) 25 thousand 92   (c) 37 thousand

(d) 8 million 356 thousand 3   (e) 75 million 25 thousand   (f) 170 million

(g) 5 billion 210 million 34 thousand 2   (h) 900 billion

PROBLEM 1-3   Write the place-value name of each underlined digit:

(a) 35<u>2</u>,167,859,203   (b) 658,214,0<u>3</u>9,715   (c) 102,314,876,92<u>5</u>   (d) 971,0<u>5</u>8,030,219

(e) 764,302,125,<u>8</u>93   (f) 561,45<u>8</u>,215,613   (g) <u>2</u>87,050,923,000   (h) 476,451,873,0<u>2</u>9

(i) 876,915,<u>2</u>54,003   (j) 301,210,58<u>2</u>,907   (k) 9<u>6</u>1,785,859,619   (l) 234,5<u>6</u>9,213,458

Solution   Recall that to help identify the correct place-value name of a given digit, you can use a place-value [see Example 1-6]:

(a) billions   (b) ten thousands   (c) ones   (d) hundred millions

(e) hundreds   (f) millions   (g) hundred billions   (h) tens

(i) hundred thousands   (j) thousands   (k) ten billions   (l) ten millions

PROBLEM 1-4   Write the value of 5 in each whole number:

(a) 213,456,201,891   (b) 340,678,203,152   (c) 503,789,621,487   (d) 413,469,521,380

(e) 935,867,004,918   (f) 613,782,495,812   (g) 718,789,247,315   (h) 819,582,413,760

(i) 759,003,018,000   (j) 619,005,783,416   (k) 118,234,658,139   (l) 218,648,231,580

Solution   Recall that to write the value of a given digit, you first identify the place-value name of that digit [see Example 1-7]:

(a) 50,000,000   (b) 50   (c) 500,000,000,000   (d) 500,000   (e) 5,000,000,000   (f) 5000

(g) 5   (h) 500,000,000   (i) 50,000,000,000   (j) 5,000,000   (k) 50,000   (l) 500

PROBLEM 1-5   Write the correct whole number for each group of one-digit values:

(a) 6 tens 5 ones   (b) 7 tens   (c) 8 ones   (d) 4 hundreds 3 tens 2 ones

(e) 2 hundreds 9 ones   (f) 1 hundred 6 tens   (g) 8 hundreds   (h) 7 thousands 9 hundreds

Solution   Recall that to write a whole number when one-digit values are given, you must be careful to write a zero (0) each time a one-digit value is missing [see Example 1-8]:

(a) 65   (b) 70   (c) 8   (d) 432   (e) 209   (f) 160   (g) 800   (h) 7900

PROBLEM 1-6   Write the correct whole number for each whole-number word name:

(a) eight   (b) eleven   (c) twenty   (d) one hundred

(e) fifty-four   (f) three hundred sixteen   (g) one thousand eighty   (h) seven hundred six

Solution   Recall that to write a whole number when the whole-number name is given, you may find it helpful to first write one-digit values [see Example 1-9]:

(a) 8   (b) 11   (c) 20   (d) 100   (e) 54   (f) 316   (g) 1080   (h) 706

PROBLEM 1-7   Write the correct whole-number word name for each whole number:

(a) 2   (b) 12   (c) 18   (d) 40   (e) 500   (f) 679   (g) 405   (h) 3090

Solution   Recall that to write the whole-number word name when the whole number is given, you may find it helpful to first write one-digit values [see Example 1-10]:

(a) two   **(b)** twelve   **(c)** eighteen   **(d)** forty

(e) five hundred   **(f)** six hundred seventy-nine   **(g)** four hundred five   **(h)** three thousand ninety

PROBLEM 1-8   Write the largest whole number of:

(a) 24 and 25   **(b)** 200 and 199   **(c)** 1020 and 1100

(d) 306 and 360   **(e)** 5150 and 5151   **(f)** 25, 430, and 26,430

(g) 356,000,001 and 356,000,000   **(h)** 728,591,346,218 and 728,591,436,218

Solution   Recall that to compare two whole numbers, you may find it helpful to write one number under the other while aligning like place values [see Example 1-13]:

(a) 25 **(b)** 200 **(c)** 1100 **(d)** 360 **(e)** 5151 **(f)** 26,430 **(g)** 356,000,001 **(h)** 728,591,436,218

PROBLEM 1-9   List each group of whole numbers from largest to smallest:

(a) 580, 579, 597, 508   **(b)** 2031, 2000, 1999, 2301, 2130   **(c)** 28, 297, 1023, 300, 1022, 1000

Solution   Recall that to order whole numbers from largest to smallest, you write the largest number first, then the next largest, and so on, until you have listed all the given numbers [see Example 1-14]:

(a) 597, 580, 579, 508   **(b)** 2301, 2130, 2031, 2000, 1999   **(c)** 1023, 1022, 1000, 300, 297, 28

PROBLEM 1-10   List each group of whole numbers from smallest to largest:

(a) 321, 123, 231, 132, 213, 312   **(b)** 5199, 5200, 5100, 4999, 5099

(c) 481, 2, 53, 1035, 480, 1036, 8   **(d)** 987, 988, 978, 789, 989

Solution   Recall that to order numbers from smallest to largest, you just reverse the largest to smallest order [see Example 1-15]:

(a) 123, 132, 213, 231, 312, 321   **(b)** 4999, 5099, 5100, 5199, 5200

(c) 2, 8, 53, 480, 481, 1035, 1036   **(d)** 789, 978, 987, 988, 989

PROBLEM 1-11   Round each whole number to the given place:

(a) 785 to the nearest hundred   **(b)** 2043 to the nearest thousand

(c) 85,641 to the nearest ten thousand   **(d)** 39 to the nearest ten

(e) 731,258,413 to the nearest million   **(f)** 619,315,208 to the nearest hundred thousand

(g) 258,319,648,215 to the nearest billion   **(h)** 981,674,321,863 to the nearest hundred million

(i) 299 to the nearest ten   **(j)** 39,999 to the nearest hundred

Solution   Recall that if the digit to the right of the digit to be rounded is 5 or more, you round up. If it's less than 5, you round down [see Examples 1-16 and 1-17]:

(a) 800   **(b)** 2000   **(c)** 90,000   **(d)** 40   **(e)** 731,000,000

(f) 619,300,000   **(g)** 258,000,000,000   **(h)** 981,700,000,000   **(i)** 300   **(j)** 40,000

PROBLEM 1-12   Add two whole numbers with no renaming:

(a) 32 + 56   **(b)** 92 + 83   **(c)** 201 + 25   **(d)** 325 + 413

(e) 729 + 830   **(f)** 2135 + 41   **(g)** 561 + 3218   **(h)** 9316 + 9270

Solution   Recall that to add whole numbers in horizontal form, you first write the numbers in vertical form [see Example 1-21]:

(a)
```
   3 2
 +5 6
 ----
   8 8
```
(b)
```
   9 2
 +8 3
 ----
 1 7 5
```
(c)
```
   2 0 1
 +   2 5
 ------
   2 2 6
```
(d)
```
   3 2 5
 +4 1 3
 ------
   7 3 8
```

(e)
```
   7 2 9
 +8 3 0
 ------
 1 5 5 9
```
(f)
```
   2 1 3 5
 +     4 1
 --------
   2 1 7 6
```
(g)
```
   3 2 1 8
 +   5 6 1
 --------
   3 7 7 9
```
(h)
```
   9 3 1 6
 +9 2 7 0
 --------
 1 8,5 8 6
```

PROBLEM 1-13   Add two whole numbers with renaming:

(a) 37 + 25   (b) 78 + 69   (c) 315 + 206   (d) 273 + 546

(e) 685 + 249   (f) 976 + 878   (g) 2859 + 729   (h) 6858 + 7578

Solution   Recall that when adding, you must rename each 2-digit value unless it is the last step [see Example 1-23]:

(a)
```
   1
   3 7
 +2 5
 ----
   6 2
```
(b)
```
   1
   7 8
 +6 9
 ----
 1 4 7
```
(c)
```
     1
   3 1 5
 +2 0 6
 ------
   5 2 1
```
(d)
```
     1
   2 7 3
 +5 4 6
 ------
   8 1 9
```

(e)
```
   1 1
   6 8 5
 +2 4 9
 ------
   9 3 4
```
(f)
```
   1 1
   9 7 6
 +8 7 8
 ------
 1 8 5 4
```
(g)
```
     1   1
   2 8 5 9
 +   7 2 9
 --------
   3 5 8 8
```
(h)
```
   1 1 1
   6 8 5 8
 +7 5 7 8
 --------
 1 4,4 3 6
```

PROBLEM 1-14   Add more than two whole numbers:

(a) 25 + 36 + 58   (b) 72 + 8 + 315   (c) 617 + 85 + 719

(d) 858 + 972 + 639   (e) 1035 + 483 + 5619   (f) 2514 + 34 + 8

(g) 12,502 + 2489 + 3488 + 25,869   (h) 8799 + 672 + 8431 + 52,789 + 482

Solution   Recall that to add more than two whole numbers, add as you would with two whole numbers [see Example 1-24]:

(a)
```
   1
   2 5
   3 6
 +5 8
 ----
 1 1 9
```
(b)
```
 1
   7 2
     8
 +3 1 5
 ------
   3 9 5
```
(c)
```
   1 2
   6 1 7
     8 5
 +7 1 9
 ------
 1 4 2 1
```
(d)
```
   1 1
   8 5 8
   9 7 2
 +6 3 9
 ------
 2 4 6 9
```

(e)
```
   1 1 1
   1 0 3 5
     4 8 3
 +5 6 1 9
 --------
   7 1 3 7
```
(f)
```
       1
   2 5 1 4
     3 4
 +     8
 --------
   2 5 5 6
```
(g)
```
   1 2 2 2
   1 2,5 0 2
     2 4 8 9
     3 4 8 8
 +2 5,8 6 9
 ----------
 4 4,3 4 8
```
(h)
```
   2 3 3 2
   8 7 9 9
     6 7 2
   8 4 3 1
   5 2,7 8 9
 +     4 8 2
 ----------
 7 1,1 7 3
```

PROBLEM 1-15   Subtract the following whole numbers with no renaming:

(a) 86 − 63   (b) 725 − 10   (c) 599 − 203   (d) 855 − 45

(e) 7883 − 23   (f) 9873 − 8820   (g) 87,759 − 77,015   (h) 529,569 − 4342

Solution   Recall that to subtract whole numbers given horizontal form, you first rename in vertical form [see Example 1-28]:

(a)
$$\begin{array}{r} 8\ 6 \\ -6\ 3 \\ \hline 2\ 3 \end{array}$$

(b)
$$\begin{array}{r} 7\ 2\ 5 \\ -\ \ 1\ 0 \\ \hline 7\ 1\ 5 \end{array}$$

(c)
$$\begin{array}{r} 5\ 9\ 9 \\ -2\ 0\ 3 \\ \hline 3\ 9\ 6 \end{array}$$

(d)
$$\begin{array}{r} 8\ 5\ 5 \\ -\ \ 4\ 5 \\ \hline 8\ 1\ 0 \end{array}$$

(e)
$$\begin{array}{r} 7\ 8\ 8\ 3 \\ -\ \ \ \ 2\ 3 \\ \hline 7\ 8\ 6\ 0 \end{array}$$

(f)
$$\begin{array}{r} 9\ 8\ 7\ 3 \\ -8\ 8\ 2\ 0 \\ \hline 1\ 0\ 5\ 3 \end{array}$$

(g)
$$\begin{array}{r} 8\ 7{,}7\ 5\ 9 \\ -7\ 7{,}0\ 1\ 5 \\ \hline 1\ 0{,}7\ 4\ 4 \end{array}$$

(h)
$$\begin{array}{r} 5\ 2\ 9{,}5\ 6\ 9 \\ -\ \ \ \ \ 4\ 3\ 4\ 2 \\ \hline 5\ 2\ 5{,}2\ 2\ 7 \end{array}$$

PROBLEM 1-16 Subtract the following whole numbers with one renaming:

(a) $91 - 75$ (b) $390 - 62$ (c) $374 - 119$ (d) $407 - 395$

(e) $661 - 90$ (f) $1539 - 729$ (g) $9166 - 1465$ (h) $37{,}024 - 4104$

Solution Recall that when a digit in any given place-value column cannot be subtracted, you rename to get more of that place-value in the minuend (top number) [see Examples 1-31 and 1-32]:

(a)
$$\begin{array}{r} {}^{8}\cancel{9}\ {}^{1}1 \\ -7\ 5 \\ \hline 1\ 6 \end{array}$$

(b)
$$\begin{array}{r} 3\ {}^{8}\cancel{9}\ {}^{1}0 \\ -\ \ 6\ 2 \\ \hline 3\ 2\ 8 \end{array}$$

(c)
$$\begin{array}{r} 3\ {}^{6}\cancel{7}\ {}^{1}4 \\ -1\ 1\ 9 \\ \hline 2\ 5\ 5 \end{array}$$

(d)
$$\begin{array}{r} {}^{3}\cancel{4}\ {}^{1}0\ 7 \\ -3\ 9\ 5 \\ \hline 1\ 2 \end{array}$$

(e)
$$\begin{array}{r} {}^{5}\cancel{6}\ {}^{1}6\ 1 \\ -\ \ 9\ 0 \\ \hline 5\ 7\ 1 \end{array}$$

(f)
$$\begin{array}{r} {}^{0}\cancel{1}\ {}^{1}5\ 3\ 9 \\ -\ \ 7\ 2\ 9 \\ \hline 8\ 1\ 0 \end{array}$$

(g)
$$\begin{array}{r} {}^{8}\cancel{9}\ {}^{1}1\ 6\ 6 \\ -1\ 4\ 6\ 5 \\ \hline 7\ 7\ 0\ 1 \end{array}$$

(h)
$$\begin{array}{r} 3\ {}^{6}\cancel{7}\ {}^{1}{,}0\ 2\ 4 \\ -\ \ 4\ 1\ 0\ 4 \\ \hline 3\ 2{,}9\ 2\ 0 \end{array}$$

PROBLEM 1-17 Subtract the following whole numbers with two or more renamings:

(a) $941 - 569$ (b) $3884 - 937$ (c) $1853 - 1654$ (d) $5621 - 2863$

(e) $6930 - 6198$ (f) $9743 - 4885$ (g) $57{,}071 - 17{,}628$ (h) $51{,}337 - 5789$

Solution Recall that to subtract with two or more renamings, you should write each renaming very neatly to avoid errors [see Example 1-33]:

(a)
$$\begin{array}{r} {}^{8}\cancel{9}\ {}^{13}\cancel{4}\ {}^{1}1 \\ -5\ 6\ 9 \\ \hline 3\ 7\ 2 \end{array}$$

(b)
$$\begin{array}{r} {}^{2}\cancel{3}\ {}^{1}8\ {}^{7}\cancel{8}\ {}^{1}4 \\ -\ \ 9\ 3\ 7 \\ \hline 2\ 9\ 4\ 7 \end{array}$$

(c)
$$\begin{array}{r} 1\ {}^{7}\cancel{8}\ {}^{14}\cancel{5}\ {}^{1}3 \\ -1\ 6\ 5\ 4 \\ \hline 1\ 9\ 9 \end{array}$$

(d)
$$\begin{array}{r} {}^{4}\cancel{5}\ {}^{15}\cancel{6}\ {}^{11}\cancel{2}\ {}^{1}1 \\ -2\ 8\ 6\ 3 \\ \hline 2\ 7\ 5\ 8 \end{array}$$

(e)
$$\begin{array}{r} 6\ {}^{8}\cancel{9}\ {}^{12}\cancel{3}\ {}^{1}0 \\ -6\ 1\ 9\ 8 \\ \hline 7\ 3\ 2 \end{array}$$

(f)
$$\begin{array}{r} {}^{8}\cancel{9}\ {}^{16}\cancel{7}\ {}^{13}\cancel{4}\ {}^{1}3 \\ -4\ 8\ 8\ 5 \\ \hline 4\ 8\ 5\ 8 \end{array}$$

(g)
$$\begin{array}{r} {}^{4}\cancel{5}\ {}^{16}\cancel{7}{,}{}^{1}0\ {}^{6}\cancel{7}\ {}^{1}1 \\ -1\ 7{,}6\ 2\ 8 \\ \hline 3\ 9{,}4\ 4\ 3 \end{array}$$

(h)
$$\begin{array}{r} {}^{4}\cancel{5}\ {}^{10}1{,}{}^{12}\cancel{3}\ {}^{12}\cancel{3}\ {}^{1}7 \\ -\ \ 5\ 7\ 8\ 9 \\ \hline 4\ 5{,}5\ 4\ 8 \end{array}$$

PROBLEM 1-18 Subtract the following whole numbers by renaming across one or more zeros:

(a) $406 - 287$ (b) $503 - 158$ (c) $900 - 841$ (d) $8005 - 287$

(e) $6100 - 3619$ (f) $70{,}002 - 2405$ (g) $60{,}201 - 34{,}285$ (h) $40{,}100 - 2639$

Solution Recall that when the minuend (top number) has one or more zeros, you will usually have to rename across those zeros [see Examples 1-34 and 1-35]:

(a)
$$\begin{array}{r} {}^{3}\cancel{4}\ {}^{9}\cancel{0}\ {}^{1}6 \\ -2\ 8\ 7 \\ \hline 1\ 1\ 9 \end{array}$$

(b)
$$\begin{array}{r} {}^{4}\cancel{5}\ {}^{9}\cancel{0}\ {}^{1}3 \\ -1\ 5\ 8 \\ \hline 3\ 4\ 5 \end{array}$$

(c)
$$\begin{array}{r} {}^{8}\cancel{9}\ {}^{9}\cancel{0}\ {}^{1}0 \\ -8\ 4\ 1 \\ \hline 5\ 9 \end{array}$$

(d)
$$\begin{array}{r} {}^{7}\cancel{8}\ {}^{9}\cancel{0}\ {}^{9}\cancel{0}\ {}^{1}5 \\ -\ \ \ 2\ 8\ 7 \\ \hline -7\ 7\ 1\ 8 \end{array}$$

(e) $\overset{5\ 10\ 9\ \ }{\cancel{6}\,\cancel{1}\,\cancel{0}\,0}$
 (f) $\overset{6\ 9\ 9\ 9\ 1}{\cancel{7}\,0,\cancel{0}\,0\,2}$
 (g) $\overset{5\ 9\ 11\ 9\ 1}{\cancel{6}\,\cancel{0},\cancel{2}\,\cancel{0}\,1}$
 (h) $\overset{3\ 9\ 10\ 9\ 1}{\cancel{4}\,\cancel{0},\cancel{1}\,\cancel{0}\,0}$

  $\underline{-3\ 6\ 1\ 9}$ $\underline{-\ \ 2\ 4\ 0\ 5}$ $\underline{-3\ 4,2\ 8\ 5}$ $\underline{-\ \ 2\ 6\ 3\ 9}$

  $2\ 4\ 8\ 1$ $6\ 7,5\ 9\ 7$ $2\ 5,9\ 1\ 6$ $3\ 7,4\ 6\ 1$

PROBLEM 1-19 Write each of the following repeated addends as a multiplication problem:

(a) $3 + 3$ (b) $4 + 4$ (c) $5 + 5 + 5$ (d) $9 + 9 + 9$

(e) $2 + 2 + 2 + 2$ (f) $7 + 7 + 7 + 7$ (g) $8 + 8 + 8 + 8 + 8$ (h) $1 + 1 + 1 + 1 + 1$

Solution Recall that you can think of $4 + 4 + 4$ as 3 equal amounts of 4 or 3×4 [see Example 1-36]:

(a) 2×3 (b) 2×4 (c) 3×5 (d) 3×9

(e) 4×2 (f) 4×7 (g) 5×8 (h) 5×1

PROBLEM 1-20 Multiply given vertical form:

(a) $\begin{array}{r} 34 \\ \times\ 2 \\ \hline \end{array}$ (b) $\begin{array}{r} 21 \\ \times\ 6 \\ \hline \end{array}$ (c) $\begin{array}{r} 70 \\ \times\ 5 \\ \hline \end{array}$ (d) $\begin{array}{r} 61 \\ \times\ 9 \\ \hline \end{array}$ (e) $\begin{array}{r} 132 \\ \times\ 3 \\ \hline \end{array}$

(f) $\begin{array}{r} 412 \\ \times\ 4 \\ \hline \end{array}$ (g) $\begin{array}{r} 1687 \\ \times\ 1 \\ \hline \end{array}$ (h) $\begin{array}{r} 3010 \\ \times\ 8 \\ \hline \end{array}$ (i) $\begin{array}{r} 40,132 \\ \times\ 2 \\ \hline \end{array}$ (j) $\begin{array}{r} 96,584 \\ \times\ 0 \\ \hline \end{array}$

Solution Recall that to multiply by one digit, you first multiply ones, then tens, then hundreds, and so on [see Example 1-40]:

(a) $\begin{array}{r} 34 \\ \times\ 2 \\ \hline 68 \end{array}$ (b) $\begin{array}{r} 21 \\ \times\ 6 \\ \hline 126 \end{array}$ (c) $\begin{array}{r} 70 \\ \times\ 5 \\ \hline 350 \end{array}$ (d) $\begin{array}{r} 61 \\ \times\ 9 \\ \hline 549 \end{array}$ (e) $\begin{array}{r} 132 \\ \times\ 3 \\ \hline 396 \end{array}$

(f) $\begin{array}{r} 412 \\ \times\ 4 \\ \hline 1648 \end{array}$ (g) $\begin{array}{r} 1687 \\ \times\ 1 \\ \hline 1687 \end{array}$ (h) $\begin{array}{r} 3010 \\ \times\ 8 \\ \hline 24{,}080 \end{array}$ (i) $\begin{array}{r} 40{,}132 \\ \times\ 2 \\ \hline 80{,}264 \end{array}$ (j) $\begin{array}{r} 96{,}584 \\ \times\ 0 \\ \hline 0 \end{array}$

PROBLEM 1-21 Multiply given horizontal form:

(a) 3×12 (b) 2×34 (c) 9×81 (d) 7×60 (e) 4×102

(f) 2×824 (g) 1×8519 (h) 0×7258 (i) $2 \times 24{,}013$ (j) $3 \times 91{,}203$

Solution Recall that to multiply given horizontal form, you first change to vertical form [see Example 1-41].

(a) $\begin{array}{r} 12 \\ \times\ 3 \\ \hline 36 \end{array}$ (b) $\begin{array}{r} 34 \\ \times\ 2 \\ \hline 68 \end{array}$ (c) $\begin{array}{r} 81 \\ \times\ 9 \\ \hline 729 \end{array}$ (d) $\begin{array}{r} 60 \\ \times\ 7 \\ \hline 420 \end{array}$ (e) $\begin{array}{r} 102 \\ \times\ 4 \\ \hline 408 \end{array}$

(f) $\begin{array}{r} 824 \\ \times\ 2 \\ \hline 1648 \end{array}$ (g) $\begin{array}{r} 8519 \\ \times\ 1 \\ \hline 8519 \end{array}$ (h) $\begin{array}{r} 7258 \\ \times\ 0 \\ \hline 0 \end{array}$ (i) $\begin{array}{r} 24{,}013 \\ \times\ 2 \\ \hline 48{,}026 \end{array}$ (j) $\begin{array}{r} 91{,}203 \\ \times\ 3 \\ \hline 273{,}609 \end{array}$

PROBLEM 1-22 Multiply when renaming is necessary:

(a) 3×25 (b) 2×319 (c) 8×591 (d) 6×487 (e) 4×2018 (f) 7×6051

(g) 9×2900 (h) 2×5186 (i) 3×4815 (j) 5×6480 (k) 4×3859 (l) $5 \times 20{,}419$

Solution Recall that you must rename each two-digit value when multiplying unless it is the last step [see Example 1-42]:

(a)
$$\begin{array}{r} \overset{1}{25} \\ \times\ 3 \\ \hline 75 \end{array}$$
(b)
$$\begin{array}{r} \overset{1}{319} \\ \times\ 2 \\ \hline 638 \end{array}$$
(c)
$$\begin{array}{r} \overset{7}{591} \\ \times\ 8 \\ \hline 4728 \end{array}$$
(d)
$$\begin{array}{r} \overset{54}{487} \\ \times\ 6 \\ \hline 2922 \end{array}$$
(e)
$$\begin{array}{r} \overset{3}{2018} \\ \times\ 4 \\ \hline 8072 \end{array}$$
(f)
$$\begin{array}{r} \overset{3}{6051} \\ \times\ 7 \\ \hline 42,357 \end{array}$$

(g)
$$\begin{array}{r} \overset{8}{2900} \\ \times\ 9 \\ \hline 26,100 \end{array}$$
(h)
$$\begin{array}{r} \overset{11}{5186} \\ \times\ 2 \\ \hline 10,372 \end{array}$$
(i)
$$\begin{array}{r} \overset{2\ 1}{4815} \\ \times\ 3 \\ \hline 14,445 \end{array}$$
(j)
$$\begin{array}{r} \overset{24}{6480} \\ \times\ 5 \\ \hline 32,400 \end{array}$$
(k)
$$\begin{array}{r} \overset{323}{3859} \\ \times\ 4 \\ \hline 15,436 \end{array}$$
(l)
$$\begin{array}{r} \overset{2\ 4}{20,419} \\ \times\ 5 \\ \hline 102,095 \end{array}$$

PROBLEM 1-23 Multiply by two digits:

(a) 12×34
(b) 15×23
(c) 41×58
(d) 53×26
(e) 23×132

(f) 14×513
(g) 72×645
(h) 87×905
(i) 96×513
(j) 75×486

Solution   Recall that you first multiply by ones and then by tens [see Example 1-43]:

(a)
$$\begin{array}{r} 34 \\ \times 12 \\ \hline 68 \\ 34\ \ \\ \hline 408 \end{array}$$
(b)
$$\begin{array}{r} \overset{1}{23} \\ \times 15 \\ \hline 115 \\ 23\ \ \\ \hline 345 \end{array}$$
(c)
$$\begin{array}{r} \overset{3}{58} \\ \times 41 \\ \hline 58 \\ 232\ \ \\ \hline 2378 \end{array}$$
(d)
$$\begin{array}{r} \overset{3}{\overset{1}{26}} \\ \times 53 \\ \hline 78 \\ 130\ \ \\ \hline 1378 \end{array}$$
(e)
$$\begin{array}{r} 132 \\ \times\ 23 \\ \hline 396 \\ 264\ \ \\ \hline 3036 \end{array}$$

(f)
$$\begin{array}{r} \overset{1}{513} \\ \times\ 14 \\ \hline 2052 \\ 513\ \ \\ \hline 7182 \end{array}$$
(g)
$$\begin{array}{r} \overset{3\ 3}{\overset{1}{645}} \\ \times\ 72 \\ \hline 1290 \\ 4515\ \ \\ \hline 46,440 \end{array}$$
(h)
$$\begin{array}{r} \overset{4}{\overset{3}{905}} \\ \times\ 87 \\ \hline 6335 \\ 7240\ \ \\ \hline 78,735 \end{array}$$
(i)
$$\begin{array}{r} \overset{1\ 2}{\overset{1}{513}} \\ \times\ 96 \\ \hline 3078 \\ 4617\ \ \\ \hline 49,248 \end{array}$$
(j)
$$\begin{array}{r} \overset{6\ 4}{\overset{4\ 3}{486}} \\ \times\ 75 \\ \hline 2430 \\ 3402\ \ \\ \hline 36,450 \end{array}$$

PROBLEM 1-24 Multiply by more than two digits:

(a) 213×123
(b) 312×234
(c) 412×287
(d) 351×386

(e) 485×285
(f) 649×782
(g) 321×3021
(h) 328×4719

Solution   Recall that to multiply by more than two digits, you first find each partial product and then add the partial products [see Example 1-44]:

(a)
$$\begin{array}{r} 123 \\ \times 213 \\ \hline 369 \\ 123\ \ \\ 246\ \ \ \\ \hline 26,199 \end{array}$$
(b)
$$\begin{array}{r} \overset{1\ 1}{234} \\ \times 312 \\ \hline 468 \\ 234\ \ \\ 702\ \ \ \\ \hline 73,008 \end{array}$$
(c)
$$\begin{array}{r} \overset{3\ 2}{\overset{1\ 1}{287}} \\ \times 412 \\ \hline 574 \\ 287\ \ \\ 1148\ \ \ \\ \hline 118,244 \end{array}$$
(d)
$$\begin{array}{r} \overset{2\ 1}{\overset{4\ 3}{386}} \\ \times 351 \\ \hline 386 \\ 1930\ \ \\ 1158\ \ \ \\ \hline 135,486 \end{array}$$

(e)
$$\begin{array}{r} \overset{3\ 2}{} \\ \overset{6\ 4}{} \\ \overset{4\ 2}{285} \\ \times 485 \\ \hline 1425 \\ 2280\ \ \\ 1140\ \ \ \\ \hline 138,225 \end{array}$$
(f)
$$\begin{array}{r} \overset{4\ 1}{} \\ \overset{3}{} \\ \overset{7\ 1}{782} \\ \times 649 \\ \hline 7038 \\ 3128\ \ \\ 4692\ \ \ \\ \hline 507,518 \end{array}$$
(g)
$$\begin{array}{r} 3021 \\ \times\ 321 \\ \hline 3021 \\ 6042\ \ \\ 9063\ \ \ \\ \hline 969,741 \end{array}$$
(h)
$$\begin{array}{r} \overset{2\ 2}{} \\ \overset{1\ 1}{} \\ \overset{5\ 1\ 7}{4719} \\ \times\ 328 \\ \hline 37752 \\ 9438\ \ \\ 14157\ \ \ \\ \hline 1,547,832 \end{array}$$

PROBLEM 1-25 Multiply using a shortcut:

(a) 20×36 (b) 45×80 (c) 60×90 (d) 300×58 (e) 700×320

(f) 90×400 (g) 500×800 (h) 7000×258 (i) 900×3500 (j) 3000×6000

Solution Recall that when a whole-number factor ends in one or more zeros, you can use a shortcut by first grouping and bringing down the zeros before multiplying [see Example 1-45]:

(a)
$$
\begin{array}{r}
\overset{1}{3\,6} \\
\times\ 2\,0 \\
\hline
7\,2\,0
\end{array}
$$

(b)
$$
\begin{array}{r}
\overset{4}{4\,5} \\
\times\ 8\,0 \\
\hline
3\,6\,0\,0
\end{array}
$$

(c)
$$
\begin{array}{r}
9\,0 \\
\times 6\,0 \\
\hline
5\,4\,0\,0
\end{array}
$$

(d)
$$
\begin{array}{r}
\overset{2}{5\,8} \\
\times\ 3\,0\,0 \\
\hline
1\,7{,}4\,0\,0
\end{array}
$$

(e)
$$
\begin{array}{r}
\overset{1}{3\,2\,0} \\
\times\ 7\,0\,0 \\
\hline
2\,2\,4{,}0\,0\,0
\end{array}
$$

(f)
$$
\begin{array}{r}
4\,0\,0 \\
\times 9\,0 \\
\hline
3\,6{,}0\,0\,0
\end{array}
$$

(g)
$$
\begin{array}{r}
8\,0\,0 \\
\times 5\,0\,0 \\
\hline
4\,0\,0{,}0\,0\,0
\end{array}
$$

(h)
$$
\begin{array}{r}
\overset{4\;5}{2\,5\,8} \\
\times\ 7\,0\,0\,0 \\
\hline
1{,}8\,0\,6{,}0\,0\,0
\end{array}
$$

(i)
$$
\begin{array}{r}
\overset{4}{3\,5\,0\,0} \\
\times\ 9\,0\,0 \\
\hline
3{,}1\,5\,0{,}0\,0\,0
\end{array}
$$

(j)
$$
\begin{array}{r}
6\,0\,0\,0 \\
\times 3\,0\,0\,0 \\
\hline
1\,8{,}0\,0\,0{,}0\,0\,0
\end{array}
$$

PROBLEM 1-26 Multiply using a shortcut:

(a) 203×345 (b) 502×420 (c) 601×500 (d) 2043×3271 (e) 3208×4526

(f) 4005×8216 (g) 5007×9250 (h) 6002×4500 (i) 8003×2000 (j) $87{,}256 \times 100{,}001$

Solution Recall that when a whole number factor has one or more zeros in the middle, you can use a shortcut by first writing that factor on the bottom as the multiplier [see Example 1-46]:

(a)
$$
\begin{array}{r}
\overset{1}{\overset{1\;1}{3\,4\,5}} \\
\times 2\,0\,3 \\
\hline
1\,0\,3\,5 \\
6\,9\,0\,0 \\
\hline
7\,0{,}0\,3\,5
\end{array}
$$

(b)
$$
\begin{array}{r}
\overset{1}{4\,2\,0} \\
\times 5\,0\,2 \\
\hline
8\,4\,0 \\
2\,1\,0\,0 \\
\hline
2\,1\,0{,}8\,4\,0
\end{array}
$$

(c)
$$
\begin{array}{r}
6\,0\,1 \\
\times\ \ 5\,0\,0 \\
\hline
3\,0\,0{,}5\,0\,0
\end{array}
$$

(d)
$$
\begin{array}{r}
\overset{1}{\overset{1\,2}{\overset{2}{3\,2\,7\,1}}} \\
\times 2\,0\,4\,3 \\
\hline
9\,8\,1\,3 \\
1\,3\,0\,8\,4 \\
6\,5\,4\,2\,0 \\
\hline
6{,}6\,8\,2{,}6\,5\,3
\end{array}
$$

(e)
$$
\begin{array}{r}
\overset{1\;\;1}{\overset{1\;\;1}{\overset{4\,2\,4}{4\,5\,2\,6}}} \\
\times 3\,2\,0\,8 \\
\hline
3\,6\,2\,0\,8 \\
9\,0\,5\,2\,0 \\
\hline
1\,3{,}5\,7\,8 \\
\hline
1\,4{,}5\,1\,9{,}4\,0\,8
\end{array}
$$

(f)
$$
\begin{array}{r}
\overset{2}{\overset{1\ \ 3}{8\,2\,1\,6}} \\
\times 4\,0\,0\,5 \\
\hline
4\,1\,0\,8\,0 \\
3\,2\,8\,6\,4\,0\,0 \\
\hline
3\,2{,}9\,0\,5{,}0\,8\,0
\end{array}
$$

(g)
$$
\begin{array}{r}
\overset{1\,2}{\overset{1\,3}{9\,2\,5\,0}} \\
\times 5\,0\,0\,7 \\
\hline
6\,4\,7\,5\,0 \\
4\,6\,2\,5\,0\,0 \\
\hline
4\,6{,}3\,1\,4{,}7\,5\,0
\end{array}
$$

(h)
$$
\begin{array}{r}
\overset{3}{\overset{1}{4\,5\,0\,0}} \\
\times 6\,0\,0\,2 \\
\hline
9\,0\,0\,0 \\
2\,7\,0\,0\,0 \\
\hline
2\,7{,}0\,0\,9{,}0\,0\,0
\end{array}
$$

(i)
$$
\begin{array}{r}
8\,0\,0\,3 \\
\times\ \ \ 2\,0\,0\,0 \\
\hline
1\,6{,}0\,0\,6{,}0\,0\,0
\end{array}
$$

(j)
$$
\begin{array}{r}
8\,7{,}2\,5\,6 \\
\times 1\,0\,0{,}0\,0\,1 \\
\hline
8\,7\,2\,5\,6 \\
8\,7\,2\,5\,6\,0\,0\,0\,0 \\
\hline
8{,}7\,2\,5{,}6\,8\,7{,}2\,5\,6
\end{array}
$$

PROBLEM 1-27 Write the related missing-factor problem for each division problem:

(a) $21 \div 7 = ?$ (b) $48 \div 6 = ?$ (c) $0 \div 4 = ?$ (d) $54 \div 6 = ?$ (e) $72 \div 9 = ?$

(f) $24 \div 3 = ?$ (g) $6 \div 6 = ?$ (h) $9 \div 1 = ?$ (i) $10 \div 5 = ?$ (j) $14 \div 2 = ?$

(k) $56 \div 7 = ?$ (l) $81 \div 9 = ?$ (m) $64 \div 8 = ?$ (n) $63 \div 7 = ?$

Solution Recall that the division problem $6 \div 3 = ?$ and the related missing factor problem $? \times 3 = 6$ are two different ways of stating the same problem [see Example 1-53]:

(a) $21 \div 7 = ?$ means $? \times 7 = 21$

(b) $48 \div 6 = ?$ means $? \times 6 = 48$

(c) $0 \div 4 = ?$ means $? \times 4 = 0$

(d) $54 \div 6 = ?$ means $? \times 6 = 54$

(e) $72 \div 9 = ?$ means $? \times 9 = 72$

(f) $24 \div 3 = ?$ means $? \times 3 = 24$

(g) $6 \div 6 = ?$ means $? \times 6 = 6$

(h) $9 \div 1 = ?$ means $? \times 1 = 9$

(i) $10 \div 5 = ?$ means $? \times 5 = 10$ **(j)** $14 \div 2 = ?$ means $? \times 2 = 14$

(k) $56 \div 7 = ?$ means $? \times 7 = 56$ **(l)** $81 \div 9 = ?$ means $? \times 9 = 81$

(m) $64 \div 8 = ?$ means $? \times 8 = 64$ **(n)** $63 \div 7 = ?$ means $? \times 7 = 63$

PROBLEM 1-28 Divide by a one-digit divisor to find a one-digit quotient with remainder.

(a) $19 \div 2$ **(b)** $61 \div 8$ **(c)** $28 \div 5$ **(d)** $5 \div 2$ **(e)** $10 \div 6$

(f) $14 \div 5$ **(g)** $18 \div 4$ **(h)** $13 \div 2$ **(i)** $62 \div 7$ **(j)** $29 \div 3$

(k) $31 \div 4$ **(l)** $39 \div 7$ **(m)** $41 \div 6$ **(n)** $25 \div 3$ **(o)** $38 \div 9$

Solution Recall that when a divisor does not divide a dividend exactly, then there will always be a nonzero remainder [see Example 1-55]:

$$
\begin{array}{ll}
\text{(a)} & 2\overline{)19} \quad 9\,\text{R}1 \\
& \underline{-18} \\
& 1
\end{array}
$$

(a) $2\,)\,19$ → 9 R1; -18; 1

(b) $8\,)\,61$ → 7 R5; -56; 5

(c) $5\,)\,28$ → 5 R3; -25; 3

(d) $2\,)\,5$ → 2 R1; -4; 1

(e) $6\,)\,10$ → 1 R4; -6; 4

(f) $5\,)\,14$ → 2 R4; -10; 4

(g) $4\,)\,18$ → 4 R2; -16; 2

(h) $2\,)\,13$ → 6 R1; -12; 1

(i) $7\,)\,62$ → 8 R6; -56; 6

(j) $3\,)\,29$ → 9 R2; -27; 2

(k) $4\,)\,31$ → 7 R3; -28; 3

(l) $7\,)\,39$ → 5 R4; -35; 4

(m) $6\,)\,41$ → 6 R5; -36; 5

(n) $3\,)\,25$ → 8 R1; -24; 1

(o) $9\,)\,38$ → 4 R2; -36; 2

PROBLEM 1-29 Divide by a one-digit divisor to find two or more digits in the quotient.

(a) $96 \div 8$ **(b)** $28 \div 2$ **(c)** $375 \div 7$ **(d)** $325 \div 9$ **(e)** $359 \div 4$

(f) $3367 \div 7$ **(g)** $3138 \div 6$ **(h)** $2000 \div 3$ **(i)** $7070 \div 9$ **(j)** $197 \div 2$

(k) $7239 \div 3$ **(l)** $21{,}144 \div 6$ **(m)** $17{,}410 \div 4$ **(n)** $39{,}349 \div 5$ **(o)** $193{,}080 \div 8$

Solution Recall that to divide when there are two or more digits in the quotient, you must bring down one or more digits from the dividend [see Example 1-59]:

(a) $8\,)\,96$ → 12; -8; 16; -16; 0

(b) $2\,)\,28$ → 14; -2; 08; -8; 0

(c) $7\,)\,375$ → 53 R4; -35; 25; -21; 4

(d) $9\,)\,325$ → 36 R1; -27; 55; -54; 1

(e) $4\,)\,359$ → 89 R3; -32; 39; -36; 3

(f) $7\,)\,3367$ → 481; -28; 56; -56; 07; -7; 0

(g) $6\,)\,3138$ → 523; -30; 13; -12; 18; -18; 0

(h) $3\,)\,2000$ → 666 R2; -18; 20; -18; 20; -18; 2

(i) $9\,)\,7070$ → 785 R5; -63; 77; -72; 50; -45; 5

(j) $2\,)\,197$ → 98 R1; -18; 17; -16; 1

(k) $3\overline{)7239}$ quotient 2413
-6
12
-12
03
-3
09
-9
0

(l) $6\overline{)21{,}144}$ quotient 3524
-18
31
-30
14
-12
24
-24
0

(m) $4\overline{)17{,}410}$ quotient 4352 R2
-16
14
-12
21
-20
10
-8
2

(n) $5\overline{)39{,}349}$ quotient 7869 R4
-35
43
-40
34
-30
49
-45
4

(o) $8\overline{)193{,}080}$ quotient 24,135
-16
33
-32
10
-8
28
-24
40
-40
0

PROBLEM 1-30 Divide by a one-digit divisor to find zeros in the quotient.

(a) $40 \div 4$ **(b)** $187 \div 9$ **(c)** $309 \div 3$ **(d)** $1240 \div 6$ **(e)** $2100 \div 7$

(f) $42{,}324 \div 6$ **(g)** $44{,}140 \div 7$ **(h)** $18{,}420 \div 2$ **(i)** $24{,}400 \div 8$ **(j)** $72{,}070 \div 9$

Solution Recall that when you cannot divide after bringing down a digit from the dividend, write a zero above that digit in the quotient [see Example 1-60]:

(a) $4\overline{)40}$ quotient 10
-4
00

(b) $9\overline{)187}$ quotient 20 R7
-18
07

(c) $3\overline{)309}$ quotient 103
-3
009
-9
0

(d) $6\overline{)1240}$ quotient 206 R4
-12
040
-36
4

(e) $7\overline{)2100}$ quotient 300
-21
000

(f) $6\overline{)42{,}324}$ quotient 7054
-42
032
-30
24
-24
0

(g) $7\overline{)44{,}140}$ quotient 6305 R5
-42
21
-21
040
-35
5

(h) $2\overline{)18{,}420}$ quotient 9210
-18
04
-4
02
-2
00

(i) $8\overline{)24{,}400}$ quotient 3050
-24
040
-40
00

(j) $9\overline{)72{,}070}$ quotient 8007 R7
-72
0070
-63
7

PROBLEM 1-31 Divide by a two-digit divisor to find a one-digit quotient.

(a) $480 \div 60$ **(b)** $600 \div 80$ **(c)** $155 \div 31$ **(d)** $199 \div 22$ **(e)** $80 \div 45$ **(f)** $450 \div 56$

Solution Recall that to divide by a two-digit divisor to find a one-digit quotient, you can use the first digit of the divisor to find an estimate for the one-digit quotient.

Caution: When the estimate is too large, you must decrease the estimate until you can subtract [see Examples 1-61 and 1-62]:

(a) $\boxed{6\overline{)48} \to 8}$ $60\overline{)480}$ quotient 8
$-480 \longleftarrow 8 \times 60$
0

(b) $\boxed{\text{about } 7 \quad 8\overline{)60}}$ $80\overline{)600}$ quotient 7 R40
$-560 \longleftarrow 7 \times 80$
40

(c) $\boxed{3\overline{)15} \to 5}$ $31\overline{)155}$ quotient 5
$-155 \longleftarrow 5 \times 31$
0

(d) $\boxed{\text{about } 9 \quad 2\overline{)19}}$ $22\overline{)199}$ quotient 9 R1
$-198 \longleftarrow 9 \times 22$
1

(e)
$$\boxed{4\overline{)8}\;\;2}$$
$$45\overline{)80}$$
$$-90 \longleftarrow 2 \times 45$$
$$\text{cannot}$$
$$\text{subtract}$$

$$\overset{1\,\text{R}35}{45\overline{)80}}$$
$$-45 \longleftarrow 1 \times 45$$
$$35$$

(f)
$$\boxed{5\overline{)45}\;\;9}$$
$$56\overline{)450}$$
$$-504 \longleftarrow 9 \times 56$$
$$\text{cannot}$$
$$\text{subtract}$$

$$\overset{8\,\text{R}2}{56\overline{)450}}$$
$$-448 \longleftarrow 8 \times 56$$
$$2$$

PROBLEM 1-32   Divide by a two-digit divisor to find two or more digits in the quotient.

(a) $1000 \div 40$ (b) $1730 \div 50$ (c) $1281 \div 21$

(d) $2200 \div 31$ (e) $2350 \div 11$ (f) $5472 \div 12$

(g) $22{,}720 \div 25$ (h) $55{,}480 \div 76$ (i) $348{,}900 \div 84$

Solution   Recall that to divide by a two-digit divisor, you can use the first digit of the divisor to find an estimate for each digit in the quotient.

Caution: When the estimate is too large, you must decrease the estimate until you can subtract [see Example 1-63]:

(a)
$$\overset{25}{40\overline{)1000}}$$
$$-80 \longleftarrow 2 \times 40$$
$$200$$
$$-200 \longleftarrow 5 \times 40$$
$$0$$

(b)
$$\overset{34\,\text{R}30}{50\overline{)1730}}$$
$$-150 \longleftarrow 3 \times 50$$
$$230$$
$$-200 \longleftarrow 4 \times 50$$
$$30$$

(c)
$$\overset{61}{21\overline{)1281}}$$
$$-126 \longleftarrow 6 \times 21$$
$$21$$
$$-21 \longleftarrow 1 \times 21$$
$$0$$

(d)
$$\overset{70\,\text{R}30}{31\overline{)2200}}$$
$$-217 \longleftarrow 7 \times 31$$
$$30$$

(e)
$$\overset{213\,\text{R}7}{11\overline{)2350}}$$
$$-22 \longleftarrow 2 \times 11$$
$$15$$
$$-11 \longleftarrow 1 \times 11$$
$$40$$
$$-33 \longleftarrow 3 \times 11$$
$$7$$

(f)
$$\overset{456}{12\overline{)5472}}$$
$$-48 \longleftarrow 4 \times 12$$
$$67$$
$$-60 \longleftarrow 5 \times 12$$
$$72$$
$$-72 \longleftarrow 6 \times 12$$
$$0$$

(g)
$$\overset{9\,0\,8\,\text{R}20}{25\overline{)22{,}720}}$$
$$-225 \longleftarrow 9 \times 25$$
$$220$$
$$-200 \longleftarrow 8 \times 25$$
$$20$$

(h)
$$\overset{7\,3\,0}{76\overline{)55{,}480}}$$
$$-532 \longleftarrow 7 \times 76$$
$$228$$
$$-228 \longleftarrow 3 \times 76$$
$$0$$

(i)
$$\overset{4\,1\,5\,3\,\text{R}48}{84\overline{)348{,}900}}$$
$$-336 \longleftarrow 4 \times 84$$
$$129$$
$$-84 \longleftarrow 1 \times 84$$
$$450$$
$$-420 \longleftarrow 5 \times 84$$
$$300$$
$$-252 \longleftarrow 3 \times 84$$
$$48$$

PROBLEM 1-33   Divide by divisors containing three or more digits.

(a) $1200 \div 300$ (b) $7300 \div 200$ (c) $3360 \div 420$ (d) $48{,}000 \div 510$

(e) $1645 \div 235$ (f) $10{,}565 \div 123$ (g) $199{,}424 \div 608$ (h) $357{,}000 \div 709$

Solution Recall that to divide by three or more digits, you can use the first digit of the divisor to find an estimate for each digit in the quotient [see Example 1-64]:

(a)

$$
\begin{array}{r} 4 \\ 3\overline{)12} \end{array}
\qquad
\begin{array}{r} 4 \\ 300\overline{)1200} \\ -1200 \quad\longleftarrow\ 4\times300 \\ \hline 0 \end{array}
$$

(b)

about 3
$2\overline{)7}$

about 6
$2\overline{)13}$

$$
\begin{array}{r} 36\ \text{R}100 \\ 200\overline{)7300} \\ -600 \quad\longleftarrow\ 3\times200 \\ \hline 1300 \\ -1200 \quad\longleftarrow\ 6\times200 \\ \hline 100 \end{array}
$$

(c)

about 8
$4\overline{)33}$

$$
\begin{array}{r} 8 \\ 420\overline{)3360} \\ -3360 \quad\longleftarrow\ 8\times420 \\ \hline 0 \end{array}
$$

(d)

about 9
$5\overline{)48}$

about 4
$5\overline{)21}$

$$
\begin{array}{r} 94\ \text{R}60 \\ 510\overline{)48{,}000} \\ -4590 \quad\longleftarrow\ 9\times510 \\ \hline 2100 \\ -2040 \quad\longleftarrow\ 4\times510 \\ \hline 60 \end{array}
$$

(e)

$8 \rightarrow 7$
$2\overline{)16}$

$$
\begin{array}{r} 7 \\ 235\overline{)1645} \\ -1645 \quad\longleftarrow\ 7\times235 \\ \hline 0 \end{array}
$$

(f)

$10 \rightarrow 9 \rightarrow 8$
$1\overline{)10}$

$7 \rightarrow 6 \rightarrow 5$
$1\overline{)7}$

$$
\begin{array}{r} 85\ \text{R}110 \\ 123\overline{)10{,}565} \\ -984 \quad\longleftarrow\ 8\times123 \\ \hline 725 \\ -615 \quad\longleftarrow\ 5\times123 \\ \hline 110 \end{array}
$$

(g)

about 3
$6\overline{)19}$

about 2
$6\overline{)17}$

8
$6\overline{)48}$

$$
\begin{array}{r} 328 \\ 608\overline{)199{,}424} \\ -1824 \quad\longleftarrow\ 3\times608 \\ \hline 1702 \\ -1216 \quad\longleftarrow\ 2\times608 \\ \hline 4864 \\ -4864 \quad\longleftarrow\ 8\times608 \\ \hline 0 \end{array}
$$

(h)

5
$7\overline{)35}$

about 3
$7\overline{)25}$

$$
\begin{array}{r} 503\ \text{R}373 \\ 709\overline{)357{,}000} \\ -3545 \quad\longleftarrow\ 5\times709 \\ \hline 2500 \\ -2127 \quad\longleftarrow\ 3\times709 \\ \hline 373 \end{array}
$$

PROBLEM 1-34 Find all the whole-number factors of each given number.

(a) 1 (b) 2 (c) 3 (d) 4 (e) 5 (f) 6 (g) 7 (h) 8 (i) 9 (j) 10

(k) 11 (l) 12 (m) 13 (n) 14 (o) 15 (p) 16 (q) 18 (r) 20 (s) 24 (t) 30

Solution Recall a given whole number is a factor of a second whole number if the given whole number divides the given whole number evenly [see Example 1-66]:

(a) 1 (b) 1, 2 (c) 1, 3 (d) 1, 2, 4

(e) 1, 5 (f) 1, 2, 3, 6 (g) 1, 7 (h) 1, 2, 4, 8

(i) 1, 3, 9 (j) 1, 2, 5, 10 (k) 1, 11 (l) 1, 2, 3, 4, 6, 12

(m) 1, 13 (n) 1, 2, 7, 14 (o) 1, 3, 5, 15 (p) 1, 2, 4, 8, 16

(q) 1, 2, 3, 6, 9, 18 (r) 1, 2, 4, 5, 10, 20 (s) 1, 2, 3, 4, 6, 8, 12, 24 (t) 1, 2, 3, 5, 6, 10, 15, 30

PROBLEM 1-35 Circle each prime number. Cross out each composite number.

0	1	2	3	4	5	6	7	8	9
10	11	12	13	14	15	16	17	18	19
20	21	22	23	24	25	26	27	28	29
30	31	32	33	34	35	36	37	38	39

40	41	42	43	44	45	46	47	48	49
50	51	52	53	54	55	56	57	58	59
60	61	62	63	64	65	66	67	68	69
70	71	72	73	74	75	76	77	78	79
80	81	82	83	84	85	86	87	88	89
90	91	92	93	94	95	96	97	98	99

Solution Recall that a prime number has exactly two different factors and a composite number has more than two different factors [see Examples 1-67 and 1-71]:

0	1	(2)	(3)	4	(5)	6	(7)	8	9
10	(11)	12	(13)	14	15	16	(17)	18	(19)
20	21	22	(23)	24	25	26	27	28	(29)
30	(31)	32	33	34	35	36	(37)	38	39
40	(41)	42	(43)	44	45	46	(47)	48	49
50	51	52	(53)	54	55	56	57	58	(59)
60	(61)	62	63	64	65	66	(67)	68	69
70	(71)	72	(73)	74	75	76	77	78	(79)
80	81	82	(83)	84	85	86	87	88	(89)
90	91	92	93	94	95	96	(97)	98	99

PROBLEM 1-36 Identify each whole number as odd or even:

(a) 0   **(b)** 1   **(c)** 5   **(d)** 8   **(e)** 107   **(f)** 2469   **(g)** 3792

Solution Recall that even numbers end in 0, 2, 4, 6, or 8 and odd numbers end in 1, 3, 5, 7, or 9 [see Example 1-68]:

(a) even   **(b)** odd   **(c)** odd   **(d)** even   **(e)** odd   **(f)** odd   **(g)** even

PROBLEM 1-37 Factor each whole number in as many different ways as possible:

(a) 0   **(b)** 1   **(c)** 2   **(d)** 4   **(e)** 18   **(f)** 100

Solution Recall that to factor a whole number, you write it as a product of primes and/or composite numbers [see Example 1-72]:

(a) $0 \times 1, 0 \times 2, 0 \times 3, \ldots$ (and infinitely many more different ways)

(b) 1×1 **(c)** 1×2 **(d)** $1 \times 4, 2 \times 2$ **(e)** $1 \times 18, 2 \times 9, 3 \times 6, 2 \times 3 \times 3$

(f) $1 \times 100, 2 \times 50, 4 \times 25, 5 \times 20, 10 \times 10$
   $2 \times 2 \times 25, 2 \times 5 \times 10, 4 \times 5 \times 5$
   $2 \times 2 \times 5 \times 5$

PROBLEM 1-38 Factor each composite number as a product of primes using division:

(a) 4   **(b)** 8   **(c)** 10   **(d)** 16   **(e)** 21   **(f)** 24   **(g)** 26   **(h)** 28   **(i)** 32

Solution [see Examples 1-74 and 1-75]:

(a) $2\lfloor 4,\ 2 \times 2$
2

(b) $2\lfloor 8,\ 2 \times 2 \times 2$
$2\lfloor 4$
2

(c) $2\lfloor 10,\ 2 \times 5$
5

(d) $2\lfloor 16,\ 2 \times 2 \times 2 \times 2$
$2\lfloor 8$
$2\lfloor 4$
2

(e) $3\lfloor 21,\ 3 \times 7$
7

(f) $2\lfloor 24,\ 2 \times 2 \times 2 \times 3$
$2\lfloor 12$
$2\lfloor 6$
3

(g) 2 | 26,   2 × 13
     13

(h) 2 | 28,   2 × 2 × 7
    2 | 14
       7

(i) 2 | 32,   2 × 2 × 2 × 2 × 2
    2 | 16
    2 | 8
    2 | 4
       2

PROBLEM 1-39   Factor each composite number as a product of primes using a factor tree:

(a) 6   **(b)** 9   **(c)** 14   **(d)** 20   **(e)** 22   **(f)** 25

(g) 27   **(h)** 33   **(i)** 35   **(j)** 38   **(k)** 40   **(l)** 44

Solution   [see Examples 1-76 and 1-77]:

(a) 6
  2 × 3

(b) 9
  3 × 3

(c) 14
  2 × 7

(d) 20
  2 × 10
  2 × 2 × 5

(e) 22
  2 × 11

(f) 25
  5 × 5

(g) 27
  3 × 9
  3 × 3 × 3

(h) 33
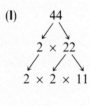
  3 × 11

(i) 35
  5 × 7

(j) 38
  2 × 19

(k) 40
  2 × 20
  2 × 2 × 10
  2 × 2 × 2 × 5

(l) 44
  2 × 22
  2 × 2 × 11

PROBLEM 1-40   Decide what to do (add, subtract, multiply, or divide) and then solve the problem:

(a) Donald weighed 200 pounds. He lost 27 pounds. How much did Donald weigh then?

(b) Karen weighed 108 pounds. She gained 17 pounds. How much did Karen weigh then?

(c) Warren sold 125 books last week. Each book cost $25. How many dollars worth of books did Warren sell last week?

(d) Marlene sold $125 worth of books last week. Each book cost $25. How many books did Marlene sell last week?

(e) Charles sold 125 books last week. His sales amounted to $2500. How much did each book cost?

(f) Susan sold 125 books last week and Paul 25 books. What was the difference between the number of books they sold?

(g) Leslie sold 125 books last week and Bill sold 25 books. How many books did they sell together?

(h) There are 144 eggs in one flat. There are 12 eggs in one dozen. How many dozen eggs are in one flat?

(i) There are 144 eggs in one flat. A restaurant orders 12 flats. How many eggs were ordered?

(j) A board is 255 cm long. It is cut so that each piece is 15 cm in length. How many pieces are there?

(k) A board is 364 cm in length. It is cut in 28 pieces of equal length. How long is each piece?

(l) Ann purchased 18 gallons of gasoline for $2 per gallon. How much did she pay for the gasoline?

(m) Bob purchased 18 liters of gasoline for $9 (900 cents). How much did Bob pay for each liter?

(n) There are 216 rolls in 27 bags. If there are the same number of rolls in each bag, how many rolls are in each bag?

(o) Aaron worked 7 hours each of 5 days last week. He earned $245 for the week. How much does Aaron earn per hour?

(p) Natalie worked 40 hours at $8 per hour. Then she worked 26 hours at double her hourly rate. How much did Natalie earn in all?

Solution   [see Examples 1-78, 1-79, 1-80, 1-81, 1-82, and 1-83]:

(a) subtract: $200 - 27 = 173$ (pounds)

(b) add: $108 + 17 = 125$ (pounds)

(c) multiply: $125 \times 25 = 3125$ (dollars)

(d) divide: $125 \div 25 = 5$ (books)

(e) divide: $2500 \div 125 = 20$ (dollars)

(f) subtract: $125 - 25 = 100$ (books)

(g) add: $125 + 25 = 150$ (books)

(h) divide: $144 \div 12 = 12$ (dozen)

(i) multiply: $144 \times 12 = 1728$ (eggs)

(j) divide: $255 \div 15 = 17$ (pieces)

(k) divide: $364 \div 28 = 13$ (cm)

(l) multiply: $18 \times 2 = 36$ (dollars)

(m) divide: $900 \div 18 = 50$ (cents)

(n) divide: $216 \div 27 = 8$ (rolls)

(o) multiply: $7 \times 5 = 35$;
then divide: $245 \div 35 = 7$ (dollars)

(p) multiply three times:
$8 \times 40 = 320, 8 \times 2 = 16, 26 \times 16 = 416$;
then add: $320 + 416 = 736$ (dollars)

Supplementary Exercises

PROBLEM 1-41   Draw a line under the digit in each whole number that has the given place value:

(a) thousands: 658,432,196,285

(b) ten millions: 269,015,347,250

(c) billions: 100,580,239,611

(d) tens: 591,036,549,872

(e) ten thousands: 300,000,000,000

(f) ten billions: 721,834,659,023

(g) ones: 487,291,635,205

(h) hundred billions: 721,834,659,023

(i) millions: 825,916,430,017

(j) hundreds: 200,310,508,200

(k) hundred millions: 534,816,729,885

(l) hundred thousands: 916,832,005,713

PROBLEM 1-42   Write each of the following whole numbers using period names instead of commas:

(a) 18,546

(b) 83,005

(c) 42,000

(d) 723,058

(e) 850,000

(f) 3,512,000

(g) 25,063,002

(h) 970,250,000,070

PROBLEM 1-43   Write the place-value name of each underlined digit:

(a) 285,316,487,915

(b) 817,639,854,203

(c) 105,213,698,540

(d) 903,142,586,200

(e) 305,002,913,617

(f) 785,413,826,918

(g) 463,218,915,834

(h) 692,413,876,518

(i) 588,765,432,017

(j) 261,823,415,286

(k) 199,381,418,275

(l) 960,108,235,872

PROBLEM 1-44   Write the value of 2 in each whole number:

(a) 862,153,089,451

(b) 615,843,569,205

(c) 318,792,013,459

(d) 213,085,619,318

(e) 186,753,048,629

(f) 734,206,518,349

(g) 963,825,619,004

(h) 425,871,643,019

(i) 563,871,942,018

(j) 618,594,231,075

(k) 185,619,348,752

(l) 791,843,621,587

PROBLEM 1-45   Write the correct whole number for each group of one-digit values:

(a) 1 ten   5 ones   **(b)** 8 ones   **(c)** 3 tens

(d) 2 hundreds   9 tens   7 ones   **(e)** 4 hundreds   6 ones   **(f)** 5 hundreds   1 ten

(g) 9 hundreds   **(h)** 6 thousands   4 tens   2 ones   **(i)** 9 thousands

PROBLEM 1-46   Write the correct whole number for each whole-number word name:

(a) five   **(b)** twelve   **(c)** thirty

(d) two hundred   **(e)** forty-three   **(f)** four hundred ten

(g) two thousand seventy-two   **(h)** seven thousand six hundred   **(i)** five hundred eleven

(j) fifteen   **(k)** sixty-nine   **(l)** eighteen

(m) ninety-one   **(n)** nineteen   **(o)** eight hundred seventeen

(p) nine hundred eight

PROBLEM 1-47   Write the correct whole-number word name for each whole number:

(a) 8   **(b)** 11   **(c)** 20   **(d)** 100   **(e)** 54   **(f)** 316   **(g)** 1080   **(h)** 706

(i) 2005   **(j)** 10,000 **(k)** 231   **(l)** 12   **(m)** 47   **(n)** 13   **(o)** 815   **(p)** 692

(q) 518   **(r)** 910   **(s)** 14   **(t)** 4017   **(u)** 19

PROBLEM 1-48   Write the smallest whole number of:

(a) 36 and 35   **(b)** 101 and 110   **(c)** 4000 and 3999   **(d)** 7201 and 7210

(e) 4103 and 4113   **(f)** 256,001 and 257,000   **(g)** 100,000 and 99,999   **(h)** 999,999 and 100,000

PROBLEM 1-49   List each group of whole numbers from largest to smallest:

(a) 63,   27,   82,   59   **(b)** 285,   582,   825,   258,   528,   852   **(c)** 3105,   3200,   3104,   3199,   2999

PROBLEM 1-50   List each group of whole numbers from smallest to largest:

(a) 58,   59,   61,   53,   60   **(b)** 1001,   1100,   1010,   1000,   1011,   1110,   1101

(c) 657,   576,   765,   675,   567,   756   **(d)** 213,   315,   428,   99,   5,   2153

PROBLEM 1-51   Round each whole number to the nearest ten:

(a) 53   **(b)** 95   **(c)** 69   **(d)** 870   **(e)** 2042   **(f)** 3927

PROBLEM 1-52   Round each whole number to the nearest hundred:

(a) 649   **(b)** 863   **(c)** 985   **(d)** 418   **(e)** 2950   **(f)** 59,987

PROBLEM 1-53   Round each whole number to the nearest thousand:

(a) 5369   **(b)** 4520   **(c)** 69,212   **(d)** 25,600

(e) 100,700   **(f)** 299,999   **(g)** 5,832,169   **(h)** 25,000,000

PROBLEM 1-54   Add whole numbers:

(a) 13 + 82   **(b)** 75 + 98   **(c)** 248 + 309   **(d)** 785 + 697

(e) 4028 + 3170   **(f)** 5816 + 4809   **(g)** 7019 + 8285   **(h)** 6487 + 9875

(i) 25 + 87 + 69   **(j)** 89 + 92 + 78 + 65   **(k)** 503 + 491 + 392   **(l)** 785 + 649 + 827 + 538

PROBLEM 1-55   Subtract whole numbers:

(a) 78 − 23   (b) 82 − 59   (c) 987 − 256   (d) 863 − 129

(e) 528 − 130   (f) 825 − 739   (g) 4816 − 382   (h) 5291 − 1834

(i) 7214 − 685   (j) 98,514 − 23,802   (k) 302 − 89   (l) 4002 − 1235

PROBLEM 1-56   Multiply by one digit:

(a) 2 × 43   (b) 9 × 38   (c) 2 × 80,143   (d) 5 × 64,980   (e) 4 × 21

(f) 3 × 408   (g) 7 × 10,231   (h) 46,105 × 8   (i) 6 × 51   (j) 671 × 5

(k) 5847 × 6   (l) 7 × 80   (m) 5 × 211   (n) 9 × 352   (o) 501 × 3

(p) 7891 × 4   (q) 8 × 3102   (r) 1 × 7685   (s) 5810 × 6   (t) 7 × 7059

PROBLEM 1-57   Multiply by two or more digits:

(a) 32 × 41   (b) 312 × 132   (c) 63 × 75   (d) 314 × 483   (e) 29 × 580   (f) 485 × 639

(g) 53 × 978   (h) 725 × 8169   (i) 18 × 39   (j) 238 × 432   (k) 14 × 512   (l) 693 × 528

(m) 45 × 803   (n) 213 × 3102   (o) 581 × 657   (p) 16 × 715   (q) 512 × 3018   (r) 85 × 608

PROBLEM 1-58   Multiply using shortcuts:

(a) 30 × 45   (b) 304 × 581   (c) 62 × 70   (d) 340 × 602   (e) 40 × 50

(f) 705 × 400   (g) 600 × 72   (h) 3058 × 6143   (i) 900 × 450   (j) 5106 × 2432

(k) 70 × 800   (l) 6003 × 5193   (m) 500 × 300   (n) 5820 × 4002   (o) 357 × 2000

(p) 3600 × 5002   (q) 2700 × 600   (r) 6000 × 7009   (s) 8000 × 4000   (t) 200,005 × 35,600

PROBLEM 1-59   Divide by a one-digit divisor:

(a) 15 ÷ 2   (b) 35 ÷ 4   (c) 23 ÷ 6   (d) 50 ÷ 8

(e) 45 ÷ 3   (f) 78 ÷ 5   (g) 609 ÷ 7   (h) 990 ÷ 9

(i) 246 ÷ 2   (j) 1069 ÷ 5   (k) 1740 ÷ 4   (l) 3070 ÷ 7

(m) 4248 ÷ 6   (n) 7475 ÷ 9   (o) 19,896 ÷ 8   (p) 24,230 ÷ 3

PROBLEM 1-60   Divide by a two-digit divisor:

(a) 100 ÷ 20   (b) 250 ÷ 30   (c) 33 ÷ 11   (d) 200 ÷ 21

(e) 912 ÷ 38   (f) 1050 ÷ 29   (g) 2250 ÷ 45   (h) 1089 ÷ 18

(i) 15,872 ÷ 64   (j) 27,900 ÷ 72   (k) 6885 ÷ 17   (l) 15,139 ÷ 28

(m) 251,906 ÷ 31   (n) 127,090 ÷ 42   (o) 78,033 ÷ 19   (p) 195,350 ÷ 39

PROBLEM 1-61   Divide by three or more digits:

(a) 2400 ÷ 300   (b) 1900 ÷ 200   (c) 2460 ÷ 410   (d) 4000 ÷ 520

(e) 13,041 ÷ 621   (f) 10,852 ÷ 312   (g) 30,972 ÷ 534   (h) 43,000 ÷ 653

(i) 266,875 ÷ 875   (j) 62,500 ÷ 297   (k) 151,200 ÷ 189   (l) 143,500 ÷ 685

(m) 6000 ÷ 2000   (n) 23,000 ÷ 4000   (o) 54,180 ÷ 3010   (p) 128,000 ÷ 5008

PROBLEM 1-62   Factor each whole number in as many different ways as possible:

(a) 3   (b) 5   (c) 6   (d) 8   (e) 10   (f) 20   (g) 36

PROBLEM 1-63 Factor each composite number as a product of primes:

(a) 12 (b) 82 (c) 15 (d) 85 (e) 18 (f) 87 (g) 88 (h) 90

(i) 91 (j) 92 (k) 93 (l) 94 (m) 95 (n) 96 (o) 98 (p) 99

(q) 100 (r) 102 (s) 86 (t) 56 (u) 84 (v) 120 (w) 81 (x) 144

PROBLEM 1-64 Solve the following word problems $(+, -, \times, \div)$:

(a) A certain car traveled 440 miles in 8 hours at a constant speed. What was the constant speed?

(b) A certain car traveled 375 miles on 15 gallons of gasoline. How many miles per gallon (mpg) did the car get?

(c) A certain train traveled 5 hours at 65 miles per hour (mph). How many miles did the train travel?

(d) A certain train traveled 315 miles at a constant rate of 45 mph. How long did the trip take?

(e) A certain bus traveled 720 miles while averaging 15 mpg. How many gallons of fuel were used?

(f) A certain bus used 50 gallons of fuel while averaging 16 mpg. How far did the bus travel?

(g) A certain plane traveled 825 miles on the first part of a trip and 975 miles on the second part. How far did the plane travel in all?

(h) The air distance between two cities is 1285 km. The driving distance between the same two cities is 1487 km. How much farther is the driving distance than the air distance?

(i) Linda earns $9 per hour. She worked 35 hours this week. How much did Linda earn this week?

(j) Paul earned $392 this week. He earns $8 per hour. How many hours did Paul work this week?

(k) Pat earned $504 this week while working 42 hours. How much does she earn per hour?

(l) A certain TV set costs $600. If there are 12 equal payments, how much is each payment for the TV set?

(m) A certain clothes dryer costs $500. Each payment is $25. How many payments must be made in all?

(n) A certain car is purchased for no money down and $285 per month for 36 months. How much will the car cost?

(o) There are 365 days in a normal year. How many *complete* 7-day weeks are in a normal year?

(p) There are 30 days in a business month. There are 12 business months in a business year. How many days in a business year?

(q) Beth worked 8 hours on Monday, 6 hours on Tuesday, 8 hours on Wednesday, 7 hours on Thursday, and 5 hours on Friday. She earns $15 per hour. How much did Beth earn Monday through Friday?

(r) Marcia earns $25 per hour. Frank earns $18 per hour. How much more does Marcia earn during a forty-hour work week than Frank?

Answers to Supplementary Exercises

(1-41) (a) 658,432,196,285 (b) 269,015,347,250 (c) 100,580,239,611 (d) 591,036,549,872
(e) 300,000,000,000 (f) 721,834,659,023 (g) 487,291,635,205 (h) 721,834,659,023
(i) 825,916,430,017 (j) 200,310,508,200 (k) 534,816,729,885 (l) 916,832,005,713

(1-42) (a) 18 thousand 546 (b) 83 thousand 5 (c) 42 thousand
(d) 723 thousand 58 (e) 850 thousand (f) 3 million 512 thousand
(g) 25 million 63 thousand 2 (h) 970 billion 250 million 70

(1-43) **(a)** billions **(b)** hundreds **(c)** ten billions **(d)** ten millions **(e)** millions **(f)** hundred thousands **(g)** hundred millions **(h)** ten thousands **(i)** thousands **(j)** ones **(k)** hundred billions **(l)** tens

(1-44) **(a)** 2,000,000,000 **(b)** 200 **(c)** 2,000,000 **(d)** 200,000,000,000 **(e)** 20 **(f)** 200,000,000 **(g)** 20,000,000 **(h)** 20,000,000,000 **(i)** 2000 **(j)** 200,000 **(k)** 2 **(l)** 20,000

(1-45) **(a)** 15 **(b)** 8 **(c)** 30 **(d)** 297 **(e)** 406 **(f)** 510 **(g)** 900 **(h)** 6042 **(i)** 9000

(1-46) **(a)** 5 **(b)** 12 **(c)** 30 **(d)** 200 **(e)** 43 **(f)** 410 **(g)** 2072 **(h)** 7600 **(i)** 511 **(j)** 15 **(k)** 69 **(l)** 18 **(m)** 91 **(n)** 19 **(o)** 817 **(p)** 908

(1-47) **(a)** eight **(b)** eleven **(c)** twenty **(d)** one hundred **(e)** fifty-four **(f)** three hundred sixteen **(g)** one thousand eighty **(h)** seven hundred six **(i)** two thousand five **(j)** ten thousand **(k)** two hundred thirty-one **(l)** twelve **(m)** forty-seven **(n)** thirteen **(o)** eight hundred fifteen **(p)** six hundred ninety-two **(q)** five hundred eighteen **(r)** nine hundred ten **(s)** fourteen **(t)** four thousand seventeen **(u)** nineteen

(1-48) **(a)** 35 **(b)** 101 **(c)** 3999 **(d)** 7201 **(e)** 4103 **(f)** 256,001 **(g)** 99,999 **(h)** 100,000

(1-49) **(a)** 82, 63, 59, 27 **(b)** 852, 825, 582, 528, 285, 258 **(c)** 3200, 3199, 3105, 3104, 2999

(1-50) **(a)** 53, 58, 59, 60, 61 **(b)** 1000, 1001, 1010, 1011, 1100, 1101, 1110 **(c)** 567, 576, 657, 675, 756, 765 **(d)** 5, 99, 213, 315, 428, 2153

(1-51) **(a)** 50 **(b)** 100 **(c)** 70 **(d)** 870 **(e)** 2040 **(f)** 3930

(1-52) **(a)** 600 **(b)** 900 **(c)** 1000 **(d)** 400 **(e)** 3000 **(f)** 60,000

(1-53) **(a)** 5000 **(b)** 5000 **(c)** 69,000 **(d)** 26,000 **(e)** 101,000 **(f)** 300,000 **(g)** 5,832,000 **(h)** 25,000,000

(1-54) **(a)** 95 **(b)** 173 **(c)** 557 **(d)** 1482 **(e)** 7198 **(f)** 10,625 **(g)** 15,304 **(h)** 16,362 **(i)** 181 **(j)** 324 **(k)** 1386 **(l)** 2799

(1-55) **(a)** 55 **(b)** 23 **(c)** 731 **(d)** 734 **(e)** 398 **(f)** 86 **(g)** 4434 **(h)** 3457 **(i)** 6529 **(j)** 74,712 **(k)** 213 **(l)** 2767

(1-56) **(a)** 86 **(b)** 342 **(c)** 160,286 **(d)** 324,900 **(e)** 84 **(f)** 1224 **(g)** 71,617 **(h)** 368,840 **(i)** 306 **(j)** 3355 **(k)** 35,082 **(l)** 560 **(m)** 1055 **(n)** 3168 **(o)** 1503 **(p)** 31,564 **(q)** 24,816 **(r)** 7685 **(s)** 34,860 **(t)** 49,413

(1-57) **(a)** 1312 **(b)** 41,184 **(c)** 4725 **(d)** 151,662 **(e)** 16,820 **(f)** 309,915 **(g)** 51,834 **(h)** 5,922,525 **(i)** 702 **(j)** 102,816 **(k)** 7168 **(l)** 365,904 **(m)** 36,135 **(n)** 660,726 **(o)** 381,717 **(p)** 11,440 **(q)** 1,545,216 **(r)** 51,680

(1-58) **(a)** 1350 **(b)** 176,624 **(c)** 4340 **(d)** 204,680 **(e)** 2000 **(f)** 282,000 **(g)** 43,200 **(h)** 18,785,294 **(i)** 405,000 **(j)** 12,417,792 **(k)** 56,000 **(l)** 31,173,579 **(m)** 150,000 **(n)** 23,291,640 **(o)** 714,000 **(p)** 18,007,200 **(q)** 1,620,000 **(r)** 42,054,000 **(s)** 32,000,000 **(t)** 7,120,178,000

(1-59) **(a)** 7 R1 **(b)** 8 R3 **(c)** 3 R5 **(d)** 6 R2 **(e)** 15 **(f)** 15 R3 **(g)** 87 **(h)** 110 **(i)** 123 **(j)** 213 R4 **(k)** 435 **(l)** 438 R4 **(m)** 708 **(n)** 830 R5 **(o)** 2487 **(p)** 8076 R2

(1-60) **(a)** 5   **(b)** 8 R10   **(c)** 3   **(d)** 9 R11   **(e)** 24   **(f)** 36 R6   **(g)** 50   **(h)** 60 R9
       (i) 248   **(j)** 387 R36   **(k)** 405   **(l)** 540 R19   **(m)** 8126   **(n)** 3025 R40   **(o)** 4107   **(p)** 5008 R38

(1-61) **(a)** 8   **(b)** 9 R100   **(c)** 6   **(d)** 7 R360   **(e)** 21   **(f)** 34 R244   **(g)** 58   **(h)** 65 R555
       (i) 305   **(j)** 210 R130   **(k)** 800   **(l)** 209 R335   **(m)** 3   **(n)** 5 R3000   **(o)** 18   **(p)** 25 R2800

(1-62) **(a)** 1×3 **(b)** 1×5 **(c)** $1 \times 6, 2 \times 3$
       (d) $1 \times 8, 2 \times 4$ **(e)** $1 \times 10, 2 \times 5$ **(f)** $1 \times 20, 2 \times 10, 4 \times 5, 2 \times 2 \times 5$
       (g) $1 \times 36, 2 \times 18, 3 \times 12, 4 \times 9, 6 \times 6, 2 \times 2 \times 9, 2 \times 3 \times 6, 3 \times 3 \times 4, 2 \times 2 \times 3 \times 3$

(1-63) **(a)** $2 \times 2 \times 3$ **(b)** 2×41 **(c)** 3×5
       (d) 5×17 **(e)** $2 \times 3 \times 3$ **(f)** 3×29
       (g) $2 \times 2 \times 2 \times 11$ **(h)** $2 \times 3 \times 3 \times 5$ **(i)** 7×13
       (j) $2 \times 2 \times 23$ **(k)** 3×31 **(l)** 2×47
       (m) 5×19 **(n)** $2 \times 2 \times 2 \times 2 \times 2 \times 3$ **(o)** $2 \times 7 \times 7$
       (p) $3 \times 3 \times 11$ **(q)** $2 \times 2 \times 5 \times 5$ **(r)** $2 \times 3 \times 17$
       (s) 2×43 **(t)** $2 \times 2 \times 2 \times 7$ **(u)** $2 \times 2 \times 3 \times 7$
       (v) $2 \times 2 \times 2 \times 3 \times 5$ **(w)** $3 \times 3 \times 3 \times 3$ **(x)** $2 \times 2 \times 2 \times 2 \times 3 \times 3$

(1-64) **(a)** 55 mph   **(b)** 25 mpg   **(c)** 325 miles   **(d)** 7 hours   **(e)** 48 gallons
       (f) 800 miles   **(g)** 1800 miles   **(h)** 202 km   **(i)** $315   **(j)** 49 hours
       (k) $12 per hour   **(l)** $50   **(m)** 20 payments   **(n)** $10,260   **(o)** 52 complete weeks
       (p) 360 days   **(q)** $510   **(r)** $280

2 INTEGERS

THIS CHAPTER IS ABOUT

☑ Identifying Integers
☑ Comparing Integers
☑ Finding Absolute Values
☑ Computing with Integers
☑ Using the Order of Operations

2-1. Identifying Integers

A. Using the number line

Recall: The whole numbers are listed as 0, 1, 2, 3, ...

To graph whole numbers, you can use a **number line.**

The number line is a line divided by points that represent numbers.
On the number line **(a)** the point that represents zero is called the **origin.**
(b) all points to the right of the origin represent **positive numbers.**
(c) all points to the left of the origin represent **negative numbers.**

EXAMPLE 2-1   Graph whole numbers on a number line.

Solution

graph of whole numbers

0 1 2 3 4 5 6 . . .   number line

whole numbers

Note: To write a positive number, you do *not* have to write a **positive sign** $(+)$, although you may include one.

EXAMPLE 2-2   Write some positive numbers.

Solution $+1$ or 1, $+2$ or 2, $+3$ or 3, . . .

B. Identifying positive and negative integers

The positive numbers listed in Example 2-2 are called the **positive integers:**

$$+1, +2, +3, . . . \quad \text{or just: } 1, 2, 3, . . .$$

Note: The natural numbers (see Chapter 1) and the positive integers are listed in the same way:

$$1, 2, 3, . . .$$

But only the positive integers are listed as:

$$+1, +2, +3, \ldots$$

On a number line, two numbers that are on opposite sides of the origin and that are the same distance from the origin are called **opposites.**

EXAMPLE 2-3   Graph the opposites of 1, 2, 3, 4, and 5.

Solution
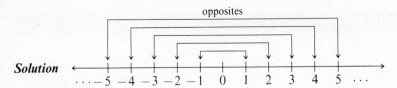

Negative numbers are the opposites of positive numbers. To write a negative number, you *must* use a negative sign $(-)$.

EXAMPLE 2-4   Write some negative numbers.

Solution $-1, -2, -3, -4, -5, -6, \ldots$

The negative numbers listed in Example 2-4 are called the **negative integers:**

$$-1, -2, -3, \ldots$$

or $$\ldots -3, -2, -1$$

The collection of all the negative integers, positive integers, and zero is called the **set of integers.**

EXAMPLE 2-5   List the set of integers.

Solution $\ldots -3, -2, -1, 0, 1, 2, 3, \ldots \leftarrow$ integers

Note: The ellipsis notation $(\ldots)$ indicates that the positive integers continue on forever in a positive counting pattern, and that the negative integers continue on forever in a negative counting pattern.

EXAMPLE 2-6   List **(a)** the first ten positive integers and **(b)** the first ten negative integers.

Solution $1, \quad 2, \quad 3, \quad 4, \quad 5, \quad 6, \quad 7, \quad 8, \quad 9, \quad 10 \longleftarrow$ first ten positive integers

$-10, -9, -8, -7, -6, -5, -4, -3, -2, -1 \longleftarrow$ first ten negative integers

2-2. Comparing Integers

To compare two numbers a and b using a number line, you write

(a) $a < b$ if a is to the *left* of b;

(b) $a > b$ if a is to the *right* of b;

(c) $a = b$ if a and b are represented by the *same* point.

EXAMPLE 2-7   Use $<, >,$ or $=$ to compare each of the following pairs of integers.

$$\textbf{(a)} \;-5 \text{ and } 2 \quad \textbf{(b)} \;-3 \text{ and } -4 \quad \textbf{(c)} \;+5 \text{ and } 5$$

Solution

(a) $-5 < 2$ because every negative number is to the left of every positive number on a number line.

(b) $-3 > -4$ because -3 is to the right of -4 on a number line.

(c) $+5 = 5$ because $+5$ and 5 are symbols for the same number, positive 5.

2-3. Finding Absolute Values

The distance between a given number and the origin (0) on a number line is called its **absolute value.** The **absolute value symbol** is | |. Read $|-5|$ as "the absolute value of negative five."

EXAMPLE 2-8   Find the absolute value of **(a)** $|+5|$, **(b)** $|-5|$, and **(c)** $|0|$.

Solution

(a) $|+5| = 5$ because $+5$ (or just 5) is 5 units from 0 on a number line.

(b) $|-5| = 5$ because -5 is 5 units from 0 on a number line.

(c) $|0| = 0$ because 0 is 0 units from 0 on a number line.

Note: The absolute value of any number is *never* negative.

2-4. Computing with Integers

A. Adding integers

To add two or more integers with like signs, you use the following addition rules.

Addition Rules for Two or More Numbers with Like Signs

(1) Find the sum of the absolute values of the addends.

(2) Write the same like sign on the sum.

EXAMPLE 2-9   Add the following integers with like signs:

$$\textbf{(a)} \ +5 + (+4) \qquad \textbf{(b)} \ -4 + (-3) \qquad \textbf{(c)} \ 8 + 5 + 9 + 7$$

Solution

absolute values

(a) $+5 + (+4) = ?\,9$ *Think:* $5 + 4 = 9$ ⟵ sum of absolute values

$\qquad\qquad\quad = +9 \text{ or } 9$ *Think:* Because both addends are positive, the sum is positive.

absolute values

(b) $-4 + (-3) = ?\,7$ *Think:* $4 + 3 = 7$ ⟵ sum of absolute values

$\qquad\qquad\quad = -7$ *Think:* Because both addends are negative, the sum is negative.

(c) $8 + 5 + 9 + 7 = ?\,29$ *Think:* $8 + 5 + 9 + 7 = 29$

$\qquad\qquad\qquad = +29 \text{ or } 29$ *Think:* No sign on a number means it is positive.

To add two integers with unlike signs, you can use the following addition rules.

Addition Rules for Two Numbers with Unlike Signs

(1) Find the difference between the absolute values of the addends.

(2) Write the sign of the number with the larger absolute value on the sum.

EXAMPLE 2-10   Add the following integers with unlike signs:

$$\textbf{(a)} \ -3 + 7 \qquad \textbf{(b)} \ 2 + (-8) \qquad \textbf{(c)} \ +6 + (-6)$$

Solution

larger absolute value

(a) $-3 + 7 = ?\,4$ *Think:* $7 - 3 = 4$ ⟵ difference between absolute values

$\qquad\qquad = +4 \text{ or } 4$ *Think:* Because 7 has the larger absolute value, the sum is positive.

larger absolute value
↓

(b) $2 + (-8) = ?\ 6$ *Think:* $8 - 2 = 6$ ←— difference between absolute values

　　　　　$= -6$ *Think:* Because -8 has the larger absolute value, the sum is negative.

(c) $+6 + (-6) = 0$ *Think:* The sum of two numbers that are opposites is always 0.

Note: The sum of any given integer and zero is always the given integer.

EXAMPLE 2-11 Add the following numbers:

$$\textbf{(a)}\ 0 + 7 \qquad \textbf{(b)}\ 0 + (-8) \qquad \textbf{(c)}\ 5 + 0 \qquad \textbf{(d)}\ -2 + 0$$

Solution

(a) $0 + 7 = 7$ 　　　**(b)** $0 + (-8) = -8$ 　　　**(c)** $5 + 0 = 5$ 　　　**(d)** $-2 + 0 = -2$

B. Subtracting integers

To subtract two integers, you can use the following subtraction rules.

Subtraction Rules for Two Numbers

(1) Change subtraction to addition and write the opposite of the subtrahend.

(2) Follow the rules for adding numbers.

(3) Check by adding the proposed difference to the original subtrahend to see if you get the original minuend.

EXAMPLE 2-12 Subtract the following integers:

$$\textbf{(a)}\ 5 - 8 \qquad \textbf{(b)}\ -2 - 7 \qquad \textbf{(c)}\ 6 - (-4) \qquad \textbf{(d)}\ -3 - (-9)$$

Solution

　　　change to +

(a) $5 - 8 = 5 + (-8)$ 　　　Add the opposite of the subtrahend.

　　　opposite of 8

　　　　$= -3$ 　　　Add numbers with unlike signs [see Example 2-10].

Check: $-3 + 8 = +5$ ←— -3 checks

　　　change to +

(b) $-2 - 7 = -2 + (-7)$ 　　　Add the opposite of the subtrahend.

　　　opposite of 7

　　　　$= -9$ 　　　Add numbers with like signs [see Example 2-9].

Check: $-9 + 7 = -2$ ←— -9 checks

　　　change to +

(c) $6 - (-4) = 6 + (+4)$ 　　　Add the opposite of the subtrahend.

　　　opposite of -4

　　　　$= +10$ or 10 　　　Add numbers with like signs.

Check: $10 + (-4) = 6$ ←— 10 checks

　　　change to +

(d) $-3 - (-9) = -3 + 9$ 　　　Add the opposite of the subtrahend.

　　　opposite of -9

　　　　$= 6$ 　　　Add numbers with unlike signs.

Check: $6 + (-9) = -3$ ←— 6 checks

C. Multiplying integers

Recall: There are several different ways to write a multiplication problem like "3 times 4" [see Example 1-39].

$$
\left.\begin{array}{l}
3 \text{ times } 4 = 3 \times 4 \\
\qquad\quad = 3 \cdot 4 \\
\qquad\quad = 3(4) \\
\qquad\quad = (3)4 \\
\qquad\quad = (3)(4)
\end{array}\right\} \text{all equivalent ways to write "3 times 4"}
$$

To multiply two integers, you can use the following multiplication rules.

Multiplication Rules for Two Numbers

(1) Find the product of the absolute values.

(2) Make the product (a) *positive* if the two factors have *like* signs;
  (b) *negative* if the two factors have *unlike* signs;
  (c) *zero* if either one of the factors is zero.

EXAMPLE 2-13 Multiply the following pairs of integers:

(a) $9 \times (-7)$ (b) $-5 \cdot 2$ (c) $(9)(3)$ (d) $(-2)0$ (e) $-4(-6)$

Solution

(a) $9 \times (-7) = ? \, 63$ *Think:* $9 \times 7 = 63 \longleftarrow$ product of the absolute values
  $\qquad\qquad\quad = -63$ The factors 9 and -7 have unlike signs so the product is negative.

(b) $-5 \cdot 2 = ? \, 10$ *Think:* $5 \times 2 = 10$
  $\qquad\qquad = -10$ The factors -5 and 2 have unlike signs so the product is negative.

(c) $(9)(3) = ? \, 27$ *Think:* $9 \times 3 = 27$
  $\qquad\quad = +27 \text{ or } 27$ The factors 9 and 3 have like signs so the product is positive.

(d) $(-2)0 = 0$ *Think:* When one of the factors is zero the product is always zero.

(e) $-4(-6) = ? \, 24$ *Think:* $4 \times 6 = 24$
  $\qquad\qquad = +24 \text{ or } 24$ *Think:* The factors -4 and -6 have like signs so the product is positive.

Note: The product of two negative integers is always positive!

To multiply more than two integers you can use the following multiplication rules.

Multiplication Rules for Two or More Numbers

(1) Find the product of all the absolute values.

(2) Make the product (a) *positive* if there are an *even number* (0, 2, 4, . . .) of negative factors;
  (b) *negative* if there are an *odd number* (1, 3, 5, . . .) of negative factors;
  (c) *zero* if one or more factors are zero.

EXAMPLE 2-14 Multiply the following groups of integers:

(a) $5 \cdot 3 \cdot 2$ (b) $4(-1)3$ (c) $-5(-2)2$ (d) $-1(-6)(-3)$ (e) $2(-3)(-1)(0)(-3)(-2)$

Solution

(a) $5 \cdot 3 \cdot 2 = ?\ 30$ *Think:* $5 \times 3 \times 2 = 30$ ⟵ product of the absolute values

            $= +30$ or 30 There is an even number (0) of negative signs so the product is positive.

(b) $4(-1)3 = ?\ 12$ *Think:* $4 \times 1 \times 3 = 12$

            $= -12$ There is an odd number (1) of negative signs so the product is negative.

(c) $-5(-2)2 = ?\ 20$ *Think:* $5 \times 2 \times 2 = 20$

            $= +20$ or 20 There is an even number (2) of negative signs so the product is positive.

(d) $-1(-6)(-3) = ?\ 18$ *Think:* $1 \times 6 \times 3 = 18$

            $= -18$ There is an odd number (3) of negative signs so the product is negative.

(e) $2(-3)(-1)(0)(-3)(-2) = 0$ *Think:* When one or more factors are zero, the product is always zero.

D. Dividing integers

Recall: There are several different ways to write the division problem "8 divided by 3" [see Example 1-50].

$$
\left.
\begin{aligned}
8 \text{ divided by } 3 &= 8 \div 3 \\
&= 8/3 \\
&= \frac{8}{3} \\
&= 8 \cdot \frac{1}{3} \\
&= \frac{1}{3} \cdot 8 \\
&= 3\overline{)8}
\end{aligned}
\right\} \text{all equivalent ways of writing "8 divided by 3"}
$$

To divide two integers, you can use the following division rules.

Division Rules for Two Numbers

(1) Find the quotient of the absolute values.

(2) Use the sign rules for multiplication to write the correct sign on the quotient.

(3) Check by multiplying the proposed quotient by the original divisor to see if you get the original dividend.

Recall: Zero divided by any nonzero number is always zero.
        Division by zero is not defined.

EXAMPLE 2-15 Divide the following pairs of integers:

(a) $15 \div 3$ (b) Divide -10 by 2.   (c) $\dfrac{8}{-2}$ (d) $-3\overline{)-18}$ (e) $\dfrac{0}{3}$ (f) $\dfrac{3}{0}$ (g) $\dfrac{0}{0}$

Solution

(a) $15 \div 3 = +5$ or 5 Use the sign rules for multiplication.

    Check: $5 \times 3 = 15$ ⟵ 5 checks

(b) Divide -10 by $2 = -10 \div 2$ Find the quotient of the absolute values.

                     $= -5$ Use the sign rules for multiplication.

    Check: $-5 \times 2 = -10$ ⟵ -5 checks

(c) $\dfrac{8}{-2} = 8 \div (-2)$ Find the quotient of the absolute values.

            $= -4$ Use the sign rules for multiplication.

    Check: $-4(-2) = +8$ or 8 ⟵ -4 checks

(d) $-3\overline{)-18} = -18 \div (-3)$ Find the quotient of the absolute values.

             $= +6$ or 6 Use the sign rules for multiplication.

    Check: $6(-3) = -18$ ⟵ 6 checks

(e) $\dfrac{0}{3} = 0$ Zero divided by any nonzero real number is always zero.

    Check: $0 \times 3 = 0$ ⟵ 0 checks

(f) $\dfrac{3}{0}$ is not defined.

(g) $\dfrac{0}{0}$ is not defined.

2-5. Using the Order of Operations

To avoid getting a wrong answer when computing with integers, you must use the following rules:

Order of Operations

- First, *clear grouping symbols* such as ( ), [ ], and { } by performing the operations inside of them.
- Next, *multiply or divide* in order *from left to right.*
- Last, *add or subtract* in order *from left to right.*

EXAMPLE 2-16   Evaluate the expressions **(a)** $3 + 5 \cdot 2$ and **(b)** $(3 + 5)2$.

Solution

(a) $3 + 5 \cdot 2 = 3 + 10$ When there are no grouping symbols, multiply and divide before you add

           $= 13$ or subtract.

(b) $(3 + 5)2 = 8 \cdot 2$ When there are grouping symbols, always clear grouping symbols first.

          $= 16$

Caution: To avoid getting a wrong answer when there are no grouping symbols present, always multiply and divide *before* adding and subtracting.

EXAMPLE 2-17   Evaluate the expressions **(a)** $8 + 6 \div 2$ and **(b)** $18 - 2 \cdot 3$.

Solution

Correct Method	Wrong Method

(a) $8 + 6 \div 2 = 8 + 3$ Divide first.

$\qquad\qquad = 11$ ⟵ correct answer

(a) $8 + 6 \div 2 = 14 \div 2$ No! Divide first.

$\qquad\qquad = 7$ ⟵ wrong answer

(b) $18 - 2 \cdot 3 = 18 - 6$ Multiply first.

$\qquad\qquad = 12$ ⟵ correct answer

(b) $18 - 2 \cdot 3 = 16 \cdot 3$ No! Multiply first.

$\qquad\qquad = 48$ ⟵ wrong answer

Caution: Be sure to work from left to right when adding, subtracting, multiplying, or dividing.

EXAMPLE 2-18   Evaluate the expressions **(a)** $3 - 5 + 7$ and **(b)** $12 \div 2 \times 3$.

Solution

Correct Method	Wrong Method

(a) $3 - 5 + 7 = -2 + 7$ Work in order from left to right.

$\qquad\qquad = 5$ ⟵ correct answer

(a) $3 - 5 + 7 = 3 - 12$ No! Never add and subtract from right to left.

$\qquad\qquad = -9$ ⟵ wrong answer

(b) $12 \div 2 \times 3 = 6 \times 3$ Work in order from left to right.

$\qquad\qquad = 18$ ⟵ correct answer

(b) $12 \div 2 \times 3 = 12 \div 6$ No! Never multiply and divide from right to left.

$\qquad\qquad = 2$ ⟵ wrong answer

SOLVED PROBLEMS

PROBLEM 2-1   Use $<$, $>$, or $=$ to compare each pair of numbers:

(a) $2, 5$ **(b)** $2, -5$ **(c)** $-2, 5$ **(d)** $-2, -5$ **(e)** $0, \dfrac{0}{2}$ **(f)** $+4, 4$

Solution   Recall that to compare two numbers a and b using a number line, you write $a < b$ if a is to the left of b, $a > b$ if a is to the right of b, and $a = b$ if a and b represent the same point [see Example 2-7]:

(a) $2 < 5$ because 2 is to the left of 5 on a number line.

(b) $2 > -5$ because every positive number is to the right of every negative number.

(c) $-2 < 5$ because every negative number is to the left of every positive number.

(d) $-2 > -5$ because -2 is to the right of -5 on a number line.

(e) $0 = \dfrac{0}{2}$ because 0 and $\dfrac{0}{2}$ are symbols for the same number.

(f) $+4 = 4$ because $+4$ and 4 are symbols for the same number.

PROBLEM 2-2   Find the absolute values of **(a)** $|+2|$, **(b)** $|-6|$, and **(c)** $|0|$.

Solution   Recall that the distance that a given number is from the origin (0) on a number line is the given number's absolute value [see Example 2-8]:

(a) $|+2| = 2$ because 2 is 2 units from 0 on a number line.

(b) $|-6| = 6$ because -6 is 6 units from 0 on a number line.

(c) $|0| = 0$ because 0 units from 0 on a number line.

PROBLEM 2-3 Add integers with like signs:

(a) $+5 + (+8)$ (b) $-6 + (-3)$ (c) $4 + 7$ (d) $+8 + (+7) + (+2)$

Solution Recall that to add two or more integers with like signs, you first find the sum of the absolute values and then write the same like sign on the sum [see Example 2-9]:

same like sign

sum of the absolute values

(a) $+5 + (+8) = +(5 + 8) = +13 \text{ or } 13$ (b) $-6 + (-3) = -(6 + 3) = -9$

(c) $4 + 7 = 11 \text{ or } +11$ (d) $+8 + (+7) + (+2) = +(8 + 7 + 2) = +17 \text{ or } 17$

PROBLEM 2-4 Add integers with unlike signs:

(a) $+8 + (-6)$ (b) $2 + (-9)$ (c) $-7 + (+8)$ (d) $-8 + 1$

(e) $0 + 6$ (f) $0 + (-7)$ (g) $+9 + 0$ (h) $-4 + 0$

Solution Recall that to add two integers with unlike signs, you first find the difference between the absolute values and then write the sign of the number with the larger absolute value on the sum [see Examples 2-10 and 2-11]:

sign of the number with the larger absolute value

difference of the absolute values

(a) $+8 + (-6) = +(8 - 6) = +2 \text{ or } 2$ (b) $2 + (-9) = -(9 - 2) = -7$

(c) $-7 + (+8) = +(8 - 7) = +1 \text{ or } 1$ (d) $-8 + 1 = -(8 - 1) = -7$

(e) $0 + 6 = 6$ (f) $0 + (-7) = -7$ (g) $+9 + 0 = +9 \text{ or } 9$ (h) $-4 + 0 = -4$

PROBLEM 2-5 Subtract integers:

(a) $+5 - (+2)$ (b) $6 - 8$ (c) $-6 - (+4)$ (d) $-2 - 9$

(e) $+8 - (-3)$ (f) $5 - (-8)$ (g) $-5 - (-2)$ (h) $-3 - (-7)$

Solution Recall that to subtract two integers, you first change subtraction to addition while writing the opposite of the subtrahend and then follow the rules for adding real numbers [see Example 2-12]:

change to addition

write opposite of subtrahend

(a) $+5 - (+2) = +5 + (-2) = +3 \text{ or } 3$ (b) $6 - 8 = 6 + (-8) = -2$

(c) $-6 - (+4) = -6 + (-4) = -10$ (d) $-2 - 9 = -2 + (-9) = -11$

(e) $+8 - (-3) = +8 + (+3) = +11 \text{ or } 11$ (f) $5 - (-8) = 5 + (+8) = +13 \text{ or } 13$

(g) $-5 - (-2) = -5 + (+2) = -3$ (h) $-3 - (-7) = -3 + (+7) = +4 \text{ or } 4$

PROBLEM 2-6 Multiply integers:

(a) $5 \cdot 2$ (b) $+4(+3)$ (c) $(-2)3$ (d) $5(-3)$

(e) $+7(-4)$ (f) $(-5)(+6)$ (g) $-2(-8)$ (h) $(-3)(-2)(-5)$

Solution Recall that to multiply two or more numbers, you first find the product of all the absolute values and then make the product positive if there are an even number of negative factors; negative if there are an odd number of negative factors; zero if one or more factors are zero [see Examples 2-13 and 2-14]:

even number (0) of negative factors

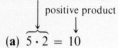
positive product

(a) $5 \cdot 2 = 10$

(d) $5(-3) = -15$

(g) $-2(-8) = +16$ or 16

(b) $+4(+3) = +12$ or 12

(e) $+7(-4) = -28$

(h) $(-3)(-2)(-5) = -30$

odd number (1) of negative factors

negative product

(c) $(-2)3 = -6$

(f) $(-5)(+6) = -30$

PROBLEM 2-7 Divide integers:

(a) $12 \div 3$ **(b)** $-18 \div 2$ **(c)** $20 \div (-5)$ **(d)** $-10 \div (-2)$ **(e)** Divide $+8$ by $+2$

(f) $\dfrac{+15}{-3}$ **(g)** $-6\overline{)-12}$ **(h)** $0 \div (-4)$ **(i)** $-4 \div 0$ **(j)** $0 \div 0$

Solution Recall that to divide two integers, you first find the quotient of the absolute values and then use the sign rules for multiplication to write the correct sign on the product [see Example 2-15]:

two positive numbers

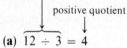
positive quotient

(a) $12 \div 3 = 4$

one negative number

negative quotient

(b) $-18 \div 2 = -9$

two negative numbers

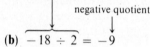
positive quotient

(c) $20 \div (-5) = -4$

(e) Divide $+8$ by $+2 = +8 \div (+2) = +4$ or 4

(g) $-6\overline{)-12} = -12 \div (-6) = +2$ or 2

(i) $-4 \div 0$ is not defined.

(d) $-10 \div (-2) = +5$ or 5

(f) $\dfrac{+15}{-3} = +15 \div (-3) = -5$

(h) $0 \div (-4) = 0$

(j) $0 \div 0$ is not defined.

PROBLEM 2-8 Evaluate using the Order of Operations:

(a) $8 - 6 + 2$ **(b)** $8 \div 2 \cdot 4$ **(c)** $9 - 3 \cdot 5$ **(d)** $9 + 3 \div 3$ **(e)** $4 + 6(-2)$ **(f)** $(4 + 6) \div 2$

(g) $5 - [8 \div (1 + 3)]3 + 6$ **(h)** $2 + [6 - (-4)(-2 + 8)] \div 6$

(i) $-3 + \{-2[16 + (-4 - (-3))] \div (-5)\}(-2)$

Solution Recall that to evaluate using the Order of Operations, you:

(1) Perform operations inside grouping symbols such as ( ), [ ], { }.

(2) Multiply or divide in order from left to right.

(3) Add or subtract in order from left to right. [See Examples 2-16, 2-17, and 2-18]:

(a) $8 - 6 + 2 = 2 + 2 = 4$ **(b)** $8 \div 2 \cdot 4 = 4 \cdot 4 = 16$ **(c)** $9 - 3 \cdot 5 = 9 - 15 = -6$

(d) $9 + 3 \div 3 = 9 + \dfrac{3}{3} = 10$ **(e)** $4 + 6(-2) = 4 + (-12) = -8$ **(f)** $(4 + 6) \div 2 = 10 \div 2 = 5$

(g) $5 - [8 \div (1 + 3)]3 + 6 = 5 - [8 \div 4]3 + 6 = 5 - 2 \cdot 3 + 6 = 5 - 6 + 6 = -1 + 6 = 5$

(h) $2 + [6 - (-4)(-2 + 8)] \div 6 = 2 + [6 - (-4)(6)] \div 6$

$$= 2 + [6 - (-24)] \div 6$$
$$= 2 + 30 \div 6$$
$$= 2 + 5$$
$$= 7$$

(i) $-3 + \{-2[16 + (-4 - (-3))] \div (-5)\}(-2) = -3 + \{-2[16 + (-1)] \div (-5)\}(-2)$

$$= -3 + \{-2[15] \div (-5)\}(-2)$$
$$= -3 + \{-30 \div (-5)\}(-2)$$
$$= -3 + \{6\}(-2)$$
$$= -3 + (-12)$$
$$= -15$$

Supplementary Exercises

PROBLEM 2-9   Use $<$, $>$, or $=$ to compare each pair of numbers:

(a) $5, 8$ **(b)** $-3, 4$ **(c)** $5, -9$ **(d)** $-6, -8$

(e) $0, 7$ **(f)** $0, -5$ **(g)** $2, +2$ **(h)** $\dfrac{0}{-5}, 0$

PROBLEM 2-10   Find each absolute value:

(a) $|2|$ **(b)** $|-9|$ **(c)** $|+7|$ **(d)** $|0|$ **(e)** $|1|$ **(f)** $|-1|$

PROBLEM 2-11   Compute with integers:

(a) $+4 + (+2)$ **(b)** $7 + (-7)$ **(c)** $4 - (-7)$ **(d)** $-6 + 5$

(e) $-8 + (-9)$ **(f)** $-8 - 6$ **(g)** $-7 - (-8)$ **(h)** $2 - 9$

(i) $2 + 3 + 5$ **(j)** $-3 + 5 - 4$ **(k)** $4 - (-3) + 5$ **(l)** $-1 - 3 + (-4)$

(m) $2 - 3 + 1 + (-2)$ **(n)** $-2 - 3 + (-4) - (-3)$ **(o)** 4×8 **(p)** $36 \div 6$

(q) $-6(7)$ **(r)** $-21 \div 3$ **(s)** $8(-3)$ **(t)** $\dfrac{18}{6}$

(u) $-7(-7)$ **(v)** $\dfrac{-35}{-7}$ **(w)** $5 \cdot 3 \cdot 2$ **(x)** $50 \div 5 \div (-2)$

PROBLEM 2-12   Evaluate using the Order of Operations:

(a) $4 \cdot 6(-3)$ **(b)** $\dfrac{-4}{-6} + 3$ **(c)** $4 - 6 \div (-3)$

(d) $-4(-6) - 3$ **(e)** $[4 + (-6)](-3)$ **(f)** $(-4 - 6) \div 5$

(g) $-5[-7 + 2(-3)]$ **(h)** $-20 \div [10 - (-3)(-2)]$ **(i)** $7[-4 - (-10) \div 2] + 2$

(j) $5 - [12 + (-2)(-4)] \div (-5)$

Answers to Supplementary Exercises

(2-9) (a) $5 < 8$ **(b)** $-3 < 4$ **(c)** $5 > -9$ **(d)** $-6 > -8$

     (e) $0 < 7$ **(f)** $0 > -5$ **(g)** $2 = +2$ **(h)** $\dfrac{0}{-5} = 0$

(2-10) (a) 2   **(b)** 9   **(c)** 7   **(d)** 0   **(e)** 1   **(f)** 1

(2-11) (a) 6   **(b)** 0   **(c)** 11   **(d)** -1 **(e)** -17 **(f)** -14 **(g)** 1   **(h)** -7

     (i) 10   **(j)** -2 **(k)** 12   **(l)** -8 **(m)** -2 **(n)** -6 **(o)** 32   **(p)** 6

     (q) -42 **(r)** -7 **(s)** -24 **(t)** 3   **(u)** 49   **(v)** 5   **(w)** 30   **(x)** -5

(2-12) (a) -72 **(b)** 5   **(c)** 6   **(d)** 21   **(e)** 6

     (f) 2   **(g)** 65   **(h)** -5 **(i)** 9   **(j)** 9

3 FRACTIONS AND MIXED NUMBERS

THIS CHAPTER IS ABOUT

- ☑ Understanding Fractions and Mixed Numbers
- ☑ Renaming Fractions and Mixed Numbers
- ☑ Simplifying Fractions and Mixed Numbers
- ☑ Multiplying with Fractions and Mixed Numbers
- ☑ Dividing with Fractions and Mixed Numbers
- ☑ Adding Fractions and Mixed Numbers
- ☑ Subtracting Fractions and Mixed Numbers
- ☑ Solving Word Problems Involving Fractions and Mixed Numbers

3-1. Understanding Fractions and Mixed Numbers

A. Understanding fractions

To understand fractions, you can think of them as *divisions of whole numbers*. If a and b are whole numbers ($b \neq 0$), then $\frac{a}{b}$ is called a **common fraction,** or just a **fraction.**

Note: Fractions can also be written as a/b or $\dfrac{a}{b}$.

EXAMPLE 3-1   Use the whole numbers 2 and 3 to make two different fractions.

Solution $\dfrac{2}{3}$ and $\dfrac{3}{2}$ ⟵ fractions

In the fraction $\frac{2}{3}$, 2 (the dividend) is called the **numerator,** 3 (the divisor) is called the **denominator,** and the horizontal bar that separates 2 and 3 (and indicates division) is called the **fraction bar.**

EXAMPLE 3-2   Name the three different parts of the fraction $\frac{0}{5}$.

Solution
$$\begin{array}{l} 0 \longleftarrow \text{numerator} \\ \overline{} \longleftarrow \text{fraction bar} \\ 5 \longleftarrow \text{denominator} \end{array}$$

Note: If $\frac{a}{b}$ is a fraction, then by definition the denominator b cannot be zero ($b \neq 0$).

EXAMPLE 3-3   (a) Can $\frac{0}{5}$ be defined as a fraction? (b) Can $\frac{5}{0}$?

Solution

(a) $\frac{0}{5}$ is defined as a fraction because 0 and 5 are whole numbers and $5 \neq 0$.

(b) $\frac{5}{0}$ is not defined because the denominator is zero.

When the numerator is less than the denominator, the fraction is called a **proper fraction.**

EXAMPLE 3-4   Which of the following are proper fractions?

$$\text{(a) } \frac{3}{4} \quad \text{(b) } \frac{9}{9} \quad \text{(c) } \frac{5}{2}$$

Solution

(a) $\frac{3}{4}$ is a proper fraction because 3 is less than 4 (3 < 4).

(b) $\frac{9}{9}$ is not a proper fraction because 9 is not less than 9 (9 = 9).

(c) $\frac{5}{2}$ is not a proper fraction because 5 is not less than 2 (5 > 2).

To name part of a whole, you can use a proper fraction.

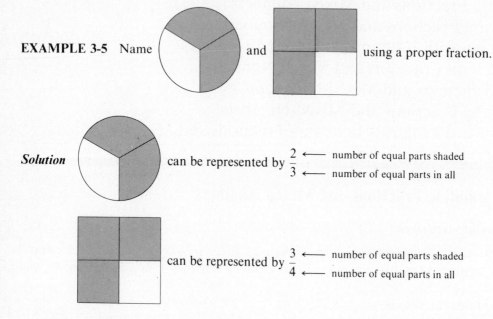

EXAMPLE 3-5   Name [circle] and [square] using a proper fraction.

Solution   [circle] can be represented by $\frac{2}{3}$ ⟵ number of equal parts shaded / ⟵ number of equal parts in all

[square] can be represented by $\frac{3}{4}$ ⟵ number of equal parts shaded / ⟵ number of equal parts in all

When the numerator is greater than or equal to the denominator, the fraction is called an **improper fraction.**

EXAMPLE 3-6   Which of the following are improper fractions?

$$\text{(a) } \frac{2}{3} \quad \text{(b) } \frac{5}{5} \quad \text{(c) } \frac{4}{1}$$

Solution

(a) $\frac{2}{3}$ is not an improper fraction because 2 is not greater than or equal to 3 (2 < 3).

(b) $\frac{5}{5}$ is an improper fraction because 5 is greater than or equal to 5 (5 = 5).

(c) $\frac{4}{1}$ is an improper fraction because 4 is greater than or equal to 1 (4 > 1).

To read or write the word name of a fraction, you use the whole-number name of the numerator, plus the name of the denominator, as follows:

(a) **wholes** if the denominator is 1;   (f) **sixths** if the denominator is 6;

(b) **halves** if the denominator is 2;   (g) **sevenths** if the denominator is 7;

(c) **thirds** if the denominator is 3;   (h) **eighths** if the denominator is 8;

(d) **fourths** if the denominator is 4;   (i) **ninths** if the denominator is 9;

(e) **fifths** if the denominator is 5;   (j) **tenths** if the denominator is 10;

and so on.

EXAMPLE 3-7   How would you read or write the following fractions?

$$\frac{9}{1}, \frac{5}{2}, \frac{2}{3}, \frac{3}{4}, \frac{7}{8}, \frac{11}{10}$$

Solution   Read or write $\frac{9}{1}$ as "nine-wholes."   Read or write $\frac{5}{2}$ as "five-halves."

Read or write $\frac{2}{3}$ as "two-thirds."   Read or write $\frac{3}{4}$ as "three-fourths."

Read or write $\frac{7}{8}$ as "seven-eighths."   Read or write $\frac{11}{10}$ as "eleven-tenths."

Caution: When writing the word name for a fraction, always place a hyphen between the word name for the numerator and the word name for the denominator.

Note: When the numerator of the fraction is 1, the denominator is singular, and should be read or written that way.

EXAMPLE 3-8   How would you read or write the following fractions?

$$\frac{1}{2}, \frac{1}{3}, \frac{1}{4}, \frac{1}{5}, \frac{1}{6}$$

Solution   Read or write $\frac{1}{2}$ as "one-half."   ⟵ *half* is the singular version of *halves*

Read or write $\frac{1}{3}$ as "one-third."   Read or write $\frac{1}{4}$ as "one-fourth."

Read or write $\frac{1}{5}$ as "one-fifth."   Read or write $\frac{1}{6}$ as "one-sixth."

B. Understanding mixed numbers

A whole number joined together with a fraction is called a **mixed number.**

EXAMPLE 3-9   Use 2 and $\frac{3}{4}$ to write a mixed number.

Solution   2 and $\frac{3}{4}$ can be written as the mixed number $2\frac{3}{4}$.

In the mixed number $2\frac{3}{4}$, 2 is called the **whole-number part** and $\frac{3}{4}$ is called the **fraction part.** To name one or more wholes and part of a whole together, you can use a mixed number.

EXAMPLE 3-10   Use a mixed number to name

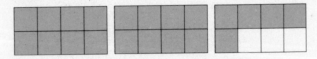

Solution

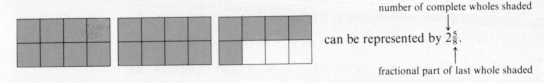

number of complete wholes shaded

can be represented by $2\frac{5}{8}$.

fractional part of last whole shaded

Caution: Always include the word *and* between the whole-number part and the fraction part when reading or writing a mixed number.

EXAMPLE 3-11   How would you read or write the word name for $2\frac{3}{4}$?

Solution   Read or write $2\frac{3}{4}$ as "two and three-fourths."

Note: Mixed numbers are a short way to write the sum of a whole number and a fraction.

EXAMPLE 3-12 Rename $2\frac{3}{4}$ as a sum of a whole number and a fraction.

Solution
$$2\frac{3}{4} = 2 + \frac{3}{4} \longleftarrow \text{sum}$$

EXAMPLE 3-13 Rename $5 + \frac{2}{3}$ as a mixed number.

Solution
$$5 + \frac{2}{3} = 5\frac{2}{3} \longleftarrow \text{mixed number}$$

3-2. Renaming Fractions and Mixed Numbers

A. Identifying equal fractions

Two fractions that represent the same part of a whole are called **equal fractions.**

EXAMPLE 3-14 Are $\frac{3}{4}$ and $\frac{6}{8}$ equal fractions?

Solution $\frac{3}{4}$ and $\frac{6}{8}$ are equal fractions because $\frac{3}{4}$ and $\frac{6}{8}$ represent the same part of a whole:

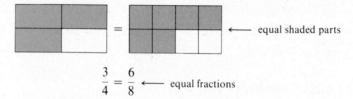

Two given fractions are equal if their **cross products** are equal.

EXAMPLE 3-15 Are $\frac{12}{18}$ and $\frac{10}{15}$ equal fractions? Are $\frac{6}{8}$ and $\frac{8}{10}$?

Solution $\frac{12}{18}$ and $\frac{10}{15}$ are equal fractions because their cross products are equal:

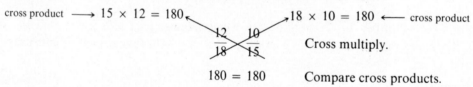

$\frac{6}{8}$ and $\frac{8}{10}$ are not equal fractions because their cross products are not equal:

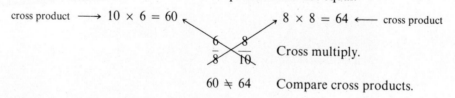

B. Renaming division problems as fractions

Every division problem can be renamed as a fraction.

To rename a division problem as a fraction, you follow these steps:

(1) Write the dividend as the numerator.

(2) Write the division symbol ($\div$) as the fraction bar (—).

(3) Write the divisor as the denominator.

EXAMPLE 3-16 Rename $8 \div 3$ as a fraction.

Solution
$$\frac{8}{3} \begin{array}{l} \longleftarrow \text{dividend} \\ \longleftarrow \text{divided by} \\ \longleftarrow \text{divisor} \end{array}$$

To rename a fraction as a division problem, you just reverse the previous steps:

(1) Write the numerator as the dividend.

(2) Write the fraction bar as the division symbol.

(3) Write the denominator as the divisor.

EXAMPLE 3-17   Rename $\frac{5}{2}$ as a division problem.

Solution

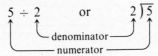

C. Renaming whole numbers as equal fractions

To rename a whole number as an equal fraction, you can write the whole number over 1.

EXAMPLE 3-18   Rename 5 as an equal fraction.

Solution
$$5 = 5 \div 1 = \frac{5}{1} \longleftarrow \text{ a fraction that is equal to 5}$$

There are many ways to rename a whole number as a fraction.

EXAMPLE 3-19   Rename 5 as an equal fraction in several different ways.

Solution
$$5 = 5 \div 1 = \frac{5}{1}$$
$$5 = 10 \div 2 = \frac{10}{2}$$
$$5 = 15 \div 3 = \frac{15}{3} \Big\} \text{ fractions that are equal to 5}$$
$$5 = 20 \div 4 = \frac{20}{4}$$
and so on

> *Agreement:* Unless otherwise stated, "rename a whole number as an equal fraction" means you rename the whole number as an equal fraction with a denominator of 1.

To rename a fraction with a denominator of 1 as an equal whole number, you write just the numerator.

EXAMPLE 3-20   Rename $\frac{3}{1}$ as an equal whole number.

Solution
$$\frac{3}{1} = 3 \div 1 = 3 \longleftarrow \text{ whole number that is equal to } \frac{3}{1}$$

D. Rename mixed numbers as fractions

Every mixed number can be renamed as an equal improper fraction.

The numerator is the only number that changes when a mixed number is renamed; the denominator always remains the same. To find the numerator of an equal improper fraction when renaming a mixed number, follow these steps:

(1) Multiply the denominator of the fraction part by the whole-number part.

(2) Add that product to the numerator of the fraction part.

EXAMPLE 3-21 Rename the mixed numbers **(a)** $2\frac{3}{4}$ and **(b)** $5\frac{7}{8}$ as equal improper fractions.

Solution

(a) $2\frac{3}{4} = \dfrac{11}{4}$ ⟵ $4 \times 2 + 3 = 8 + 3 = 11$

⟵ same denominator

(b) $5\frac{7}{8} = \dfrac{47}{8}$ ⟵ $8 \times 5 + 7 = 40 + 7 = 47$

⟵ same denominator

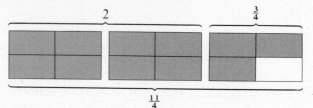

To rename an improper fraction as an equal mixed number using division, you divide the numerator by the denominator. Then

- The quotient becomes the whole-number part of the mixed number.
- The remainder is the numerator of the fraction part.
- The divisor is the denominator of the fraction part, which always remains the same.

EXAMPLE 3-22 Rename the improper fractions **(a)** $\frac{23}{4}$ and **(b)** $\frac{14}{3}$ as equal mixed numbers using division.

Solution

(a) $\dfrac{23}{4} = 5\frac{3}{4}$ because: $\dfrac{23}{4} = 23 \div 4$ or $4\overline{)\,23\,}$

$$
\begin{array}{r}
5\tfrac{3}{4} \leftarrow \text{remainder} \\
\leftarrow \text{divisor} \\
4\overline{)\;23\;} \\
-20 \\
\hline
3
\end{array}
$$

(b) $\dfrac{14}{3} = 4\frac{2}{3}$ because: $\dfrac{14}{3} = 14 \div 3$ or $3\overline{)\,14\,}$

$$
\begin{array}{r}
4\tfrac{2}{3} \leftarrow \text{remainder} \\
\leftarrow \text{divisor} \\
3\overline{)\;14\;} \\
-12 \\
\hline
2
\end{array}
$$

Note: The answer to the division problem $23 \div 4$ can be written in two forms:

(1) remainder form: 5 R3   **(2)** mixed-number form: $5\frac{3}{4}$

3-3. Simplifying Fractions and Mixed Numbers

A. Simplifying proper fractions

A fraction in which the numerator and denominator do not share a common factor other than 1 is called a **fraction in lowest terms**.

EXAMPLE 3-23 Which of the fractions $\frac{3}{4}$ and $\frac{4}{6}$ is in lowest terms?

Solution $\frac{3}{4}$ is in lowest terms because 3 and 4 do not share a common factor other than 1.

$$\frac{3}{4} = \frac{3}{2 \times 2} \quad\text{no common factors other than 1}$$

$\frac{4}{6}$ is not in lowest terms because 4 and 6 do share a common factor of 2.

$$\frac{4}{6} = \frac{2 \times 2}{3 \times 2} \underset{\longleftarrow}{\overset{\longleftarrow}{\rule{1.5cm}{0pt}}} \text{common factor of 2}$$

When you add, subtract, multiply, or divide and get a fraction or a mixed number for an answer, you should always **simplify** that answer by writing it in **simplest form.** The simplest form of a fraction or mixed number can be

(1) a proper fraction in lowest terms,

(2) a whole number, or

(3) a mixed number when the fraction part is a proper fraction in lowest terms

To simplify a proper fraction that is not in lowest terms, you reduce the fraction to lowest terms using the following rule:

Fundamental Rule for Fractions

If a, b, and c are whole numbers ($b \neq 0$ and $c \neq 0$), then $\dfrac{a \times c}{b \times c} = \dfrac{a}{b}$.

Note: According to the Fundamental Rule for Fractions, the value of a fraction will not change when you divide both the numerator and denominator by the same nonzero number.

To simplify a fraction that is not in lowest terms, you first factor the numerator and denominator, then eliminate all the common factors.

EXAMPLE 3-24   Simplify the fractions (a) $\frac{6}{8}$, (b) $\frac{6}{9}$, and (c) $\frac{4}{8}$.

Solution

(a) $\dfrac{6}{8} = \dfrac{3 \times \cancel{2}}{4 \times \cancel{2}} = \dfrac{3}{4} \longleftarrow \text{simplest form}$ **(b)** $\dfrac{6}{9} = \dfrac{2 \times \cancel{3}}{3 \times \cancel{3}} = \dfrac{2}{3} \longleftarrow \text{simplest form}$ **(c)** $\dfrac{4}{8} = \dfrac{1 \times \cancel{4}}{2 \times \cancel{4}} = \dfrac{1}{2} \longleftarrow \text{simplest form}$

Caution: A fraction is not in simplest form until all common whole-number factors other than 1 have been eliminated.

EXAMPLE 3-25   Simplify the fraction $\frac{12}{18}$.

Solution

Wrong Method

$\dfrac{12}{18} = \dfrac{6 \times \cancel{2}}{9 \times \cancel{2}} = \dfrac{6}{9} \longleftarrow \begin{array}{l}\text{not in lowest terms}\\ \text{(6 and 9 have a}\\ \text{common factor 3)}\end{array}$

Correct Method

$\dfrac{12}{18} = \dfrac{2 \times \cancel{6}}{3 \times \cancel{6}} = \dfrac{2}{3} \longleftarrow \text{lowest terms}$

Caution: If all factors are eliminated in either the numerator or denominator, put the number 1 in place of the eliminated factors.

EXAMPLE 3-26   Simplify the fraction $\frac{2}{6}$.

Solution

Wrong Method

$\dfrac{2}{6} = \dfrac{\cancel{2}}{\cancel{2} \times 3} = \dfrac{0}{3} \longleftarrow \begin{array}{l}\text{Wrong! Always replace the}\\ \text{eliminated factors with 1.}\end{array}$

Correct Method

$\dfrac{2}{6} = \dfrac{\cancel{2}}{\cancel{2} \times 3} = \dfrac{1}{3}$ or $\dfrac{2}{6} = \dfrac{\cancel{2} \times 1}{\cancel{2} \times 3} = \dfrac{1}{3} \longleftarrow \text{correct}$

Caution: To eliminate a factor in the numerator of a fraction, you must eliminate a like factor in the denominator. Similarly, to eliminate a factor in the denominator, you must eliminate a like factor in the numerator.

EXAMPLE 3-27 Simplify the fraction $\frac{16}{36}$.

Solution

Wrong Method

$$\frac{16}{36} = \frac{4 \times \cancel{4}}{3 \times 3 \times 4} \longleftarrow \text{Wrong! One factor must be from the denominator.}$$

Wrong Method

$$\frac{16}{36} = \frac{4 \times 4}{\cancel{3} \times \cancel{3} \times 4} \longleftarrow \text{Wrong! One factor must be from the numerator.}$$

Correct Method

$$\frac{16}{36} = \frac{4 \times \cancel{4}}{3 \times 3 \times \cancel{4}}$$

$$= \frac{4}{3 \times 3} \quad \text{or} \quad \frac{4}{9} \longleftarrow \text{correct}$$

B. Simplifying improper fractions

To **simplify an improper fraction,** you first reduce to lowest terms (when possible), then rename the improper fraction as a mixed number in lowest terms or as a whole number.

EXAMPLE 3-28 Simplify the fractions (**a**) $\frac{10}{6}$ and (**b**) $\frac{45}{15}$.

Solution

(**a**) $\dfrac{10}{6} = \dfrac{5 \times \cancel{2}}{3 \times \cancel{2}} = \dfrac{5}{3} = 1\frac{2}{3} \longleftarrow$ simplest form

(**b**) $\dfrac{45}{15} = \dfrac{3 \times \cancel{15}}{1 \times \cancel{15}} = \dfrac{3}{1} = 3 \longleftarrow$ simplest form

C. Simplifying mixed numbers

To **simplify a mixed number** that is not in simplest form, you simplify the fraction part.

EXAMPLE 3-29 Simplify the mixed numbers (**a**) $5\frac{8}{10}$, (**b**) $4\frac{11}{3}$, and (**c**) $3\frac{18}{2}$.

Solution

(**a**) $5\frac{8}{10} = 5 + \dfrac{8}{10} = 5 + \dfrac{4 \times \cancel{2}}{5 \times \cancel{2}} = 5 + \dfrac{4}{5} = 5\frac{4}{5} \longleftarrow$

(**b**) $4\frac{11}{3} = 4 + \dfrac{11}{3} = 4 + 3\frac{2}{3} = 4 + 3 + \dfrac{2}{3} = 7 + \dfrac{2}{3} = 7\frac{2}{3} \longleftarrow$ simplest form

(**c**) $3\frac{18}{2} = 3 + \dfrac{18}{2} = 3 + \dfrac{9 \times \cancel{2}}{1 \times \cancel{2}} = 3 + \dfrac{9}{1} = 3 + 9 = 12 \longleftarrow$

3-4. Multiplying with Fractions and Mixed Numbers

To find a fractional part of an amount, you **multiply by a fraction.**

EXAMPLE 3-30 Rename $\frac{5}{6}$ of $\frac{3}{4}$ as multiplication.

Solution

$$\frac{5}{6} \text{ of } \frac{3}{4} = \frac{5}{6} \times \frac{3}{4} \longleftarrow \text{multiplication}$$

Note: When the word *of* appears between two numbers, you can *usually* replace the word with a multiplication symbol (×).

To join equal fractional parts together, you also multiply with fractions.

EXAMPLE 3-31 Write $\frac{2}{3} + \frac{2}{3} + \frac{2}{3} + \frac{2}{3}$ as multiplication.

Solution Think of $\frac{2}{3} + \frac{2}{3} + \frac{2}{3} + \frac{2}{3}$ as 4 equal amounts of $\frac{2}{3}$, or

$$4 \times \frac{2}{3} \longleftarrow \text{multiplication}$$

A. Multiplying with fractions

To multiply with fractions, you can use the following rule:

Multiplication Rule for Fractions

If $\frac{a}{b}$ and $\frac{c}{d}$ are fractions, then

$$\frac{a}{b} \times \frac{c}{d} = \frac{a \times c}{b \times d}$$

Note: To multiply fractions, you multiply the numerators first and then the denominators.

EXAMPLE 3-32 Multiply $\frac{5}{6} \times \frac{3}{4}$.

Solution

$$\frac{5}{6} \times \frac{3}{4} = \frac{5 \times 3}{6 \times 4} \qquad \text{Use the Multiplication Rule for Fractions.}$$

$$= \frac{15}{24} \qquad \begin{array}{l}\text{Multiply the numerators.} \\ \text{Multiply the denominators.}\end{array}$$

$$= \frac{5 \times \cancel{3}}{8 \times \cancel{3}} \qquad \text{Eliminate common factors.}$$

$$= \frac{5}{8} \longleftarrow \text{simplest form}$$

Note: If you factor first and eliminate all common factors before multiplying, the product will always be in lowest terms after multiplying.

EXAMPLE 3-33 Eliminate common factors before multiplying $\frac{5}{6} \times \frac{3}{4}$.

Solution

$$\frac{5}{6} \times \frac{3}{4} = \frac{5}{2 \times 3} \times \frac{3}{2 \times 2} \qquad \text{Factor.}$$

$$= \frac{5}{2 \times \cancel{3}} \times \frac{\cancel{3}}{4} \qquad \text{Eliminate all common factors.}$$

$$= \frac{5 \times 1}{2 \times 4} \qquad \text{Multiply fractions.}$$

$$= \frac{5}{8} \longleftarrow \text{simplest form}$$

To multiply two or more fractions, you eliminate common factors, then multiply the numerators and denominators, simplifying when possible.

EXAMPLE 3-34 Multiply $\frac{2}{3} \times \frac{3}{5} \times \frac{15}{4}$.

Solution

$$\frac{2}{3} \times \frac{3}{5} \times \frac{15}{4} = \frac{\cancel{2}}{3} \times \frac{3}{5} \times \frac{3 \times 5}{\cancel{2} \times 2}$$

$$= \frac{1}{\cancel{3}} \times \frac{\cancel{3}}{5} \times \frac{3 \times 5}{2}$$

Eliminate all common factors.

$$= \frac{1}{1} \times \frac{1}{\cancel{5}} \times \frac{3 \times \cancel{5}}{2}$$

$$= \frac{1}{1} \times \frac{1}{1} \times \frac{3}{2}$$

$$= \frac{1 \times 1 \times 3}{1 \times 1 \times 2}$$

Multiply the numerators.
Multiply the denominators.

$$= \frac{3}{2}$$

Simplify.

$$= 1\frac{1}{2} \longleftarrow \text{ simplest form}$$

To multiply with fractions and whole numbers, rename each whole number as an equal fraction.

EXAMPLE 3-35 Multiply $4 \times \frac{2}{3}$.

Solution

$$4 \times \frac{2}{3} = \frac{4}{1} \times \frac{2}{3}$$

Rename: $4 = \frac{4}{1}$

$$= \frac{4 \times 2}{1 \times 3}$$

Multiply fractions as before.

$$= \frac{8}{3}$$

Simplify.

$$= 2\frac{2}{3} \longleftarrow \text{ simplest form}$$

B. Multiplying with mixed numbers

To multiply with mixed numbers, you rename each mixed number as an equal fraction.

EXAMPLE 3-36 Multiply $2\frac{1}{2} \times 4 \times \frac{3}{5}$.

Solution

$$2\frac{1}{2} \times 4 \times \frac{3}{5} = \frac{5}{2} \times \frac{4}{1} \times \frac{3}{5}$$

Rename: $2\frac{1}{2} = \frac{5}{2}$ and $4 = \frac{4}{1}$

$$= \frac{\cancel{5}}{\cancel{2}} \times \frac{\cancel{2} \times 2}{1} \times \frac{3}{\cancel{5}}$$

Eliminate all common factors.

$$= \frac{1 \times 2 \times 3}{1 \times 1 \times 1}$$

Multiply fractions.

$$= \frac{6}{1}$$

Simplify.

$$= 6 \longleftarrow \text{ simplest form}$$

3-5. Dividing with Fractions and Mixed Numbers

When the product of two fractions is 1, each fraction is called a **reciprocal** of the other.

EXAMPLE 3-37 Interpret the equation $\frac{3}{4} \times \frac{4}{3} = 1$ in terms of reciprocals.

Solution $\frac{3}{4} \times \frac{4}{3} = 1$ means that $\frac{3}{4}$ is the reciprocal of $\frac{4}{3}$ and that $\frac{4}{3}$ is the reciprocal of $\frac{3}{4}$.

When you divide with fractions, you need to find the reciprocal of the divisor.

A. Finding the reciprocal of a fraction

To find the reciprocal of a fraction, you interchange the whole numbers in the numerator and denominator.

EXAMPLE 3-38 Find the reciprocals of (a) $\frac{2}{3}$ and (b) $\frac{8}{5}$.

Solution

(a) The reciprocal of $\frac{2}{3}$ is $\frac{3}{2}$ (or $1\frac{1}{2}$) because $\frac{2}{3} \times \frac{3}{2} = 1$.

(b) The reciprocal of $\frac{8}{5}$ is $\frac{5}{8}$ because $\frac{8}{5} \times \frac{5}{8} = 1$.

B. Finding the reciprocal of a whole number

To find the reciprocal of a whole number, you rename the whole number as an equal fraction, then interchange the numbers in the numerator and denominator.

EXAMPLE 3-39 Find the reciprocals of the whole numbers (a) 3 and (b) 0.

Solution

(a) The reciprocal of 3 (or $\frac{3}{1}$) is $\frac{1}{3}$ because $3 \times \frac{1}{3} = \frac{3}{1} \times \frac{1}{3} = 1$.

(b) The reciprocal of 0 (or $\frac{0}{1}$) is not defined because $\frac{1}{0}$ (division by zero) is not defined.

Note: Zero is the only whole number that does not have a reciprocal.

C. Finding the reciprocal of a mixed number

To find the reciprocal of a mixed number, you rename the mixed number as an equal fraction, then interchange the numbers in the numerator and denominator.

EXAMPLE 3-40 Find the reciprocals of the mixed numbers (a) $3\frac{1}{4}$ and (b) $2\frac{5}{8}$.

Solution

(a) The reciprocal of $3\frac{1}{4}$ (or $\frac{13}{4}$) is $\frac{4}{13}$ because $3\frac{1}{4} \times \frac{4}{13} = \frac{13}{4} \times \frac{4}{13} = 1$.

(b) The reciprocal of $2\frac{5}{8}$ (or $\frac{21}{8}$) is $\frac{8}{21}$ because $2\frac{5}{8} \times \frac{8}{21} = \frac{21}{8} \times \frac{8}{21} = 1$.

D. Dividing with fractions

Given a fractional part of some amount, you can find what that amount is if you **divide by a fraction**.

EXAMPLE 3-41 Write the following as a division problem with fractions:

$$\frac{1}{8} \text{ is } \frac{1}{4} \text{ of an amount}$$

Solution $\frac{1}{8}$ is $\frac{1}{4}$ of an amount means

$$\frac{1}{8} \div \frac{1}{4} = \text{the amount} \quad \text{or} \quad \text{the amount} = \frac{1}{8} \div \frac{1}{4} \longleftarrow \text{division}$$

Note: When the word *is* appears between two fractions, you can *usually* replace the word with a division symbol ($\div$).

To separate a fractional amount into equal parts, you also divide by a fraction.

EXAMPLE 3-42 Write the following as a division problem with fractions:

$$\frac{3}{4} \text{ separated into 8 equal parts}$$

Solution $\frac{3}{4}$ separated into 8 equal parts means

$$\frac{3}{4} \div 8 \quad \text{or} \quad \frac{3}{4} \div \frac{8}{1} \longleftarrow \text{division}$$

To divide with fractions, you can use the following rule:

Division Rule for Fractions

If $\frac{a}{b}$ and $\frac{c}{d}$ are fractions ($c \neq 0$), then

$$\frac{a}{b} \div \frac{c}{d} = \frac{a}{b} \times \frac{d}{c}$$

Note: To divide fractions, you change the divisor to its reciprocal and then multiply.

EXAMPLE 3-43   Divide $\frac{1}{8} \div \frac{1}{4}$.

change to multiplication

Solution $\dfrac{1}{8} \div \dfrac{1}{4} = \dfrac{1}{8} \times \dfrac{4}{1}$ Use the Division Rule for Fractions.

write the reciprocal of the divisor

$$= \frac{1}{2 \times 4} \times \frac{4 \times 1}{1} \qquad \text{Multiply as before [see Example 3-29].}$$

$$= \frac{1}{2} \times \frac{1}{1}$$

$$= \frac{1}{2} \longleftarrow \text{simplest form}$$

Note: $\frac{1}{8} \div \frac{1}{4} = \frac{1}{2}$ means $\frac{1}{8}$ is $\frac{1}{4}$ of $\frac{1}{2}$.

To check division, you multiply the proposed quotient by the original divisor to see if you get the original dividend.

EXAMPLE 3-44   Check the division problem $\frac{1}{8} \div \frac{1}{4} = \frac{1}{2}$.

original divisor   proposed quotient   original dividend

Solution $\dfrac{1}{4} \times \dfrac{1}{2} = \dfrac{1 \times 1}{4 \times 2} = \dfrac{1}{8} \longleftarrow \frac{1}{2}$ checks

Caution: Do not try to eliminate common factors before changing a division problem to a multiplication problem.

EXAMPLE 3-45   Divide $\frac{3}{5} \div \frac{2}{3}$.

Solution

   Wrong Method

$$\frac{3}{5} \div \frac{2}{3} = \frac{3}{5} \div \frac{2}{3} \longleftarrow Wrong! \qquad \text{Never eliminate common factors while there is a division symbol in the problem.}$$

$$= \frac{1}{5} \div \frac{2}{1}$$

$$= \frac{1}{5} \times \frac{1}{2}$$

$$= \frac{1}{10} \longleftarrow \text{wrong answer}$$

Check: $\quad \frac{2}{3} \times \frac{1}{10} = \frac{2}{3} \times \frac{1}{2 \times 5}$

$$= \frac{1}{15} \longleftarrow \text{not the original dividend } \frac{3}{5} \left(\text{so } \frac{1}{10} \text{ does not check}\right)$$

Correct Method

$$\frac{3}{5} \div \frac{2}{3} = \frac{3}{5} \times \frac{3}{2} \qquad \text{Change to multiplication and invert the divisor.}$$

$$= \frac{3 \times 3}{5 \times 2}$$

$$= \frac{9}{10} \longleftarrow \text{correct answer}$$

Check: $\quad \frac{2}{3} \times \frac{9}{10} = \frac{2}{3} \times \frac{3 \times 3}{2 \times 5}$

$$= \frac{3}{5} \longleftarrow \text{original dividend } \left(\frac{9}{10} \text{ checks}\right)$$

To divide a fraction by a whole number, you rename the whole number as an equal fraction.

EXAMPLE 3-46 Divide $\frac{3}{4} \div 8$.

Solution $\quad \frac{3}{4} \div 8 = \frac{3}{4} \div \frac{8}{1} \qquad\qquad\qquad$ Rename: $8 = \frac{8}{1}$

$$= \frac{3}{4} \times \frac{1}{8} \qquad\qquad\qquad \text{Multiply by the reciprocal of the divisor.}$$

$$= \frac{3 \times 1}{4 \times 8} \qquad\qquad\qquad \text{Multiply as before.}$$

$$= \frac{3}{32} \longleftarrow \text{proposed answer}$$

Check: $\quad 8 \times \frac{3}{32} = \frac{8}{1} \times \frac{3}{32} \qquad\qquad$ Rename.

$$= \frac{8}{1} \times \frac{3}{4 \times 8} \qquad\qquad \text{Multiply as before.}$$

$$= \frac{3}{4} \longleftarrow \text{original dividend } \left(\frac{3}{32} \text{ checks}\right)$$

To divide a whole number by a fraction, you rename the whole number as an equal fraction.

EXAMPLE 3-47 Divide $6 \div \frac{2}{3}$.

Solution $6 \div \dfrac{2}{3} = \dfrac{6}{1} \div \dfrac{2}{3}$ Rename.

$= \dfrac{6}{1} \times \dfrac{3}{2}$ Multiply by the reciprocal
of the divisor.

$= \dfrac{\cancel{2} \times 3}{1} \times \dfrac{3}{\cancel{2}}$ Multiply as before.

$= \dfrac{9}{1}$

$= 9$ ←—— proposed answer

Check: $\dfrac{2}{3} \times 9 = \dfrac{2}{3} \times \dfrac{9}{1}$ Rename.

$= \dfrac{2}{\cancel{3}} \times \dfrac{\cancel{3} \times 3}{1}$ Multiply as before.

$= \dfrac{6}{1}$

$= 6$ ←—— original dividend (9 checks)

E. Dividing with mixed numbers

To divide with mixed numbers, you rename each mixed number as an equal fraction.

EXAMPLE 3-48   Divide the mixed numbers $2\frac{2}{5} \div 1\frac{1}{5}$.

Solution $2\frac{2}{5} \div 1\frac{1}{5} = \dfrac{12}{5} \div \dfrac{6}{5}$ Rename: $2\frac{2}{5} = \dfrac{12}{5}$ and $1\frac{1}{5} = \dfrac{6}{5}$

$= \dfrac{12}{5} \times \dfrac{5}{6}$ Multiply by the reciprocal
of the divisor.

$= \dfrac{2 \times \cancel{6}}{\cancel{5}} \times \dfrac{\cancel{5}}{\cancel{6}}$ Multiply as before.

$= \dfrac{2}{1}$

$= 2$ ←—— proposed answer

Check: $1\frac{1}{5} \times 2 = \dfrac{6}{5} \times \dfrac{2}{1}$ Rename and multiply as before.

$= \dfrac{12}{5}$

$= 2\frac{2}{5}$ ←—— original dividend (2 checks)

3-6. Adding Fractions and Mixed Numbers

To join two or more fractional amounts together, you **add fractions.**

EXAMPLE 3-49   Rename the following as addition:

$$\dfrac{2}{8} \text{ joined together with } \dfrac{3}{8}$$

Solution $\frac{2}{8}$ joined together with $\frac{3}{8}$ means

$$\frac{2}{8} + \frac{3}{8} \longleftarrow \text{addition}$$

A. Adding like fractions

Fractions with the same denominator are called **like fractions.**

EXAMPLE 3-50 Write two like fractions.

Solution $\frac{2}{8}$ and $\frac{3}{8}$ are two like fractions because their denominators are the same: $8 = 8$.

To add like fractions, you can use the following rule:

Addition Rule for Like Fractions

If $\frac{a}{c}$ and $\frac{b}{c}$ are any two like fractions, then

$$\frac{a}{c} + \frac{b}{c} = \frac{a + b}{c}$$

Note: To add two like fractions, you add the two numerators and then write the same denominator.

EXAMPLE 3-51 Add $\frac{2}{8} + \frac{3}{8}$.

Solution
$$\frac{2}{8} + \frac{3}{8} = \frac{2 + 3}{8}$$
Add the two numerators.
Write the same denominator.

$$= \frac{5}{8}$$
Add in the numerator: $2 + 3 = 5$

The following example can help you understand addition of two like fractions.

EXAMPLE 3-52 Translate addition of like fractions to words and pictures.

Solution

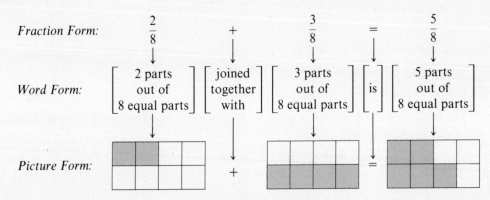

Caution: Always simplify a fraction sum when possible.

EXAMPLE 3-53 Add $\frac{3}{12} + \frac{5}{12}$.

Solution $\dfrac{3}{12} + \dfrac{5}{12} = \dfrac{3+5}{12}$ Add like fractions.

$= \dfrac{8}{12}$

$= \dfrac{2 \times \cancel{4}}{3 \times \cancel{4}}$ Simplify [see Section 3-3].

$= \dfrac{2}{3}$ simplest form

To add more than two like fractions, you add all the numerators and write the same denominator.

EXAMPLE 3-54   Add $\frac{3}{4} + \frac{1}{4} + \frac{5}{4} + \frac{1}{4}$.

Solution $\dfrac{3}{4} + \dfrac{1}{4} + \dfrac{5}{4} + \dfrac{1}{4} = \dfrac{3+1+5+1}{4}$ Add all the numerators.
Write the same denominator.

$= \dfrac{10}{4}$ Add in the numerator.

$= \dfrac{\cancel{2} \times 5}{\cancel{2} \times 2}$ Simplify.

$= \dfrac{5}{2}$

$= 2\frac{1}{2}$ ⟵ simplest form

B. Finding the least common denominator (LCD)

Fractions with different denominators are called **unlike fractions.** To add unlike fractions, the first step is to find the **least common denominator (LCD).** The least common denominator for two or more fractions is the smallest nonzero whole number that all of the denominators divide into *evenly* (with a zero remainder). When the smaller denominator divides the larger denominator evenly, the larger denominator is the LCD.

EXAMPLE 3-55   Find the LCD for the fractions $\frac{3}{4}$ and $\frac{5}{8}$.

Solution   The LCD for $\frac{3}{4}$ and $\frac{5}{8}$ is 8 because 8 is the smallest nonzero whole number that both 4 and 8 will divide into evenly.

Note: In this example the larger denominator 8 is the LCD because 4 divides 8 evenly: $8 \div 4 = 2$ R0, or just 2.

When the denominators do not share any common prime factors, the product of the denominators is the LCD.

EXAMPLE 3-56   Find the LCD for the fractions $\frac{2}{3}$ and $\frac{3}{4}$.

Solution   The LCD for $\frac{2}{3}$ and $\frac{3}{4}$ is 12 because 12 ($= 3 \times 4$) is the smallest nonzero whole number that both 3 and 4 will divide into evenly.

Note: In this example you can multiply the denominators 3 and 4 to find the LCD 12 because 3 and 4 do not share a common prime factor: $3 = 1 \times 3$ and $4 = 1 \times 4$ or 2×2.

If you have trouble finding the LCD for two fractions, you can use the following method:

Factoring Method for Finding the LCD

(1) Factor each denominator as a product of primes.

(2) Identify the greatest number of times that each prime factor occurs in any single factorization from step 1.

(3) Write the LCD as a product of the factors found in step 2.

EXAMPLE 3-57   Find the LCD for the fractions $\frac{7}{12}$ and $\frac{3}{10}$.

Solution

$12 = 2 \times 2 \times 3 \qquad 10 = 2 \times 5$ Factor each denominator as a product of primes.

$12 = \boxed{2 \times 2} \times \boxed{3} \qquad 10 = 2 \times \boxed{5}$ Circle the greatest number of times that each different prime factor occurs in any single factorization.

$\text{LCD} = 2 \times 2 \times 3 \times 5 = 60$ Write the LCD as a product of the circled factors.

Note 1: The LCD for $\frac{7}{12}$ and $\frac{3}{10}$ is 60 because 60 is the smallest nonzero whole number that both 10 and 12 will divide into evenly.

Note 2: There are two 2s in the LCD because the greatest number of times that 2 occurs as a factor in any single factorization is twice in $2 \times 2 \times 3$ (2 occurs only once in 2×5).

The following shortcut can save you time and effort in finding the LCD for two unlike fractions in almost every practical application.

Shortcut: To find the LCD for two fractions when the difference between their denominators divides each denominator evenly:

(1) Find the difference between the denominators.

(2) Divide the difference found in step 1 into one of the denominators.

(3) Multiply the quotient from step 2 times the other denominator to get the LCD.

EXAMPLE 3-58   Find the LCD for $\frac{3}{10}$ and $\frac{7}{12}$ using a shortcut.

Solution   The LCD for $\frac{3}{10}$ and $\frac{7}{12}$ is 60 because

$$12 - 10 = 2 \longrightarrow 2\overline{)12} \quad 6 \times 10 = 60 \qquad \text{or} \qquad 12 - 10 = 2 \longrightarrow 2\overline{)10} \quad 5 \times 12 = 60$$

When each smaller denominator divides the larger denominator evenly, the larger denominator is the LCD.

EXAMPLE 3-59   Find the LCD for the fractions $\frac{1}{6}, \frac{5}{12}$, and $\frac{3}{4}$.

Solution   The LCD for $\frac{1}{6}, \frac{5}{12}$, and $\frac{3}{4}$ is 12 because 12 is the smallest nonzero whole number that 6, 12, and 4 will divide into evenly.

Note: The larger denominator 12 is the LCD because 4, 6, and 12 all divide into 12 evenly: $12 \div 4 = 3$, $12 \div 6 = 2$, and $12 \div 12 = 1$.

When no two denominators share a common prime factor, the product of all the denominators is the LCD.

EXAMPLE 3-60   Find the LCD for the fractions $\frac{1}{2}, \frac{2}{3}$, and $\frac{3}{5}$.

Solution   The LCD for $\frac{1}{2}, \frac{2}{3}$, and $\frac{3}{5}$ is 30 because 30 ($= 2 \times 3 \times 5$) is the smallest nonzero whole number that 2, 3, and 5 will all divide into evenly.

Note: You can multiply the denominators 2, 3, and 5 together to find the LCD because no two denominators share a common prime factor: $2 = 1 \times 2$, $3 = 1 \times 3$, and $5 = 1 \times 5$.

When you have trouble finding the LCD for more than two fractions, you can always use the Factoring Method for Finding the LCD presented in Example 3-57.

EXAMPLE 3-61   Find the LCD for the fractions $\frac{7}{18}$, $\frac{5}{6}$, and $\frac{3}{4}$.

Solution $18 = 2 \times \boxed{3 \times 3}$ Factor each denominator as a product of primes.

$6 = 2 \times 3$ Circle the greatest number of times that each different prime factor occurs in any single factorization.

$4 = \boxed{2 \times 2}$

$LCD = 2 \times 2 \times 3 \times 3$ Write the LCD as a product of the circled primes.

$= 4 \times 9$

$= 36 \longleftarrow LCD$

Note: The LCD for $\frac{7}{18}$, $\frac{5}{6}$, and $\frac{3}{4}$ is 36 because 36 is the smallest nonzero whole number that 18, 6, and 4 will all divide into evenly.

C. Building up fractions

Recall the Fundamental Rule for Fractions from Section 3-3:

If a, b, and c are whole numbers ($b \neq 0$ and $c \neq 0$), then $\dfrac{a}{b} = \dfrac{a \times c}{b \times c}$.

EXAMPLE 3-62   Show that $\dfrac{2}{3} = \dfrac{2 \times 4}{3 \times 4}$.

Solution $\dfrac{2}{3} = \dfrac{2}{3} \times 1$ *Think:* Any number times 1 is that number.

$= \dfrac{2}{3} \times \dfrac{4}{4}$ Substitute: $1 = \dfrac{4}{4}$

$= \dfrac{2 \times 4}{3 \times 4}$ Multiply fractions.

Recall that to add unlike fractions, the first step is to find the least common denominator (LCD). The second step is to build up to get like fractions using the LCD. To build up a given fraction as an equal fraction with a given denominator, you first compare denominators to find the correct **building factor**.

EXAMPLE 3-63   Build up $\frac{2}{3}$ as an equal fraction with a denominator of 12.

Solution $\dfrac{2}{3} = \dfrac{?}{12}$ *Think:* 3 times what building factor is 12?

$3 \times 4 = 12$ *Think:* The correct building factor is 4 because $3 \times 4 = 12$.

$\dfrac{2}{3} = \dfrac{2 \times 4}{3 \times 4}$ Use the Fundamental Rule for Fractions.

$= \dfrac{8}{12} \longleftarrow \frac{2}{3}$ built up as an equal fraction with a denominator of 12

Note: To build up $\frac{2}{3}$ as an equal fraction with a denominator of 12, you multiply both the numerator and denominator of $\frac{2}{3}$ by the building factor 4.

To rename two unlike fractions as equivalent like fractions using the LCD, you first find the LCD and then build up to get like fractions using the LCD.

EXAMPLE 3-64 Rename the fractions $\frac{1}{2}$ and $\frac{2}{3}$ as equivalent like fractions using the LCD.

Solution The LCD of $\frac{1}{2}$ and $\frac{2}{3}$ is 6.

$$\frac{1}{2} = \frac{?}{6} \longleftarrow 2 \times \mathbf{3} = 6 \text{ means the building factor for } \frac{1}{2} \text{ is 3}$$

$$\frac{2}{3} = \frac{?}{6} \longleftarrow 3 \times \mathbf{2} = 6 \text{ means the building factor for } \frac{2}{3} \text{ is 2}$$

$$\frac{1}{2} = \frac{1 \times 3}{2 \times 3} = \frac{3}{6} \longleftarrow \frac{1}{2} \text{ built up as an equal fraction using the LCD 6}$$

$$\frac{2}{3} = \frac{2 \times 2}{3 \times 2} = \frac{4}{6} \longleftarrow \frac{2}{3} \text{ built up as an equal fraction using the LCD 6}$$

Renamed as equivalent like fractions using the LCD, $\frac{1}{2}$ and $\frac{2}{3}$ are $\frac{3}{6}$ and $\frac{4}{6}$, respectively.

To rename more than two unlike fractions as equivalent like fractions using the LCD, you rename as with two unlike fractions.

EXAMPLE 3-65 Rename the fractions $\frac{5}{6}$, $\frac{1}{3}$, and $\frac{7}{12}$ as equivalent like fractions using the LCD.

Solution The LCD of $\frac{5}{6}$, $\frac{1}{3}$, and $\frac{7}{12}$ is 12.

$$\frac{5}{6} = \frac{?}{12} \longleftarrow 6 \times \mathbf{2} = 12 \text{ means the building factor for } \frac{5}{6} \text{ is 2}$$

$$\frac{1}{3} = \frac{?}{12} \longleftarrow 3 \times \mathbf{4} = 12 \text{ means the building factor for } \frac{1}{3} \text{ is 4}$$

$$\frac{7}{12} = \frac{7}{12} \longleftarrow \text{no building factor needed because 12 is the LCD}$$

$$\frac{5}{6} = \frac{5 \times 2}{6 \times 2} = \frac{10}{12} \longleftarrow \frac{5}{6} \text{ built up as an equal fraction using the LCD 12}$$

$$\frac{1}{3} = \frac{1 \times 4}{3 \times 4} = \frac{4}{12} \longleftarrow \frac{1}{3} \text{ built up as an equal fraction using the LCD 12}$$

$$\frac{7}{12} = \frac{7}{12} \longleftarrow \frac{7}{12} \text{ is already a fraction with the LCD 12 for a denominator.}$$

Renamed as equivalent fractions using the LCD, $\frac{5}{6}$, $\frac{1}{3}$, and $\frac{7}{12}$ are $\frac{10}{12}$, $\frac{4}{12}$, and $\frac{7}{12}$, respectively.

D. Adding unlike fractions

To add unlike fractions, you can use the following rules:

Addition Rules for Unlike Fractions

(1) Find the LCD for the unlike fractions.
(2) Build up to get like fractions using the LCD from step 1.
(3) Substitute the like fractions for the original unlike fractions.
(4) Add the like fractions from step 3.
(5) Simplify the sum from step 4 when possible.

Note: To add two or more unlike fractions, you first rename as like fractions using the LCD and then add the like fractions.

To add two unlike fractions, you use the Addition Rules for Unlike Fractions.

EXAMPLE 3-66 Add $\frac{1}{2} + \frac{2}{3}$.

Solution The LCD for $\frac{1}{2}$ and $\frac{2}{3}$ is 6.

$$\frac{1}{2} = \frac{1 \times 3}{2 \times 3} = \frac{3}{6}$$ Build up to get like fractions using the LCD.

$$\frac{2}{3} = \frac{2 \times 2}{3 \times 2} = \frac{4}{6}$$ LCD

$$\frac{1}{2} + \frac{2}{3} = \frac{3}{6} + \frac{4}{6} \longleftarrow \text{like fractions}$$ Substitute the like fractions for the original unlike fractions.

$$= \frac{3 + 4}{6}$$ Add like fractions as before [see Example 3-49].

$$= \frac{7}{6}$$

$$= 1\frac{1}{6}$$ Simplify [see Section 3-3].

Note: $\frac{1}{2} + \frac{2}{3} = 1\frac{1}{6} \longleftarrow$ simplest form

To add more than two unlike fractions, you use the Addition Rules for Unlike Fractions.

EXAMPLE 3-67 Add $\frac{11}{15} + \frac{3}{5} + \frac{1}{6}$.

Solution The LCD for $\frac{11}{15}, \frac{3}{5}$, and $\frac{1}{6}$ is 30.

$$\frac{11}{15} = \frac{11 \times 2}{15 \times 2} = \frac{22}{30}$$ Build up to get like fractions.

$$\frac{3}{5} = \frac{3 \times 6}{5 \times 6} = \frac{18}{30}$$ LCD

$$\frac{1}{6} = \frac{1 \times 5}{6 \times 5} = \frac{5}{30}$$

$$\frac{11}{15} + \frac{3}{5} + \frac{1}{6} = \frac{22}{30} + \frac{18}{30} + \frac{5}{30} \longleftarrow \text{like fractions}$$ Substitute.

$$= \frac{22 + 18 + 5}{30}$$ Add like fractions.

$$= \frac{45}{30}$$

$$= \frac{3 \times \cancel{15}}{2 \times \cancel{15}}$$ Simplify.

$$= \frac{3}{2}$$

$$= 1\frac{1}{2}$$

Note: $\frac{11}{15} + \frac{3}{5} + \frac{1}{6} = 1\frac{1}{2} \longleftarrow$ simplest form

E. Adding mixed numbers

To add two or more mixed numbers, you can use the following rules:

Addition Rules for Mixed Numbers

(1) Write in vertical addition form.
(2) Build up to get like-fraction parts when necessary.
(3) Add the like-fraction parts to find the fraction part of the sum.
(4) Add the whole-number parts to find the whole-number part of the sum.
(5) Simplify the sum from steps 3 and 4 when possible.

To add a mixed number to one or more whole numbers, you add the whole numbers and then write the same fraction part.

EXAMPLE 3-68   Add $5\frac{1}{2} + 3$.

Solution $\qquad 5\frac{1}{2} + 3 = 8\frac{1}{2}$ Add the whole-number parts: $5 + 3 = 8$
$\qquad\qquad\qquad\qquad\qquad$ Write the same fraction part.

Note: $5\frac{1}{2} + 3 = 8\frac{1}{2}$ because $5\frac{1}{2} + 3 = 5 + \dfrac{1}{2} + 3$

$$= (5 + 3) + \frac{1}{2}$$

$$= 8 + \frac{1}{2}$$

$$= 8\frac{1}{2}$$

To add a mixed number and one or more fractions, you add the fractions and then write the same whole-number part.

EXAMPLE 3-69   Add $\frac{5}{6} + 2\frac{1}{3}$.

Solution $\quad \dfrac{5}{6} + 2\frac{1}{3} = \dfrac{5}{6} + 2 + \dfrac{1}{3}$ *Think:* The LCD of $\dfrac{5}{6}$ and $\dfrac{1}{3}$ is 6.

$$= 2 + \frac{5}{6} + \frac{2}{6} \qquad \text{Build up to get like fractions using the LCD.}$$

$$= 2 + \frac{7}{6} \qquad\qquad \text{Add like fractions.}$$

$$= 2 + 1\frac{1}{6} \qquad\qquad \text{Simplify.}$$

$$= 2 + 1 + \frac{1}{6}$$

$$= 3 + \frac{1}{6}$$

$$= 3\frac{1}{6}$$

Note: $\frac{5}{6} + 2\frac{1}{3} = 3\frac{1}{6}$ ⟵ simplest form

To add two or more mixed numbers, you first write the mixed numbers in vertical form, then add the fraction parts, and then add the whole-number parts.

Note: It is usually easier to add mixed numbers in vertical form.

EXAMPLE 3-70   Add $2\frac{3}{4} + 3\frac{1}{2}$.

Solution   The LCD for $\frac{3}{4}$ and $\frac{1}{2}$ is 4.

$$2\frac{3}{4} = 2\frac{3}{4}$$
$$+3\frac{1}{2} = 3\frac{2}{4}$$ LCD   Write in vertical form.
Build up to get like-fraction parts.

$$5\frac{5}{4} = 5 + \frac{5}{4}$$ Add the like-fraction parts: $\frac{3}{4} + \frac{2}{4} = \frac{5}{4}$
Add the whole-number parts: $2 + 3 = 5$

$$= 5 + 1\frac{1}{4}$$ Simplify.

$$= 6\frac{1}{4}$$

Note: $2\frac{3}{4} + 3\frac{1}{2} = 6\frac{1}{4}$ ⟵ simplest form

3-7. Subtracting Fractions and Mixed Numbers

To take one fractional amount away from another fractional amount, you **subtract fractions.**

EXAMPLE 3-71   Rename $\frac{5}{8}$ take away $\frac{2}{8}$ as subtraction.

Solution $\frac{5}{8}$ take away $\frac{2}{8}$ means

$$\frac{5}{8} - \frac{2}{8}$$ ⟵ subtraction

A. Subtracting like fractions

To subtract like fractions, you can use the following rule:

Subtraction Rule for Like Fractions

If $\frac{a}{c}$ and $\frac{b}{c}$ are any two like fractions, then

$$\frac{a}{c} - \frac{b}{c} = \frac{a - b}{c}.$$

Note: To subtract like fractions, you subtract the numerators and then write the same denominator.

EXAMPLE 3-72   Subtract the fractions $\frac{5}{8} - \frac{2}{8}$.

Solution $\dfrac{5}{8} - \dfrac{2}{8} = \dfrac{5 - 2}{8}$ Subtract the numerators.
Write the same denominator.

$$= \frac{3}{8}$$ Subtract in the numerator: $5 - 2 = 3$

To check a subtraction answer, you add the proposed difference to the original subtrahend to see if you get the original minuend.

EXAMPLE 3-73   Check the subtraction problem $\frac{5}{8} - \frac{2}{8} = \frac{3}{8}$.

Solution

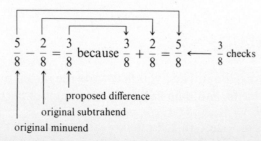

$$\frac{5}{8} - \frac{2}{8} = \frac{3}{8} \text{ because } \frac{3}{8} + \frac{2}{8} = \frac{5}{8} \longleftarrow \frac{3}{8} \text{ checks}$$

proposed difference
original subtrahend
original minuend

The following example can help you understand subtraction of like fractions.

EXAMPLE 3-74 Translate the subtraction of like fractions $\frac{5}{8} - \frac{2}{8} = \frac{3}{8}$ to words and pictures:

Solution

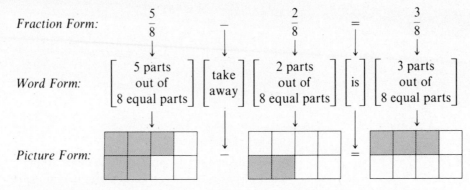

Caution: Always simplify a fractional difference when possible [see Section 3-3].

EXAMPLE 3-75 Subtract the fractions $\frac{11}{12} - \frac{5}{12}$.

Solution
$$\frac{11}{12} - \frac{5}{12} = \frac{11 - 5}{12} \qquad \text{Subtract like fractions.}$$

$$= \frac{6}{12}$$

$$= \frac{1 \times \cancel{6}}{2 \times \cancel{6}} \qquad \text{Simplify [see Section 3-3].}$$

$$= \frac{1}{2} \longleftarrow \text{simplest form}$$

B. Subtracting unlike fractions

To subtract unlike fractions, you can use the following method:

Subtracting Rules for Unlike Fractions

(1) Find the LCD for the unlike fractions.

(2) Build up to get like fractions using the LCD from step 1.

(3) Substitute the like fractions from step 2 for the original unlike fractions.

(4) Subtract the like fractions from step 3.

(5) Simplify the difference from step 4 when possible.

(6) Check by adding the simplest form of the proposed difference from step 4 or 5 to the original subtrahend to see if you get the original minuend.

Note: To subtract unlike fractions, you first rename as equivalent like fractions using the LCD and then subtract the like fractions.

EXAMPLE 3-76 Subtract the fractions $\frac{11}{12} - \frac{1}{6}$.

Solution The LCD for $\frac{11}{12}$ and $\frac{1}{6}$ is 12.

$$\frac{11}{12} = \frac{11}{12} \qquad \text{Build up to get like fractions using the LCD.}$$

$$\frac{1}{6} = \frac{1 \times 2}{6 \times 2} = \frac{2}{12} \Bigg\rangle \text{LCD}$$

$$\frac{11}{12} - \frac{1}{6} = \frac{11}{12} - \frac{2}{12} \quad \longleftarrow \text{ like fractions}$$

Substitute the like fractions for the original unlike fractions.

$$= \frac{11 - 2}{12}$$

Subtract the like fractions.

$$= \frac{9}{12}$$

$$= \frac{\cancel{3} \times 3}{\cancel{3} \times 4}$$

Simplify.

$$= \frac{3}{4} \quad \longleftarrow \text{ proposed difference}$$

$$\frac{3}{4} + \frac{1}{6} = \frac{9}{12} + \frac{2}{12}$$

Check by adding the proposed difference to the original subtrahend to see if you get the original minuend.

$$= \frac{9 + 2}{12}$$

$$= \frac{11}{12} \quad \longleftarrow \frac{3}{4} \text{ checks}$$

Note: $\frac{11}{12} - \frac{1}{6} = \frac{3}{4} \longleftarrow$ simplest form

C. Subtracting with mixed numbers

To subtract a whole number from a mixed number, you subtract the whole numbers and then write the same fraction part.

EXAMPLE 3-77   Subtract $7\frac{2}{3} - 3$.

Solution

$$7\frac{2}{3} - 3 = 4\frac{2}{3} \quad \text{or} \quad \begin{array}{r} 7\frac{2}{3} \\ -3 \\ \hline 4\frac{2}{3} \end{array}$$

Subtract the whole numbers.
Write the same fractional part.

Note: $7\frac{2}{3} - 3 = 4\frac{2}{3}$ because $7\frac{2}{3} - 3 = 7 + \frac{2}{3} - 3$

$$= (7 - 3) + \frac{2}{3}$$
$$= 4 + \frac{2}{3}$$
$$= 4\frac{2}{3}$$

To subtract a fraction from a mixed number, you first write the numbers in vertical form, then subtract the fractions and write the same whole-number part.

EXAMPLE 3-78   Subtract $5\frac{2}{3} - \frac{1}{2}$.

Solution

$$\begin{array}{r} 5\frac{2}{3} = 5\frac{4}{6} \\ -\frac{1}{2} = \frac{3}{6} \\ \hline 5\frac{1}{6} \end{array}$$

Write in vertical subtraction form.
Build up like fractions using the LCD.
Subtract like fractions and write the same whole-number part.

$$\begin{array}{r} 5\frac{1}{6} = 5\frac{1}{6} \\ +\frac{1}{2} = \frac{3}{6} \\ \hline 5\frac{4}{6} = 5\frac{2}{3} \quad \longleftarrow 5\frac{1}{6} \text{ checks} \end{array}$$

Check the proposed difference.

Note: $5\frac{2}{3} - \frac{1}{2} = 5\frac{1}{6} \longleftarrow$ simplest form

To subtract two mixed numbers, you first write the numbers in vertical form, then subtract the fraction parts, and then subtract the whole-number parts.

EXAMPLE 3-79 Subtract $9\frac{5}{8} - 2\frac{1}{6}$.

Solution   The LCD for $\frac{5}{8}$ and $\frac{1}{6}$ is 24.

$$4 \times 6 = 24$$
Think: $8 - 6 = 2$ and $2\overline{)8}$

$$
\begin{aligned}
9\frac{5}{8} &= 9\frac{15}{24} \\
-2\frac{1}{6} &= 2\frac{4}{24}
\end{aligned}
\Bigg\rangle \text{LCD}
$$

$$7\frac{11}{24} \longleftarrow \text{proposed difference}$$

Write in vertical form.
Build up to get like fraction parts.
Subtract the like fraction parts: $\frac{15}{24} - \frac{4}{24} = \frac{11}{24}$
Subtract the whole-number parts: $9 - 2 = 7$

$$
\begin{aligned}
7\frac{11}{24} &= 7\frac{11}{24} \\
+2\frac{1}{6} &= 2\frac{4}{24}
\end{aligned}
$$

Check by adding the proposed difference to the original subtrahend to see if you get the original minuend.

$$
\begin{aligned}
9\frac{15}{24} &= 9 + \frac{15}{24} \\
&= 9 + \frac{\not{3} \times 5}{\not{3} \times 8} \\
&= 9 + \frac{5}{8} \\
&= 9\frac{5}{8} \longleftarrow 7\frac{11}{24} \text{ checks}
\end{aligned}
$$

Note: $9\frac{5}{8} - 2\frac{1}{6} = 7\frac{11}{24} \longleftarrow$ simplest form

D. Renaming to subtract

To subtract a fraction or a mixed number from a whole number, you first rename the whole number as an equal mixed number so that the fraction parts are like fractions.

EXAMPLE 3-80 Subtract $8 - 2\frac{3}{4}$.

Solution   First rename the whole number (8) as an equal mixed number.

$$
\begin{aligned}
8^{\downarrow} &= 7\frac{4}{4} \\
-2\frac{3}{4} &= 2\frac{3}{4}
\end{aligned}
\Bigg\rangle \text{like fractions}
$$

missing fraction part

$$5\frac{1}{4} \longleftarrow \text{proposed difference}$$

Think: $8 = 7 + 1 = 7 + \frac{4}{4} = 7\frac{4}{4}$

$$
\begin{aligned}
5\frac{1}{4} \\
+2\frac{3}{4}
\end{aligned}
$$

Check the proposed difference.

$$
\begin{aligned}
7\frac{4}{4} &= 7 + \frac{4}{4} \\
&= 7 + 1 \\
&= 8 \longleftarrow 5\frac{1}{4} \text{ checks}
\end{aligned}
$$

Caution: Recall that if you're subtracting a whole number from a mixed number, you subtract the whole-number parts and then write the fraction part. But you can't use this method when subtracting a mixed number from a whole number. This means you can't just move a fraction in a mixed-number subtrahend to the whole number in the minuend. For example, see what happens when the fraction $\frac{3}{4}$ is moved from the subtrahend to the minuend in $8 - 2\frac{3}{4}$:

$$8 - 2\frac{3}{4} = 7\frac{4}{4} - 2\frac{3}{4} \qquad\qquad\qquad 8\frac{3}{4} - 2 = (8 - 2) + \frac{3}{4}$$

$$= (7 - 2) + \left(\frac{4}{4} - \frac{3}{4}\right) \qquad\qquad = 6 + \frac{3}{4}$$

$$= 5 + \frac{1}{4} \qquad\qquad\qquad\qquad = 6\frac{3}{4}$$

$$= 5\frac{1}{4}$$

You can see that $5\frac{1}{4} \neq 6\frac{3}{4}$, so $8 - 2\frac{3}{4} \neq 8\frac{3}{4} - 2$.

Always rename the whole number as an equal mixed number when subtracting a mixed number from a whole number.

To avoid making an error when subtracting whole numbers and fractions or mixed numbers, you should always use vertical subtraction form.

EXAMPLE 3-81 Subtract the following whole numbers and fractions:

$$\textbf{(a)}\ 5 - \tfrac{1}{2} \qquad \textbf{(b)}\ 6 - 3\tfrac{5}{8} \qquad \textbf{(c)}\ 6\tfrac{5}{8} - 3$$

Solution

$$\textbf{(a)} \quad \begin{array}{r} 5 = 4\frac{2}{2} \\ -\frac{1}{2} = \frac{1}{2} \\ \hline 4\frac{1}{2} \end{array} \qquad \textbf{(b)} \quad \begin{array}{r} 6 = 5\frac{8}{8} \\ -3\frac{5}{8} = 3\frac{5}{8} \\ \hline 2\frac{3}{8} \end{array} \qquad \textbf{(c)} \quad \begin{array}{r} 6\frac{5}{8} \\ -3 \\ \hline 3\frac{5}{8} \end{array} \quad \text{Check as before.}$$

To subtract two mixed numbers when the minuend of the fraction part is too small, you first rename the minuend to get an equal mixed number with a larger fraction part.

EXAMPLE 3-82 Subtract the mixed numbers $9\frac{1}{6} - 1\frac{5}{6}$.

Solution

$$\overset{\substack{\text{too} \\ \text{small}}}{\downarrow} \quad \overset{\text{more sixths}}{\downarrow}$$

$$9\tfrac{1}{6} = 8\tfrac{7}{6} \qquad \textit{Think: } 9\tfrac{1}{6} = 9 + \tfrac{1}{6} = 8 + 1 + \tfrac{1}{6} = 8 + \tfrac{6}{6} + \tfrac{1}{6} = 8 + \tfrac{7}{6} = 8\tfrac{7}{6} \text{ or}$$

$$\begin{array}{r} -1\frac{5}{6} = 1\frac{5}{6} \\ \hline 7\frac{2}{6} \end{array} \qquad\qquad \overset{\overset{\displaystyle 1 + 6}{\frown}}{9\tfrac{1}{6} = 8\tfrac{7}{6}} \quad \text{(The new numerator is the sum of the old numerator and denominator.)}$$

$$\underset{\underset{\displaystyle 9 - 1}{\smile}}{}$$

$$7\tfrac{2}{6} = 7\tfrac{1}{3} \longleftarrow \text{proposed difference} \qquad \textit{Think: } \frac{2}{6} = \frac{\cancel{2} \times 1}{\cancel{2} \times 3} = \frac{1}{3}$$

$$\begin{array}{r} 7\frac{1}{3} = 7\frac{2}{6} \\ +1\frac{5}{6} = 1\frac{5}{6} \\ \hline 8\frac{7}{6} \end{array} \quad \text{Check the proposed difference.}$$

$$8\tfrac{7}{6} = 8 + 1\tfrac{1}{6}$$

$$= 9\tfrac{1}{6} \longleftarrow 7\tfrac{1}{3} \text{ checks}$$

Caution: To avoid making an error when subtracting a fraction from a mixed number, you should always use vertical subtraction form.

EXAMPLE 3-83 Subtract $5\frac{1}{2} - \frac{3}{4}$.

$$\text{too small} \quad \text{more fourths}$$
$$\downarrow \qquad\quad \downarrow$$

Solution $\quad 5\frac{1}{2} = 5\frac{2}{4} = 4\frac{6}{4} \qquad$ *Think:* $5\frac{2}{4} = 5 + \frac{2}{4} = 4 + 1 + \frac{2}{4} = 4 + \frac{4}{4} + \frac{2}{4} = 4 + \frac{6}{4} = 4\frac{6}{4}$

$$-\frac{3}{4} = \frac{3}{4} = \frac{3}{4}$$
$$\overline{\qquad\qquad\qquad 4\frac{3}{4}} \longleftarrow \text{proposed difference}$$

$\quad 4\frac{3}{4} \qquad$ Check the proposed difference.
$+\frac{3}{4}$
$\overline{\quad 4\frac{6}{4}} = 4 + \dfrac{3}{2}$

$$= 4 + 1\frac{1}{2}$$

$$= 5\frac{1}{2} \longleftarrow 4\frac{3}{4} \text{ checks}$$

3-8. Solving Word Problems Involving Fractions and Mixed Numbers

A. Solving word problems containing fractions

To solve *addition* word problems containing fractions, you find how much two or more fractional parts are all together.

To solve *subtraction* word problems containing fractions, you find

(a) how much is left;

(b) how much more is needed; or

(c) the difference between two fractional parts.

To solve *multiplication* word problems containing fractions, you find

(a) a fractional part of a given amount;

(b) how much two or more equal fractional parts are all together; or

(c) the product of one fractional part times another fractional part.

To solve *division* word problems containing fractions, you find

(a) an amount given a fractional part of it;

(b) how many equal fractional parts there are in all;

(c) how much is in each equal fractional part; or

(d) how many times larger one fractional part is than another fractional part.

EXAMPLE 3-84 Solve the following word problem containing fractions:

Lillian and George were painting. Lillian used half ($\frac{1}{2}$) of the paint. George used one-fourth ($\frac{1}{4}$) of the paint. **(a)** How much paint did they use together? **(b)** How much of the paint was left?

Solution

(1) *Identify:* Lillian used ($\frac{1}{2}$ of the paint). ⟵ — Circle each fact.

George used ($\frac{1}{4}$ of the paint). ⟵

(a) How much of the paint did they use together? ⟵ Underline the questions.

(b) How much of the paint was left? ⟵

(2) *Understand:* **(a)** The first part of the problem asks you to join $\frac{1}{2}$ and $\frac{1}{4}$ together.

(b) The second part of the problem asks you to subtract the sum of $\frac{1}{4}$ and $\frac{1}{2}$ from the total amount of paint.

(3) *Decide:*   To join fractional parts together, you *add.* | To take a fractional part away, you *subtract.*

(4) *Compute:* $\frac{1}{2} + \frac{1}{4} = \frac{2}{4} + \frac{1}{4} = \frac{3}{4}$ | $1 - \frac{3}{4} = \frac{4}{4} - \frac{3}{4} = \frac{1}{4}$

(5) *Interpret:* $\frac{3}{4}$ means they used $\frac{3}{4}$ *of the paint.* | $\frac{1}{4}$ means $\frac{1}{4}$ *of the paint is left.*

(6) *Check:*   Is $\frac{3}{4}$ of the paint (the amount used) and $\frac{1}{4}$ of the paint (the amount left) equal to all of the paint? Yes: $\frac{3}{4} + \frac{1}{4} = \frac{4}{4} = 1$ ◀—— the whole amount of the paint

EXAMPLE 3-85   Solve the following word problems containing fractions:

(a) Ralph was awake two-thirds ($\frac{2}{3}$) of the day. He worked half ($\frac{1}{2}$) of the time he was awake. What part of a day did Ralph work?

(b) Norma worked half ($\frac{1}{2}$) of the day. The time she worked was two-thirds ($\frac{2}{3}$) of the time she was awake. What part of the day was Norma awake?

Solution   **(a)** | **(b)**

(1) *Identify:*   Ralph was awake $\frac{2}{3}$ of the day. He worked $\frac{1}{2}$ of the time he was awake. <u>What part of a day did Ralph work?</u>

(1) *Identify:*   Norma worked $\frac{1}{2}$ of the day. The time she worked was $\frac{2}{3}$ of the time she was awake. <u>What part of the day was Norma awake?</u>

(2) *Understand:*   The question asks you to find $\frac{1}{2}$ *of* $\frac{2}{3}$.

(2) *Understand:*   The question asks you to find what part of $\frac{2}{3}$ *is* $\frac{1}{2}$.

(3) *Decide:*   To find a fractional part of an amount, you *multiply.*

(3) *Decide:*   To find an amount given a fractional part of it, you *divide.*

(4) *Compute:* $\frac{1}{2}$ of $\frac{2}{3}$

$$\frac{1}{2} \times \frac{2}{3} = \frac{1}{\cancel{2}} \times \frac{\cancel{2}}{3}$$

$$= \frac{1}{3}$$

(4) *Compute:* $\frac{1}{2}$ is $\frac{2}{3}$ of what part?
↓
$$\frac{1}{2} \div \frac{2}{3} = \frac{1}{2} \times \frac{3}{2}$$

$$= \frac{3}{4}$$

(5) *Picture:*

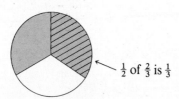

← $\frac{1}{2}$ of $\frac{2}{3}$ is $\frac{1}{3}$

(5) *Picture:*

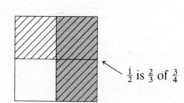

← $\frac{1}{2}$ is $\frac{2}{3}$ of $\frac{3}{4}$

(6) *Interpret:* $\frac{1}{3}$ means that Ralph worked $\frac{1}{3}$ *of the day.*

(6) *Interpret:* $\frac{3}{4}$ means that Norma was awake $\frac{3}{4}$ *of the day.*

(7) *Check:*   Is the time that Ralph was awake ($\frac{2}{3}$ of the day) twice as long as the time he worked ($\frac{1}{3}$ of the day)? Yes: $2 \times \frac{1}{3} = \frac{2}{3}$

(7) *Check:*   Is the time Norma worked ($\frac{2}{3}$ of $\frac{3}{4}$ of the day) equal to $\frac{1}{2}$ of the day? Yes: $\frac{2}{3} \times \frac{3}{4} = \frac{1}{2}$

B. Solving word problems containing mixed numbers

To solve a word problem containing mixed numbers, it is usually helpful to first solve a similar problem containing easier whole numbers.

EXAMPLE 3-86   Solve the following word problem containing mixed numbers:

Margaret needs $3\frac{1}{3}$ yards of material for each curtain panel. She needs $3\frac{3}{4}$ panels in all. How much material should Margaret buy?

Solution

(1) *Identify:*   Margaret needs $\left(3\frac{1}{3} \text{ yards}\right)$ of material — Circle each fact.

for each curtain panel. She needs

$\left(3\frac{3}{4} \text{ panels}\right)$ in all. How much material ← Underline the question.

should Margaret buy?

(2) *Think:*   Margaret needs *3* yards of material ← Use a similar problem containing easier whole numbers.

for each curtain panel. She needs

4 panels in all. How much material

should Margaret buy?

(3) *Understand:*   The question asks you to join 4 equal amounts of 3 together.

(4) *Decide:*   To join equal amounts together, you *multiply*.

(5) *Compute using whole numbers:* $4 \times 3 = 12$ ← approximate answer

(6) *Compute using mixed numbers:*

$$3\frac{3}{4} \times 3\frac{1}{3} = \frac{15}{4} \times \frac{10}{3}$$ ← Replace the easier whole numbers with the original mixed numbers and compute as before.

$$= \frac{\cancel{3} \times 5}{\cancel{4} \times 2} \times \frac{\cancel{2} \times 5}{\cancel{3}}$$

$$= \frac{5}{2} \times \frac{5}{1}$$

$$= \frac{25}{2}$$

$$= 12\frac{1}{2}$$ ← exact answer

(7) *Interpret:* $12\frac{1}{2}$ means that making $3\frac{3}{4}$ panels with $3\frac{1}{3}$ yards of material each requires $12\frac{1}{2}$ yards of material.

EXAMPLE 3-87   Solve the following word problem containing mixed numbers:

Connie had to visit her brother, Mac. Connie used $22\frac{1}{4}$ gallons of gas to drive $311\frac{1}{2}$ miles. How many miles per gallon (mpg) did Connie get on her trip?

Solution

(1) *Identify:*   Connie used $\left(22\frac{1}{4} \text{ gallons}\right)$ ← Circle each fact.

of gas to drive $\left(311\frac{1}{2} \text{ miles}\right)$.

How many miles per gallon (mpg) ← Underline the question.

did Connie get on this trip?

(2) *Think:*   Connie used *20* gallons of gas ← Use a similar problem containing easier whole numbers.

to drive *300* miles. How many

miles per gallon (mpg) did

Connie get on this trip?

(3) *Understand:*   The question asks you to find how many equal amounts of *20* there are in *300*.

(4) *Decide:*   To find how many equal amounts of 20 there are in 300, you *divide*.

(5) *Compute using
   whole numbers:* $300 \div 20 = 15$ ⟵ approximate answer

(6) *Compute using
   mixed numbers:* $311\frac{1}{2} \div 22\frac{1}{4} = \dfrac{623}{2} \div \dfrac{89}{4}$ ⟵ Replace the easier whole numbers with the original
                                                                                         mixed numbers and compute as before.

$$= \dfrac{623}{2} \times \dfrac{4}{89}$$

$$= \dfrac{7 \times \cancel{89}}{\cancel{2}} \times \dfrac{\cancel{2} \times 2}{\cancel{89}}$$

$$= \dfrac{7}{1} \times \dfrac{2}{1}$$

$$= 14 \ ⟵ \text{ exact answer}$$

(7) *Interpret:*   14 means that Connie got *14 mpg on this trip:* $311\frac{1}{2}$ miles on $22\frac{1}{4}$ gallons of gas.

SOLVED PROBLEMS

PROBLEM 3-1   Identify each fraction as proper, improper, or not defined:

(a) $\frac{2}{3}$ **(b)** $\frac{3}{2}$ **(c)** $\frac{4}{4}$ **(d)** $\frac{0}{4}$ **(e)** $\frac{4}{0}$ **(f)** $\frac{8}{1}$ **(g)** $\frac{6}{6}$ **(h)** $\frac{1}{0}$ **(i)** $\frac{0}{1}$ **(j)** $\frac{3}{100}$

Solution   Recall that a proper fraction has a numerator that is less than the denominator, an improper fraction has a numerator that is greater than or equal to the denominator, and a fraction is not defined when the denominator is zero [see Examples 3-3, 3-4, and 3-6]:

(a) proper   **(b)** improper   **(c)** improper   **(d)** proper   **(e)** not defined

(f) improper   **(g)** improper   **(h)** not defined   **(i)** proper   **(j)** proper

PROBLEM 3-2   Name the shaded part of each figure using a proper fraction:

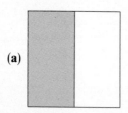 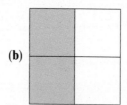 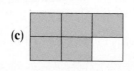 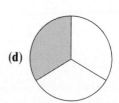

Solution   Recall that the numerator names the number of equal parts that are shaded and the denominator names the number of equal parts in all [see Example 3-5]:

(a) $\dfrac{1}{2}$ **(b)** $\dfrac{2}{4}$ **(c)** $\dfrac{5}{6}$ **(d)** $\dfrac{1}{3}$

PROBLEM 3-3   Write the correct word name for each fraction:

(a) $\frac{1}{2}$ **(b)** $\frac{4}{3}$ **(c)** $\frac{3}{4}$ **(d)** $\frac{6}{5}$ **(e)** $\frac{5}{6}$ **(f)** $\frac{1}{8}$ **(g)** $\frac{8}{9}$ **(h)** $\frac{7}{10}$ **(i)** $\frac{11}{12}$ **(j)** $\frac{29}{100}$

Solution   Recall that to read or write the word name of a fraction, you use the whole-number word name of the numerator followed by: *whole(s)* if the denominator is 1; *half* or *halves* if the denominator is 2; *third(s)* if the denominator is 3; and so on [see Examples 3-7 and 3-8]:

(a) one-half   **(b)** four-thirds   **(c)** three-fourths   **(d)** six-fifths   **(e)** five-sixths

(f) one-eighth **(g)** eight-ninths **(h)** seven-tenths   **(i)** eleven-twelfths **(j)** twenty-nine-hundredths

PROBLEM 3-4 Name each group of shaded figures using a mixed number:

 (a)   (b) 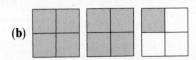   (c)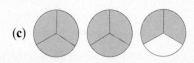

Solution Each completely shaded whole represents one [see Example 3-10]:

(a) $1\frac{1}{2}$ (b) $2\frac{1}{4}$ (c) $2\frac{2}{3}$

PROBLEM 3-5 Write the correct word name for each mixed number:

(a) $1\frac{1}{3}$ (b) $2\frac{1}{4}$ (c) $8\frac{3}{2}$ (d) $5\frac{7}{6}$ (e) $3\frac{5}{8}$ (f) $9\frac{3}{10}$ (g) $4\frac{13}{12}$ (h) $10\frac{2}{9}$ (i) $25\frac{4}{15}$

Solution Recall that you always write the word *and* between the whole-number part and the fraction part of the mixed number [see Example 3-11]:

(a) one and one-third   (b) two and one-fourth   (c) eight and three-halves

(d) five and seven-sixths   (e) three and five-eighths   (f) nine and three-tenths

(g) four and thirteen-twelfths   (h) ten and two-ninths   (i) twenty-five and four-fifteenths

PROBLEM 3-6 Rename each mixed number as a sum:

(a) $5\frac{1}{2}$ (b) $3\frac{1}{4}$ (c) $2\frac{5}{4}$ (d) $4\frac{9}{5}$ (e) $9\frac{3}{5}$ (f) $11\frac{7}{8}$ (g) $15\frac{5}{9}$ (h) $20\frac{3}{10}$ (i) $5\frac{1}{25}$ (j) $16\frac{39}{100}$

Solution Recall that mixed numbers are just a short way to write the sum of a whole number and a fraction [see Example 3-12]:

(a) $5 + \frac{1}{2}$ (b) $3 + \frac{1}{4}$ (c) $2 + \frac{5}{4}$ (d) $4 + \frac{9}{5}$ (e) $9 + \frac{3}{5}$

(f) $11 + \frac{7}{8}$ (g) $15 + \frac{5}{9}$ (h) $20 + \frac{3}{10}$ (i) $5 + \frac{1}{25}$ (j) $16 + \frac{39}{100}$

PROBLEM 3-7 Rename each sum as a mixed number:

(a) $4 + \frac{3}{4}$ (b) $2 + \frac{5}{2}$ (c) $8 + \frac{1}{3}$ (d) $6 + \frac{4}{5}$ (e) $8 + \frac{3}{7}$

(f) $10 + \frac{11}{6}$ (g) $25 + \frac{9}{10}$ (h) $100 + \frac{97}{100}$ (i) $1 + \frac{1}{4}$ (j) $99 + \frac{999}{1000}$

Solution Recall that the sum of a whole number and a fraction is just another way to write a mixed number [see Example 3-13]:

(a) $4\frac{3}{4}$ (b) $2\frac{5}{2}$ (c) $8\frac{1}{3}$ (d) $6\frac{4}{5}$ (e) $8\frac{3}{7}$

(f) $10\frac{11}{6}$ (g) $25\frac{9}{10}$ (h) $100\frac{97}{100}$ (i) $1\frac{1}{4}$ (j) $99\frac{999}{1000}$

PROBLEM 3-8 Are the following pairs of fractions equal or not equal?

(a) $\frac{1}{2}$ and $\frac{2}{4}$ (b) $\frac{1}{2}$ and $\frac{2}{3}$ (c) $\frac{0}{3}$ and $\frac{0}{5}$ (d) $\frac{1}{4}$ and $\frac{4}{8}$

(e) $\frac{1}{4}$ and $\frac{2}{8}$ (f) $\frac{3}{5}$ and $\frac{6}{10}$ (g) $\frac{3}{4}$ and $\frac{6}{8}$ (h) $\frac{5}{8}$ and $\frac{15}{16}$

Solution Recall that two given fractions are equal only when their cross products are equal [see Examples 3-14 and 3-15]:

(a) $4 = 4$; equal   (b) $3 \neq 4$; not equal   (c) $0 = 0$; equal   (d) $8 \neq 16$; not equal

(e) $8 = 8$; equal   (f) $30 = 30$; equal   (g) $24 = 24$; equal   (h) $80 \neq 120$; not equal

PROBLEM 3-9 Rename each division problem as a fraction:

(a) $2 \div 3$ (b) $3 \div 2$ (c) $8 \div 5$ (d) $1 \div 2$ (e) $0 \div 6$

(f) $15 \div 4$ (g) $99 \div 100$ (h) $7 \div 10$ (i) $1 \div 8$ (j) $243 \div 17$

Solution Recall to rename a division problem as a fraction, you write the dividend as the numerator, write the division symbol ($\div$) as the fraction bar (—), and write the divisor as the denominator [see

Example 3-16]:

(a) $\frac{2}{3}$ **(b)** $\frac{3}{2}$ **(c)** $\frac{8}{5}$ **(d)** $\frac{1}{2}$ **(e)** $\frac{0}{6}$ **(f)** $\frac{15}{4}$ **(g)** $\frac{99}{100}$ **(h)** $\frac{7}{10}$ **(i)** $\frac{1}{8}$ **(j)** $\frac{243}{17}$

PROBLEM 3-10 Rename each fraction as a division problem:

(a) $\frac{7}{10}$ **(b)** $\frac{1}{2}$ **(c)** $\frac{3}{5}$ **(d)** $\frac{8}{3}$ **(e)** $\frac{5}{4}$ **(f)** $\frac{17}{8}$ **(g)** $\frac{1}{10}$ **(h)** $\frac{100}{11}$ **(i)** $\frac{49}{100}$ **(j)** $\frac{357}{25}$

Solution Recall that to rename a fraction as a division problem, you write the numerator as the dividend, write the fractor bar (—) as the division symbol (÷), and write the denominator as the divisor [see Example 3-17]:

(a) $7 \div 10$ **(b)** $1 \div 2$ **(c)** $3 \div 5$ **(d)** $8 \div 3$ **(e)** $5 \div 4$

(f) $17 \div 8$ **(g)** $1 \div 10$ **(h)** $100 \div 11$ **(i)** $49 \div 100$ **(j)** $357 \div 25$

PROBLEM 3-11 Rename each whole number as an equal fraction:

(a) 2 **(b)** 8 **(c)** 1 **(d)** 0 **(e)** 25

Solution Recall that to rename a whole number as an equal fraction, you write the whole number over 1 [see Example 3-18]:

(a $\frac{2}{1}$ **(b)** $\frac{8}{1}$ **(c)** $\frac{1}{1}$ **(d)** $\frac{0}{1}$ **(e)** $\frac{25}{1}$

PROBLEM 3-12 Rename each mixed number as an equal fraction:

(a) $1\frac{1}{2}$ **(b)** $4\frac{2}{3}$ **(c)** $5\frac{3}{4}$ **(d)** $3\frac{1}{4}$ **(e)** $6\frac{3}{8}$ **(f)** $2\frac{6}{5}$ **(g)** $7\frac{5}{2}$ **(h)** $3\frac{9}{10}$ **(i)** $2\frac{49}{50}$

Solution Recall that to find the numerator of an equal fraction when renaming a mixed number, you multiply the whole-number part by the denominator of the fraction part and then add that product to the numerator of the fraction part [see Example 3-21]:

(a) $2 \times 1 + 1 = 3; \frac{3}{2}$ **(b)** $3 \times 4 + 2 = 14; \frac{14}{3}$ **(c)** $4 \times 5 + 3 = 23; \frac{23}{4}$

(d) $4 \times 3 + 1 = 13; \frac{13}{4}$ **(e)** $8 \times 6 + 3 = 51; \frac{51}{8}$ **(f)** $5 \times 2 + 6 = 16; \frac{16}{5}$

(g) $2 \times 7 + 5 = 19; \frac{19}{2}$ **(h)** $10 \times 3 + 9 = 39; \frac{39}{10}$ **(i)** $50 \times 2 + 49 = 149; \frac{149}{50}$

PROBLEM 3-13 Rename each improper fraction as an equal mixed number:

(a) $\frac{9}{2}$ **(b)** $\frac{16}{3}$ **(c)** $\frac{7}{4}$ **(d)** $\frac{14}{5}$ **(e)** $\frac{70}{6}$ **(f)** $\frac{11}{8}$ **(g)** $\frac{47}{9}$ **(h)** $\frac{87}{10}$ **(i)** $\frac{25}{12}$ **(j)** $\frac{347}{100}$

Solution Recall that to rename an improper fraction as an equal mixed number using division, divide the numerator by the denominator. The quotient then becomes the whole-number part of the mixed number, the remainder is the numerator of the fraction part, and the divisor is the denominator of the fraction part, which never changes [see Example 3-22]:

(a) $2\,\overline{)\,9}\,;4\frac{1}{2}$ **(b)** $3\,\overline{)\,16}\,;5\frac{1}{3}$ **(c)** $4\,\overline{)\,7}\,;1\frac{3}{4}$ **(d)** $5\,\overline{)\,14}\,;2\frac{4}{5}$ **(e)** $6\,\overline{)\,70}\,;11\frac{4}{6}=11\frac{2}{3}$
$\quad\;\; \frac{-8}{1}$ $\quad\;\; \frac{-15}{1}$ $\quad\;\; \frac{-4}{3}$ $\quad\;\; \frac{-10}{4}$ $\quad\;\; \frac{-6}{10}$
$\qquad\qquad\qquad\qquad\qquad\qquad\qquad\qquad\qquad\qquad\qquad\qquad\qquad\qquad\qquad \frac{-6}{4}$

(f) $8\,\overline{)\,11}\,;1\frac{3}{8}$ **(g)** $9\,\overline{)\,47}\,;5\frac{2}{9}$ **(h)** $10\,\overline{)\,87}\,;8\frac{7}{10}$ **(i)** $12\,\overline{)\,25}\,;2\frac{1}{12}$ **(j)** $100\,\overline{)\,347}\,;3\frac{47}{100}$
$\quad\;\; \frac{-8}{3}$ $\quad\;\; \frac{-45}{2}$ $\quad\;\; \frac{-80}{7}$ $\quad\;\; \frac{-24}{1}$ $\quad\;\; \frac{-300}{47}$

PROBLEM 3-14 Simplify each proper fraction:

(a) $\frac{2}{4}$ **(b)** $\frac{4}{6}$ **(c)** $\frac{3}{12}$ **(d)** $\frac{5}{15}$ **(e)** $\frac{6}{8}$ **(f)** $\frac{4}{20}$ **(g)** $\frac{3}{24}$ **(h)** $\frac{4}{10}$

Solution Recall that to simplify a proper fraction that is not in lowest terms, you reduce the fraction to lowest terms using the Fundamental Rule for Fractions [see Examples 3-24, 3-25, 3-26, and 3-27]:

(a) $\dfrac{2}{4} = \dfrac{1 \times \cancel{2}}{2 \times \cancel{2}} = \dfrac{1}{2}$ **(b)** $\dfrac{4}{6} = \dfrac{\cancel{2} \times 2}{\cancel{2} \times 3} = \dfrac{2}{3}$ (c) $\dfrac{3}{12} = \dfrac{1 \times \cancel{3}}{\cancel{3} \times 4} = \dfrac{1}{4}$ (d) $\dfrac{5}{15} = \dfrac{1 \times \cancel{5}}{3 \times \cancel{5}} = \dfrac{1}{3}$

(e) $\dfrac{6}{8} = \dfrac{\cancel{2} \times 3}{\cancel{2} \times 4} = \dfrac{3}{4}$ (f) $\dfrac{4}{20} = \dfrac{1 \times \cancel{4}}{\cancel{4} \times 5} = \dfrac{1}{5}$ (g) $\dfrac{3}{24} = \dfrac{1 \times \cancel{3}}{\cancel{3} \times 8} = \dfrac{1}{8}$ (h) $\dfrac{4}{10} = \dfrac{\cancel{2} \times 2}{\cancel{2} \times 5} = \dfrac{2}{5}$

PROBLEM 3-15 Simplify each improper fraction:

(a) $\dfrac{7}{2}$ **(b)** $\dfrac{7}{3}$ (c) $\dfrac{14}{8}$ **(d)** $\dfrac{42}{9}$

Solution Recall that to simplify an improper fraction, you first reduce to lowest terms when possible and then rename the improper fraction in lowest terms as a mixed number [see Example 3-28]:

(a) $\dfrac{7}{2} = 2\overline{)7}\;;\; 3\tfrac{1}{2}$ $\dfrac{-6}{\,1}$

(b) $\dfrac{7}{3} = 3\overline{)7}\;;\; 2\tfrac{1}{3}$ $\dfrac{-6}{\,1}$

(c) $\dfrac{14}{8} = \dfrac{\cancel{2} \times 7}{\cancel{2} \times 4} = \dfrac{7}{4};\; 4\overline{)7}\;;\; 1\tfrac{3}{4}$ $\dfrac{-4}{\,3}$

(d) $\dfrac{42}{9} = \dfrac{\cancel{3} \times 14}{\cancel{3} \times 3} = \dfrac{14}{3};\; 3\overline{)14}\;;\; 4\tfrac{2}{3}$ $\dfrac{-12}{\;\,2}$

PROBLEM 3-16 Simplify each mixed number:

(a) $3\tfrac{4}{8}$ **(b)** $5\tfrac{6}{9}$ (c) $4\tfrac{3}{2}$ **(d)** $8\tfrac{5}{3}$ (e) $1\tfrac{8}{4}$ **(f)** $2\tfrac{15}{5}$ (g) $6\tfrac{24}{30}$

Solution Recall that to simplify a mixed number that is not in simplest form, you simplify the fraction part [see Example 3-29]:

(a) $3\tfrac{4}{8} = 3 + \dfrac{4}{8} = 3 + \dfrac{1}{2} = 3\tfrac{1}{2}$ (b) $5\tfrac{6}{9} = 5 + \dfrac{6}{9} = 5 + \dfrac{2}{3} = 5\tfrac{2}{3}$

(c) $4\tfrac{3}{2} = 4 + \dfrac{3}{2} = 4 + 1\tfrac{1}{2} = 4 + 1 + \dfrac{1}{2} = 5 + \dfrac{1}{2} = 5\tfrac{1}{2}$

(d) $8\tfrac{5}{3} = 8 + \dfrac{5}{3} = 8 + 1\tfrac{2}{3} = 8 + 1 + \dfrac{2}{3} = 9 + \dfrac{2}{3} = 9\tfrac{2}{3}$ (e) $1\tfrac{8}{4} = 1 + \dfrac{8}{4} = 1 + 2 = 3$

(f) $2\tfrac{15}{5} = 2 + \dfrac{15}{5} = 2 + 3 = 5$ (g) $6\tfrac{24}{30} = 6 + \dfrac{24}{30} = 6 + \dfrac{4}{5} = 6\tfrac{4}{5}$

PROBLEM 3-17 Rename each fractional statement as multiplication:

(a) $\tfrac{1}{2}$ of $\tfrac{3}{4}$ **(b)** $\tfrac{2}{3}$ of $\tfrac{5}{8}$ (c) $\tfrac{1}{3}$ of 5 **(d)** $\tfrac{1}{4}$ of 10 (e) $\tfrac{7}{10}$ of $2\tfrac{1}{2}$ **(f)** $\tfrac{2}{5}$ of $2\tfrac{3}{4}$

Solution Recall that to find a fractional part of an amount you multiply. Also, you can usually replace the word *of* with a multiplication symbol ($\times$) [see Example 3-30]:

(a) $\dfrac{1}{2} \times \dfrac{3}{4}$ **(b)** $\dfrac{2}{3} \times \dfrac{5}{8}$ (c) $\dfrac{1}{3} \times 5$ **(d)** $\dfrac{1}{4} \times 10$ (e) $\dfrac{7}{10} \times 2\tfrac{1}{2}$ **(f)** $\dfrac{2}{5} \times 2\tfrac{3}{4}$

PROBLEM 3-18 Rename each repeated addition problem as multiplication:

(a) $\tfrac{1}{3} + \tfrac{1}{3}$ **(b)** $\tfrac{1}{2} + \tfrac{1}{2} + \tfrac{1}{2}$

Solution Recall that to join equal fractional parts together, you multiply [see Example 3-31]:

(a) 2 equal amounts of $\tfrac{1}{3}$, or $2 \times \tfrac{1}{3}$ (b) 3 equal amounts of $\tfrac{1}{2}$, or $3 \times \tfrac{1}{2}$

PROBLEM 3-19   Multiply with fractions:

(a) $\frac{2}{3} \times \frac{1}{5}$ (b) $\frac{1}{2} \times \frac{3}{4}$ (c) $\frac{2}{5} \times \frac{5}{9}$ (d) $\frac{1}{3} \times \frac{6}{5}$ (e) $\frac{2}{3} \times \frac{9}{2}$

(f) $\frac{8}{3} \times \frac{3}{4}$ (g) $\frac{5}{3} \times 2$ (h) $\frac{5}{8} \times 8$ (i) $5 \times \frac{1}{2}$ (j) $6 \times \frac{2}{3}$

Solution   Recall that if you eliminate all common factors before you multiply fractions, then the product will always be in lowest terms after multiplying [see Examples 3-32 through 3-35]:

(a) $\dfrac{2}{3} \times \dfrac{1}{5} = \dfrac{2 \times 1}{3 \times 5} = \dfrac{2}{15}$

(b) $\dfrac{1}{2} \times \dfrac{3}{4} = \dfrac{1 \times 3}{2 \times 4} = \dfrac{3}{8}$

(c) $\dfrac{2}{\cancel{3}} \times \dfrac{\cancel{5}}{9} = \dfrac{2}{1} \times \dfrac{1}{9} = \dfrac{2}{9}$

(d) $\dfrac{1}{3} \times \dfrac{6}{5} = \dfrac{1}{\cancel{3}} \times \dfrac{2 \times \cancel{3}}{5} = \dfrac{1}{1} \times \dfrac{2}{5} = \dfrac{2}{5}$

(e) $\dfrac{\cancel{2}}{3} \times \dfrac{9}{\cancel{2}} = \dfrac{1}{\cancel{3}} \times \dfrac{\cancel{3} \times 3}{1} = \dfrac{1}{1} \times \dfrac{3}{1} = \dfrac{3}{1} = 3$

(f) $\dfrac{8}{\cancel{3}} \times \dfrac{\cancel{3}}{4} = \dfrac{2 \times 4}{1} \times \dfrac{1}{4} = \dfrac{2}{1} \times \dfrac{1}{1} = \dfrac{2}{1} = 2$

(g) $\dfrac{5}{3} \times 2 = \dfrac{5}{3} \times \dfrac{2}{1} = \dfrac{10}{3} = 3\frac{1}{3}$

(h) $\dfrac{5}{8} \times 8 = \dfrac{5}{\cancel{8}} \times \dfrac{\cancel{8}}{1} = \dfrac{5}{1} \times \dfrac{1}{1} = \dfrac{5}{1} = 5$

(i) $5 \times \dfrac{1}{2} = \dfrac{5}{1} \times \dfrac{1}{2} = \dfrac{5}{2} = 2\frac{1}{2}$

(j) $6 \times \dfrac{2}{3} = \dfrac{6}{1} \times \dfrac{2}{3} = \dfrac{2 \times \cancel{3}}{1} \times \dfrac{2}{\cancel{3}} = \dfrac{2}{1} \times \dfrac{2}{1} = \dfrac{4}{1} = 4$

PROBLEM 3-20   Multiply with mixed numbers:

(a) $2\frac{1}{4} \times \frac{5}{6}$ (b) $1\frac{1}{2} \times \frac{1}{3}$ (c) $\frac{2}{5} \times 1\frac{3}{4}$ (d) $1\frac{1}{2} \times \frac{1}{2}$

(e) $\frac{3}{8} \times 2\frac{1}{3}$ (f) $\frac{3}{5} \times 7\frac{1}{2}$ (g) $4\frac{1}{2} \times \frac{2}{3}$ (h) $7\frac{1}{2} \times 1\frac{3}{5}$

Solution   Recall that to multiply with mixed numbers, you first rename each mixed number and whole number as an equal fraction and then multiply [see Example 3-36]:

(a) $2\frac{1}{4} \times \dfrac{5}{6} = \dfrac{9}{4} \times \dfrac{5}{6} = \dfrac{3 \times \cancel{3}}{4} \times \dfrac{5}{2 \times \cancel{3}} = \dfrac{15}{8} = 1\frac{7}{8}$

(b) $1\frac{1}{2} \times \dfrac{1}{3} = \dfrac{\cancel{3}}{2} \times \dfrac{1}{\cancel{3}} = \dfrac{1}{2}$

(c) $\dfrac{2}{5} \times 1\frac{3}{4} = \dfrac{2}{5} \times \dfrac{7}{4} = \dfrac{\cancel{2}}{5} \times \dfrac{7}{\cancel{2} \times 2} = \dfrac{7}{10}$

(d) $1\frac{1}{2} \times \dfrac{1}{2} = \dfrac{3}{2} \times \dfrac{1}{2} = \dfrac{3}{4}$

(e) $\dfrac{3}{8} \times 2\frac{1}{3} = \dfrac{\cancel{3}}{8} \times \dfrac{7}{\cancel{3}} = \dfrac{7}{8}$

(f) $\dfrac{3}{5} \times 7\frac{1}{2} = \dfrac{3}{5} \times \dfrac{15}{2} = \dfrac{3}{\cancel{5}} \times \dfrac{3 \times \cancel{5}}{2} = \dfrac{9}{2} = 4\frac{1}{2}$

(g) $4\frac{1}{2} \times \dfrac{2}{3} = \dfrac{9}{\cancel{2}} \times \dfrac{\cancel{2}}{3} = \dfrac{\cancel{3} \times 3}{1} \times \dfrac{1}{\cancel{3}} = 3$

(h) $7\frac{1}{2} \times 1\frac{3}{5} = \dfrac{15}{2} \times \dfrac{8}{5} = \dfrac{3 \times \cancel{5}}{\cancel{2}} \times \dfrac{\cancel{2} \times 4}{\cancel{5}} = 12$

PROBLEM 3-21   Find and check the reciprocal of each fraction:

(a) $\frac{1}{2}$ (b) $\frac{2}{3}$ (c) $\frac{5}{4}$ (d) $\frac{0}{1}$

Solution   Recall that to find the reciprocal of a fraction, you interchange the whole numbers in the numerator and denominator. To check the reciprocal of a fraction, you multiply the two fractions to see if the product is 1 [see Example 3-38]:

(a) The reciprocal of $\frac{1}{2}$ is $\frac{2}{1}$ (or 2) because $\frac{1}{2} \times \frac{2}{1} = 1$.

(b) The reciprocal of $\frac{2}{3}$ is $\frac{3}{2}$ because $\frac{2}{3} \times \frac{3}{2} = 1$.

(c) The reciprocal of $\frac{5}{4}$ is $\frac{4}{5}$ because $\frac{5}{4} \times \frac{4}{5} = 1$.

(d) The reciprocal of $\frac{0}{1}$ is not defined because the denominator of a fraction can never be zero.

PROBLEM 3-22   Find and check the reciprocal of each whole number:

(a) 2 (b) 6 (c) 0 (d) 15

Solution   Recall that to find the reciprocal of a whole number, you first rename the whole number as an equal fraction and then find the reciprocal of that fraction [see Example 3-39]:

(a) The reciprocal of 2 (or $\frac{2}{1}$) is $\frac{1}{2}$ because $2 \times \frac{1}{2} = \frac{2}{1} \times \frac{1}{2} = 1$.

(b) The reciprocal of 6 (or $\frac{6}{1}$) is $\frac{1}{6}$ because $6 \times \frac{1}{6} = \frac{6}{1} \times \frac{1}{6} = 1$.

(c) The reciprocal of 0 is not defined because $0 = \frac{0}{1}$ and $\frac{1}{0}$ (its reciprocal) is not defined.

(d) The reciprocal of 15 (or $\frac{15}{1}$) is $\frac{1}{15}$ because $15 \times \frac{1}{15} = \frac{15}{1} \times \frac{1}{15} = 1$.

PROBLEM 3-23   Find and check the reciprocal of each mixed number:

(a) $2\frac{1}{3}$ 　　　　　　　　　　　(b) $3\frac{1}{2}$ 　　　　　　　　　　　(c) $2\frac{3}{4}$

Solution   Recall that to find the reciprocal of a mixed number, you first rename the mixed number as an equal fraction and then find the reciprocal of that fraction [see Example 3-40]:

(a) The reciprocal of $2\frac{1}{3}$ (or $\frac{7}{3}$) is $\frac{3}{7}$ because $2\frac{1}{3} \times \frac{3}{7} = \frac{7}{3} \times \frac{3}{7} = 1$.

(b) The reciprocal of $3\frac{1}{2}$ (or $\frac{7}{2}$) is $\frac{2}{7}$ because $3\frac{1}{2} \times \frac{2}{7} = \frac{7}{2} \times \frac{2}{7} = 1$.

(c) The reciprocal of $2\frac{3}{4}$ (or $\frac{11}{4}$) is $\frac{4}{11}$ because $2\frac{3}{4} \times \frac{4}{11} = \frac{11}{4} \times \frac{4}{11} = 1$.

PROBLEM 3-24   Rename each fractional statement as division:

(a) $\frac{2}{3}$ is $\frac{1}{4}$ of an amount 　　(b) $\frac{2}{3}$ of an amount is $\frac{1}{2}$ 　　(c) 2 is $\frac{5}{8}$ of an amount

(d) $\frac{1}{2}$ of an amount is 5 　　(e) $3\frac{1}{4}$ is $\frac{1}{3}$ of an amount 　　(f) $\frac{1}{10}$ of an amount is $1\frac{1}{2}$

Solution   Recall that to find an amount given a fractional part of it, you divide. Also, when the word *is* appears between two numbers, you can usually replace the word with a division symbol ($\div$) [see Example 3-41]:

(a) $\frac{2}{3} \div \frac{1}{4}$ 　　　　　　　　(b) $\frac{1}{2}$ is $\frac{2}{3}$ of amount, or $\frac{1}{2} \div \frac{2}{3}$

(c) $2 \div \frac{5}{8}$ 　　　　　　　　(d) 5 is $\frac{1}{2}$ of an amount, or $5 \div \frac{1}{2}$

(e) $3\frac{1}{4} \div \frac{1}{3}$ 　　　　　　　　(f) $1\frac{1}{2}$ is $\frac{1}{10}$ of an amount, or $1\frac{1}{2} \div \frac{1}{10}$

PROBLEM 3-25   Divide the following fractions.

(a) $\frac{3}{4} \div \frac{2}{3}$ 　(b) $\frac{3}{8} \div \frac{2}{3}$ 　(c) $\frac{2}{3} \div \frac{5}{6}$ 　(d) $\frac{1}{6} \div \frac{2}{5}$ 　(e) $\frac{1}{2} \div \frac{1}{3}$ 　(f) $\frac{1}{3} \div \frac{1}{2}$

(g) $\frac{3}{5} \div \frac{3}{5}$ 　(h) $\frac{0}{3} \div \frac{7}{9}$ 　(i) $\frac{5}{6} \div 3$ 　(j) $\frac{2}{3} \div 4$ 　(k) $2 \div \frac{1}{2}$ 　(l) $9 \div \frac{3}{4}$

Solution   Recall that to divide with fractions, you change the divisor to its reciprocal and then multiply. To check division, you multiply the proposed quotient by the original divisor to see if you get the original dividend [see Examples 3-43 through 3-47]:

(a) $\dfrac{3}{4} \div \dfrac{2}{3} = \dfrac{3}{4} \times \dfrac{3}{2} = \dfrac{9}{8} = 1\frac{1}{8}$ 　　　　(b) $\dfrac{3}{8} \div \dfrac{2}{3} = \dfrac{3}{8} \times \dfrac{3}{2} = \dfrac{9}{16}$

(c) $\dfrac{2}{3} \div \dfrac{5}{6} = \dfrac{2}{3} \times \dfrac{6}{5} = \dfrac{2}{\cancel{3}} \times \dfrac{2 \times \cancel{3}}{5} = \dfrac{4}{5}$ 　　　　(d) $\dfrac{1}{6} \div \dfrac{2}{5} = \dfrac{1}{6} \times \dfrac{5}{2} = \dfrac{5}{12}$

(e) $\dfrac{1}{2} \div \dfrac{1}{3} = \dfrac{1}{2} \times \dfrac{3}{1} = \dfrac{3}{2} = 1\frac{1}{2}$ 　　　　(f) $\dfrac{1}{3} \div \dfrac{1}{2} = \dfrac{1}{3} \times \dfrac{2}{1} = \dfrac{2}{3}$

(g) $\dfrac{3}{5} \div \dfrac{3}{5} = \dfrac{\cancel{3}}{\cancel{5}} \times \dfrac{\cancel{5}}{\cancel{3}} = 1$ 　　　　(h) $\dfrac{0}{3} \div \dfrac{7}{9} = 0 \div \dfrac{7}{9} = 0$

(i) $\dfrac{5}{6} \div 3 = \dfrac{5}{6} \div \dfrac{3}{1} = \dfrac{5}{6} \times \dfrac{1}{3} = \dfrac{5}{18}$

(j) $\dfrac{2}{3} \div 4 = \dfrac{2}{3} \div \dfrac{4}{1} = \dfrac{2}{3} \times \dfrac{1}{4} = \dfrac{\cancel{2}}{3} \times \dfrac{1}{\cancel{2} \times 2} = \dfrac{1}{6}$

(k) $2 \div \dfrac{1}{2} = \dfrac{2}{1} \div \dfrac{1}{2} = \dfrac{2}{1} \times \dfrac{2}{1} = \dfrac{4}{1} = 4$

(l) $9 \div \dfrac{3}{4} = \dfrac{9}{1} \div \dfrac{3}{4} = \dfrac{9}{1} \times \dfrac{4}{3} = \dfrac{\cancel{3} \times 3}{1} \times \dfrac{4}{\cancel{3}} = \dfrac{12}{1} = 12$

PROBLEM 3-26   Divide with mixed numbers.

(a) $2\frac{1}{2} \div \frac{1}{4}$ **(b)** $4\frac{1}{2} \div 2$ **(c)** $\frac{3}{4} \div 1\frac{1}{2}$ **(d)** $1\frac{1}{2} \div \frac{2}{3}$ **(e)** $2\frac{1}{2} \div 1\frac{1}{2}$ **(f)** $3\frac{3}{4} \div 4\frac{1}{2}$

(g) $9 \div 2\frac{1}{4}$ **(h)** $4\frac{1}{4} \div 6$ **(i)** $2\frac{5}{8} \div 3\frac{1}{2}$ **(j)** $4\frac{1}{2} \div \frac{3}{4}$ **(k)** $0 \div 5\frac{3}{4}$ **(l)** $6\frac{7}{8} \div 1$

Solution   Recall that to divide with mixed numbers, you first rename each mixed number (and each whole number) as an equal fraction and then divide. To check division, you multiply the proposed quotient by the original divisor to see if you get the original dividend [see Example 3-48]:

(a) $2\frac{1}{2} \div \dfrac{1}{4} = \dfrac{5}{2} \div \dfrac{1}{4} = \dfrac{5}{2} \times \dfrac{4}{1} = \dfrac{5}{\cancel{2}} \times \dfrac{\cancel{2} \times 2}{1} = \dfrac{10}{1} = 10$

(b) $4\frac{1}{2} \div 2 = \dfrac{9}{2} \div \dfrac{2}{1} = \dfrac{9}{2} \times \dfrac{1}{2} = \dfrac{9}{4} = 2\frac{1}{4}$

(c) $\dfrac{3}{4} \div 1\frac{1}{2} = \dfrac{3}{4} \div \dfrac{3}{2} = \dfrac{\cancel{3}}{4} \times \dfrac{2}{\cancel{3}} = \dfrac{1}{\cancel{2} \times 2} \times \dfrac{\cancel{2}}{1} = \dfrac{1}{2}$

(d) $1\frac{1}{2} \div \dfrac{2}{3} = \dfrac{3}{2} \div \dfrac{2}{3} = \dfrac{3}{2} \times \dfrac{3}{2} = \dfrac{9}{4} = 2\frac{1}{4}$

(e) $2\frac{1}{2} \div 1\frac{1}{2} = \dfrac{5}{2} \div \dfrac{3}{2} = \dfrac{5}{\cancel{2}} \times \dfrac{\cancel{2}}{3} = \dfrac{5}{3} = 1\frac{2}{3}$

(f) $3\frac{3}{4} \div 4\frac{1}{2} = \dfrac{15}{4} \div \dfrac{9}{2} = \dfrac{15}{4} \times \dfrac{2}{9} = \dfrac{\cancel{3} \times 5}{\cancel{2} \times 2} \times \dfrac{\cancel{2}}{\cancel{3} \times 3} = \dfrac{5}{6}$

(g) $9 \div 2\frac{1}{4} = \dfrac{9}{1} \div \dfrac{9}{4} = \dfrac{\cancel{9}}{1} \times \dfrac{4}{\cancel{9}} = \dfrac{4}{1} = 4$

(h) $4\frac{1}{4} \div 6 = \dfrac{17}{4} \div \dfrac{6}{1} = \dfrac{17}{4} \times \dfrac{1}{6} = \dfrac{17}{24}$

(i) $2\frac{5}{8} \div 3\frac{1}{2} = \dfrac{21}{8} \div \dfrac{7}{2} = \dfrac{21}{8} \times \dfrac{2}{7} = \dfrac{3 \times \cancel{7}}{\cancel{2} \times 4} \times \dfrac{\cancel{2}}{\cancel{7}} = \dfrac{3}{4}$

(j) $4\frac{1}{2} \div \dfrac{3}{4} = \dfrac{9}{2} \div \dfrac{3}{4} = \dfrac{9}{2} \times \dfrac{4}{3} = \dfrac{\cancel{3} \times 3}{\cancel{2}} \times \dfrac{\cancel{2} \times 2}{\cancel{3}} = \dfrac{6}{1} = 6$

(k) $0 \div 5\frac{3}{4} = 0$ (0 divided by any nonzero number is 0.)

(l) $6\frac{7}{8} \div 1 = 6\frac{7}{8}$ (Any number divided by 1 equals that number.)

PROBLEM 3-27   Add two like fractions; simplify your answers into lowest terms.

(a) $\frac{1}{2} + \frac{1}{2}$ **(b)** $\frac{1}{2} + \frac{3}{2}$ **(c)** $\frac{1}{4} + \frac{1}{4}$ **(d)** $\frac{3}{4} + \frac{1}{4}$ **(e)** $\frac{3}{4} + \frac{3}{4}$ **(f)** $\frac{1}{8} + \frac{1}{8}$ **(g)** $\frac{3}{8} + \frac{1}{8}$

(h) $\frac{3}{8} + \frac{5}{8}$ **(i)** $\frac{5}{8} + \frac{5}{8}$ **(j)** $\frac{1}{10} + \frac{1}{10}$ **(k)** $\frac{3}{10} + \frac{1}{10}$ **(l)** $\frac{1}{10} + \frac{7}{10}$ **(m)** $\frac{7}{10} + \frac{7}{10}$ **(n)** $\frac{1}{12} + \frac{1}{12}$

Solution   Recall that to add two like fractions, you add the numerators and then write the same denominator [see Examples 3-51 and 3-53]:

(a) $\dfrac{1}{2} + \dfrac{1}{2} = \dfrac{1+1}{2} = \dfrac{2}{2} = 1$

(b) $\dfrac{1}{2} + \dfrac{3}{2} = \dfrac{1+3}{2} = \dfrac{4}{2} = 2$

(c) $\dfrac{1}{4} + \dfrac{1}{4} = \dfrac{1+1}{4} = \dfrac{2}{4} = \dfrac{1}{2}$

(d) $\dfrac{3}{4} + \dfrac{1}{4} = \dfrac{3+1}{4} = \dfrac{4}{4} = 1$

(e) $\frac{3}{4} + \frac{3}{4} = \frac{3+3}{4} = \frac{6}{4} = \frac{3}{2} = 1\frac{1}{2}$

(f) $\frac{1}{8} + \frac{1}{8} = \frac{1+1}{8} = \frac{2}{8} = \frac{1}{4}$

(g) $\frac{3}{8} + \frac{1}{8} = \frac{3+1}{8} = \frac{4}{8} = \frac{1}{2}$

(h) $\frac{3}{8} + \frac{5}{8} = \frac{3+5}{8} = \frac{8}{8} = 1$

(i) $\frac{5}{8} + \frac{5}{8} = \frac{5+5}{8} = \frac{10}{8} = \frac{5}{4} = 1\frac{1}{4}$

(j) $\frac{1}{10} + \frac{1}{10} = \frac{1+1}{10} = \frac{2}{10} = \frac{1}{5}$

(k) $\frac{3}{10} + \frac{1}{10} = \frac{3+1}{10} = \frac{4}{10} = \frac{2}{5}$

(l) $\frac{1}{10} + \frac{7}{10} = \frac{1+7}{10} = \frac{8}{10} = \frac{4}{5}$

(m) $\frac{7}{10} + \frac{7}{10} = \frac{7+7}{10} = \frac{14}{10} = \frac{7}{5} = 1\frac{2}{5}$

(n) $\frac{1}{12} + \frac{1}{12} = \frac{1+1}{12} = \frac{2}{12} = \frac{1}{6}$

PROBLEM 3-28 Add more than two like fractions; simplify your answers.

(a) $\frac{1}{2} + \frac{1}{2} + \frac{1}{2}$ (b) $\frac{1}{3} + \frac{1}{3} + \frac{1}{3}$ (c) $\frac{1}{4} + \frac{1}{4} + \frac{1}{4}$ (d) $\frac{1}{5} + \frac{2}{5} + \frac{1}{5}$ (e) $\frac{1}{6} + \frac{5}{6} + \frac{5}{6}$

(f) $\frac{2}{7} + \frac{1}{7} + \frac{3}{7}$ (g) $\frac{5}{8} + \frac{1}{8} + \frac{3}{8}$ (h) $\frac{7}{9} + \frac{8}{9} + \frac{5}{9}$ (i) $\frac{1}{10} + \frac{3}{10} + \frac{7}{10}$ (j) $\frac{5}{12} + \frac{7}{12} + \frac{1}{12}$

Solution Recall that to add more than two like fractions, you add all of the numerators and then write the same denominator [see Example 3-54]:

(a) $\frac{1}{2} + \frac{1}{2} + \frac{1}{2} = \frac{1+1+1}{2} = \frac{3}{2} = 1\frac{1}{2}$

(b) $\frac{1}{3} + \frac{1}{3} + \frac{1}{3} = \frac{1+1+1}{3} = \frac{3}{3} = 1$

(c) $\frac{1}{4} + \frac{1}{4} + \frac{1}{4} = \frac{1+1+1}{4} = \frac{3}{4}$

(d) $\frac{1}{5} + \frac{2}{5} + \frac{1}{5} = \frac{1+2+1}{5} = \frac{4}{5}$

(e) $\frac{1}{6} + \frac{5}{6} + \frac{5}{6} = \frac{1+5+5}{6} = \frac{11}{6} = 1\frac{5}{6}$

(f) $\frac{2}{7} + \frac{1}{7} + \frac{3}{7} = \frac{2+1+3}{7} = \frac{6}{7}$

(g) $\frac{5}{8} + \frac{1}{8} + \frac{3}{8} = \frac{5+1+3}{8} = \frac{9}{8} = 1\frac{1}{8}$

(h) $\frac{7}{9} + \frac{8}{9} + \frac{5}{9} = \frac{7+8+5}{9} = \frac{20}{9} = 2\frac{2}{9}$

(i) $\frac{1}{10} + \frac{3}{10} + \frac{7}{10} = \frac{1+3+7}{10} = \frac{11}{10} = 1\frac{1}{10}$

(j) $\frac{5}{12} + \frac{7}{12} + \frac{1}{12} = \frac{5+7+1}{12} = \frac{13}{12} = 1\frac{1}{12}$

PROBLEM 3-29 Find the LCD for two unlike fractions when the smaller denominator divides the larger denominator evenly:

(a) $\frac{1}{2}$ and $\frac{3}{4}$ (b) $\frac{1}{2}$ and $\frac{1}{6}$ (c) $\frac{5}{8}$ and $\frac{1}{2}$ (d) $\frac{3}{10}$ and $\frac{1}{2}$ (e) $\frac{1}{3}$ and $\frac{5}{6}$

(f) $\frac{2}{3}$ and $\frac{7}{9}$ (g) $\frac{5}{12}$ and $\frac{2}{3}$ (h) $\frac{1}{4}$ and $\frac{7}{8}$ (i) $\frac{3}{4}$ and $\frac{7}{12}$ (j) $\frac{1}{5}$ and $\frac{7}{10}$

Solution Recall that when the smaller denominator divides the larger denominator evenly, the larger denominator is the LCD [see Example 3-55]:

(a) The LCD for $\frac{1}{2}$ and $\frac{3}{4}$ is 4 because 2 divides 4 evenly ($4 \div 2 = 2$).

(b) The LCD for $\frac{1}{2}$ and $\frac{1}{6}$ is 6 because 2 divides 6 evenly ($6 \div 2 = 3$).

(c) The LCD for $\frac{5}{8}$ and $\frac{1}{2}$ is 8 because 2 divides 8 evenly ($8 \div 2 = 4$).

(d) The LCD for $\frac{3}{10}$ and $\frac{1}{2}$ is 10 because 2 divides 10 evenly ($10 \div 2 = 5$).

(e) The LCD for $\frac{1}{3}$ and $\frac{5}{6}$ is 6 because 3 divides 6 evenly ($6 \div 3 = 2$).

(f) The LCD for $\frac{2}{3}$ and $\frac{7}{9}$ is 9 because 3 divides 9 evenly ($9 \div 3 = 3$).

(g) The LCD for $\frac{5}{12}$ and $\frac{2}{3}$ is 12 because 3 divides 12 evenly ($12 \div 3 = 4$).

(h) The LCD for $\frac{1}{4}$ and $\frac{7}{8}$ is 8 because 4 divides 8 evenly ($8 \div 4 = 2$).

(i) The LCD for $\frac{3}{4}$ and $\frac{7}{12}$ is 12 because 4 divides 12 evenly ($12 \div 4 = 3$).

(j) The LCD for $\frac{1}{5}$ and $\frac{7}{10}$ is 10 because 5 divides 10 evenly ($10 \div 5 = 2$).

PROBLEM 3-30 Find the LCD for two fractions when the denominators do not share a common prime factor:

(a) $\frac{1}{2}$ and $\frac{2}{3}$ (b) $\frac{1}{2}$ and $\frac{3}{5}$ (c) $\frac{5}{7}$ and $\frac{1}{2}$ (d) $\frac{2}{9}$ and $\frac{1}{2}$ (e) $\frac{1}{3}$ and $\frac{3}{4}$

(f) $\frac{2}{3}$ and $\frac{1}{5}$ (g) $\frac{1}{4}$ and $\frac{4}{5}$ (h) $\frac{1}{9}$ and $\frac{3}{4}$ (i) $\frac{3}{5}$ and $\frac{5}{6}$ (j) $\frac{1}{8}$ and $\frac{2}{5}$

Solution Recall that when two denominators do not share a common prime factor, the product of the denominators is the LCD [see Example 3-52]:

(a) The LCD of $\frac{1}{2}$ and $\frac{2}{3}$ is 6 ($= 2 \times 3$) because 2 and 3 do not share a common prime factor.

(b) The LCD of $\frac{1}{2}$ and $\frac{3}{5}$ is 10 ($= 2 \times 5$) because 2 and 5 do not share a common prime factor.

(c) The LCD of $\frac{5}{7}$ and $\frac{1}{2}$ is 14 ($= 7 \times 2$) because 7 and 2 do not share a common prime factor.

(d) The LCD of $\frac{2}{9}$ and $\frac{1}{2}$ is 18 ($= 9 \times 2$) because 9 and 2 do not share a common prime factor.

(e) The LCD of $\frac{1}{3}$ and $\frac{3}{4}$ is 12 ($= 3 \times 4$) because 3 and 4 do not share a common prime factor.

(f) The LCD of $\frac{2}{3}$ and $\frac{1}{5}$ is 15 ($= 3 \times 5$) because 3 and 5 do not share a common prime factor.

(g) The LCD of $\frac{1}{4}$ and $\frac{4}{5}$ is 20 ($= 4 \times 5$) because 4 and 5 do not share a common prime factor.

(h) The LCD of $\frac{1}{9}$ and $\frac{3}{4}$ is 36 ($= 9 \times 4$) because 9 and 4 do not share a common prime factor.

(i) The LCD of $\frac{3}{5}$ and $\frac{5}{6}$ is 30 ($= 5 \times 6$) because 5 and 6 do not share a common prime factor.

(j) The LCD of $\frac{1}{8}$ and $\frac{2}{5}$ is 40 ($= 8 \times 5$) because 8 and 5 do not share a common prime factor.

PROBLEM 3-31 Find the LCD of two fractions when the difference between the denominators divides each denominator evenly:

(a) $\frac{3}{4}$ and $\frac{5}{6}$ (b) $\frac{3}{8}$ and $\frac{1}{6}$ (c) $\frac{1}{8}$ and $\frac{7}{10}$ (d) $\frac{2}{9}$ and $\frac{5}{12}$

Solution Recall that when the difference between the two denominators divides each denominator evenly, you divide the difference into one of the denominators and then multiply that quotient times the other denominator to get the LCD [see Example 3-58]:

(a) The LCD for $\frac{3}{4}$ and $\frac{5}{6}$ is 12 because: $6 - 4 = 2$ and $2\overline{)6}$ gives $3 \times 4 = 12$ or $2\overline{)4}$ gives $2 \times 6 = 12$

(b) The LCD for $\frac{3}{8}$ and $\frac{1}{6}$ is 24 because: $8 - 6 = 2$ and $2\overline{)8}$ gives $4 \times 6 = 24$ or $2\overline{)6}$ gives $3 \times 8 = 24$

(c) The LCD for $\frac{1}{8}$ and $\frac{7}{10}$ is 40 because: $10 - 8 = 2$ and $2\overline{)10}$ gives $5 \times 8 = 40$ or $2\overline{)8}$ gives $4 \times 10 = 40$

(d) The LCD for $\frac{2}{9}$ and $\frac{5}{12}$ is 36 because: $12 - 9 = 3$ and $3\overline{)12}$ gives $4 \times 9 = 36$ or $3\overline{)9}$ gives $3 \times 12 = 36$

PROBLEM 3-32 Find the LCD of two fractions using the Factoring Method for finding the LCD:

(a) $\frac{1}{6}$ and $\frac{3}{10}$ (b) $\frac{1}{10}$ and $\frac{3}{4}$ (c) $\frac{1}{20}$ and $\frac{11}{12}$ (d) $\frac{5}{12}$ and $\frac{1}{21}$ (e) $\frac{4}{15}$ and $\frac{2}{9}$ (f) $\frac{3}{14}$ and $\frac{5}{6}$

Solution Recall that when you have trouble finding the LCD, you can use the Factoring Method for finding the LCD [see Example 3-57]:

(a) The LCD for $\frac{1}{6}$ and $\frac{3}{10}$ is 30 because $6 = ②\times③$ and $10 = 2 \times ⑤$ means:
LCD $= 2 \times 3 \times 5 = 30$.

(b) The LCD for $\frac{1}{10}$ and $\frac{3}{4}$ is 20 because $10 = 2 \times ⑤$ and $4 = ②\times②$ means:
LCD $= 5 \times 2 \times 2 = 20$.

(c) The LCD for $\frac{1}{20}$ and $\frac{11}{12}$ is 60 because $20 = ②\times②\times⑤$ and $12 = 2 \times 2 \times ③$ means:
LCD $= 2 \times 2 \times 5 \times 3 = 60$.

(d) The LCD for $\frac{5}{12}$ and $\frac{1}{21}$ is 84 because $12 = ②\times②\times③$ and $21 = 3 \times ⑦$ means:
LCD $= 2 \times 2 \times 3 \times 7 = 84$.

(e) The LCD for $\frac{4}{15}$ and $\frac{2}{9}$ is 45 because $15 = 3 \times \boxed{5}$ and $9 = \boxed{3 \times 3}$ means: LCD $= 5 \times 3 \times 3 = 45$.

(f) The LCD for $\frac{3}{14}$ and $\frac{5}{6}$ is 42 because $14 = \boxed{2} \times \boxed{7}$ and $6 = 2 \times \boxed{3}$ means: LCD $= 2 \times 7 \times 3 = 42$.

PROBLEM 3-33 Find the LCD for more than two fractions:

(a) $\frac{1}{2}, \frac{3}{4}$, and $\frac{5}{8}$ **(b)** $\frac{5}{6}, \frac{2}{3}$, and $\frac{7}{12}$ **(c)** $\frac{3}{10}, \frac{1}{2}$, and $\frac{2}{5}$ **(d)** $\frac{11}{25}, \frac{49}{100}$, and $\frac{3}{20}$ **(e)** $\frac{1}{2}, \frac{2}{3}$, and $\frac{1}{5}$

(f) $\frac{1}{3}, \frac{1}{4}$, and $\frac{3}{5}$ **(g)** $\frac{7}{8}, \frac{4}{5}$, and $\frac{2}{3}$ **(h)** $\frac{1}{2}, \frac{3}{4}$, and $\frac{1}{6}$ **(i)** $\frac{1}{4}, \frac{7}{10}$, and $\frac{5}{6}$ **(j)** $\frac{1}{12}, \frac{3}{8}$, and $\frac{5}{9}$

Solution Recall that to find the LCD for more than two fractions you find the smallest nonzero whole number that each denominator will divide into evenly [see Examples 3-59, 3-60, and 3-61]:

(a) The LCD for $\frac{1}{2}, \frac{3}{4}$, and $\frac{5}{8}$ is 8 because 2, 4, and 8 all divide 8 evenly.

(b) The LCD for $\frac{5}{6}, \frac{2}{3}$, and $\frac{7}{12}$ is 12 because 6, 3, and 12 all divide 12 evenly.

(c) The LCD for $\frac{3}{10}, \frac{1}{2}$, and $\frac{2}{5}$ is 10 because 10, 2, and 5 all divide 10 evenly.

(d) The LCD for $\frac{11}{25}, \frac{49}{100}$, and $\frac{3}{20}$ is 100 because 25, 100, and 20 all divide 100 evenly.

(e) The LCD for $\frac{1}{2}, \frac{2}{3}$, and $\frac{1}{5}$ is 30 ($= 2 \times 3 \times 5$) because no two of 2, 3, and 5 share a common prime factor.

(f) The LCD for $\frac{1}{3}, \frac{1}{4}$, and $\frac{3}{5}$ is 60 ($= 3 \times 4 \times 5$) because no two of 3, 4, and 5 share a common prime factor.

(g) The LCD for $\frac{7}{8}, \frac{4}{5}$, and $\frac{2}{3}$ is 120 ($= 3 \times 5 \times 8$) because no two of 8, 5, and 3 share a common prime factor.

(h) The LCD for $\frac{1}{2}, \frac{3}{4}$, and $\frac{1}{6}$ is 12 because $2 = 2, 4 = \boxed{2 \times 2}, 6 = 2 \times \boxed{3}$; thus $2 \times 2 \times 3 = 12$.

(i) The LCD for $\frac{1}{4}, \frac{7}{10}$, and $\frac{5}{6}$ is 60 because $4 = \boxed{2 \times 2}, 10 = 2 \times \boxed{5}, 6 = 2 \times \boxed{3}$; thus $2 \times 2 \times 5 \times 3 = 60$.

(j) The LCD for $\frac{1}{12}, \frac{3}{8}$, and $\frac{5}{9}$ is 72 because $12 = 2 \times 2 \times 3, 8 = \boxed{2 \times 2 \times 2}, 9 = \boxed{3 \times 3}$; thus $2 \times 2 \times 2 \times 3 \times 3 = 72$.

PROBLEM 3-34 Build up each fraction as an equal fraction with the given denominator:

(a) $\frac{1}{2} = \frac{?}{6}$ **(b)** $\frac{1}{2} = \frac{?}{8}$ **(c)** $\frac{1}{3} = \frac{?}{9}$ **(d)** $\frac{2}{3} = \frac{?}{15}$ **(e)** $\frac{1}{4} = \frac{?}{8}$ **(f)** $\frac{3}{4} = \frac{?}{16}$

Solution Recall that to build up a given fraction as an equal fraction with a given denominator, you first compare denominators to find the correct building factor and then multiply both the numerator and denominator by that building factor [see Examples 3-62 and 3-63]:

(a) The building factor is 3 ($2 \times 3 = 6$) and $\dfrac{1}{2} = \dfrac{1 \times 3}{2 \times 3} = \dfrac{3}{6}$.

(b) The building factor is 4 ($2 \times 4 = 8$) and $\dfrac{1}{2} = \dfrac{1 \times 4}{2 \times 4} = \dfrac{4}{8}$.

(c) The building factor is 3 ($3 \times 3 = 9$) and $\dfrac{1}{3} = \dfrac{1 \times 3}{3 \times 3} = \dfrac{3}{9}$.

(d) The building factor is 5 ($3 \times 5 = 15$) and $\dfrac{2}{3} = \dfrac{2 \times 5}{3 \times 5} = \dfrac{10}{15}$.

(e) The building factor is 2 ($4 \times 2 = 8$) and $\dfrac{1}{4} = \dfrac{1 \times 2}{4 \times 2} = \dfrac{2}{8}$.

(f) The building factor is 4 ($4 \times 4 = 16$) and $\dfrac{3}{4} = \dfrac{3 \times 4}{4 \times 4} = \dfrac{12}{16}$.

PROBLEM 3-35   Rename unlike fractions as equivalent like fractions using the LCD:

(a) $\frac{1}{2}$ and $\frac{3}{4}$ (b) $\frac{2}{3}$ and $\frac{5}{6}$ (c) $\frac{5}{8}$ and $\frac{1}{4}$ (d) $\frac{1}{2}$ and $\frac{1}{3}$ (e) $\frac{2}{3}$ and $\frac{3}{4}$

(f) $\frac{4}{5}$ and $\frac{1}{2}$ (g) $\frac{1}{4}$ and $\frac{1}{6}$ (h) $\frac{5}{6}$ and $\frac{7}{8}$ (i) $\frac{1}{4}, \frac{7}{8}$, and $\frac{1}{2}$ (j) $\frac{2}{3}, \frac{1}{2}$, and $\frac{3}{5}$

Solution   Recall that to rename unlike fractions as like fractions using the LCD, you first find the LCD and then build up to get like fractions using the LCD [see Examples 3-64 and 3-65]:

(a) The LCD of $\frac{1}{2}$ and $\frac{3}{4}$ is 4: $\frac{1}{2} = \frac{1 \times 2}{2 \times 2} = \frac{2}{4}$ and $\frac{3}{4} = \frac{3}{4}$.

(b) The LCD of $\frac{2}{3}$ and $\frac{5}{6}$ is 6: $\frac{2}{3} = \frac{2 \times 2}{3 \times 2} = \frac{4}{6}$ and $\frac{5}{6} = \frac{5}{6}$.

(c) The LCD of $\frac{5}{8}$ and $\frac{1}{4}$ is 8: $\frac{5}{8} = \frac{5}{8}$ and $\frac{1}{4} = \frac{1 \times 2}{4 \times 2} = \frac{2}{8}$.

(d) The LCD of $\frac{1}{2}$ and $\frac{1}{3}$ is 6: $\frac{1}{2} = \frac{1 \times 3}{2 \times 3} = \frac{3}{6}$ and $\frac{1}{3} = \frac{1 \times 2}{3 \times 2} = \frac{2}{6}$.

(e) The LCD of $\frac{2}{3}$ and $\frac{3}{4}$ is 12: $\frac{2}{3} = \frac{2 \times 4}{3 \times 4} = \frac{18}{12}$ and $\frac{3}{4} = \frac{3 \times 3}{4 \times 3} = \frac{9}{12}$.

(f) The LCD of $\frac{4}{5}$ and $\frac{1}{2}$ is 10: $\frac{4}{5} = \frac{4 \times 2}{5 \times 2} = \frac{8}{10}$ and $\frac{1}{2} = \frac{1 \times 5}{2 \times 5} = \frac{5}{10}$.

(g) The LCD of $\frac{1}{4}$ and $\frac{1}{6}$ is 12: $\frac{1}{4} = \frac{1 \times 3}{4 \times 3} = \frac{3}{12}$ and $\frac{1}{6} = \frac{1 \times 2}{6 \times 2} = \frac{2}{12}$.

(h) The LCD of $\frac{5}{6}$ and $\frac{7}{8}$ is 24: $\frac{5}{6} = \frac{5 \times 4}{6 \times 4} = \frac{20}{24}$ and $\frac{7}{8} = \frac{7 \times 3}{8 \times 3} = \frac{21}{24}$.

(i) The LCD of $\frac{1}{4}, \frac{7}{8}$, and $\frac{1}{2}$ is 8: $\frac{1}{4} = \frac{1 \times 2}{4 \times 2} = \frac{2}{8}$, $\frac{7}{8} = \frac{7}{8}$, and $\frac{1}{2} = \frac{1 \times 4}{2 \times 4} = \frac{4}{8}$.

(j) The LCD of $\frac{2}{3}, \frac{1}{2}$, and $\frac{3}{5}$ is 30: $\frac{2}{3} = \frac{2 \times 10}{3 \times 10} = \frac{20}{30}$, $\frac{1}{2} = \frac{1 \times 15}{2 \times 15} = \frac{15}{30}$, and $\frac{3}{5} = \frac{3 \times 6}{5 \times 6} = \frac{18}{30}$.

PROBLEM 3-36   Add two unlike fractions:

(a) $\frac{1}{2} + \frac{1}{4}$ (b) $\frac{1}{4} + \frac{1}{8}$ (c) $\frac{1}{6} + \frac{1}{3}$ (d) $\frac{1}{2} + \frac{1}{3}$ (e) $\frac{1}{3} + \frac{1}{4}$

(f) $\frac{2}{5} + \frac{1}{2}$ (g) $\frac{1}{4} + \frac{1}{6}$ (h) $\frac{1}{6} + \frac{3}{8}$ (i) $\frac{3}{10} + \frac{1}{8}$ (j) $\frac{5}{12} + \frac{3}{8}$

Solution   Recall that to add two unlike fractions, first rename as equivalent like fractions using the LCD and then add the like fractions [see Example 3-66]:

(a) The LCD for $\frac{1}{2}$ and $\frac{1}{4}$ is 4: $\frac{1}{2} + \frac{1}{4} = \frac{1 \times 2}{2 \times 2} + \frac{1}{4} = \frac{2}{4} + \frac{1}{4} = \frac{2 + 1}{4} = \frac{3}{4}$.

(b) The LCD for $\frac{1}{4}$ and $\frac{1}{8}$ is 8: $\frac{1}{4} + \frac{1}{8} = \frac{1 \times 2}{4 \times 2} + \frac{1}{8} = \frac{2}{8} + \frac{1}{8} = \frac{3}{8}$.

(c) The LCD for $\frac{1}{6}$ and $\frac{1}{3}$ is 6: $\frac{1}{6} + \frac{1}{3} = \frac{1}{6} + \frac{1 \times 2}{3 \times 2} = \frac{1}{6} + \frac{2}{6} = \frac{3}{6} = \frac{1 \times \cancel{3}}{2 \times \cancel{3}} = \frac{1}{2}$.

(d) The LCD for $\frac{1}{2}$ and $\frac{1}{3}$ is 6: $\frac{1}{2} + \frac{1}{3} = \frac{1 \times 3}{2 \times 3} + \frac{1 \times 2}{3 \times 2} = \frac{3}{6} + \frac{2}{6} = \frac{5}{6}$.

(e) The LCD for $\frac{1}{3}$ and $\frac{1}{4}$ is 12: $\frac{1}{3} + \frac{1}{4} = \frac{1 \times 4}{3 \times 4} + \frac{1 \times 3}{4 \times 3} = \frac{4}{12} + \frac{3}{12} = \frac{7}{12}$.

(f) The LCD for $\frac{2}{5}$ and $\frac{1}{2}$ is 10: $\frac{2}{5} + \frac{1}{2} = \frac{2 \times 2}{5 \times 2} + \frac{1 \times 5}{2 \times 5} = \frac{4}{10} + \frac{5}{10} = \frac{9}{10}$.

(g) The LCD for $\frac{1}{4}$ and $\frac{1}{6}$ is 12: $\frac{1}{4} + \frac{1}{6} = \frac{1 \times 3}{4 \times 3} + \frac{1 \times 2}{6 \times 2} = \frac{3}{12} + \frac{2}{12} = \frac{5}{12}$.

(h) The LCD for $\frac{1}{6}$ and $\frac{3}{8}$ is 24: $\frac{1}{6} + \frac{3}{8} = \frac{1 \times 4}{6 \times 4} + \frac{3 \times 3}{8 \times 3} = \frac{4}{24} + \frac{9}{24} = \frac{13}{24}$.

(i) The LCD for $\frac{3}{10}$ and $\frac{1}{8}$ is 40: $\frac{3}{10} + \frac{1}{8} = \frac{3 \times 4}{10 \times 4} + \frac{1 \times 5}{8 \times 5} = \frac{12}{40} + \frac{5}{40} = \frac{17}{40}$.

(j) The LCD for $\frac{5}{12}$ and $\frac{3}{8}$ is 24: $\frac{5}{12} + \frac{3}{8} = \frac{5 \times 2}{12 \times 2} + \frac{3 \times 3}{8 \times 3} = \frac{10}{24} + \frac{9}{24} = \frac{19}{24}$.

PROBLEM 3-37 Add more than two unlike fractions:

(a) $\frac{1}{4} + \frac{1}{2} + \frac{1}{8}$ **(b)** $\frac{5}{6} + \frac{1}{2} + \frac{2}{3}$ **(c)** $\frac{1}{3} + \frac{1}{2} + \frac{1}{5}$ **(d)** $\frac{3}{4} + \frac{4}{5} + \frac{2}{3}$ **(e)** $\frac{3}{8} + \frac{5}{6} + \frac{3}{10}$ **(f)** $\frac{7}{12} + \frac{9}{10} + \frac{5}{8}$

Solution To add more than two unlike fractions, you proceed in the same way as you did with two unlike fractions [see Example 3-67]:

(a) The LCD for $\frac{1}{4}, \frac{1}{2}$, and $\frac{1}{8}$ is 8:

$$\frac{1}{4} + \frac{1}{2} + \frac{1}{8} = \frac{1 \times 2}{4 \times 2} + \frac{1 \times 4}{2 \times 4} + \frac{1}{8} = \frac{2}{8} + \frac{4}{8} + \frac{1}{8} = \frac{7}{8}.$$

(b) The LCD for $\frac{5}{6}, \frac{1}{2}$, and $\frac{2}{3}$ is 6:

$$\frac{5}{6} + \frac{1}{2} + \frac{2}{3} = \frac{5}{6} + \frac{1 \times 3}{2 \times 3} + \frac{2 \times 2}{3 \times 2} = \frac{5}{6} + \frac{3}{6} + \frac{4}{6} = \frac{12}{6} = \frac{2}{1} = 2.$$

(c) The LCD for $\frac{1}{3}, \frac{1}{2}$, and $\frac{1}{5}$ is 30:

$$\frac{1}{3} + \frac{1}{2} + \frac{1}{5} = \frac{1 \times 10}{3 \times 10} + \frac{1 \times 15}{2 \times 15} + \frac{1 \times 6}{5 \times 6} = \frac{10}{30} + \frac{15}{30} + \frac{6}{30} = \frac{31}{30} = 1\frac{1}{30}.$$

(d) The LCD for $\frac{3}{4}, \frac{4}{5}$, and $\frac{2}{3}$ is 60:

$$\frac{3}{4} + \frac{4}{5} + \frac{2}{3} = \frac{3 \times 15}{4 \times 15} + \frac{4 \times 12}{5 \times 12} + \frac{2 \times 20}{3 \times 20} = \frac{45}{60} + \frac{48}{60} + \frac{40}{60} = \frac{133}{60} = 2\frac{13}{60}.$$

(e) The LCD for $\frac{3}{8}, \frac{5}{6}$, and $\frac{3}{10}$ is 120:

$$\frac{3}{8} + \frac{5}{6} + \frac{3}{10} = \frac{3 \times 15}{8 \times 15} + \frac{5 \times 20}{6 \times 20} + \frac{3 \times 12}{10 \times 12} = \frac{45}{120} + \frac{100}{120} + \frac{36}{120} = \frac{181}{120} = 1\frac{61}{120}.$$

(f) The LCD for $\frac{7}{12}, \frac{9}{10}$, and $\frac{5}{8}$ is 120:

$$\frac{7}{12} + \frac{9}{10} + \frac{5}{8} = \frac{7 \times 10}{12 \times 10} + \frac{9 \times 12}{10 \times 12} + \frac{5 \times 15}{8 \times 15} = \frac{70}{120} + \frac{108}{120} + \frac{75}{120} = \frac{253}{120} = 2\frac{13}{120}.$$

PROBLEM 3-38 Add mixed numbers and whole numbers:

(a) $2 + 5\frac{1}{3}$ **(b)** $1 + 3\frac{3}{4}$ **(c)** $4\frac{1}{5} + 8$ **(d)** $6\frac{5}{8} + 9$ **(e)** $4 + 1\frac{2}{3}$

(f) $7 + 2\frac{5}{6}$ **(g)** $1\frac{1}{4} + 2$ **(h)** $5\frac{1}{2} + 1$ **(i)** $3 + 3\frac{2}{5}$ **(j)** $3 + 5\frac{1}{8}$

(k) $2\frac{7}{10} + 4$ **(l)** $4\frac{1}{6} + 3$ **(m)** $4 + 5\frac{3}{10} + 2$ **(n)** $1\frac{3}{5} + 3 + 6$ **(o)** $5 + 8\frac{7}{8} + 2 + 7$

Solution Recall that to add a mixed number to one or more whole numbers, you add the whole numbers and then write the same fraction part [see Example 3-68]:

(a) $2 + 5\frac{1}{3} = 7\frac{1}{3}$ (b) $1 + 3\frac{3}{4} = 4\frac{3}{4}$ (c) $4\frac{1}{5} + 8 = 12\frac{1}{5}$

(d) $6\frac{5}{8} + 9 = 15\frac{5}{8}$ (e) $4 + 1\frac{2}{3} = 5\frac{2}{3}$ (f) $7 + 2\frac{5}{6} = 9\frac{5}{6}$

(g) $1\frac{1}{4} + 2 = 3\frac{1}{4}$ (h) $5\frac{1}{2} + 1 = 6\frac{1}{2}$ (i) $3 + 3\frac{2}{5} = 6\frac{2}{5}$

(j) $3 + 5\frac{1}{8} = 8\frac{1}{8}$ (k) $2\frac{7}{10} + 4 = 6\frac{7}{10}$ (l) $4\frac{1}{6} + 3 = 7\frac{1}{6}$

(m) $4 + 5\frac{3}{10} + 2 = 11\frac{3}{10}$ (n) $1\frac{3}{5} + 3 + 6 = 10\frac{3}{5}$ (o) $5 + 8\frac{7}{8} + 2 + 7 = 22\frac{7}{8}$

PROBLEM 3-39 Add mixed numbers and fractions:

(a) $\frac{1}{4} + 3\frac{1}{4}$ (b) $8\frac{1}{2} + \frac{1}{2}$ (c) $1\frac{1}{3} + \frac{2}{3}$ (d) $\frac{1}{2} + 7\frac{1}{4}$ (e) $\frac{1}{6} + 2\frac{1}{3}$

(f) $4\frac{5}{8} + \frac{3}{4}$ (g) $6\frac{4}{5} + \frac{9}{10}$ (h) $2\frac{5}{6} + \frac{1}{2} + \frac{2}{3}$ (i) $\frac{3}{5} + 8\frac{1}{2} + \frac{2}{3}$ (j) $\frac{1}{8} + \frac{1}{2} + 3\frac{3}{4} + \frac{5}{16}$

Solution Recall that to add a mixed number and one or more fractions, you add the fractions and then write the same whole-number part [see Example 3-69]:

(a)
$$3\frac{1}{4}$$
$$+\ \frac{1}{4}$$
$$\overline{3\frac{2}{4}} = 3\frac{1}{2}$$

(b)
$$8\frac{1}{2}$$
$$+\ \frac{1}{2}$$
$$\overline{8\frac{2}{2}} = 8 + \frac{2}{2} = 8 + 1 = 9$$

(c)
$$1\frac{1}{3}$$
$$+\ \frac{2}{3}$$
$$\overline{1\frac{3}{3}} = 1 + \frac{3}{3} = 1 + 1 = 2$$

(d)
$$7\frac{1}{4} = 7\frac{1}{4}$$
$$+\ \frac{1}{2} = \frac{2}{4}$$
$$\overline{7\frac{3}{4}}$$

(e)
$$2\frac{1}{3} = 2\frac{2}{6}$$
$$+\ \frac{1}{6} = \frac{1}{6}$$
$$\overline{2\frac{3}{6}} = 2\frac{1}{2}$$

(f)
$$4\frac{5}{8} = 4\frac{5}{8}$$
$$+\ \frac{3}{4} = \frac{6}{8}$$
$$\overline{4\frac{11}{8}} = 4 + \frac{11}{8} = 4 + 1\frac{3}{8} = 5\frac{3}{8}$$

(g)
$$6\frac{4}{5} = 6\frac{8}{10}$$
$$+\ \frac{9}{10} = \frac{9}{10}$$
$$\overline{6\frac{17}{10}} = 6 + \frac{17}{10} = 6 + 1\frac{7}{10} = 7\frac{7}{10}$$

(h)
$$2\frac{5}{6} = 2\frac{5}{6}$$
$$\frac{1}{2} = \frac{3}{6}$$
$$+\ \frac{2}{3} = \frac{4}{6}$$
$$\overline{2\frac{12}{6}} = 2 + \frac{12}{6} = 2 + \frac{2}{1} = 4$$

(i)
$$8\frac{1}{2} = 8\frac{15}{30}$$
$$\frac{3}{5} = \frac{18}{30}$$
$$+\ \frac{2}{3} = \frac{20}{30}$$
$$\overline{8\frac{53}{30}} = 8 + \frac{53}{30} = 8 + 1\frac{23}{30} = 9\frac{23}{30}$$

(j)
$$3\frac{3}{4} = 3\frac{12}{16}$$
$$\frac{1}{8} = \frac{2}{16}$$
$$\frac{1}{2} = \frac{8}{16}$$
$$+\ \frac{5}{16} = \frac{5}{16}$$
$$\overline{3\frac{27}{16}} = 3 + \frac{27}{16} = 3 + 1\frac{11}{16} = 4\frac{11}{16}$$

PROBLEM 3-40 Add mixed numbers:

(a) $1\frac{1}{6} + 1\frac{1}{6}$ (b) $2\frac{1}{3} + 3\frac{1}{3}$ (c) $5\frac{3}{4} + 2\frac{1}{4}$ (d) $3\frac{5}{8} + 5\frac{7}{8}$

(e) $2\frac{1}{4} + 2\frac{1}{2}$ (f) $3\frac{1}{3} + 4\frac{1}{6}$ (g) $3\frac{3}{4} + 3\frac{5}{8}$ (h) $4\frac{1}{2} + 5\frac{1}{3}$

(i) $2\frac{1}{3} + 4\frac{1}{4}$ (j) $3\frac{1}{2} + 1\frac{1}{2} + 2\frac{1}{2}$ (k) $4\frac{1}{2} + 3\frac{3}{5} + 5\frac{7}{10}$ (l) $2\frac{11}{16} + 1\frac{5}{8} + 3\frac{1}{2} + 8\frac{3}{4}$

Solution Recall that to add two or more mixed numbers, you first write the numbers in vertical form, then add the fraction parts, and then add the whole-number parts [see Example 3-70]:

(a)
$$1\frac{1}{6}$$
$$+1\frac{1}{6}$$
$$\overline{2\frac{2}{6}} = 2\frac{1}{3}$$

(b)
$$2\frac{1}{3}$$
$$+3\frac{1}{3}$$
$$\overline{5\frac{2}{3}}$$

(c)
$$5\frac{3}{4}$$
$$+2\frac{1}{4}$$
$$\overline{7\frac{4}{4}} = 7 + \frac{4}{4} = 7 + 1 = 8$$

(d)
$$3\frac{5}{8}$$
$$+5\frac{7}{8}$$
$$\overline{8\frac{12}{8}} = 8\frac{3}{2} = 8 + \frac{3}{2}$$
$$= 8 + 1\frac{1}{2} = 9\frac{1}{2}$$

(e)
$$2\frac{1}{4} = 2\frac{1}{4}$$
$$+2\frac{1}{2} = 2\frac{2}{4}$$
$$\overline{4\frac{3}{4}}$$

(f)
$$3\frac{1}{3} = 3\frac{2}{6}$$
$$+4\frac{1}{6} = 4\frac{1}{6}$$
$$\overline{7\frac{3}{6}} = 7\frac{1}{2}$$

(g) $3\frac{3}{4} = 3\frac{6}{8}$
$\underline{+3\frac{5}{8} = 3\frac{5}{8}}$
$6\frac{11}{8} = 6 + \frac{11}{8}$
$= 6 + 1\frac{3}{8}$
$= 7\frac{3}{8}$

(h) $4\frac{1}{2} = 4\frac{3}{6}$
$\underline{+5\frac{1}{3} = 5\frac{2}{6}}$
$9\frac{5}{6}$

(i) $2\frac{1}{3} = 2\frac{4}{12}$
$\underline{+4\frac{1}{4} = 4\frac{3}{12}}$
$6\frac{7}{12}$

(j) $3\frac{1}{2}$
$1\frac{1}{2}$
$\underline{+2\frac{1}{2}}$
$6\frac{3}{2} = 6 + \frac{3}{2}$
$= 6 + 1\frac{1}{2}$
$= 7\frac{1}{2}$

(k) $4\frac{1}{2} = 4\frac{5}{10}$
$3\frac{3}{5} = 3\frac{6}{10}$
$\underline{+5\frac{7}{10} = 5\frac{7}{10}}$
$12\frac{18}{10} = 12\frac{9}{5} = 12 + \frac{9}{5}$
$= 12 + 1\frac{4}{5}$
$= 13\frac{4}{5}$

(l) $2\frac{11}{16} = 2\frac{11}{16}$
$1\frac{5}{8} = 1\frac{10}{16}$
$3\frac{1}{2} = 3\frac{8}{16}$
$\underline{+8\frac{3}{4} = 8\frac{12}{16}}$
$14\frac{41}{16} = 14 + \frac{41}{16}$
$= 14 + 2\frac{9}{16}$
$= 16\frac{9}{16}$

PROBLEM 3-41 Subtract like fractions:

(a) $\frac{2}{3} - \frac{1}{3}$ **(b)** $\frac{7}{8} - \frac{1}{8}$ **(c)** $\frac{1}{2} - \frac{1}{2}$ **(d)** $\frac{9}{10} - \frac{3}{10}$ **(e)** $\frac{3}{4} - \frac{1}{4}$

(f) $\frac{11}{12} - \frac{1}{12}$ **(g)** $\frac{4}{5} - \frac{3}{5}$ **(h)** $\frac{8}{9} - \frac{4}{9}$ **(i)** $\frac{5}{6} - \frac{1}{6}$ **(j)** $\frac{6}{7} - \frac{3}{7}$

Solution Recall that to subtract like fractions, you subtract the numerators and then write the same denominator [see Examples 3-72, 3-73, and 3-75]:

(a) $\frac{2}{3} - \frac{1}{3} = \frac{2-1}{3} = \frac{1}{3}$

(b) $\frac{7}{8} - \frac{1}{8} = \frac{7-1}{8} = \frac{6}{8} = \frac{3}{4}$

(c) $\frac{1}{2} - \frac{1}{2} = \frac{1-1}{2} = \frac{0}{2} = 0$

(d) $\frac{9}{10} - \frac{3}{10} = \frac{9-3}{10} = \frac{6}{10} = \frac{3}{5}$

(e) $\frac{3}{4} - \frac{1}{4} = \frac{3-1}{4} = \frac{2}{4} = \frac{1}{2}$

(f) $\frac{11}{12} - \frac{1}{12} = \frac{11-1}{12} = \frac{10}{12} = \frac{5}{6}$

(g) $\frac{4}{5} - \frac{3}{5} = \frac{4-3}{5} = \frac{1}{5}$

(h) $\frac{8}{9} - \frac{4}{9} = \frac{8-4}{9} = \frac{4}{9}$

(i) $\frac{5}{6} - \frac{1}{6} = \frac{5-1}{6} = \frac{4}{6} = \frac{2}{3}$

(j) $\frac{6}{7} - \frac{3}{7} = \frac{6-3}{7} = \frac{3}{7}$

PROBLEM 3-42 Subtract unlike fractions:

(a) $\frac{3}{4} - \frac{1}{2}$ **(b)** $\frac{1}{2} - \frac{1}{6}$ **(c)** $\frac{1}{2} - \frac{1}{3}$ **(d)** $\frac{3}{4} - \frac{1}{3}$ **(e)** $\frac{1}{4} - \frac{1}{6}$ **(f)** $\frac{5}{6} - \frac{5}{8}$ **(g)** $\frac{9}{10} - \frac{7}{12}$ **(h)** $\frac{24}{25} - \frac{3}{20}$

Solution Recall that to subtract unlike fractions, first rename as equivalent like fractions using the LCD and then subtract the like fractions [see Example 3-76]:

(a) The LCD for $\frac{3}{4}$ and $\frac{1}{2}$ is 4: $\frac{3}{4} - \frac{1}{2} = \frac{3}{4} - \frac{2}{4} = \frac{1}{4}$

(b) The LCD for $\frac{1}{2}$ and $\frac{1}{6}$ is 6: $\frac{1}{2} - \frac{1}{6} = \frac{3}{6} - \frac{1}{6} = \frac{2}{6} = \frac{1}{3}$

(c) The LCD for $\frac{1}{2}$ and $\frac{1}{3}$ is 6: $\frac{1}{2} - \frac{1}{3} = \frac{3}{6} - \frac{2}{6} = \frac{1}{6}$

(d) The LCD for $\frac{3}{4}$ and $\frac{1}{3}$ is 12: $\frac{3}{4} - \frac{1}{3} = \frac{9}{12} - \frac{4}{12} = \frac{5}{12}$

(e) The LCD for $\frac{1}{4}$ and $\frac{1}{6}$ is 12: $\frac{1}{4} - \frac{1}{6} = \frac{3}{12} - \frac{2}{12} = \frac{1}{12}$

(f) The LCD for $\dfrac{5}{6}$ and $\dfrac{5}{8}$ is 24: $\quad \dfrac{5}{6} - \dfrac{5}{8} = \dfrac{20}{24} - \dfrac{15}{24} = \dfrac{5}{24}$

(g) The LCD for $\dfrac{9}{10}$ and $\dfrac{7}{12}$ is 60: $\quad \dfrac{9}{10} - \dfrac{7}{12} = \dfrac{54}{60} - \dfrac{35}{60} = \dfrac{19}{60}$

(h) The LCD for $\dfrac{24}{25}$ and $\dfrac{3}{20}$ is 100: $\quad \dfrac{24}{25} - \dfrac{3}{20} = \dfrac{96}{100} - \dfrac{15}{100} = \dfrac{81}{100}$

PROBLEM 3-43   Subtract whole numbers from mixed numbers:

(a) $7\frac{3}{8} - 2$ **(b)** $8\frac{1}{4} - 1$ **(c)** $4\frac{1}{2} - 3$ **(d)** $6\frac{2}{3} - 5$ **(e)** $9\frac{1}{5} - 4$ **(f)** $1\frac{1}{6} - 1$

(g) $2\frac{4}{9} - 1$ **(h)** $3\frac{3}{10} - 1$ **(i)** $8\frac{11}{12} - 5$ **(j)** $7\frac{3}{4} - 6$ **(k)** $9\frac{1}{3} - 9$ **(l)** $6\frac{5}{6} - 1$

(m) $11\frac{2}{5} - 2$ **(n)** $23\frac{7}{10} - 15$ **(o)** $5\frac{2}{9} - 1$ **(p)** $15\frac{5}{12} - 11$ **(q)** $10\frac{1}{10} - 1$ **(r)** $8\frac{4}{5} - 8$

Solution   Recall that to subtract a whole number from a mixed number, you subtract the whole numbers and then write the fraction part [see Example 3-77]:

(a) $\begin{array}{r} 7\frac{3}{8} \\ -2 \\ \hline 5\frac{3}{8} \end{array}$ **(b)** $\begin{array}{r} 8\frac{1}{4} \\ -1 \\ \hline 7\frac{1}{4} \end{array}$ **(c)** $\begin{array}{r} 4\frac{1}{2} \\ -3 \\ \hline 1\frac{1}{2} \end{array}$ **(d)** $\begin{array}{r} 6\frac{2}{3} \\ -5 \\ \hline 1\frac{2}{3} \end{array}$ **(e)** $\begin{array}{r} 9\frac{1}{5} \\ -4 \\ \hline 5\frac{1}{5} \end{array}$ **(f)** $\begin{array}{r} 1\frac{1}{6} \\ -1 \\ \hline \frac{1}{6} \end{array}$

(g) $\begin{array}{r} 2\frac{4}{9} \\ -1 \\ \hline 1\frac{4}{9} \end{array}$ **(h)** $\begin{array}{r} 3\frac{3}{10} \\ -1 \\ \hline 2\frac{3}{10} \end{array}$ **(i)** $\begin{array}{r} 8\frac{11}{12} \\ -5 \\ \hline 3\frac{11}{12} \end{array}$ **(j)** $\begin{array}{r} 7\frac{3}{4} \\ -6 \\ \hline 1\frac{3}{4} \end{array}$ **(k)** $\begin{array}{r} 9\frac{1}{3} \\ -9 \\ \hline \frac{1}{3} \end{array}$ **(l)** $\begin{array}{r} 6\frac{5}{6} \\ -1 \\ \hline 5\frac{5}{6} \end{array}$

(m) $\begin{array}{r} 11\frac{2}{5} \\ -\ 2 \\ \hline 9\frac{2}{5} \end{array}$ **(n)** $\begin{array}{r} 23\frac{7}{10} \\ -15 \\ \hline 8\frac{7}{10} \end{array}$ **(o)** $\begin{array}{r} 5\frac{2}{9} \\ -1 \\ \hline 4\frac{2}{9} \end{array}$ **(p)** $\begin{array}{r} 15\frac{5}{12} \\ -11 \\ \hline 4\frac{5}{12} \end{array}$ **(q)** $\begin{array}{r} 10\frac{1}{10} \\ -\ 1 \\ \hline 9\frac{1}{10} \end{array}$ **(r)** $\begin{array}{r} 8\frac{4}{5} \\ -8 \\ \hline \frac{4}{5} \end{array}$

PROBLEM 3-44   Subtract fractions from mixed numbers:

(a) $7\frac{5}{6} - \frac{1}{6}$ **(b)** $2\frac{7}{8} - \frac{1}{8}$ **(c)** $9\frac{3}{4} - \frac{1}{4}$ **(d)** $1\frac{9}{10} - \frac{3}{10}$ **(e)** $8\frac{2}{3} - \frac{1}{3}$

(f) $3\frac{11}{12} - \frac{1}{12}$ **(g)** $5\frac{1}{2} - \frac{1}{2}$ **(h)** $4\frac{4}{5} - \frac{1}{5}$ **(i)** $6\frac{1}{2} - \frac{3}{8}$ **(j)** $12\frac{9}{10} - \frac{1}{2}$

Solution   Recall that to subtract a fraction from a mixed number, you first write the numbers in vertical form, then subtract fractions, and then write the same whole-number part [see Example 3-78]:

(a) $\begin{array}{r} 7\frac{5}{6} \\ -\ \frac{1}{6} \\ \hline 7\frac{4}{6} = 7\frac{2}{3} \end{array}$ **(b)** $\begin{array}{r} 2\frac{7}{8} \\ -\ \frac{1}{8} \\ \hline 2\frac{6}{8} = 2\frac{3}{4} \end{array}$ **(c)** $\begin{array}{r} 9\frac{3}{4} \\ -\ \frac{1}{4} \\ \hline 9\frac{2}{4} = 9\frac{1}{2} \end{array}$ **(d)** $\begin{array}{r} 1\frac{9}{10} \\ -\ \frac{3}{10} \\ \hline 1\frac{6}{10} = 1\frac{3}{5} \end{array}$ **(e)** $\begin{array}{r} 8\frac{2}{3} \\ -\ \frac{1}{3} \\ \hline 8\frac{1}{3} \end{array}$

(f) $\begin{array}{r} 3\frac{11}{12} \\ -\ \frac{1}{12} \\ \hline 3\frac{10}{12} = 3\frac{5}{6} \end{array}$ **(g)** $\begin{array}{r} 5\frac{1}{2} \\ -\ \frac{1}{2} \\ \hline 5 \end{array}$ **(h)** $\begin{array}{r} 4\frac{4}{5} \\ -\ \frac{1}{5} \\ \hline 4\frac{3}{5} \end{array}$ **(i)** $\begin{array}{r} 6\frac{1}{2} = 6\frac{4}{8} \\ -\ \frac{3}{8} = \frac{3}{8} \\ \hline 6\frac{1}{8} \end{array}$ **(j)** $\begin{array}{r} 12\frac{9}{10} = 12\frac{9}{10} \\ -\ \frac{1}{2} = \frac{5}{10} \\ \hline 12\frac{4}{10} = 12\frac{2}{5} \end{array}$

PROBLEM 3-45   Subtract two mixed numbers:

(a) $9\frac{3}{4} - 5\frac{1}{4}$ **(b)** $2\frac{2}{3} - 1\frac{1}{3}$ **(c)** $5\frac{11}{12} - 5\frac{7}{12}$ **(d)** $8\frac{1}{2} - 7\frac{1}{2}$ **(e)** $6\frac{5}{6} - 1\frac{1}{6}$

(f) $4\frac{3}{8} - 3\frac{1}{8}$ **(g)** $8\frac{7}{10} - 8\frac{3}{10}$ **(h)** $7\frac{2}{5} - 2\frac{1}{5}$ **(i)** $6\frac{1}{2} - 4\frac{1}{4}$ **(j)** $10\frac{11}{12} - 2\frac{1}{3}$

Solution   Recall that to subtract two mixed numbers, you first write the numbers in vertical form, then subtract the fraction parts, and then subtract the whole-number parts [see Example 3-79]:

(a) $\begin{array}{r} 9\frac{3}{4} \\ -5\frac{1}{4} \\ \hline 4\frac{2}{4} = 4\frac{1}{2} \end{array}$ **(b)** $\begin{array}{r} 2\frac{2}{3} \\ -1\frac{1}{3} \\ \hline 1\frac{1}{3} \end{array}$ **(c)** $\begin{array}{r} 5\frac{11}{12} \\ -5\frac{7}{12} \\ \hline \frac{4}{12} = \frac{1}{3} \end{array}$ **(d)** $\begin{array}{r} 8\frac{1}{2} \\ -7\frac{1}{2} \\ \hline 1 \end{array}$ **(e)** $\begin{array}{r} 6\frac{5}{6} \\ -1\frac{1}{6} \\ \hline 5\frac{4}{6} = 5\frac{2}{3} \end{array}$

(f) $\quad 4\frac{3}{8}$
$\quad -3\frac{1}{8}$
$\quad \overline{1\frac{2}{8}} = 1\frac{1}{4}$

(g) $\quad 8\frac{7}{10}$
$\quad -8\frac{3}{10}$
$\quad \overline{\frac{4}{10}} = \frac{2}{5}$

(h) $\quad 7\frac{2}{5}$
$\quad -2\frac{1}{5}$
$\quad \overline{5\frac{1}{5}}$

(i) $\quad 6\frac{1}{2} = 6\frac{2}{4}$
$\quad -4\frac{1}{4} = 4\frac{1}{4}$
$\quad \overline{\phantom{-4\frac{1}{4} = }2\frac{1}{4}}$

(j) $\quad 10\frac{11}{12} = 10\frac{11}{12}$
$\quad -\ 2\frac{1}{3} = 2\frac{4}{12}$
$\quad \overline{\phantom{-\ 2\frac{1}{3} = }8\frac{7}{12}}$

PROBLEM 3-46 Subtract mixed numbers from whole numbers:

(a) $8 - 2\frac{1}{2}$ **(b)** $3 - 1\frac{3}{4}$ **(c)** $12 - 3\frac{7}{8}$ **(d)** $9 - 1\frac{1}{3}$ **(e)** $2 - 1\frac{5}{6}$

Solution Recall that to subtract a mixed number from a whole number, you first rename the whole number as an equal mixed number so that the fraction parts are like fractions [see Example 3-80]:

(a) $\quad 8 = 7\frac{2}{2}$
$\quad -2\frac{1}{2} = 2\frac{1}{2}$
$\quad \overline{\phantom{-2\frac{1}{2} = }5\frac{1}{2}}$

(b) $\quad 3 = 2\frac{4}{4}$
$\quad -1\frac{3}{4} = 1\frac{3}{4}$
$\quad \overline{\phantom{-1\frac{3}{4} = }1\frac{1}{4}}$

(c) $\quad 12 = 11\frac{8}{8}$
$\quad -\ 3\frac{7}{8} = 3\frac{7}{8}$
$\quad \overline{\phantom{-\ 3\frac{7}{8} = }8\frac{1}{8}}$

(d) $\quad 9 = 8\frac{3}{3}$
$\quad -1\frac{1}{3} = 1\frac{1}{3}$
$\quad \overline{\phantom{-1\frac{1}{3} = }7\frac{2}{3}}$

(e) $\quad 2 = 1\frac{6}{6}$
$\quad -1\frac{5}{6} = 1\frac{5}{6}$
$\quad \overline{\phantom{-1\frac{5}{6} = }\frac{1}{6}}$

PROBLEM 3-47 Subtract fractions from whole numbers:

(a) $2 - \frac{3}{8}$ **(b)** $10 - \frac{1}{5}$ **(c)** $7 - \frac{1}{10}$ **(d)** $1 - \frac{1}{12}$ **(e)** $12 - \frac{1}{2}$

Solution Recall that to subtract a fraction from a whole number, you first rename the whole number as an equal mixed number so that the fractions are like fractions [see Example 3-81]:

(a) $\quad 2 = 1\frac{8}{8}$
$\quad -\frac{3}{8} = \frac{3}{8}$
$\quad \overline{\phantom{-\frac{3}{8} = }1\frac{5}{8}}$

(b) $\quad 10 = 9\frac{5}{5}$
$\quad -\frac{1}{5} = \frac{1}{5}$
$\quad \overline{\phantom{-\frac{1}{5} = }9\frac{4}{5}}$

(c) $\quad 7 = 6\frac{10}{10}$
$\quad -\frac{1}{10} = \frac{1}{10}$
$\quad \overline{\phantom{-\frac{1}{10} = }6\frac{9}{10}}$

(d) $\quad 1 = \frac{12}{12}$
$\quad -\frac{1}{12} = \frac{1}{12}$
$\quad \overline{\phantom{-\frac{1}{12} = }\frac{11}{12}}$

(e) $\quad 12 = 11\frac{2}{2}$
$\quad -\frac{1}{2} = \frac{1}{2}$
$\quad \overline{\phantom{-\frac{1}{2} = }11\frac{1}{2}}$

PROBLEM 3-48 Subtract two mixed numbers when the fraction part of the minuend is too small:

(a) $8\frac{1}{3} - 3\frac{2}{3}$ **(b)** $9\frac{2}{5} - 2\frac{3}{5}$ **(c)** $7\frac{1}{8} - 4\frac{7}{8}$ **(d)** $10\frac{1}{4} - 1\frac{3}{4}$ **(e)** $6\frac{1}{6} - 5\frac{5}{6}$ **(f)** $8\frac{3}{10} - 2\frac{7}{10}$

(g) $11\frac{1}{2} - 1\frac{3}{4}$ **(h)** $6\frac{1}{5} - 4\frac{3}{10}$ **(i)** $7\frac{1}{2} - 2\frac{2}{3}$ **(j)** $8\frac{2}{5} - 1\frac{1}{2}$ **(k)** $6\frac{1}{8} - 3\frac{5}{12}$ **(l)** $9\frac{1}{4} - 3\frac{5}{6}$

Solution Recall that to subtract two mixed numbers when the minuend of the fraction part is too small, you first rename the minuend to get an equal mixed number with a larger fraction part [see Example 3-82]:

(a) $\quad 8\frac{1}{3} = 7\frac{4}{3}$
$\quad -3\frac{2}{3} = 3\frac{2}{3}$
$\quad \overline{\phantom{-3\frac{2}{3} = }4\frac{2}{3}}$

(b) $\quad 9\frac{2}{5} = 8\frac{7}{5}$
$\quad -2\frac{3}{5} = 2\frac{3}{5}$
$\quad \overline{\phantom{-2\frac{3}{5} = }6\frac{4}{5}}$

(c) $\quad 7\frac{1}{8} = 6\frac{9}{8}$
$\quad -4\frac{7}{8} = 4\frac{7}{8}$
$\quad \overline{\phantom{-4\frac{7}{8} = }2\frac{2}{8}} = 2\frac{1}{4}$

(d) $\quad 10\frac{1}{4} = 9\frac{5}{4}$
$\quad -\ 1\frac{3}{4} = 1\frac{3}{4}$
$\quad \overline{\phantom{-\ 1\frac{3}{4} = }8\frac{2}{4}} = 8\frac{1}{2}$

(e) $\quad 6\frac{1}{6} = 5\frac{7}{6}$
$\quad -5\frac{5}{6} = 5\frac{5}{6}$
$\quad \overline{\phantom{-5\frac{5}{6} = }\frac{2}{6}} = \frac{1}{3}$

(f) $\quad 8\frac{3}{10} = 7\frac{13}{10}$
$\quad -2\frac{7}{10} = 2\frac{7}{10}$
$\quad \overline{\phantom{-2\frac{7}{10} = }5\frac{6}{10}} = 5\frac{3}{5}$

(g) $\quad 11\frac{1}{2} = 11\frac{2}{4} = 10\frac{6}{4}$
$\quad -\ 1\frac{3}{4} = \ 1\frac{3}{4} = \ 1\frac{3}{4}$
$\quad \overline{\phantom{-\ 1\frac{3}{4} = \ 1\frac{3}{4} = \ }9\frac{3}{4}}$

(h) $\quad 6\frac{1}{5} = 6\frac{2}{10} = 5\frac{12}{10}$
$\quad -4\frac{3}{10} = 4\frac{3}{10} = 4\frac{3}{10}$
$\quad \overline{\phantom{-4\frac{3}{10} = 4\frac{3}{10} = }1\frac{9}{10}}$

(i) $\quad 7\frac{1}{2} = 7\frac{3}{6} = 6\frac{9}{6}$
$\quad -2\frac{2}{3} = 2\frac{4}{6} = 2\frac{4}{6}$
$\quad \overline{\phantom{-2\frac{2}{3} = 2\frac{4}{6} = }4\frac{5}{6}}$

(j) $\quad 8\frac{2}{5} = 8\frac{4}{10} = 7\frac{14}{10}$
$\quad -1\frac{1}{2} = 1\frac{5}{10} = 1\frac{5}{10}$
$\quad \overline{\phantom{-1\frac{1}{2} = 1\frac{5}{10} = }6\frac{9}{10}}$

(k) $\quad 6\frac{1}{8} = 6\frac{3}{24} = 5\frac{27}{24}$
$\quad -3\frac{5}{12} = 3\frac{10}{24} = 3\frac{10}{24}$
$\quad \overline{\phantom{-3\frac{5}{12} = 3\frac{10}{24} = }2\frac{17}{24}}$

(l) $\quad 9\frac{1}{4} = 9\frac{3}{12} = 8\frac{15}{12}$
$\quad -3\frac{5}{6} = 3\frac{10}{12} = 3\frac{10}{12}$
$\quad \overline{\phantom{-3\frac{5}{6} = 3\frac{10}{12} = }5\frac{5}{12}}$

PROBLEM 3-49 Subtract fractions from mixed numbers when the minuend of the fraction part is too small:

(a) $2\frac{2}{9} - \frac{7}{9}$ **(b)** $5\frac{1}{3} - \frac{2}{3}$ **(c)** $1\frac{3}{8} - \frac{5}{8}$

(d) $9\frac{1}{10} - \frac{3}{10}$ **(e)** $4\frac{1}{2} - \frac{5}{6}$ **(f)** $6\frac{1}{8} - \frac{1}{4}$

(g) $10\frac{1}{4} - \frac{1}{3}$ **(h)** $12\frac{1}{3} - \frac{2}{5}$ **(i)** $2\frac{1}{8} - \frac{1}{6}$

Solution   Recall that to subtract a fraction from a mixed number when the minuend of the fraction part is too small, you first rename the mixed number as an equal mixed number with a larger fraction part [see Example 3-83]:

(a)
$$2\frac{2}{9} = 1\frac{11}{9}$$
$$-\frac{7}{9} \quad \frac{7}{9}$$
$$\overline{\quad 1\frac{4}{9}}$$

(b)
$$5\frac{1}{3} = 4\frac{4}{3}$$
$$-\frac{2}{3} \quad \frac{2}{3}$$
$$\overline{\quad 4\frac{2}{3}}$$

(c)
$$1\frac{3}{8} = \frac{11}{8}$$
$$-\frac{5}{8} = \frac{5}{8}$$
$$\overline{\quad \frac{6}{8}} = \frac{3}{4}$$

(d)
$$9\frac{1}{10} = 8\frac{11}{10}$$
$$-\frac{3}{10} = \frac{3}{10}$$
$$\overline{\quad 8\frac{8}{10}} = 8\frac{4}{5}$$

(e)
$$4\frac{1}{2} = 4\frac{3}{6} = 3\frac{9}{6}$$
$$-\frac{5}{6} = \frac{5}{6} = \frac{5}{6}$$
$$\overline{\qquad 3\frac{4}{6}} = 3\frac{2}{3}$$

(f)
$$6\frac{1}{8} = 6\frac{1}{8} = 5\frac{9}{8}$$
$$-\frac{1}{4} = \frac{2}{8} = \frac{2}{8}$$
$$\overline{\qquad 5\frac{7}{8}}$$

(g)
$$10\frac{1}{4} = 10\frac{3}{12} = 9\frac{15}{12}$$
$$-\frac{1}{3} = \frac{4}{12} = \frac{4}{12}$$
$$\overline{\qquad 9\frac{11}{12}}$$

(h)
$$12\frac{1}{3} = 12\frac{5}{15} = 11\frac{20}{15}$$
$$-\frac{2}{5} = \frac{6}{15} = \frac{6}{15}$$
$$\overline{\qquad 11\frac{14}{15}}$$

(i)
$$2\frac{1}{8} = 2\frac{3}{24} = 1\frac{27}{24}$$
$$-\frac{1}{6} = \frac{4}{24} = \frac{4}{24}$$
$$\overline{\qquad 1\frac{23}{24}}$$

PROBLEM 3-50   Decide what to do (add, subtract, multiply, or divide with fractions) and then solve the following problems:

(a) A weekend is $\frac{2}{7}$ of a week. What fractional part of a week are the weekdays, not including the weekend?

(b) Each lap around the track is $\frac{1}{4}$ mile. How many laps must be completed to run 5 miles?

(c) George spent $\frac{1}{2}$ hour studying math and $\frac{3}{4}$ hour studying English. How long did he study in all?

(d) Pat practices the piano $\frac{3}{4}$ hour each day. How long does she practice in a 7-day week?

(e) A recipe calls for $\frac{3}{4}$ cup sugar. Richard has only $\frac{1}{3}$ cup on hand. How much more sugar is needed?

(f) Mom gave Tom $\frac{2}{3}$ of the pie. He ate $\frac{1}{2}$ of that portion. What fractional part of the whole pie did he eat?

(g) Greg studied for $\frac{1}{2}$ hour. This was only $\frac{1}{3}$ of the time he should have studied. How long should Greg have studied?

(h) Bobbi paid $\frac{1}{4}$ of her monthly pay for rent and $\frac{1}{3}$ for food. What part of her monthly pay did she pay for rent and food together?

Solution   Recall that to join parts together, you add; to take one part away from another part, you subtract; to join equal parts together, you multiply; and to separate an amount into equal parts, you divide [see Examples 3-84 and 3-85]:

(a) subtract: whole week $\longrightarrow$ 1 or $\frac{7}{7}$; $\frac{7}{7} - \frac{2}{7} = \frac{5}{7}$ (of a week)

(b) divide: $5 \div \frac{1}{4} = \frac{5}{1} \div \frac{1}{4} = \frac{5}{1} \times \frac{4}{1} = \frac{20}{1} = 20$ (laps)

(c) add: $\frac{1}{2} + \frac{3}{4} = \frac{2}{4} + \frac{3}{4} = \frac{5}{4} = 1\frac{1}{4}$ (hours)

(d) multiply: $\frac{3}{4} \times 7 = \frac{3}{4} \times \frac{7}{1} = \frac{21}{4} = 5\frac{1}{4}$ (hours)

(e) subtract: $\frac{3}{4} - \frac{1}{3} = \frac{9}{12} - \frac{4}{12} = \frac{5}{12}$ (of a cup)

(f) multiply: $\frac{1}{2} \times \frac{2}{3} = \frac{1}{3}$ (of the pie)

(g) divide: $\frac{1}{2} \div \frac{1}{3} = \frac{1}{2} \times \frac{3}{1} = \frac{3}{2} = 1\frac{1}{2}$ (hours)

(h) add: $\frac{1}{4} + \frac{1}{3} = \frac{3}{12} + \frac{4}{12} = \frac{7}{12}$ (of her monthly pay)

PROBLEM 3-51   Decide what to do (add, subtract, multiply, or divide with mixed numbers) and then solve the following problems:

(a) Amy swam $8\frac{1}{2}$ laps. Bill swam $2\frac{3}{4}$ laps. How many more laps did Amy swim than Bill?

(b) Aaron ran $3\frac{1}{2}$ miles and then walked $4\frac{3}{4}$ miles. How far did he run and walk all together?

(c) If Ann takes $3\frac{1}{2}$ minutes to run each lap around the track, how long will it take her to run 45 laps?

(d) If Alfred takes $3\frac{1}{2}$ minutes to run each lap around the track, how many laps will he complete in 45 minutes?

(e) If each book weighs $1\frac{1}{3}$ pounds, how much will 100 books weigh?

(f) Brant types at $72\frac{1}{2}$ words per minute. About how long will it take him to type a report of 1000 words?

(g) Emil is $62\frac{1}{2}$ inches tall. Last year he was $60\frac{3}{4}$ inches tall. How much did Emil grow since last year?

(h) Brice was a $5\frac{1}{2}$-pound baby. His twin sister weighed $6\frac{3}{4}$ pounds at birth. How much was the combined weight of the twins at birth?

Solution Recall that to join amounts together, you add; to take one amount away from another, you subtract; to join equal amounts together, you multiply; and to separate into equal amounts, you divide [see Examples 3-86 and 3-87]:

(a) subtract: $8\frac{1}{2} - 2\frac{3}{4} = 5\frac{3}{4}$ (laps)

(b) add: $3\frac{1}{2} + 4\frac{3}{4} = 8\frac{1}{4}$ (miles)

(c) multiply: $3\frac{1}{2} \times 45 = 157\frac{1}{2}$ (minutes)

(d) divide: $45 \div 3\frac{1}{2} = 12\frac{6}{7}$ or 12 (complete laps)

(e) multiply: $1\frac{1}{3} \times 100 = 133\frac{1}{3}$ (pounds)

(f) divide: $1000 \div 72\frac{1}{2} = 13\frac{23}{29}$ or about 14 (minutes)

(g) subtract: $62\frac{1}{2} - 60\frac{3}{4} = 1\frac{3}{4}$ (inches)

(h) add: $5\frac{1}{2} + 6\frac{3}{4} = 12\frac{1}{4}$ (pounds)

Supplementary Exercises

PROBLEM 3-52 Are the following pairs of fractions equal or not equal?

(a) $\frac{3}{4}$ and $\frac{6}{8}$ **(b)** $\frac{2}{3}$ and $\frac{8}{12}$ **(c)** $\frac{1}{2}$ and $\frac{4}{6}$ **(d)** $\frac{4}{5}$ and $\frac{12}{15}$ **(e)** $\frac{1}{3}$ and $\frac{6}{21}$

(f) $\frac{1}{4}$ and $\frac{5}{20}$ **(g)** $\frac{1}{5}$ and $\frac{2}{10}$ **(h)** $\frac{3}{8}$ and $\frac{8}{24}$ **(i)** $\frac{7}{10}$ and $\frac{20}{30}$ **(j)** $\frac{5}{12}$ and $\frac{15}{36}$

(k) $\frac{6}{8}$ and $\frac{30}{40}$ **(l)** $\frac{4}{6}$ and $\frac{18}{24}$ **(m)** $\frac{8}{12}$ and $\frac{24}{36}$ **(n)** $\frac{4}{12}$ and $\frac{16}{60}$ **(o)** $\frac{6}{10}$ and $\frac{18}{30}$

PROBLEM 3-53 Rename each division problem as a fraction:

(a) $1 \div 2$ **(b)** $7 \div 8$ **(c)** $5 \div 3$ **(d)** $8 \div 5$ **(e)** $2 \div 9$

(f) $5 \div 6$ **(g)** $12 \div 1$ **(h)** $0 \div 4$ **(i)** $3 \div 7$ **(j)** $7 \div 10$

(k) $11 \div 12$ **(l)** $3 \div 20$ **(m)** $27 \div 50$ **(n)** $49 \div 100$ **(o)** $253 \div 1000$

PROBLEM 3-54 Rename each fraction as a division problem:

(a) $\frac{5}{8}$ **(b)** $\frac{2}{3}$ **(c)** $\frac{6}{5}$ **(d)** $\frac{7}{1}$ **(e)** $\frac{0}{9}$

(f) $\frac{3}{2}$ **(g)** $\frac{1}{6}$ **(h)** $\frac{3}{4}$ **(i)** $\frac{8}{7}$ **(j)** $\frac{3}{10}$

(k) $\frac{5}{12}$ **(l)** $\frac{7}{18}$ **(m)** $\frac{21}{25}$ **(n)** $\frac{199}{100}$ **(o)** $\frac{249}{500}$

PROBLEM 3-55 Rename each whole or mixed number as an equal fraction:

(a) 2 **(b)** 5 **(c)** $3\frac{1}{2}$ **(d)** $1\frac{2}{3}$ **(e)** $2\frac{3}{4}$

(f) $4\frac{3}{5}$ **(g)** 6 **(h)** 0 **(i)** $2\frac{1}{3}$ **(j)** $5\frac{1}{4}$

(k) $3\frac{1}{5}$ **(l)** $2\frac{7}{10}$ **(m)** 1 **(n)** 20 **(o)** $25\frac{49}{100}$

PROBLEM 3-56 Rename each improper fraction as an equal mixed number:

(a) $\frac{3}{2}$ **(b)** $\frac{5}{4}$ **(c)** $\frac{8}{5}$ **(d)** $\frac{5}{3}$ **(e)** $\frac{11}{6}$

(f) $\frac{13}{8}$ **(g)** $\frac{17}{9}$ **(h)** $\frac{13}{10}$ **(i)** $\frac{7}{2}$ **(j)** $\frac{11}{4}$

(k) $\frac{10}{3}$ **(l)** $\frac{25}{8}$ **(m)** $\frac{24}{5}$ **(n)** $\frac{37}{6}$ **(o)** $\frac{51}{8}$

PROBLEM 3-57 Simplify each fraction or mixed number:

(a) $\frac{2}{6}$ (b) $\frac{4}{8}$ (c) $\frac{8}{4}$ (d) $\frac{12}{9}$ (e) $3\frac{4}{6}$

(f) $2\frac{10}{8}$ (g) $\frac{9}{12}$ (h) $\frac{8}{12}$ (i) $\frac{10}{6}$ (j) $\frac{6}{1}$

(k) $1\frac{8}{10}$ (l) $3\frac{9}{18}$ (m) $\frac{12}{18}$ (n) $\frac{35}{20}$ (o) $5\frac{13}{3}$

PROBLEM 3-58 Multiply with fractions and mixed numbers:

(a) $\frac{3}{4} \times \frac{2}{3}$ (b) $\frac{1}{2} \times \frac{1}{4}$ (c) $\frac{2}{3} \times 2\frac{1}{4}$

(d) $2\frac{1}{2} \times \frac{4}{5}$ (e) $1\frac{3}{4} \times 2\frac{1}{8}$ (f) $1\frac{1}{2} \times 2\frac{1}{2}$

(g) $\frac{5}{8} \times \frac{1}{4}$ (h) $\frac{5}{6} \times \frac{2}{3}$ (i) $\frac{3}{8} \times 1\frac{1}{3}$

(j) $2\frac{1}{4} \times \frac{1}{3}$ (k) $1\frac{1}{2} \times \frac{3}{4} \times 2\frac{1}{3} \times 4$ (l) $\frac{5}{8} \times 2\frac{1}{2} \times \frac{3}{10} \times 2$

(m) $1\frac{1}{2} \times 1\frac{1}{3} \times 1\frac{1}{4} \times 1\frac{1}{5}$ (n) $\frac{3}{8} \times 1\frac{1}{3} \times \frac{4}{5} \times 1\frac{7}{10} \times 0 \times 5\frac{7}{10}$ (o) $\frac{1}{2} \times 1\frac{1}{2} \times \frac{3}{4} \times 1\frac{1}{3} \times \frac{4}{5} \times 1\frac{1}{4} \times \frac{5}{6} \times 1\frac{1}{5}$

PROBLEM 3-59 Find the reciprocal of each number:

(a) $\frac{1}{4}$ (b) $\frac{1}{3}$ (c) 2 (d) 1 (e) $1\frac{1}{2}$

(f) $1\frac{1}{3}$ (g) $\frac{2}{5}$ (h) $\frac{0}{8}$ (i) 0 (j) 5

(k) $1\frac{1}{5}$ (l) $2\frac{1}{4}$ (m) $\frac{7}{10}$ (n) $\frac{39}{100}$ (o) $4\frac{7}{12}$

PROBLEM 3-60 Divide with fractions and mixed numbers:

(a) $\frac{3}{4} \div \frac{1}{2}$ (b) $\frac{2}{3} \div \frac{1}{2}$ (c) $\frac{5}{8} \div \frac{5}{8}$ (d) $\frac{0}{2} \div \frac{7}{10}$ (e) $\frac{1}{3} \div 2$

(f) $3 \div \frac{2}{5}$ (g) $3\frac{1}{2} \div \frac{1}{2}$ (h) $\frac{1}{4} \div 2\frac{1}{2}$ (i) $1\frac{1}{2} \div 3$ (j) $2 \div 2\frac{3}{4}$

(k) $1\frac{1}{5} \div 1\frac{1}{3}$ (l) $2\frac{1}{2} \div 2\frac{1}{4}$ (m) $\frac{7}{12} \div \frac{5}{12}$ (n) $1\frac{3}{4} \div 1\frac{3}{4}$ (o) $\frac{5}{18} \div 1\frac{1}{9}$

PROBLEM 3-61 Add like fractions:

(a) $\frac{1}{8} + \frac{5}{8}$ (b) $\frac{3}{10} + \frac{3}{10}$ (c) $\frac{1}{12} + \frac{11}{12}$ (d) $\frac{1}{5} + \frac{2}{5}$

(e) $\frac{1}{6} + \frac{1}{6}$ (f) $\frac{1}{3} + \frac{1}{3}$ (g) $\frac{7}{8} + \frac{1}{8}$ (h) $\frac{3}{10} + \frac{7}{10}$

(i) $\frac{5}{12} + \frac{5}{12}$ (j) $\frac{3}{5} + \frac{1}{5}$ (k) $\frac{1}{6} + \frac{5}{6}$ (l) $\frac{2}{3} + \frac{2}{3}$

(m) $\frac{3}{8} + \frac{7}{8}$ (n) $\frac{7}{12} + \frac{5}{12}$ (o) $\frac{2}{5} + \frac{3}{5}$ (p) $\frac{5}{6} + \frac{5}{6}$

(q) $\frac{1}{3} + \frac{2}{3}$ (r) $\frac{7}{8} + \frac{7}{8}$ (s) $\frac{5}{12} + \frac{11}{12} + \frac{7}{12}$ (t) $\frac{3}{5} + \frac{4}{5} + \frac{1}{5}$

(u) $\frac{1}{2} + \frac{1}{2} + \frac{1}{2}$ (v) $\frac{1}{4} + \frac{3}{4} + \frac{3}{4}$ (w) $\frac{3}{8} + \frac{5}{8} + \frac{1}{8} + \frac{7}{8}$ (x) $\frac{5}{10} + \frac{7}{10} + \frac{9}{10} + \frac{9}{10}$

PROBLEM 3-62 Find the LCD for unlike fractions without using the Factoring Method for finding the LCD:

(a) $\frac{1}{2}$ and $\frac{5}{12}$ (b) $\frac{2}{3}$ and $\frac{7}{15}$ (c) $\frac{3}{4}$ and $\frac{5}{16}$ (d) $\frac{2}{5}$ and $\frac{1}{15}$ (e) $\frac{1}{6}$ and $\frac{1}{12}$

(f) $\frac{3}{8}$ and $\frac{5}{16}$ (g) $\frac{1}{3}$ and $\frac{2}{7}$ (h) $\frac{1}{4}$ and $\frac{1}{3}$ (i) $\frac{3}{4}$ and $\frac{5}{7}$ (j) $\frac{2}{5}$ and $\frac{1}{7}$

(k) $\frac{3}{8}$ and $\frac{1}{9}$ (l) $\frac{1}{4}$ and $\frac{5}{9}$ (m) $\frac{5}{6}$ and $\frac{7}{8}$ (n) $\frac{1}{8}$ and $\frac{7}{10}$ (o) $\frac{3}{10}$ and $\frac{7}{12}$

(p) $\frac{5}{9}$ and $\frac{11}{15}$ (q) $\frac{9}{10}$ and $\frac{11}{12}$ (r) $\frac{5}{12}$ and $\frac{13}{18}$ (s) $\frac{3}{4}, \frac{1}{2},$ and $\frac{5}{8}$ (t) $\frac{2}{3}, \frac{1}{6},$ and $\frac{1}{2}$

(u) $\frac{2}{3}, \frac{3}{5},$ and $\frac{1}{2}$ (v) $\frac{3}{4}, \frac{4}{5},$ and $\frac{1}{3}$ (w) $\frac{5}{6}, \frac{3}{8},$ and $\frac{7}{10}$ (x) $\frac{3}{10}, \frac{5}{18},$ and $\frac{9}{8}$ (y) $\frac{7}{18}, \frac{11}{12},$ and $\frac{10}{9}$

PROBLEM 3-63 Find the LCD for unlike fractions using the Factoring Method for finding the LCD:

(a) $\frac{1}{4}$ and $\frac{7}{10}$ (b) $\frac{1}{6}$ and $\frac{3}{14}$ (c) $\frac{5}{8}$ and $\frac{9}{14}$ (d) $\frac{2}{9}$ and $\frac{1}{15}$ (e) $\frac{7}{16}$ and $\frac{4}{9}$

(f) $\frac{3}{10}$ and $\frac{1}{14}$ (g) $\frac{3}{16}$ and $\frac{1}{10}$ (h) $\frac{3}{10}$ and $\frac{5}{18}$ (i) $\frac{1}{12}$ and $\frac{11}{20}$ (j) $\frac{5}{12}$ and $\frac{11}{21}$

(k) $\frac{1}{12}$ and $\frac{9}{22}$ (l) $\frac{5}{14}$ and $\frac{17}{18}$ (m) $\frac{19}{20}$ and $\frac{1}{14}$ (n) $\frac{13}{14}$ and $\frac{17}{21}$ (o) $\frac{23}{24}$ and $\frac{13}{14}$

(p) $\frac{15}{16}$ and $\frac{20}{21}$ (q) $\frac{17}{18}$ and $\frac{3}{25}$ (r) $\frac{1}{4}, \frac{1}{6},$ and $\frac{1}{8}$ (s) $\frac{2}{3}, \frac{5}{8},$ and $\frac{1}{6}$ (t) $\frac{3}{4}, \frac{7}{8},$ and $\frac{4}{9}$

(u) $\frac{5}{6}, \frac{1}{8},$ and $\frac{3}{10}$ (v) $\frac{7}{12}, \frac{5}{8},$ and $\frac{5}{6}$ (w) $\frac{7}{10}, \frac{11}{12},$ and $\frac{14}{15}$ (x) $\frac{1}{12}, \frac{1}{18},$ and $\frac{1}{10}$ (y) $\frac{3}{20}, \frac{5}{18},$ and $\frac{11}{16}$

PROBLEM 3-64 Rename unlike fractions as equivalent like fractions using the LCD:

(a) $\frac{1}{2}$ and $\frac{3}{8}$ 　　(b) $\frac{5}{6}$ and $\frac{1}{2}$ 　　(c) $\frac{1}{4}$ and $\frac{5}{12}$ 　　(d) $\frac{3}{5}$ and $\frac{7}{10}$ 　　(e) $\frac{1}{2}$ and $\frac{1}{4}$

(f) $\frac{2}{3}$ and $\frac{5}{9}$ 　　(g) $\frac{1}{3}$ and $\frac{3}{5}$ 　　(h) $\frac{3}{4}$ and $\frac{1}{5}$ 　　(i) $\frac{5}{6}$ and $\frac{2}{5}$ 　　(j) $\frac{4}{5}$ and $\frac{5}{8}$

(k) $\frac{1}{2}$ and $\frac{2}{3}$ 　　(l) $\frac{1}{3}$ and $\frac{7}{8}$ 　　(m) $\frac{5}{12}$ and $\frac{7}{18}$ 　　(n) $\frac{1}{15}$ and $\frac{11}{12}$ 　　(o) $\frac{4}{15}$ and $\frac{3}{20}$

(p) $\frac{17}{18}$ and $\frac{11}{15}$ 　　(q) $\frac{7}{12}$ and $\frac{9}{16}$ 　　(r) $\frac{13}{16}$ and $\frac{15}{18}$ 　　(s) $\frac{5}{6}, \frac{1}{3}$, and $\frac{1}{2}$ 　　(t) $\frac{3}{4}, \frac{5}{6}$, and $\frac{7}{12}$

(u) $\frac{1}{4}, \frac{2}{3}$, and $\frac{4}{5}$ 　　(v) $\frac{1}{2}, \frac{3}{7}$, and $\frac{2}{5}$ 　　(w) $\frac{7}{10}, \frac{5}{12}$, and $\frac{1}{9}$ 　　(x) $\frac{3}{8}, \frac{2}{15}$, and $\frac{7}{12}$ 　　(y) $\frac{15}{16}, \frac{1}{4}, \frac{1}{2}$, and $\frac{7}{8}$

PROBLEM 3-65 Add unlike fractions:

(a) $\frac{1}{2} + \frac{5}{6}$ 　　(b) $\frac{3}{8} + \frac{1}{2}$ 　　(c) $\frac{2}{3} + \frac{7}{9}$ 　　(d) $\frac{3}{4} + \frac{5}{16}$

(e) $\frac{1}{5} + \frac{7}{10}$ 　　(f) $\frac{5}{12} + \frac{1}{6}$ 　　(g) $\frac{1}{2} + \frac{3}{7}$ 　　(h) $\frac{2}{5} + \frac{5}{8}$

(i) $\frac{1}{4} + \frac{4}{5}$ 　　(j) $\frac{4}{7} + \frac{1}{5}$ 　　(k) $\frac{1}{6} + \frac{3}{5}$ 　　(l) $\frac{7}{8} + \frac{1}{3}$

(m) $\frac{5}{6} + \frac{1}{9}$ 　　(n) $\frac{1}{12} + \frac{11}{14}$ 　　(o) $\frac{15}{16} + \frac{7}{12}$ 　　(p) $\frac{1}{12} + \frac{8}{15}$

(q) $\frac{5}{18} + \frac{11}{12}$ 　　(r) $\frac{14}{15} + \frac{1}{18}$ 　　(s) $\frac{2}{15} + \frac{1}{5} + \frac{1}{3}$ 　　(t) $\frac{7}{12} + \frac{5}{6} + \frac{1}{2}$

(u) $\frac{1}{2} + \frac{1}{3} + \frac{1}{5}$ 　　(v) $\frac{5}{6} + \frac{7}{8} + \frac{9}{10}$ 　　(w) $\frac{1}{8} + \frac{3}{16} + \frac{1}{4} + \frac{1}{2}$ 　　(x) $\frac{5}{6} + \frac{1}{2} + \frac{5}{12} + \frac{1}{3}$

PROBLEM 3-66 Add with mixed numbers:

(a) $2 + 3\frac{2}{3}$ 　　(b) $5 + 8\frac{1}{2}$ 　　(c) $3\frac{1}{4} + 4$ 　　(d) $1\frac{1}{5} + 2$

(e) $3\frac{3}{4} + 1$ 　　(f) $3 + 9\frac{5}{6}$ 　　(g) $\frac{1}{2} + 4\frac{1}{2}$ 　　(h) $\frac{1}{3} + 9\frac{1}{3}$

(i) $8\frac{3}{4} + \frac{3}{4}$ 　　(j) $7\frac{5}{8} + \frac{7}{8}$ 　　(k) $\frac{1}{3} + 6\frac{1}{6}$ 　　(l) $\frac{3}{4} + 5\frac{5}{8}$

(m) $4\frac{1}{2} + \frac{3}{4}$ 　　(n) $3\frac{1}{2} + 1\frac{2}{3}$ 　　(o) $3\frac{3}{4} + 5\frac{3}{5}$ 　　(p) $4\frac{1}{6} + 1\frac{5}{8}$

(q) $5\frac{3}{10} + 4\frac{7}{12}$ 　　(r) $6\frac{5}{15} + 5\frac{4}{20}$ 　　(s) $2 + \frac{1}{3} + 8\frac{5}{9}$ 　　(t) $\frac{1}{12} + 9 + 1\frac{1}{3}$

(u) $3 + 4\frac{4}{8} + \frac{1}{9}$ 　　(v) $\frac{7}{12} + 5\frac{1}{6} + 2$ 　　(w) $6\frac{3}{12} + 7 + \frac{5}{18}$ 　　(x) $8\frac{2}{3} + \frac{4}{5} + 2$

PROBLEM 3-67 Subtract fractions:

(a) $\frac{3}{5} - \frac{1}{5}$ 　　(b) $\frac{4}{3} - \frac{1}{3}$ 　　(c) $\frac{1}{6} - \frac{1}{6}$ 　　(d) $\frac{5}{2} - \frac{1}{2}$ 　　(e) $\frac{5}{8} - \frac{3}{8}$

(f) $\frac{5}{4} - \frac{3}{4}$ 　　(g) $\frac{7}{10} - \frac{3}{10}$ 　　(h) $\frac{7}{12} - \frac{1}{12}$ 　　(i) $\frac{7}{32} - \frac{3}{32}$ 　　(j) $\frac{11}{16} - \frac{9}{16}$

(k) $\frac{1}{2} - \frac{1}{8}$ 　　(l) $\frac{9}{10} - \frac{1}{2}$ 　　(m) $\frac{3}{4} - \frac{5}{12}$ 　　(n) $\frac{7}{8} - \frac{1}{16}$ 　　(o) $\frac{4}{5} - \frac{1}{4}$

(p) $\frac{1}{2} - \frac{2}{7}$ 　　(q) $\frac{5}{8} - \frac{3}{10}$ 　　(r) $\frac{11}{12} - \frac{5}{8}$ 　　(s) $\frac{7}{10} - \frac{1}{12}$ 　　(t) $\frac{14}{15} - \frac{5}{12}$

(u) $\frac{9}{10} - \frac{11}{16}$ 　　(v) $\frac{11}{20} - \frac{13}{30}$ 　　(w) $\frac{15}{16} - \frac{11}{24}$ 　　(x) $\frac{11}{18} - \frac{11}{24}$ 　　(y) $\frac{9}{16} - \frac{5}{32}$

PROBLEM 3-68 Subtract with mixed numbers:

(a) $5\frac{3}{4} - 3$ 　　(b) $8\frac{2}{3} - 1$ 　　(c) $3\frac{5}{6} - 2$ 　　(d) $4\frac{7}{10} - 1$ 　　(e) $1\frac{2}{3} - 1$

(f) $6\frac{4}{5} - 2$ 　　(g) $2\frac{5}{9} - 1$ 　　(h) $7\frac{1}{2} - 7$ 　　(i) $9\frac{3}{10} - \frac{1}{10}$ 　　(j) $1\frac{5}{8} - \frac{1}{2}$

(k) $8\frac{3}{4} - \frac{1}{2}$ 　　(l) $2\frac{2}{3} - \frac{1}{6}$ 　　(m) $7\frac{1}{2} - \frac{1}{3}$ 　　(n) $3\frac{2}{3} - \frac{1}{4}$ 　　(o) $6\frac{5}{6} - \frac{3}{10}$

(p) $4\frac{5}{12} - \frac{1}{5}$ 　　(q) $5\frac{1}{4} - 4\frac{1}{4}$ 　　(r) $6\frac{7}{8} - 3\frac{1}{8}$ 　　(s) $10\frac{1}{2} - 2\frac{1}{6}$ 　　(t) $15\frac{2}{3} - 4\frac{5}{12}$

(u) $12\frac{1}{2} - 1\frac{2}{5}$ 　　(v) $8\frac{4}{5} - 5\frac{1}{4}$ 　　(w) $9\frac{3}{4} - 2\frac{1}{10}$ 　　(x) $11\frac{11}{12} - 8\frac{7}{18}$ 　　(y) $3\frac{11}{20} - 3\frac{5}{12}$

PROBLEM 3-69 Subtract fractions and mixed numbers from whole numbers:

(a) $5 - 1\frac{1}{2}$ 　　(b) $12 - 2\frac{1}{3}$ 　　(c) $2 - 1\frac{1}{4}$ 　　(d) $9 - 5\frac{1}{8}$ 　　(e) $15 - 6\frac{5}{6}$

(f) $16 - 2\frac{4}{5}$ 　　(g) $3 - 2\frac{7}{10}$ 　　(h) $11 - 1\frac{5}{12}$ 　　(i) $7 - 5\frac{3}{7}$ 　　(j) $4 - 1\frac{7}{9}$

(k) $15 - 3\frac{5}{16}$ 　　(l) $8 - 5\frac{5}{32}$ 　　(m) $6 - 5\frac{1}{5}$ 　　(n) $10 - 3\frac{3}{8}$ 　　(o) $2 - \frac{2}{3}$

(p) $5 - \frac{3}{4}$ 　　(q) $10 - \frac{5}{8}$ 　　(r) $3 - \frac{1}{6}$ 　　(s) $7 - \frac{2}{5}$ 　　(t) $9 - \frac{3}{10}$

(u) $6 - \frac{1}{12}$ 　　(v) $1 - \frac{1}{16}$ 　　(w) $4 - \frac{29}{32}$ 　　(x) $6 - \frac{7}{8}$ 　　(y) $14 - \frac{1}{10}$

PROBLEM 3-70 Subtract fractions and mixed numbers from mixed numbers when the minuend of the fraction part is too small:

(a) $8\frac{1}{4} - 5\frac{3}{4}$ (b) $7\frac{1}{8} - 2\frac{7}{8}$ (c) $9\frac{1}{3} - 8\frac{2}{3}$ (d) $3\frac{1}{5} - 2\frac{2}{5}$ (e) $6\frac{1}{3} - 1\frac{1}{2}$

(f) $10\frac{1}{4} - 2\frac{1}{3}$ (g) $11\frac{1}{4} - 3\frac{1}{2}$ (h) $6\frac{1}{3} - 1\frac{5}{6}$ (i) $8\frac{3}{4} - 2\frac{5}{6}$ (j) $3\frac{5}{8} - 1\frac{15}{16}$

(k) $5\frac{3}{10} - 1\frac{3}{4}$ (l) $12\frac{1}{16} - 3\frac{3}{16}$ (m) $5\frac{1}{2} - 4\frac{3}{4}$ (n) $2\frac{1}{6} - 1\frac{5}{6}$ (o) $9\frac{1}{2} - \frac{7}{10}$

(p) $1\frac{1}{4} - \frac{3}{4}$ (q) $8\frac{1}{3} - \frac{5}{8}$ (r) $2\frac{1}{6} - \frac{1}{3}$ (s) $7\frac{2}{5} - \frac{3}{5}$ (t) $3\frac{3}{10} - \frac{7}{10}$

(u) $6\frac{3}{8} - \frac{1}{2}$ (v) $4\frac{1}{12} - \frac{1}{10}$ (w) $5\frac{5}{12} - \frac{3}{4}$ (x) $10\frac{1}{16} - \frac{1}{4}$ (y) $12\frac{3}{16} - \frac{5}{8}$

PROBLEM 3-71 Solve word problems containing fractions and mixed numbers $(+, -, \times, \div)$:

(a) How many $7\frac{1}{2}$-inch boards can be cut from a 22-inch piece of stock?

(b) A board is cut into 20 equal pieces so that each piece is $3\frac{1}{2}$ inches in length. How long was the board before it was cut?

(c) A bedroom is $8\frac{1}{2}$ feet by $9\frac{1}{4}$ feet. How much baseboard is needed to go around the room if there is one doorway that is $3\frac{1}{2}$ feet wide and 7 feet high?

(d) The walls of the room in problem (c) are 8 feet high. Wallpaper is $1\frac{1}{2}$ feet wide and the length of each roll is 20 feet. How many rolls are needed to paper the room using only whole 8-feet strips except for the doorway?

(e) What is the difference between the width and length of an $8\frac{1}{2} \times 11$ inch sheet of paper?

(f) It takes $\frac{1}{2}$ pound of meat for each serving of meat loaf. (1) How much meat is needed for $7\frac{1}{2}$ servings? (2) How many complete servings can be made from $7\frac{1}{4}$ pounds of meat?

(g) Stella worked $4\frac{1}{2}$ hours Monday, $5\frac{1}{4}$ hours Tuesday, $3\frac{3}{4}$ hours Wednesday, $6\frac{1}{2}$ hours Thursday, and $7\frac{1}{4}$ hours Friday. How many hours did she work in all?

(h) Ralph purchased $3\frac{1}{2}$ pounds of potatoes for 49¢ per pound and $5\frac{1}{4}$ pounds of onions at 19¢ per pound. How much will the potatoes and onions cost? (*Hint:* Stores always round up to the next nearest cent.)

Answers to Supplementary Exercises

(3-52) (a) equal   (b) equal   (c) not equal   (d) equal   (e) not equal
       (f) equal   (g) equal   (h) not equal   (i) not equal   (j) equal
       (k) equal   (l) not equal   (m) equal   (n) not equal   (o) equal

(3-53) (a) $\frac{1}{2}$ (b) $\frac{7}{8}$ (c) $\frac{5}{3}$ (d) $\frac{8}{5}$ (e) $\frac{2}{9}$
       (f) $\frac{5}{6}$ (g) $\frac{12}{1}$ (h) $\frac{0}{4}$ (i) $\frac{3}{7}$ (j) $\frac{7}{10}$
       (k) $\frac{11}{12}$ (l) $\frac{3}{20}$ (m) $\frac{27}{50}$ (n) $\frac{49}{100}$ (o) $\frac{253}{1000}$

(3-54) (a) $5 \div 8$ (b) $2 \div 3$ (c) $6 \div 5$ (d) $7 \div 1$ (e) $0 \div 9$
       (f) $3 \div 2$ (g) $1 \div 6$ (h) $3 \div 4$ (i) $8 \div 7$ (j) $3 \div 10$
       (k) $5 \div 12$ (l) $7 \div 18$ (m) $21 \div 25$ (n) $199 \div 100$ (o) $249 \div 500$

(3-55) (a) $\frac{2}{1}$ (b) $\frac{5}{1}$ (c) $\frac{7}{2}$ (d) $\frac{5}{3}$ (e) $\frac{11}{4}$
       (f) $\frac{23}{5}$ (g) $\frac{6}{1}$ (h) $\frac{0}{1}$ (i) $\frac{7}{3}$ (j) $\frac{21}{4}$
       (k) $\frac{16}{5}$ (l) $\frac{27}{10}$ (m) $\frac{1}{1}$ (n) $\frac{20}{1}$ (o) $\frac{2549}{100}$

(3-56) (a) $1\frac{1}{2}$ (b) $1\frac{1}{4}$ (c) $1\frac{3}{5}$ (d) $1\frac{2}{3}$ (e) $1\frac{5}{6}$
       (f) $1\frac{5}{8}$ (g) $1\frac{8}{9}$ (h) $1\frac{3}{10}$ (i) $3\frac{1}{2}$ (j) $2\frac{3}{4}$
       (k) $3\frac{1}{3}$ (l) $3\frac{1}{8}$ (m) $4\frac{4}{5}$ (n) $6\frac{1}{6}$ (o) $6\frac{3}{8}$

(3-57) (a) $\frac{1}{3}$ (b) $\frac{1}{2}$ (c) 2 (d) $1\frac{1}{3}$ (e) $3\frac{2}{3}$
(f) $3\frac{1}{4}$ (g) $\frac{3}{4}$ (h) $\frac{2}{3}$ (i) $1\frac{2}{3}$ (j) 6
(k) $1\frac{4}{5}$ (l) $3\frac{1}{2}$ (m) $\frac{2}{3}$ (n) $1\frac{3}{4}$ (o) $9\frac{1}{3}$

(3-58) (a) $\frac{1}{2}$ (b) $\frac{1}{8}$ (c) $1\frac{1}{2}$ (d) 2 (e) $3\frac{23}{32}$
(f) $3\frac{3}{4}$ (g) $\frac{5}{32}$ (h) $\frac{5}{9}$ (i) $\frac{1}{2}$ (j) $\frac{3}{4}$
(k) $10\frac{1}{2}$ (l) $\frac{15}{16}$ (m) 3 (n) 0 (o) $\frac{3}{4}$

(3-59) (a) $\frac{4}{1}$ (b) $\frac{3}{1}$ (c) $\frac{1}{2}$ (d) $\frac{1}{1}$ (e) $\frac{2}{3}$
(f) $\frac{3}{4}$ (g) $\frac{5}{2}$ (h) not defined (i) not defined (j) $\frac{1}{5}$
(k) $\frac{5}{6}$ (l) $\frac{4}{9}$ (m) $\frac{10}{7}$ (n) $\frac{100}{39}$ (o) $\frac{12}{55}$

(3-60) (a) $1\frac{1}{2}$ (b) $1\frac{1}{3}$ (c) 1 (d) 0 (e) $\frac{1}{6}$
(f) $7\frac{1}{2}$ (g) 7 (h) $\frac{1}{10}$ (i) $\frac{1}{2}$ (j) $\frac{8}{11}$
(k) $\frac{9}{10}$ (l) $1\frac{1}{9}$ (m) $1\frac{2}{5}$ (n) 1 (o) $\frac{1}{4}$

(3-61) (a) $\frac{3}{4}$ (b) $\frac{3}{5}$ (c) 1 (d) $\frac{3}{5}$ (e) $\frac{1}{3}$ (f) $\frac{2}{3}$
(g) 1 (h) 1 (i) $\frac{5}{6}$ (j) $\frac{4}{5}$ (k) 1 (l) $1\frac{1}{3}$
(m) $1\frac{1}{4}$ (n) 1 (o) 1 (p) $1\frac{2}{3}$ (q) 1 (r) $1\frac{3}{4}$
(s) $1\frac{11}{12}$ (t) $1\frac{3}{5}$ (u) $1\frac{1}{2}$ (v) $1\frac{3}{4}$ (w) 2 (x) 3

(3-62) (a) 12 (b) 15 (c) 16 (d) 15 (e) 12
(f) 16 (g) 21 (h) 12 (i) 28 (j) 35
(k) 72 (l) 36 (m) 24 (n) 40 (o) 60
(p) 45 (q) 60 (r) 36 (s) 8 (t) 6
(u) 30 (v) 60 (w) 120 (x) 360 (y) 36

(3-63) (a) 20 (b) 42 (c) 56 (d) 45 (e) 144
(f) 70 (g) 80 (h) 90 (i) 60 (j) 84
(k) 132 (l) 126 (m) 140 (n) 42 (o) 168
(p) 336 (q) 450 (r) 24 (s) 24 (t) 72
(u) 120 (v) 24 (w) 60 (x) 180 (y) 720

(3-64) (a) $\frac{4}{8}$ and $\frac{3}{8}$ (b) $\frac{5}{6}$ and $\frac{3}{6}$ (c) $\frac{3}{12}$ and $\frac{5}{12}$ (d) $\frac{6}{10}$ and $\frac{7}{10}$
(e) $\frac{2}{4}$ and $\frac{1}{4}$ (f) $\frac{6}{9}$ and $\frac{5}{9}$ (g) $\frac{5}{15}$ and $\frac{9}{15}$ (h) $\frac{15}{20}$ and $\frac{4}{20}$
(i) $\frac{25}{30}$ and $\frac{12}{30}$ (j) $\frac{32}{40}$ and $\frac{25}{40}$ (k) $\frac{3}{6}$ and $\frac{4}{6}$ (l) $\frac{8}{24}$ and $\frac{21}{24}$
(m) $\frac{15}{36}$ and $\frac{14}{36}$ (n) $\frac{4}{60}$ and $\frac{55}{60}$ (o) $\frac{16}{60}$ and $\frac{9}{60}$ (p) $\frac{85}{90}$ and $\frac{66}{90}$
(q) $\frac{28}{48}$ and $\frac{27}{48}$ (r) $\frac{117}{144}$ and $\frac{120}{144}$ (s) $\frac{5}{6}, \frac{2}{6}$, and $\frac{3}{6}$ (t) $\frac{9}{12}, \frac{10}{12}$, and $\frac{7}{12}$
(u) $\frac{15}{60}, \frac{40}{60}$, and $\frac{48}{60}$ (v) $\frac{35}{70}, \frac{30}{70}$, and $\frac{28}{70}$ (w) $\frac{126}{180}, \frac{75}{180}$, and $\frac{20}{180}$ (x) $\frac{45}{120}, \frac{16}{120}$, and $\frac{70}{120}$
(y) $\frac{15}{16}, \frac{4}{16}, \frac{8}{16}$, and $\frac{14}{16}$

(3-65) (a) $1\frac{1}{3}$ (b) $\frac{7}{8}$ (c) $1\frac{4}{9}$ (d) $1\frac{1}{16}$ (e) $\frac{9}{10}$ (f) $\frac{7}{12}$
(g) $\frac{13}{14}$ (h) $1\frac{1}{40}$ (i) $1\frac{1}{20}$ (j) $\frac{27}{35}$ (k) $\frac{23}{30}$ (l) $1\frac{5}{24}$
(m) $\frac{17}{18}$ (n) $\frac{73}{84}$ (o) $1\frac{25}{48}$ (p) $\frac{37}{60}$ (q) $1\frac{7}{36}$ (r) $\frac{89}{90}$
(s) $1\frac{1}{15}$ (t) $1\frac{11}{12}$ (u) $1\frac{1}{30}$ (v) $2\frac{73}{120}$ (w) $1\frac{1}{16}$ (x) $2\frac{1}{12}$

(3-66) (a) $5\frac{2}{3}$ (b) $13\frac{1}{2}$ (c) $7\frac{1}{4}$ (d) $3\frac{1}{5}$ (e) $4\frac{3}{4}$ (f) $12\frac{5}{6}$
(g) 5 (h) $9\frac{2}{3}$ (i) $9\frac{1}{2}$ (j) $8\frac{1}{2}$ (k) $6\frac{1}{2}$ (l) $6\frac{3}{8}$
(m) $5\frac{1}{4}$ (n) $5\frac{1}{6}$ (o) $9\frac{7}{20}$ (p) $5\frac{19}{24}$ (q) $9\frac{53}{60}$ (r) $11\frac{8}{15}$
(s) $10\frac{8}{9}$ (t) $10\frac{5}{12}$ (u) $7\frac{11}{18}$ (v) $7\frac{3}{4}$ (w) $13\frac{19}{36}$ (x) $11\frac{7}{15}$

(3-67) (a) $\frac{2}{5}$ (b) 1 (c) 0 (d) 2 (e) $\frac{1}{4}$
(f) $\frac{1}{2}$ (g) $\frac{2}{5}$ (h) $\frac{1}{2}$ (i) $\frac{1}{8}$ (j) $\frac{1}{8}$

(k) $\frac{3}{8}$ (l) $\frac{2}{5}$ (m) $\frac{1}{3}$ (n) $\frac{13}{16}$ (o) $\frac{11}{20}$

(p) $\frac{3}{14}$ (q) $\frac{13}{40}$ (r) $\frac{7}{24}$ (s) $\frac{37}{60}$ (t) $\frac{31}{60}$

(u) $\frac{17}{80}$ (v) $\frac{7}{60}$ (w) $\frac{23}{48}$ (x) $\frac{11}{72}$ (y) $\frac{13}{32}$

(3-68) (a) $2\frac{3}{4}$ (b) $7\frac{2}{3}$ (c) $1\frac{5}{6}$ (d) $3\frac{7}{10}$ (e) $\frac{2}{3}$

(f) $4\frac{4}{5}$ (g) $1\frac{5}{9}$ (h) $\frac{1}{2}$ (i) $9\frac{1}{5}$ (j) $1\frac{1}{8}$

(k) $8\frac{1}{4}$ (l) $2\frac{1}{2}$ (m) $7\frac{1}{6}$ (n) $3\frac{5}{12}$ (o) $6\frac{8}{15}$

(p) $4\frac{13}{60}$ (q) 1 (r) $3\frac{3}{4}$ (s) $8\frac{1}{3}$ (t) $11\frac{1}{4}$

(u) $11\frac{1}{10}$ (v) $3\frac{11}{20}$ (w) $7\frac{13}{20}$ (x) $3\frac{19}{36}$ (y) $\frac{2}{15}$

(3-69) (a) $3\frac{1}{2}$ (b) $9\frac{2}{3}$ (c) $\frac{3}{4}$ (d) $3\frac{7}{8}$ (e) $8\frac{1}{6}$

(f) $13\frac{1}{5}$ (g) $\frac{3}{10}$ (h) $9\frac{7}{12}$ (i) $1\frac{4}{7}$ (j) $2\frac{2}{9}$

(k) $11\frac{11}{16}$ (l) $2\frac{27}{32}$ (m) $\frac{4}{5}$ (n) $6\frac{5}{8}$ (o) $1\frac{1}{3}$

(p) $4\frac{1}{4}$ (q) $9\frac{3}{8}$ (r) $2\frac{5}{6}$ (s) $6\frac{3}{5}$ (t) $8\frac{7}{10}$

(u) $5\frac{11}{12}$ (v) $\frac{15}{16}$ (w) $3\frac{3}{32}$ (x) $5\frac{1}{8}$ (y) $13\frac{9}{10}$

(3-70) (a) $2\frac{1}{2}$ (b) $4\frac{1}{4}$ (c) $\frac{2}{3}$ (d) $\frac{4}{5}$ (e) $4\frac{5}{6}$

(f) $7\frac{11}{12}$ (g) $7\frac{3}{4}$ (h) $4\frac{1}{2}$ (i) $5\frac{1}{12}$ (j) $1\frac{11}{16}$

(k) $3\frac{11}{20}$ (l) $8\frac{7}{8}$ (m) $\frac{3}{4}$ (n) $\frac{1}{3}$ (o) $8\frac{4}{5}$

(p) $\frac{1}{2}$ (q) $7\frac{17}{24}$ (r) $1\frac{5}{6}$ (s) $6\frac{4}{5}$ (t) $2\frac{3}{5}$

(u) $5\frac{7}{8}$ (v) $3\frac{59}{60}$ (w) $4\frac{2}{3}$ (x) $9\frac{13}{16}$ (y) $11\frac{9}{16}$

(3-71) (a) $35\frac{1}{5}$ boards   (b) 70 inches   (c) 32 feet   (d) 11 rolls

(e) $2\frac{1}{2}$ inches   (f) (1) $3\frac{3}{4}$ pounds (2) 14 complete servings   (g) $27\frac{1}{4}$ hours   (h) \$2.72

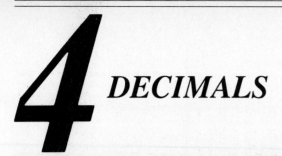

DECIMALS

4-1. Using Place Value with Decimals

A. Identifying digits in decimals

The numbers used to measure things accurately are called **decimal numbers** or just **decimals.**

EXAMPLE 4-1   List all the decimals containing the digits 0, 2, and 5 *exactly once* in each number.

Solution   The decimals containing 0, 2, and 5 exactly once are

$$52.0, \quad 25.0, \quad 50.2, \quad 20.5, \quad 05.2 = 5.20, \quad 5.02, \quad 02.5 = 2.50, \quad 2.05, \quad 0.52, \quad 0.25$$

Every decimal has a **whole-number part** and a **decimal-fraction part** separated by a **decimal point.**

EXAMPLE 4-2   Identify the three parts of 0.25.

Solution   In 0.25: 0 is the whole-number part
                       25 is the decimal-fraction part.
                       . is the decimal point.

To help locate the correct digit in a decimal given the digit's place-value, you can use a place-value chart like the one shown on page 116.

Decimal Place-Value Chart (from millions to millionths)

Whole-Number Part							Decimal-Fraction Part					
millions	hundred thousands	ten thousands	thousands	hundreds	tens	ones	tenths	hundredths	thousandths	ten thousandths	hundred thousandths	millionths
						•						

← place-value names

EXAMPLE 4-3   Draw a line under the digit that is in the thousandths place in 1.732051.

Solution

Whole-Number Part							Decimal-Fraction Part					
millions	hundred thousands	ten thousands	thousands	hundreds	tens	ones	tenths	hundredths	thousandths	ten thousandths	hundred thousandths	millionths
						1 •	7	3	2̲	0	5	1

← place-value names

You can also use a place-value chart to help identify the correct place-value name for a given digit.

EXAMPLE 4-4   Write the place-value name for the underlined digit in the decimal 3.1<u>4</u>1592.

Solution

Whole-Number Part							Decimal-Fraction Part					
millions	hundred thousands	ten thousands	thousands	hundreds	tens	ones	tenths	hundredths	thousandths	ten thousandths	hundred thousandths	millionths
						3 •	1	4̲	1	5	9	2

The place-value name of the underlined digit is hundredths.

Once you identify the place-value name for a given digit, you can write the value of that digit.

EXAMPLE 4-5   In 235.980176, what is the value of each of the following digits?

(**a**) 9   (**b**) 8   (**c**) 0   (**d**) 1   (**e**) 7   (**f**) 6

Solution In 235.980176: **(a)** the value of 9 is 9 tenths (or 0.9);

(b) the value of 8 is 8 hundredths (or 0.08);

(c) the value of 0 is 0 thousandths (or 0.000);

(d) the value of 1 is 1 ten thousandth (or 0.0001);

(e) the value of 7 is 7 hundred thousandths (or 0.00007);

(f) the value of 6 is 6 millionths (or 0.000006).

B. Writing decimals

To write the **decimal-word name** for a given decimal number, you write the word *and* for the decimal point when the whole number part is not zero.

EXAMPLE 4-6 Write the decimal-word name for the decimal number 2.5

Solution Read or write 2.5 as "two and five tenths" or "2 and 5 tenths."

Note: "2 and 5 tenths" is often referred to as the **short decimal-word name** for 2.5.

To write the correct short decimal-word name for a given decimal number, you always name the whole number, then name the decimal-fraction part with the place-value name of the last digit on the right.

EXAMPLE 4-7 Write the short decimal-word name for the following decimal numbers:

(a) 1.7   **(b)** 1.73   **(c)** 1.732   **(d)** 1.7320   **(e)** 1.73205   **(f)** 1.732051

Solution Read or write: **(a)** 1.7 as "1 and 7 tenths";

(b) 1.73 as "1 and 73 hundredths";

(c) 1.732 as "1 and 732 thousandths";

(d) 1.7320 as "1 and 7320 ten thousandths";

(e) 1.73205 as "1 and 73,205 hundred thousandths";

(f) 1.732051 as "1 and 732,051 millionths."

To write the correct decimal number given a short decimal-word name, you write the last digit of the decimal-fraction part so that it has the same value as the given short decimal-word name.

EXAMPLE 4-8 Write the decimal number for the short decimal-word name 3 and 75 thousandths.

Solution

Whole-Number Part							Decimal-Fraction Part					
millions	hundred thousands	ten thousands	thousands	hundreds	tens	ones	tenths	hundredths	thousandths	ten thousandths	hundred thousandths	millionths
						3 ·	0	7	5			

3 goes in the ones place

write a zero

"75 thousandths" means that the last digit 5 must go in the thousandths place.

and

When the whole-number part of a decimal number is zero, the decimal is called a **decimal fraction.**

EXAMPLE 4-9   Write decimal fractions for the following decimal-word names:

(a) 2 tenths
(b) 3 hundredths
(c) 7 thousandths
(d) 15 ten thousandths
(e) 9 hundred thousands
(f) 235 millionths

Solution   The decimal fraction for: (a) 2 tenths is 0.2;
            (b) 3 hundredths is 0.03;
            (c) 7 thousandths is 0.007;
            (d) 15 ten thousandths is 0.0015;
            (e) 9 hundred thousandths is 0.00009;
            (f) 235 millionths is 0.000235.

Note: A zero (0) is *usually* written for the whole-number part of a decimal fraction.

EXAMPLE 4-10   Which is the preferred form of 3 tenths, .3 or 0.3?

Solution   Both forms are correct, but 0.3 is the preferred form of 3 tenths because the zero is included in the whole-number part.

C. Comparing decimals

To **compare two decimals,** you determine whether one decimal is greater (larger) or less (smaller) than the other or whether the two decimals are equal. You can use one of the following symbols to compare two decimals

       > (is greater than)

       < (is less than)

       = (is equal to)

To compare two decimals, you may find it helpful to write one decimal under the other while aligning the decimal points and like place values.

EXAMPLE 4-11   Compare the decimals 0.09 and 0.1.

Solution

Whole-Number Part							Decimal-Fraction Part					
millions	hundred thousands	ten thousands	thousands	hundreds	tens	ones	tenths	hundredths	thousandths	ten thousandths	hundred thousandths	millionths
						0 ·	0	9				
						0 ·	1					

Write one number under the other while aligning the decimal points and like place values.

     ↑ ↑

   same different (0 tenths < 1 tenth)

0.09 is less than 0.1; that is, 0.09 < 0.1. Or 0.1 is greater than 0.09; that is, 0.1 > 0.09.

Note: The inequality symbols < and > should always point at the smaller number.

D. Ordering decimals

To **order two or more decimals,** you list them from smallest to largest or from largest to smallest.

EXAMPLE 4-12   Order the following decimals from smallest to largest:

9.99,   15.29,   83.00,   7.05,   15.30,   9.49,   5.29,   83.01,   10.00,   15.28

Solution

$$
\begin{array}{l}
5.2\ 9 \longleftarrow \text{smallest decimal} \\
7.0\ 5 \\
9.4\ 9 \\
9.9\ 9 \\
1\ 0.0\ 0 \\
1\ 5.2\ 8 \\
1\ 5.2\ 9 \\
1\ 5.3\ 0 \\
8\ 3.0\ 0 \\
8\ 3.0\ 1 \longleftarrow \text{largest decimal}
\end{array}
$$

Note: To order two or more decimals from largest to smallest, you just reverse the smallest-to-largest order.

EXAMPLE 4-13   Order the following decimals from largest to smallest:

138.49,   138.50,   138.51,   39.49,   37.99,   38.00,   8.50,   30.00,   8.49,   18.51,   0.99,   0.09,   0.1

Solution

$$
\begin{array}{l}
1\ 3\ 8.5\ 1 \longleftarrow \text{largest decimal} \\
1\ 3\ 8.5\ 0 \\
1\ 3\ 8.4\ 9 \\
3\ 9.4\ 9 \\
3\ 8.0\ 0 \\
3\ 7.9\ 9 \\
3\ 0.0\ 0 \\
1\ 8.5\ 1 \\
8.5\ 0 \\
8.4\ 9 \\
0.9\ 9 \\
0.1 \\
0.0\ 9 \longleftarrow \text{smallest decimal}
\end{array}
$$

E. Rounding numbers

When you don't need an exact answer for a problem using decimals, you may want to make the computation easier and **round decimals** before computing. The place to which you round the decimals will depend on how accurate the answer needs to be.

If the digit to the right of the digit to be rounded is 5 or more, then **round up** as follows:

(**1**) Increase the digit to be rounded by 1.
(**2**) If the rounded digit is in the whole-number part, replace the digits to the right of the rounded digit with zeros in the whole-number part.
(**3**) Omit all the digits to the right of the rounded digit in the decimal-fraction part.

EXAMPLE 4-14   Round the decimal 3.14159 to the nearest thousandth.

thousandths
↓

Solution   3.14159 = 3.14159   Draw a line under the digit to be rounded.

= 3.14159   *Think:* Is the digit to the right 5 or more? Yes: 5 = 5.

≈ 3.142   Round up: Increase 1 by 1 to get 2.
Omit 5 and 9.

If the digit to the right of the digit to be rounded is less than 5, then **round down** as follows:

(1) Leave the digit to be rounded the same.

(2) If the rounded digit is in the whole-number part, replace the digits to the right of the rounded digit by zeros in the whole-number part.

(3) Omit all the digits to the right of the rounded digit in the decimal-fraction part.

EXAMPLE 4-15   Round the decimal 963.25 to the nearest ten.

Solution   963.25 = 963.25   Draw a line under the digit to be rounded.

= 963.25   *Think:* Is the digit to the right 5 or more? No: 3 < 5

≈ 960   Round down: Leave 6 the same.
Replace 3 with 0.
Omit 2 and 5.

4-2. Renaming with Decimals

A. Renaming whole numbers as decimals

Every whole number can be renamed as an equal decimal. To do this, you write the decimal point following the whole number, then write one or more zeros following the decimal point.

EXAMPLE 4-16   Rename the whole numbers **(a)** 8, **(b)** 10, and **(c)** 258 as decimals.

Solution

whole number   decimals
↓

(a)   8 =   8.0   or   8.00 or   8.000   (and so on)

(b)   10 =   10.0   or   10.00 or   10.000   (and so on)

(c) 258 = 258.0   or   258.00 or 258.000   (and so on)

Note: When the decimal-fraction part of a decimal is zero, the decimal can always be renamed as an equal whole number.

EXAMPLE 4-17   Rename the decimals **(a)** 5.0, **(b)** 23.00, and **(c)** 1.000 as equal whole numbers.

Solution

decimal   whole number
↓   ↓

(a)   5.0   =   5

(b) 23.00   =   23

(c)   1.000 =   1

B. Renaming decimals as fractions and mixed numbers

To rename a decimal fraction as an equal proper fraction, you first write the value of the decimal fraction, and then rename using tenths, hundredths, thousandths, . . . , as necessary.

Note: A value of tenths, hundredths, thousandths, . . . means the denominator of the proper fraction is 10, 100, 1000, . . . , respectively.

EXAMPLE 4-18 Rename the decimal fractions (a) 0.5, (b) 0.25, and (c) 0.001 as equal proper fractions in simplest form.

Solution

$$
\begin{array}{ccc}
\text{value} & \text{fraction} & \text{simplest form} \\
\downarrow & \downarrow & \downarrow
\end{array}
$$

(a) $0.5 = \overbrace{5 \text{ tenths}} = \frac{5}{10} = \frac{1}{2}$

(b) $0.25 = 25 \text{ hundredths} = \frac{25}{100} = \frac{1}{4}$

(c) $0.001 = 1 \text{ thousandth} = \frac{1}{1000}$

When a fraction has a denominator of 10, or 100, or 1000, . . . , you can rename that proper fraction as a equal decimal fraction using tenths, hundredths, thousandths, . . . , respectively.

EXAMPLE 4-19 Rename the fractions (a) $\frac{7}{10}$, (b) $\frac{3}{100}$, and (c) $\frac{581}{1000}$ as equal decimals.

Solution

$$
\begin{array}{cc}
\text{value} & \text{decimal} \\
\downarrow & \downarrow
\end{array}
$$

(a) $\frac{7}{10} = \overbrace{7 \text{ tenths}} = 0.7$

(b) $\frac{3}{100} \quad 3 \text{ hundredths} = 0.03$

(c) $\frac{581}{1000} = 581 \text{ thousandths} = 0.581$

Note: When a fraction has a denominator that is not 10, 100, or 1000, . . . , you can rename that fraction as an equal decimal also. The method for renaming such a fraction as an equal decimal is presented in Section 4-8.

When the whole-number part of a decimal is not zero, the decimal can be renamed as an equal mixed number using the short decimal-word name.

EXAMPLE 4-20 Rename the decimals (a) 2.8, (b) 49.05, and (c) 8.003 as equal mixed numbers in simplest form.

Solution

$$
\begin{array}{cccc}
 & \text{proper} & \text{mixed} & \text{simplest} \\
 & \text{fraction} & \text{number} & \text{form} \\
\text{decimal-word name} & \downarrow & \downarrow & \downarrow
\end{array}
$$

(a) $2.8 = \overbrace{2 \text{ and } 8 \text{ tenths}} = 2 \text{ and } \frac{8}{10} = 2\frac{8}{10} = 2\frac{4}{5}$

(b) $49.05 = 49 \text{ and } 5 \text{ hundredths} = 49 \text{ and } \frac{5}{100} = 49\frac{5}{100} = 49\frac{1}{20}$

(c) $8.003 = 8 \text{ and } 3 \text{ thousandths} = 8 \text{ and } \frac{3}{1000} = 8\frac{3}{1000}$

When the whole-number part of a decimal is not zero, the decimal can also be renamed as an equal improper fraction by first using place values.

EXAMPLE 4-21   Rename **(a)** 2.8, **(b)** 49.05, and **(c)** 8.003 as equal improper fractions in lowest terms.

Solution

$$\begin{array}{c} \text{place} \\ \text{value} \\ \downarrow \end{array} \qquad \begin{array}{c} \text{improper} \\ \text{fraction} \\ \downarrow \end{array} \quad \begin{array}{c} \text{lowest} \\ \text{terms} \\ \downarrow \end{array}$$

(a) $2.8 \ = \ \overbrace{28 \text{ tenths}} \ = \ \frac{28}{10} = \frac{14}{5}$

(b) $49.05 \ = \ 4905 \text{ hundredths} \ = \ \frac{4905}{100} = \frac{981}{20}$

(c) $8.003 = 8003 \text{ thousandths} = \frac{8003}{1000}$

C. Renaming decimals to get more decimal places

Each digit in the decimal-fraction part of a decimal counts as one **decimal place.**

EXAMPLE 4-22   Find the number of decimal places in the decimals **(a)** 0.70, **(b)** 21.5, and **(c)** 0.001.

Solution

(a) 0.70 has 2 decimal places because there are 2 digits in the decimal-fraction part.

(b) 21.5 has 1 decimal place because there is 1 digit in the decimal-fraction part.

(c) 0.001 has 3 decimal places because there are 3 digits in the decimal-fraction part.

To rename a decimal as an equal decimal containing more decimal places, you can add the necessary number of zeros on the right-hand side of the decimal-fraction part.

EXAMPLE 4-23   Rename 1.5 as an equal decimal with **(a)** 2 decimal places, **(b)** 3 decimal places, and **(c)** 4 decimal places.

Solution   **(a)** $1.5 = 1.50$ ⟵ 2 decimal places

(b) $1.5 = 1.500$ ⟵ 3 decimal places

(c) $1.5 = 1.5000$ ⟵ 4 decimal places

Note: Any number of zeros can be written on the right-hand side of the decimal-fraction part of a decimal without changing the value of that decimal.

4-3.   Adding with Decimals

A. Adding decimals

To join two or more decimal amounts together, you **add decimals.**

To add decimals, you can first rename as equal fractions and then add fractions.

EXAMPLE 4-24   Add the decimals 0.3 + 0.4 using fractions.

Solution $0.3 + 0.4 = 3 \text{ tenths} + 4 \text{ tenths}$ Rename as equal fractions.

$$= \frac{3}{10} + \frac{4}{10} \quad \text{⟵ equal fractions}$$

$$= \frac{7}{10} \qquad\qquad \text{Add fractions.}$$

$$= 7 \text{ tenths} \qquad \text{Rename as an equal decimal.}$$

$$= 0.7 \text{ ⟵ sum of } 0.3 + 0.4$$

It's much easier, however, to add decimals by first writing in vertical form and then adding as you would for whole numbers, remembering to write the decimal point in the answer.

EXAMPLE 4-25 Add the decimals 0.3 + 0.4 in vertical form.

Solution

(1) Write in vertical form.

$$
\begin{array}{r}
0.3 \\
+0.4 \\
\hline
\uparrow
\end{array}
$$

align
decimal
points

(2) Add as you would for whole numbers.

$$
\begin{array}{r}
0.3 \\
+0.4 \\
\hline
0\ 7
\end{array}
$$

$\uparrow\ \uparrow$
$3 + 4 = 7$
$0 + 0 = 0$

(3) Write the decimal point in the answer.

$$
\begin{array}{r}
0.3 \\
+0.4 \\
\hline
0.7
\end{array}
$$

write the decimal
point directly
below the other
decimal points

When decimals do not have the same number of decimal places, you'll usually find it easier to add if you first rename to get an equal number of decimal places in each decimal.

EXAMPLE 4-26 Add the decimals 0.5 + 2.73 + 18.2 + 1.125.

Solution

$$
\begin{array}{r}
\overset{1\ \ 1}{}\\
0.5\ 0\ 0\ \longleftarrow\ 0.5\ =\ 0.500 \\
2.7\ 3\ 0\ \longleftarrow\ 2.73\ =\ 2.730 \\
1\ 8.2\ 0\ 0\ \longleftarrow\ 18.2\ =\ 18.200 \\
+\ \ 1.1\ 2\ 5 \\
\hline
2\ 2.5\ 5\ 5
\end{array}
$$

Write in vertical form.
Rename to get an equal number of decimal places.
Add as for whole numbers.

Write the decimal point in the answer.

B. Adding decimals and whole numbers

To add decimals and whole numbers, you first rename each whole number as an equal decimal with the same number of decimal places as the given decimal and then add as before.

EXAMPLE 4-27 Add 258 + 63.49.

Solution

$$
\begin{array}{r}
\overset{1\ \ 1}{}\\
2\ 5\ 8.0\ 0\ \longleftarrow\ 258\ =\ 258.\ =\ 258.00 \\
+\ \ 6\ 3.4\ 9 \\
\hline
3\ 2\ 1.4\ 9
\end{array}
$$

Rename to get the same
number of decimal places.

C. Adding with money

To **add with money,** you first rename each dollar-and-cent amount as an equal decimal with two decimal places and then add as before.

EXAMPLE 4-28 Add $54.62 + $15.76 + 87¢ + $9 + 15¢ + 9¢.

Solution

$$
\begin{array}{r}
\overset{2\ 2\ 2}{}\\
\$5\ 4.6\ 2 \\
1\ 5.7\ 6 \\
0.8\ 7\ \longleftarrow\ 87¢ = 87\ \text{hundredths} = \$0.87 \\
9.0\ 0\ \longleftarrow\ \ \ \$9 = \$9.00\ \ \ \ \ \ \ = \$9.00 \\
0.1\ 5\ \longleftarrow\ 15¢ = 15\ \text{hundredths} = \$0.15 \\
0.0\ 9\ \longleftarrow\ \ 9¢ = \ 9\ \text{hundredths} = \$0.09 \\
\hline
\$8\ 0.4\ 9
\end{array}
$$

4-4. Subtracting with Decimals

A. Subtracting decimals

To take one decimal amount away from another, you **subtract decimals.**

To subtract decimals, you can first rename the decimals as equal fractions and then subtract fractions.

EXAMPLE 4-29   Subtract the decimals $0.7 - 0.3$ using fractions.

Solution $0.7 - 0.3 = 7$ tenths $- 3$ tenths   Rename as equal fractions.

$$= \frac{7}{10} - \frac{3}{10} \longleftarrow \text{ like fractions}$$

$$= \frac{4}{10}$$ Subtract like fractions.

$$= 4 \text{ tenths}$$ Rename as an equal decimal.

$$= 0.4 \longleftarrow \text{ difference of 0.7 and 0.3}$$

It's much easier, however, to subtract decimals by first writing in vertical form, then subtracting as you would for whole numbers, remembering to write the decimal point in the answer.

EXAMPLE 4-30   Subtract the decimals $0.7 - 0.4$ in vertical form.

Solution

(1) Write in vertical form.

$$\begin{array}{r} 0.7 \\ -0.4 \\ \hline \uparrow \end{array}$$

align
decimal
points

(2) Subtract as you would for whole numbers.

$$\begin{array}{r} 0.7 \\ -0.4 \\ \hline 0\ 3 \end{array}$$

$7 - 4 = 3$
$0 - 0 = 0$

(3) Write the decimal point in the answer.

$$\begin{array}{r} 0.7 \\ -0.4 \\ \hline 0.3 \end{array}$$

write the decimal
point directly
below the other
decimal points

When the minuend has more decimal places than the subtrahend, you'll usually find it easier to subtract if you first rename the subtrahend to get an equal number of decimal places in both decimals

EXAMPLE 4-31   Subtract the decimals $25.375 - 4.8$ in vertical form.

Solution
$$\begin{array}{r} \overset{4}{2}\overset{1}{5}.3\ 7\ 5 \\ -\ 4.8\ 0\ 0 \longleftarrow \ 4.8 = 4.800 \\ \hline 2\ 0.5\ 7\ 5 \end{array}$$

Write in vertical form.
Rename to get an equal number of decimal places.
Subtract as you would for whole numbers.
Write the decimal point in the answer.

Caution: When the subtrahend has more decimal places than the minuend, you must rename to get an equal number of decimal places in order to subtract.

EXAMPLE 4-32   Subtract the decimals $45.2 - 9.125$ in vertical form.

Solution

$$
\begin{array}{r}
{}^{3}\!\!{}_{1}\;{}^{1}\;{}^{9}\!\!{}_{1} \\
\cancel{4}\,5.\cancel{2}\cancel{0}\,0 \longleftarrow 45.2 = 45.200 \\
-\quad 9.1\,2\,5 \\
\hline
3\,6.0\,7\,5
\end{array}
$$

Rename to get the same number of decimal places.

B. Subtracting decimals and whole decimals

To subtract a whole number from a decimal, you must first rename the whole number as an equal decimal with the same number of decimal places as the given decimal.

EXAMPLE 4-33   Subtract 185.36 − 78.

Solution

$$
\begin{array}{r}
{}^{7}\!\!{}_{1} \\
1\,\cancel{8}\,5.3\,6 \\
-\quad 7\,8.0\,0 \longleftarrow 78 = 78.00 \\
\hline
1\,0\,7.3\,6
\end{array}
$$

Rename to get the same number of decimal places.

Caution: To subtract a decimal from a whole number, you must first rename the whole number as an equal decimal with the same number of decimal places as the given decimal.

EXAMPLE 4-34   Subtract 5 − 0.008.

Solution

$$
\begin{array}{r}
{}^{4}\;{}^{9}\;{}^{9}\!\!{}_{1} \\
\cancel{5}.\cancel{0}\cancel{0}\,0 \longleftarrow 5 = 5.000 \\
-0.0\,0\,8 \\
\hline
4.9\,9\,2
\end{array}
$$

Rename to get the same number of decimal places in order to subtract.

C. Subtracting with money

To **subtract with money,** you first rename each dollar-and-cent amount as an equal decimal with two decimal places and then subtract as before.

EXAMPLE 4-35   Subtract $18 − 5¢.

Solution

$$
\begin{array}{r}
{}^{7}\;{}^{9}\!\!{}_{1} \\
\$1\,\cancel{8}.\cancel{0}\,0 \longleftarrow \$18 = \$18.00 \\
-\quad 0.0\,5 \qquad 5¢ = 5 \text{ hundredths} = \$0.05 \\
\hline
\$1\,7.9\,5
\end{array}
$$

4-5. Multiplying with Decimals

A. Multiplying a decimal by a whole number

To join two or more equal decimal amounts together, you **multiply the decimal** amount by the appropriate whole number.

EXAMPLE 4-36   Write "3 equal amounts of 0.4" as multiplication.

Solution   3 equal amounts of 0.4 means 0.4 + 0.4 + 0.4 or 3 × 0.4 ⟵ multiplication

To multiply a decimal by a whole number, you can first rename both as equal fractions.

EXAMPLE 4-37   Multiply 3 × 0.4 using fractions.

Solution $3 \times 0.4 = 3 \times 4$ tenths   Rename as equal fractions.

$$= \frac{3}{1} \times \frac{4}{10} \longleftarrow \text{equal fractions}$$

$$= \frac{3 \times 4}{1 \times 10}$$ Multiply fractions.

$$= \frac{12}{10}$$

$$= 12 \text{ tenths}$$

$$= 1.2 \longleftarrow 3 \times 0.4$$ Rename as an equal decimal.

It is much easier to multiply a decimal by a whole number by first writing in vertical form, multiplying as you would for whole numbers, and then writing the decimal point in the product.

Note: The number of decimal places in the product should equal the total number of decimal places contained in the two factors.

EXAMPLE 4-38   Multiply 3×0.4 in vertical form.

Solution

(1) Write in vertical form.   **(2)** Multiply as you would for whole numbers.
$3 \times 4 = 12$ **(3)** Write the decimal point in the product.

align right-hand digits

$$\begin{array}{r} 0.4 \\ \times \quad 3 \\ \hline \end{array}$$

$$\begin{array}{r} 1 \\ 0.4 \\ \times \quad 3 \\ \hline 1\,2 \end{array}$$

$$\begin{array}{r} 0.4 \longleftarrow \text{1 decimal place} \\ \times \quad 3 \longleftarrow \text{0 decimal places} \\ \hline 1.2 \longleftarrow \text{1 (} = 1 + 0\text{) decimal places} \end{array}$$

Multiplication Rule for Decimals

(1) Write in vertical form while aligning the right-hand digits.

(2) Multiply as you would for two whole numbers.

(3) Put the same number of decimal places in the product as the total number of decimal places contained in both factors.

EXAMPLE 4-39   Multiply 25×3.49.

Solution

(1) Write in vertical form.   **(2)** Multiply as you would for whole numbers.
$25 \times 349 = 8725$ **(3)** Write the decimal point in the product.

align right-hand digits

$$\begin{array}{r} 3.4\,9 \\ \times \quad 2\,5 \\ \hline \end{array}$$

$$\begin{array}{r} 1 \\ 2\,4 \\ 3.4\,9 \\ \times \quad 2\,5 \\ \hline 1\,7\,4\,5 \\ 6\,9\,8 \\ \hline 8\,7\,2\,5 \end{array}$$

$$\begin{array}{r} 3.4\,9 \longleftarrow \text{2 decimal places} \\ \times \quad 2\,5 \longleftarrow \text{0 decimal places} \\ \hline 1\,7\,4\,5 \\ 6\,9\,8 \\ \hline 8\,7.2\,5 \longleftarrow \text{2 (} = 2 + 0\text{) decimal places} \end{array}$$

Note: When you multiply a decimal by a whole number, the product will always have the same number of decimal places as the decimal factor because a whole number always has 0 decimal places.

B. Multiplying with decimals

When both factors are decimals, you must add the number of decimal places in each factor together to find the number of decimal places that belong in the product.

EXAMPLE 4-40   Multiply 5.14 × 23.6.

Solution

(1) Write in vertical form.

(2) Multiply as you would for whole numbers.
514 × 236 = 121304

(3) Write the decimal point in the product.

$$
\begin{array}{r}
2\ 3.6 \\
\times 5.1\ 4 \\
\hline
\end{array}
$$

$$
\begin{array}{r}
{}^{1\ 3} \\
{}^{1\ 2} \\
2\ 3.6 \\
\times 5.1\ 4 \\
\hline
9\ 4\ 4 \\
2\ 3\ 6 \\
1\ 1\ 8\ 0 \\
\hline
1\ 2\ 1\ 3\ 0\ 4
\end{array}
$$

$$
\begin{array}{r}
2\ 3.6 \quad \longleftarrow \ 1 \text{ decimal place} \\
\times 5.1\ 4 \quad \longleftarrow \ 2 \text{ decimal places} \\
\hline
9\ 4\ 4 \\
2\ 3\ 6 \\
1\ 1\ 8\ 0 \\
\hline
1\ 2\ 1.3\ 0\ 4 \quad \longleftarrow \ 3\,(=1+2) \text{ decimal places}
\end{array}
$$

C. Writing zeros in the product

When the number of decimal places that belong in the product is greater than the number of digits in the product, you must write zeros in the product until there are enough digits to write the decimal point in the correct place.

EXAMPLE 4-41   Multiply 0.02 × 0.03.

Solution

(1) Write in vertical form.

(2) Multiply as you would for whole numbers.
2 × 3 = 6

(3) Write the decimal point in the product

$$
\begin{array}{r}
0.0\ 3 \\
\times 0.0\ 2 \\
\hline
\end{array}
$$

$$
\begin{array}{r}
0.0\ 3 \\
\times 0.0\ 2 \\
\hline
6
\end{array}
$$

write zeros ⟶
as needed

$$
\begin{array}{r}
0.0\ 3 \quad \longleftarrow \ 2 \text{ decimal places} \\
\times 0.0\ 2 \quad \longleftarrow \ 2 \text{ decimal places} \\
\hline
0.0\ 0\ 0\ 6 \quad \longleftarrow \ 4\,(=2+2) \text{ decimal places}
\end{array}
$$

Note: In this example 3 zeros are needed to write the decimal point in the product and a 4th zero is needed to show that the whole-number part of the product is zero.

D. Multiplying by 10, 100, and 1000

To multiply a decimal by 10, you can proceed as shown in Examples 4-38 and 4-39.

EXAMPLE 4-42   Multiply the decimal 7.49 by 10.

Solution

(1) Write in vertical form.

$$
\begin{array}{r}
7.4\ 9 \\
\times \quad 1\ 0 \\
\hline
\end{array}
$$

(2) Multiply as before.

$$
\begin{array}{r}
7.4\ 9 \quad \longleftarrow \ 2 \text{ decimal places} \\
\times \quad 1\ 0 \quad \longleftarrow \ 0 \text{ decimal places} \\
\hline
7\ 4.9\ 0 \quad \longleftarrow \ 2\,(=2+0) \text{ decimal places}
\end{array}
$$

Shortcut: To multiply 7.49 by 10, you could have moved the decimal point one decimal place to the right: 10 × 7.49 = 74.9

Multiply by 10 Rule

To multiply by 10, you move the decimal point one decimal place to the right.

EXAMPLE 4-43   Multiply the decimals (**a**) 1.89, (**b** 3.2, (**c**) 0.25, and (**d**) 5 by 10.

Solution

(**a**) $10 \times 1.89 = 18.9 = 18.9$

(**b**) $10 \times 3.2 = 32. = 32$

(**c**) $10 \times 0.25 = 02.5 = 2.5$

(**d**) $10 \times 5 = 50. = 50$

To multiply a decimal by 100, you can proceed as shown in Examples 4-38 and 4-39.

EXAMPLE 4-44   Multiply the decimal 0.125 by 100.

Solution

(**1**) Write in vertical form.

$$\begin{array}{r} 0.1\,2\,5 \\ \times\quad 1\,0\,0 \\ \hline \end{array}$$

(**2**) Multiply as before.

$$\begin{array}{r} 0.1\,2\,5 \leftarrow \text{3 decimal places} \\ \times\quad 1\,0\,0 \leftarrow \text{0 decimal places} \\ \hline 1\,2.5\,0\,0 \leftarrow 3(=3+0)\ \text{decimal places)} \end{array}$$

Shortcut: To multiply 0.125 by 100, you could have moved the decimal point two decimal places to the right: $100 \times 0.125 = 012.5 = 12.5$

Multiply by 100 Rule

To multiply by 100, you move the decimal point two decimal places to the right.

EXAMPLE 4-45   Multiply the decimals (**a**) 0.059, (**b**) 0.02, (**c**) 2.5, and (**d**) 39 by 100.

Solution

(**a**) $100 \times 0.059 = 005.9 = 5.9$

(**b**) $100 \times 0.02 = 002. = 2$

(**c**) $100 \times 2.5 = 250 = 250. = 250$

(**d**) $100 \times 39 = 100 \times 39. = 3900. = 3900$

To multiply by 1000, you can proceed as shown in Examples 4-38 and 4-39.

EXAMPLE 4-46   Multiply 3.14159 by 1000.

Solution

(**1**) Write in vertical form.

$$\begin{array}{r} 3.1\,4\,1\,5\,9 \\ \times\quad 1\,0\,0\,0 \\ \hline \end{array}$$

(**2**) Multiply as before.

$$\begin{array}{r} 3.1\,4\,1\,5\,9 \quad \text{5 decimal places} \\ \times\quad 1\,0\,0\,0 \quad \text{0 decimal places} \\ \hline 3\,1\,4\,1.5\,9\,0\,0\,0 \quad 5(=5+0)\ \text{decimal places} \end{array}$$

Shortcut: To multiply 3.14159 by 1000, you could have moved the decimal point three decimal places to the right: $1000 \times 3.14159 = 3141.59$

Multiply by 1000 Rule

To multiply by 1000, you move the decimal point three decimal places to the right.

EXAMPLE 4-47   Multiply the decimals numbers (**a**) 0.2576, (**b**) 0.002, (**c**) 1.3, and (**d**) 8 by 1000.

Solution

(**a**) $1000 \times 0.2576 = 0257.6 = 257.6$

(**b**) $1000 \times 0.002 = 0002. = 0002 = 2$

(**c**) $1000 \times 1.3 = 1300. = 1300$

(**d**) $1000 \times 8 = 8000. = 8000$

The whole numbers 10, 100, 1000, . . . , are called **powers of 10.** The following rule is a basic rule for multiplying by 10, 100, 1000, or any power of 10.

Multiply by Powers of 10 Rule

To multiply by a power of 10, move the decimal point one decimal place to the right for each zero in the power of 10.

EXAMPLE 4-48 Multiply 0.75 by **(a)** 10, **(b)** 100, **(c)** 1000, and **(d)** 10,000.

Solution

(a) $10 \times 0.75 = 07.5 = 7.5$ One zero in 10 means 1 decimal place to the right.

(b) $100 \times 0.75 = 075. = 75$ Two zeros in 100 means 2 decimal places to the right.

(c) $1000 \times 0.75 = 0750. = 750$ Three zeros in 1000 means 3 decimal places to the right.

(d) $10,000 \times 0.75 = 07500. = 7500$ Four zeros in 10,000 means 4 decimal places to the right.

4-6. Dividing with Decimals

A. Dividing a decimal by a whole number

To separate a decimal amount into two or more equal amounts, you **divide the decimal amount** by the appropriate whole number.

EXAMPLE 4-49 Write "1.2 separated into 3 equal amounts" as division.

Solution 1.2 separated into 3 equal amounts means $1.2 \div 3 \longleftarrow$ division

To divide a decimal by a whole number, you can first rename the decimal and whole number as equal fractions and then divide fractions

EXAMPLE 4-50 Divide $1.2 \div 3$ using fractions.

Solution $\quad 1.2 \div 3 = 12 \text{ tenths} \div 3 \qquad$ Rename as equal fractions.

$$= \frac{12}{10} \div \frac{3}{1} \longleftarrow \text{equal fractions}$$

$$= \frac{12}{10} \times \frac{1}{3} \qquad \text{Divide fractions.}$$

$$= \frac{\cancel{3} \times 4}{10} \times \frac{1}{\cancel{3}}$$

$$= \frac{4}{10}$$

$$= 4 \text{ tenths}$$

$$= 0.4 \longleftarrow 1.2 \div 3 \qquad \text{Rename as an equal decimal.}$$

It is much easier to divide a decimal by a whole number by first writing in division box form, dividing as you would for whole numbers, and then writing the decimal point in the quotient.

EXAMPLE 4-51 Divide $1.2 \div 3$ in division box form.

Solution

(1) Write in division box form.

$$3\overline{)1.2}$$

(2) Divide as you would for whole numbers.
$$12 \div 3 = 4$$

$$\begin{array}{r} 4 \\ 3\overline{)\ 1.2} \\ -1\ 2 \\ \hline 0 \end{array}$$

(3) Write the decimal point in the quotient.

$$\begin{array}{r} 0.4 \\ 3\overline{)\ 1.2} \\ -1\ 2 \\ \hline 0 \end{array}$$ ⟶ 1 decimal place

Note: When you divide a decimal by a whole number, the number of decimal places in the quotient should equal number of decimal places in the decimal dividend.

Recall: To check division, you multiply the proposed quotient by the original divisor to see if you get the original dividend.

EXAMPLE 4-52   Check the division problem $1.2 \div 3 = 0.4$.

Solution

$$\begin{array}{r} 0.4 \\ \times\quad 3 \\ \hline 1.2 \end{array}$$

0.4 ⟵ proposed quotient
× 3 ⟵ original divisor
1.2 ⟵ original dividend (0.4 checks)

B. Dividing with decimals

To divide by a decimal, you can first rename the dividend and the divisor as equal fractions.

EXAMPLE 4-53   Divide $0.12 \div 0.3$ using fractions.

Solution $0.12 \div 0.3 = 12$ hundredths $\div 3$ tenths   Rename as equal fractions.

$$= \frac{12}{100} \div \frac{3}{10} \longleftarrow \text{equal fractions}$$

$$= \frac{12}{100} \times \frac{10}{3}$$ Divide fractions.

$$= \frac{\cancel{3} \times 4}{\cancel{10} \times 10} \times \frac{\cancel{10}}{\cancel{3}}$$

$$= \frac{4}{10}$$

$$= 4 \text{ tenths}$$

$$= 0.4 \longleftarrow 0.12 \div 0.3$$ Rename as an equal decimal.

It is much easier to divide by a decimal in division box form.

Caution: To divide by a decimal in division box form, you must first make the divisor a whole number by multiplying both the divisor and dividend by the same power of 10.

EXAMPLE 4-54   Divide $0.12 \div 0.3$ in division box form.

Solution

(1) Write in division box form.

$$0.3\overline{)0.1\ 2}$$

(2) Make the divisor a whole number.

whole-number divisor

$$0.3\overline{)0.1\,2} = 3\overline{)1.2} \longleftarrow 10 \times 0.12$$

$$10 \times 0.3$$

(3) Divide by a whole number.

$$
\begin{array}{r}
0.4 \\
3\overline{)\ 1.2} \\
-1\ 2 \\
\hline
0
\end{array}
$$ → 1 decimal place

(4) Check.

$$
\begin{array}{r}
0.4 \longleftarrow \text{proposed quotient} \\
\times\ \ 0.3 \longleftarrow \text{original divisor} \\
\hline
0.1\ 2 \longleftarrow \text{original dividend (0.4 checks)}
\end{array}
$$

Caution: If the decimal divisor has to be multiplied by a power of 10 to get a whole-number divisor, then the dividend must also be multiplied by the same power of 10. For example,

$$0.12 \div 0.3 = 1.2 \div 3 = 0.4$$

Multiplying both the divisor and dividend by the same nonzero number will not change the quotient of the original problem.

Division Rule for Decimals

(1) Write in division box form.

(2) If the divisor is a decimal, then make the divisor a whole number by multiplying both the divisor and the dividend by the same power of 10.

(3) Divide as you would for whole numbers.

(4) Put the same number of decimal places in the quotient as in the decimal dividend from Step 2.

EXAMPLE 4-55 Divide $1.71 \div 0.025$.

Solution

(1) Write in division box form.

$$0.025\overline{)1.7\ 1}$$

(2) Make the divisor a whole number.

whole-number divisor

$$0.025\overline{)1.7\ 1\ 0} = 25\overline{)1\ 7\ 1\ 0.} \longleftarrow 1000 \times 1.71$$

$$1000 \times 0.025$$

(3) Divide as you would for whole numbers.
$$1710 \div 25 = 684$$

$$
\begin{array}{r}
6\ 8\ 4 \\
25\overline{)\ 1\ 7\ 1\ 0.0} \longleftarrow 1710 = 1710.0 \\
-1\ 5\ 0 \\
\hline
2\ 1\ 0 \\
-2\ 0\ 0 \\
\hline
1\ 0\ 0 \\
-1\ 0\ 0 \\
\hline
0
\end{array}
$$

(4) Write the decimal point in the quotient.

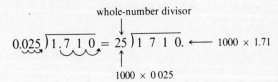

$$
\begin{array}{r}
6\ 8.4 \\
25\overline{)1\ 7\ 1\ 0.0}
\end{array}
$$ → 1 decimal place

Note: $0.025 \times 68.4 = 1.71$ (68.4 checks)

C. Writing zeros in the quotient

When the number of decimal places that belong in the quotient is greater than the number of digits in the quotient, you must write zeros in the quotient until there are enough digits to write the decimal point in the correct place.

EXAMPLE 4-56 Divide $0.0001 \div 0.02$.

Solution

(1) Write in division box form.

$$0.02 \overline{)0.0\ 0\ 0\ 1}$$

(2) Make the divisor a whole number.

whole-number divisor

$$0.02 \overline{)0.0001} = 2\overline{)0.01} \longleftarrow 100 \times 0.0001$$

100×0.02

(3) Divide as you would for whole numbers.
$$10 \div 2 = 5$$

$$\begin{array}{r} 5 \\ 2\overline{)0.0\ 1\ 0} \\ -1\ 0 \\ \hline 0 \end{array}$$

(4) Write the decimal point in the quotient and write zeros as needed.

write zeros

$$\overbrace{0.0\ 0\ 5} \\ 2\overline{)0.0\ 1\ 0} \longrightarrow 3 \text{ decimal places}$$

Note: $0.0001 \div 0.02 = 0.01 \div 2 = 0.005$ and $0.02 \times 0.005 = 0.0001$ (0.005 checks)

D. Rounding the quotient

When you divide decimals and get more digits than you want in the quotient, you will usually want to round the quotient. The place to which you round the quotient will depend on how accurate your answer needs to be.

To divide and round the quotient to a given place, you must carry out the division to one place beyond the given place and then round the quotient (see Examples 4-14 and 4-15).

EXAMPLE 4-57 Divide $0.23 \div 0.3$ and round the quotient to the nearest tenth.

Solution

(1) Make the divisor a whole number.

$$0.3 \overline{)0.2\,3} = 3\overline{)2.3}$$

(2) Divide one place beyond the given place.

hundredths place

$$\begin{array}{r} 0.7\ 6 \\ 3\overline{)\ 2.3\ 0} \\ -2\ 1 \\ \hline 2\ 0 \\ -1\ 8 \\ \hline 2 \end{array}$$

(3) Round to the given place.

nearest tenth

$$\begin{array}{r} 0.7\ 6 \approx 0.8 \\ 3\overline{)2.3\ 0} \end{array}$$

Note: $0.3 \times 0.8 = 0.24 \approx 0.23$ (0.8 checks)

E. Dividing by 10, 100, and 1000

To divide a decimal by 10, you can proceed as shown in Example 4-51.

EXAMPLE 4-58   Divide 25.8 by 10.

Solution

(1) Write in division
box form.

$$10 \overline{)2\ 5.8}$$

(2) Divide as before.

$$
10 \overline{)\ 2\ 5.8\ 0} \quad \begin{array}{l} 2.5\ 8 \leftarrow \\ \\ \end{array} \text{2 decimal places}
$$

$$
\begin{array}{r}
2.5\ 8 \\
10\,\overline{)\ 2\ 5.8\ 0} \\
-2\ 0 \\
\hline
5\ 8 \\
-5\ 0 \\
\hline
8\ 0 \\
-8\ 0 \\
\hline
0
\end{array}
$$

(3) Check.

$10 \times 2.58 = 25.8$ (2.58 checks)

Note: To divide 25.8 by 10, you could have moved the decimal point one decimal place to the left:
$25.8 \div 10 = 25.8 = 2.58$

Divide by 10 Rule

To divide by 10, you move the decimal point one decimal place to the left.

EXAMPLE 4-59   Divide **(a)** 63, **(b)** 2.35, **(c)** 50, and **(d)** 200 by 10.

Solution

(a) $63 \div 10 = 63.0 \div 10 = 6.30 = 6.3$

(b) $2.35 \div 10 = .235 = 0.235$

(c) $50 \div 10 = 5.0 = 5. = 5$ or $5\emptyset \div 1\emptyset = 5 \div 1 = 5$

(d) $200 \div 10 = 20.0 = 20. = 20$ or $20\emptyset \div 1\emptyset = 20 \div 1 = 20$

To divide a decimal by 100, you can proceed as shown in Example 4-51.

EXAMPLE 4-60   Divide 2.5 by 100.

Solution

(1) Write in division box form.

$$100 \overline{)2.5}$$

(2) Divide as before.

$$
\begin{array}{r}
0.0\ 2\ 5 \leftarrow \\
100\,\overline{)\ 2.5\ 0\ 0} \quad \text{3 decimal places} \\
-2\ 0\ 0 \\
\hline
5\ 0\ 0 \\
-5\ 0\ 0 \\
\hline
0
\end{array}
$$

(3) Check.

$100 \times 0.025 = 2.5$
(0.025 checks)

Note: To divide 2.5 by 100, you could have moved the decimal point two decimal places to the left:
$2.5 \div 100 = .0250 = 0.025$.

Divide by 100 Rule

To divide by 100, you move the decimal point two decimal places to the left.

EXAMPLE 4-61   Divide **(a)** 275, **(b)** 37.5, **(c)** 40, and **(d)** 600 by 100.

Solution

(a) $275 \div 100 = 275.0 \div 100 = 2.750 = 2.75$

(b) $37.5 \div 100 = .375 = 0.375$

(c) $40 \div 100 = 40. \div 100 = .40 = 0.4$

(d) $600 \div 100 = 600. \div 100 = 6.00 = 6. = 6$ or $6\emptyset\emptyset \div 1\emptyset\emptyset = 6 \div 1 = 6$

To divide a decimal by 1000, you can proceed as shown in Example 4-51.

EXAMPLE 4-62 Divide 635.2 by 1000 in division box form.

Solution

(1) Write in division box form.

$$1000\overline{)6\ 3\ 5.2}$$

(2) Divide as before.

$$
\begin{array}{r}
0.6\ 3\ 5\ 2 \\
1000\overline{)6\ 3\ 5.2\ 0\ 0\ 0} \\
-6\ 0\ 0\ 0 \\
\hline
3\ 5\ 2\ 0 \\
-3\ 0\ 0\ 0 \\
\hline
5\ 2\ 0\ 0 \\
-5\ 0\ 0\ 0 \\
\hline
2\ 0\ 0\ 0 \\
-2\ 0\ 0\ 0 \\
\hline
0
\end{array}
$$

4 decimal places

(3) Check.

$1000 \times 0.6352 = 635.2$

(0.6352 checks)

Note: To divide 635.2 by 1000, you could have moved the decimal point three decimal places to the left:
$635.2 \div 1000 = .6352 = 0.6352$

Divide by 1000 Rule

To divide by 1000, you move the decimal point three decimal places to the left.

EXAMPLE 4-63 Divide (a) 85.7, (b) 9, (c) 1280, and (d) 500 by 1000.

Solution

(a) $85.7 \div 1000 = .0857 = 0.0857$

(b) $9 \div 1000\ \ \ = 9. \div 1000\ \ \ = .009\ \ = 0.009$

(c) $1280 \div 1000 = 1280. \div 1000 = 1.280 = 1.28$

(d) $5000 \div 1000 = 5000. \div 1000 = 5.000 = 5$ or $5\emptyset\emptyset\emptyset \div 1\emptyset\emptyset\emptyset = 5 \div 1 = 5$

The following rule is a basic rule for dividing by 10, 100, 1000, or any power of 10.

Divide by Powers of 10 Rule

To divide by a power of 10, you move the decimal point one decimal place to the left for each zero in the power of 10.

EXAMPLE 4-64 Divide 3.8 by (a) 10, (b) 100, (c) 1000, and (d) 10,000.

Solution

(a) $3.8 \div 10\ \ \ \ = .38\ \ \ = 0.38$ One zero in 10 means 1 decimal place to the left.

(b) $3.8 \div 100\ \ \ = .038\ \ = 0.038$ Two zeros in 100 means 2 decimal places to the left.

(c) $3.8 \div 1000\ = .0038 = 0.0038$ Three zeros in 1000 means 3 decimal places to the left.

(d) $3.8 \div 10,000 = .00038 = 0.00038$ Four zeros in 10,000 means 4 decimal places to the left.

4-7. Finding the Average (Mean)

A. Finding the average (mean) of two or more given numbers

To find the **average (mean)** of two or more given numbers:

(1) Add the numbers.

(2) Divide the sum from Step 1 by the number of given numbers.

EXAMPLE 4-65   Find the average (mean) of 82, 9.1, 36, 0.5, and 1.3.

Solution

(1) Find the sum.   (2) Divide the sum by the   (3) Check.
                          number of given numbers.

$$
\begin{array}{r}
1\\
8\,2.0\\
9.1\\
3\,6.0\\
0.5\\
+\ \ 1.3\\
\hline
1\,2\,8.9
\end{array}
$$

number of given numbers $\longrightarrow$ 5) 1 2 8.9 0 $\longleftarrow$ sum, 2 5.7 8 $\longleftarrow$ average

$$
\begin{array}{r}
2\,5.7\,8\\
5\,\overline{)\,1\,2\,8.9\,0}\\
-1\,0\\
\hline
2\,8\\
-2\,5\\
\hline
3\,9\\
-3\,5\\
\hline
4\,0\\
-4\,0\\
\hline
0
\end{array}
$$

$$
\begin{array}{r}
2\ 3\ 4\\
2\ 5.7\ 8\\
\times\qquad 5\\
\hline
1\ 2\ 8.9\ 0
\end{array}
$$
$\longleftarrow$ 25.78 checks

The average (mean) of 82, 9.1, 36, 0.5, and 1.3 is 25.78.

B. Finding the average (mean) of a given number of addends, given the sum

To find the average (mean) of a given number of addends when you know the sum, you divide the sum by the number of addends.

EXAMPLE 4-66   Find the average (mean) of 4 numbers whose sum is 0.1.

0.0 2 5 $\longleftarrow$ average (mean)   Check as before.

number of addends $\longrightarrow$ 4) 0.1 0 0 $\longleftarrow$ sum

$$
\begin{array}{r}
0.0\,2\,5\\
4\,\overline{)\,0.1\,0\,0}\\
-\quad 8\\
\hline
2\,0\\
-2\,0\\
\hline
0
\end{array}
$$

4-8. Renaming Fractions as Decimals

To rename certain fractions as decimals, you can first rename the fraction with a denominator of 10, 100, 1000, . . . , and then rename as a decimal, as in Example 4-19.

EXAMPLE 4-67   Rename $\frac{2}{5}$ as a decimal by renaming first as a fraction with a denominator of 10.

Solution $\dfrac{2}{5} = \dfrac{2}{5} \times 1$ Multiply by 1.

$= \dfrac{2}{5} \times \dfrac{2}{2}$ *Think:* $1 = \dfrac{2}{2}$

$$= \frac{2 \times 2}{5 \times 2} \qquad \text{Multiply fractions.}$$

$$= \frac{4}{10}$$

$$= 4 \text{ tenths}$$

$$= 0.4 \longleftarrow \tfrac{2}{5} \qquad \text{Rename as an equal decimal [see Example 4-19].}$$

It is easier to rename a fraction as an equal decimal by first renaming the fraction as a division problem, and then dividing to get a decimal quotient.

EXAMPLE 4-68   Rename $\frac{2}{5}$ as an equal decimal using division.

Solution $\dfrac{2}{5} = 2 \div 5 \quad \text{or} \quad 5\overline{)2} \qquad \qquad \text{Rename as divison.}$

$5\overline{)2.0} \qquad \qquad \text{Rename the dividend as a decimal.}$

$\begin{array}{r} 0.4 \longleftarrow \tfrac{2}{5} \\ 5\overline{)\,2.0} \\ -2\,0 \\ \hline 2\,0 \end{array} \qquad \text{Divide to get a decimal quotient.}$

The decimal 0.4 is called a **terminating decimal** because it ends in a given place (the tenths place). A decimal that does not end in any given place is called a **nonterminating decimal.**

EXAMPLE 4-69   Which of the following decimals is a nonterminating decimal?

(**a**) 0.3   (**b**) 0.33   (**c**) 0.333   (**d**) 0.333 . . .

Solution

(**a**) 0.3 is a terminating decimal because it ends in the tenths place.

(**b**) 0.33 is a terminating decimal because it ends in the hundredths place.

(**c**) 0.333 is a terminating decimal because it ends in the thousandths place.

(**d**) 0.333 . . . is a nonterminating decimal because the ellipsis symbol (. . .) indicates that the 3s continue on forever.

A nonterminating decimal (like 0.333 . . .) that repeats one or more digits forever is called a **repeating decimal.**

EXAMPLE 4-70   (**a**) Is 1.234 . . . a repeating decimal? (**b**) Is 27.2727 . . . a repeating decimal?

Solution

(**a**) 1.234 . . . is not a repeating decimal because the counting pattern shown goes on forever, but no set of digits repeats: 1.23456789101112 13 . . .

(**b**) 27.2727 . . . is a repeating decimal because the digits 2 and 7 repeat forever in the same way.

Repeating decimals (like 0.666 . . . and 27.2727 . . .) can be written with **bar notation** instead of ellipsis notation. To write a repeating decimal using bar notation, you draw the bar over each digit in the decimal fraction part that repeats.

EXAMPLE 4-71   Write (**a**) 0.666 . . . , (**b**) 27.2727 . . . , and (**c**) 0.8333 . . . , using bar notation.

Solution

(**a**) $0.666 \ldots = 0.\overline{6}$

(**b**) $27.2727 \ldots = 27.\overline{27} \longleftrightarrow$ bar notation

(**c**) $0.8333 \ldots = 0.8\overline{3}$

Note: $0.666\ldots = \overset{\text{simplest form}}{\overline{0.6}} = 0.6\overline{6} = 0.66\overline{6} = \cdots$

Caution: $27.2727\ldots$ cannot be written as $\overline{27}$ ← wrong

because $\overline{27} = 27272727\ldots$

but $27.272727\ldots$ can be written as $27.\overline{27}$

Caution: Do not write $0.8333\ldots$ as $0.\overline{83}$ ← wrong

because $0.\overline{83} = 0.838383\ldots$

but $0.8\overline{3} = 0.8333\ldots$

Every fraction can be renamed as an equal terminating decimal or as an equal repeating decimal by dividing the denominator into the numerator to get the decimal quotient.

EXAMPLE 4-72 Rename the following fractions as equal decimals:

(a) $\dfrac{1}{2}$ **(b)** $\dfrac{3}{4}$ **(c)** $\dfrac{5}{8}$ **(d)** $\dfrac{1}{3}$ **(e)** $\dfrac{5}{6}$

Solution

(a) $\dfrac{1}{2} = 1 \div 2$ or $2\overline{)1}$:

$$
\begin{array}{r}
0.5 \quad \longleftarrow \tfrac{1}{2}\text{ as an equal decimal}\\
2\overline{)\ 1.0}\\
\underline{-1\ 0}\\
0
\end{array}
$$

(b) $\dfrac{3}{4} = 3 \div 4$ or $4\overline{)3}$:

$$
\begin{array}{r}
0.7\ 5 \quad \longleftarrow \tfrac{3}{4}\text{ as an equal decimal}\\
4\overline{)\ 3.0\ 0}\\
\underline{-2\ 8}\\
2\ 0\\
\underline{-2\ 0}\\
0
\end{array}
$$

(c) $\dfrac{5}{8} = 5 \div 8$ or $8\overline{)5}$:

$$
\begin{array}{r}
0.6\ 2\ 5 \quad \longleftarrow \tfrac{5}{8}\text{ as an equal decimal}\\
8\overline{)\ 5.0\ 0\ 0}\\
\underline{-4\ 8}\\
2\ 0\\
\underline{-1\ 6}\\
4\ 0\\
\underline{-4\ 0}\\
0
\end{array}
$$

(d) $\dfrac{1}{3} = 1 \div 3$ or $3\overline{)1}$:

$$
\begin{array}{r}
0.3\ 3\ 3\ldots = 0.\overline{3} \quad \longleftarrow \tfrac{1}{3}\text{ as an equal decimal}\\
3\overline{)\ 1.0\ 0\ 0}\\
\underline{-\ 9}\\
1\,0\\
\underline{-\ 9}\\
1\,0\\
\underline{-\ 9}\\
1
\end{array}
$$

repeating 1s

(e) $\dfrac{5}{6} = 5 \div 6$ or $6\overline{)5}$:

$$
\begin{array}{r}
0.8\ 3\ 3\ 3\ldots = 0.8\overline{3} \quad \longleftarrow \tfrac{5}{6}\text{ as an equal decimal}\\
6\overline{)\ 5.0\ 0\ 0\ 0}\\
\underline{-4\ 8}\\
2\,0\\
\underline{-1\ 8}\\
2\,0\\
\underline{-1\ 8}\\
2
\end{array}
$$

repeating 2s

4-9. Evaluating Powers and Square Roots

A. Writing power notation

Recall: Multiplication is a short way to write a sum of repeated addends.

EXAMPLE 4-73 Write $2 + 2 + 2 + 2$ as a multiplication problem.

Solution

repeated addend

number of repeated addends

$2 + 2 + 2 + 2 = 2 \times 4$ ⟵ multiplication problem (4 repeated addends of 2)

Recall: In 2×4, 2 and 4 are called factors.

Power notation or **exponential notation** is a short way to write a product of **repeated factors.**

EXAMPLE 4-74 Write $2 \times 2 \times 2 \times 2$ as a power notation.

Solution

repeated factor

number of repeated factors

$2 \times 2 \times 2 \times 2 = 2^4$ ⟵ power notation (4 repeated factors of 2)

In power notation, such as 2^4, 2 is called the **base** and 4 is called the **power** or **exponent.** The power notation 2^4 is read as "two to the *fourth power*" or "two *raised* to the fourth power."

Power Notation Rule

To write a product of repeated factors as power notation, you use the repeated factor as the base and the number of repeated factors as the power (exponent):

repeated factors ⟶ $\underbrace{a \times a \times a \times \cdots \times a}_{n \text{ repeated factors of } a} = a^n$ ⟵ power notation

EXAMPLE 4-75 Write the repeated factors **(a)** 10×10 and **(b)** $1.2 \times 1.2 \times 1.2$ using power notation.

Solution

(a) $10 \times 10 = 10^2$ Use the repeated factor 10 for the base.
Use the number of repeated factors (2) for the power (exponent).

(b) $1.2 \times 1.2 \times 1.2 = 1.2^3$ Use the repeated factor 1.2 for the base.
Use the number of repeated factors (3) for the power (exponent).

Note: The power notation 10^2 is read as "ten to the *second power*" or "ten *squared*." The power notation 1.2^3 is read as "one and two tenths to the *third power*" or "one and two tenths *cubed*."

To **write power notation** when the repeated factor is a fraction, you must write parentheses around the fraction before writing the power (exponent).

EXAMPLE 4-76 Write the repeated factors (a) $\frac{3}{4} \times \frac{3}{4}$ and (b) $\frac{3 \times 3}{4}$ using power notation.

Solution

(a) $\frac{3}{4} \times \frac{3}{4} = \left(\frac{3}{4}\right)^2$ Use parentheses to show that the whole fraction is being raised to the second power.

(b) $\frac{3 \times 3}{4} = \frac{3^2}{4}$ Do not use parentheses to show that only the numerator is being raised to the second power.

EXAMPLE 4-77 Show that $\frac{2^3}{5}$ does not mean $\left(\frac{2}{5}\right)^3$.

Solution

$$\frac{2^3}{5} = \frac{2 \times 2 \times 2}{5} = \frac{8}{5} \quad \bigg| \quad \left(\frac{2}{5}\right)^3 = \frac{2}{5} \times \frac{2}{5} \times \frac{2}{5} = \frac{8}{125}$$

B. Evaluating power notation

To **evaluate power notation,** you first write the power notation as a product of repeated factors and then multiply to get a whole number, or fraction, or decimal product.

EXAMPLE 4-78 Evaluate (a) 3^2, (b) $(\frac{2}{5})^3$, (c) 1^5, (d) 0^6, (e) 0.1^4, and (f) 10^7.

Solution

(a) In 3^2, the base 3 has a power (exponent) of 2.
A power (exponent) of 2 means the base 3 is used as a repeated factor 2 times.

$3^2 = 3 \times 3$ Write 2 repeated factors of 3.

    $= 9$ Multiply.

(b) In $\left(\frac{2}{5}\right)^3$, the base $\frac{2}{5}$ has a power (exponent) of 3.

A power (exponent) of 3 means the base $\frac{2}{5}$ is used as a repeated factor 3 times.

$\left(\frac{2}{5}\right)^3 = \frac{2}{5} \times \frac{2}{5} \times \frac{2}{5}$ Write 3 repeated factors of $\frac{2}{5}$.

        $= \frac{8}{125}$ Multiply.

(c) In 1^5, the base 1 has a power (exponent) of 5.
A power (exponent) of 5 means the base 1 is used as a repeated factor 5 times.

$1^5 = 1 \times 1 \times 1 \times 1 \times 1$ Write 5 repeated factors of 1.

    $= 1$ Multiply

(d) In 0^6, the base 0 has a power (exponent) of 6.
A power (exponent) of 6 means the base 0 is used as a repeated factor 6 times.

$0^6 = 0 \times 0 \times 0 \times 0 \times 0 \times 0$ Write 6 repeated factors of 0.

    $= 0$ Multiply.

(e) In 0.1^4, the base 0.1 has a power (exponent) of 4.
A power (exponent) of 4 means the base 0.1 is used as a repeated factor 4 times.

$$0.1^4 = 0.1 \times 0.1 \times 0.1 \times 0.1 \quad | \quad \text{Write 4 repeated factors of 0.1.}$$

$$= \underline{0.0001} \qquad\qquad\qquad | \quad \text{Multiply.}$$

$$\underbrace{\qquad}_{\text{4 zeros}}$$

(f) In 10^7, the base 10 has a power (exponent) of 7.
A power (exponent) of 7 means the base 10 is used as a repeated factor 7 times.

$$10^7 = 10 \times 10 \times 10 \times 10 \times 10 \times 10 \times 10 \qquad \text{Write 7 repeated factors of 10.}$$

$$= \underline{10,000,000} \qquad\qquad\qquad\qquad\qquad \text{Multiply.}$$

$$\underbrace{\qquad}_{\text{7 zeros}}$$

Example 4-79 illustrates some rules you should remember:

- If n is any power (exponent), then $1^n = 1$.
- If n is any nonzero power (exponent), then $0^n = 0$. (*Note:* 0^0 is not defined.)
- If n is any whole number power (exponent), then $0.1^n = \underline{0.00 \cdots 01}$.

$$\underbrace{\qquad}_{n \text{ zeros}}$$

- If n is any whole number power (exponent), then $10^n = \underline{1000 \cdots 0}$.

$$\underbrace{\qquad}_{n \text{ zeros}}$$

If you use 2 as a base with whole-number exponents, an interesting pattern occurs. As each exponent decreases by 1, the amount is *halved*. This is illustrated in the following example.

EXAMPLE 4-79   Evaluate the following powers of 2:

$$\textbf{(a)} \ 2^5 \quad \textbf{(b)} \ 2^4 \quad \textbf{(c)} \ 2^3 \quad \textbf{(d)} \ 2^2 \quad \textbf{(e)} \ 2^1 \quad \textbf{(f)} \ 2^0$$

Solution   **(a)** $2^5 = 2 \times 2 \times 2 \times 2 \times 2 = 32$

(b) $2^4 = 2 \times 2 \times 2 \times 2 = 16 \longleftarrow$ half of 32

(c) $2^3 = 2 \times 2 \times 2 = 8 \longleftarrow$ half of 16

(d) $2^2 = 2 \times 2 = 4 \longleftarrow$ half of 8

(e) $2^1 = 2 \longleftarrow$ half of 4

(f) $2^0 = 1 \longleftarrow$ half of 2

The power notation 2^1 is just another way of writing the base 2. The general rule for this power notation is

- If a is any base, then $a^1 = a$.

The power notation 2^0 is just another way of writing the whole number 1. The general rule for this power notation is

- If a is any nonzero base, then $a^0 = 1$.   (0^0 is not defined.)

EXAMPLE 4-80   Evaluate the following power notation:

(a) 5^1 **(b)** $\left(\dfrac{3}{4}\right)^1$ **(c)** 2.5^1 **(d)** 1^1 **(e)** 0^1

(f) 10^0 **(g)** $\left(\dfrac{2}{3}\right)^0$ **(h)** 0.75^0 **(i)** 1^0 **(j)** 0^0

Solution

(a) $5^1 = 5$ (b) $\left(\dfrac{3}{4}\right)^1 = \dfrac{3}{4}$ (c) $2.5^1 = 2.5$ (d) $1^1 = 1$ (e) $0^1 = 0$

(f) $10^0 = 1$ (g) $\left(\dfrac{2}{3}\right)^0 = 1$ (h) $0.75^0 = 1$ (i) $1^0 = 1$ (j) 0^0 is not defined.

C. Finding squares and square roots using paper and pencil

If a and b are numbers and $a = b^2$, then a is called the **square** of b.

EXAMPLE 4-81   Find the square of (a) 3, (b) $\frac{3}{4}$, and (c) 0.1.

Solution

(a) The square of 3 is 9 because $3^2 = 3 \times 3 = 9$.

(b) The square of $\dfrac{3}{4}$ is $\dfrac{9}{16}$ because $\left(\dfrac{3}{4}\right)^2 = \dfrac{3}{4} \times \dfrac{3}{4} = \dfrac{9}{16}$.

(c) The square of 0.1 is 0.01 because $0.1^2 = \underset{\text{2 zeros}}{\underbrace{0.01}}$

To find the square of a given number, you multiply the given number times itself.

You should know all the whole-number squares up to 100 from memory.

EXAMPLE 4-82   Name all the whole-number squares up to 100.

Solution
$$
\begin{array}{llll}
1 \text{ is a whole-number square because} & 1^2 &=& 1 \\
4 \text{ is a whole-number square because} & 2^2 &=& 4 \\
9 \text{ is a whole-number square because} & 3^2 &=& 9 \\
16 \text{ is a whole-number square because} & 4^2 &=& 16 \\
25 \text{ is a whole-number square because} & 5^2 &=& 25 \\
36 \text{ is a whole-number square because} & 6^2 &=& 36 \\
49 \text{ is a whole-number square because} & 7^2 &=& 49 \\
64 \text{ is a whole-number square because} & 8^2 &=& 64 \\
81 \text{ is a whole-number square because} & 9^2 &=& 81 \\
100 \text{ is a whole-number square because} & 10^2 &=& 100
\end{array}
$$

If a and b are numbers and $a = b^2$, then b is called the **square root** of a.

EXAMPLE 4-83   Find the square root of (a) 9, (b) $\frac{9}{16}$, and (c) 0.01.

Solution

(a) The square root of 9 is 3 because $9 = 3 \times 3 = 3^2$ [see Example 4-81(a)].

(b) The square root of $\dfrac{9}{16}$ is $\dfrac{3}{4}$ because $\dfrac{9}{16} = \dfrac{3 \times 3}{4 \times 4} = \left(\dfrac{3}{4}\right)^2$ [see Example 4-81(b)].

(c) The square root of 0.01 is 0.1 because $0.01 = 0.1 \times 0.1 = 0.1^2$ [see Example 4-81(c)].

To find the square root of a given number, you find the number that, when squared, will equal the given number.

"The square root of 9" can be written as $\sqrt{9}$. In $\sqrt{9}$, the symbol $\sqrt{}$ is called the **square root symbol** or **radical sign,** 9 is called the square or **radicand,** and $\sqrt{9}$ is called a **radical.**

To find the square root of a given whole-number square, you can divide the given whole number by various whole-number divisors until the whole-number divisor and the quotient are the same.

EXAMPLE 4-84 Find $\sqrt{196}$ using division.

Solution

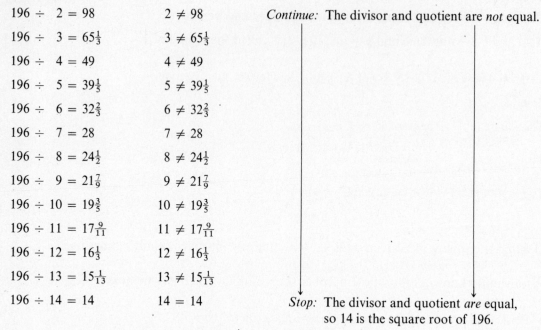

$196 \div 2 = 98$ $2 \neq 98$ *Continue:* The divisor and quotient are *not* equal.

$196 \div 3 = 65\frac{1}{3}$ $3 \neq 65\frac{1}{3}$

$196 \div 4 = 49$ $4 \neq 49$

$196 \div 5 = 39\frac{1}{5}$ $5 \neq 39\frac{1}{5}$

$196 \div 6 = 32\frac{2}{3}$ $6 \neq 32\frac{2}{3}$

$196 \div 7 = 28$ $7 \neq 28$

$196 \div 8 = 24\frac{1}{2}$ $8 \neq 24\frac{1}{2}$

$196 \div 9 = 21\frac{7}{9}$ $9 \neq 21\frac{7}{9}$

$196 \div 10 = 19\frac{3}{5}$ $10 \neq 19\frac{3}{5}$

$196 \div 11 = 17\frac{9}{11}$ $11 \neq 17\frac{9}{11}$

$196 \div 12 = 16\frac{1}{3}$ $12 \neq 16\frac{1}{3}$

$196 \div 13 = 15\frac{1}{13}$ $13 \neq 15\frac{1}{13}$

$196 \div 14 = 14$ $14 = 14$ *Stop:* The divisor and quotient *are* equal, so 14 is the square root of 196.

$\sqrt{196} = 14$ Write square root notation.

Hint: The multiplication fact $9 \times 9 = 81$ or $10 \times 10 = 100$ tells you that the square root of 196 is larger than 9 or 10. This means you should start the division process in Example 4-84 with a whole number at least as large as 11.

To find the square root of a fraction, you can write the square root of the numerator over the square root of the denominator.

EXAMPLE 4-85 Find $\sqrt{\dfrac{25}{64}}$ using division.

Solution

same

$25 \div 5 = 5$ means $\sqrt{25} = 5$

same

$64 \div 8 = 8$ means $\sqrt{64} = 8$

$\sqrt{\dfrac{25}{64}} = \dfrac{5}{8}$ ← square root of the numerator
              ← square root of the denominator

D. Finding squares and square roots using a table

To find the square of any whole number from 0 to 100, you can use Appendix Table 2 found in the back of this book. Part of Appendix Table 2 is displayed in the following example.

EXAMPLE 4-86 Find 47^2 using Appendix Table 2.

Solution

Number N	Square N²	Square Root √N
35	1225	5.916
36	1296	6
37	1369	6.083
38	1444	6.164
39	1521	6.245
40	1600	6.325
41	1681	6.403
42	1764	6.481
43	1849	6.557
44	1936	6.633
45	2025	6.708
46	2116	6.782
47	2209	6.856
48	2304	6.928
49	2401	7

(1) Place your left finger on the base 47 in the "Number N" column.

(2) Place your right finger on the top of the "Square N^2" column.

(3) Move your left finger straight across and your right finger straight down until they meet at 2209. Therefore $47^2 = 2209$.

Note: $47^2 = 47 \times 47 = 2209$.

You can also use Appendix Table 2 to find the square root of any whole number from 0 to 100. Part of Appendix Table 2 is displayed in Example 4-87.

EXAMPLE 4-87 Find $\sqrt{47}$ using Appendix Table 2.

Solution

Number N	Square N²	Square Root √N
35	1225	5.916
36	1296	6
37	1369	6.083
38	1444	6.164
39	1521	6.245
40	1600	6.325
41	1681	6.403
42	1764	6.481
43	1849	6.557
44	1936	6.633
45	2025	6.708
46	2116	6.782
47	2209	6.856
48	2304	6.928
49	2401	7

(1) Place your left finger on the radicand 47 in the "Number N" column.

(2) Place your right finger at the top of the "Square Root $\sqrt{N}$" column.

(3) Move your left finger straight across and your right finger straight down until they meet at 6.856. Therefore $\sqrt{47} \approx 6.856$.

Note: $\sqrt{47} \approx 6.856$ because $6.856 \times 6.856 = 47.004736 \approx 47$.

E. Finding squares and square roots using a calculator

To find the square of a whole number or decimal with four digits or less, you can use a standard 8-digit display **calculator.** To find the square of a whole number or decimal with 5 digits or less, you can use a standard 10-digit display calculator.

EXAMPLE 4-88 Find (a) 276^2 and (b) 915.8^2 using a calculator.

Solution	Press	Display	Interpret
(a)	2 7 6	276	276
	×	276	276 ×
	2 7 6	276	276 × 276
	=	76176	276 × 276 = 76,176
(b)	9 1 5 . 8	915.8	915.8
	×	915.8	915.8 ×
	9 1 5 . 8	915.8	915.8 × 915.8
	=	838689.64	915.8 × 915.8 = 838,689.64

You can also find (or approximate) the square root of a whole number or decimal using a display calculator. To find or approximate the square root of an 8-digit whole number or decimal with 8 digits or less, you can use a standard 8-digit calculator. To find or approximate the square root of a whole number or decimal with 10 digits or less, you can use a standard 10-digit calculator.

EXAMPLE 4-89 Approximate (a) $\sqrt{785{,}273}$ and (b) $\sqrt{14{,}906.385}$ using a calculator.

Solution	Press	Display	Interpret
(a)	7 8 5 2 7 3	785273	785,273
	√	886.15631	$\sqrt{785{,}273} \approx 886.15631$
(b)	1 4 9 0 6 . 3 8 5	14906.385	14,906.385
	√	122.09171	$\sqrt{14{,}906.385} \approx 122.09171$

To find or approximate the square root of a fraction using a calculator, you first divide the numerator by the denominator and then press the square root key.

EXAMPLE 4-90 Approximate $\sqrt{\dfrac{17}{32}}$ using a calculator.

Solution	Press	Display	Interpret
	1 7	17	17
	÷	17	17 ÷
	3 2	32	17 ÷ 32
	=	0.53125	17 ÷ 32 = 0.53125
	√	0.7288689	$\sqrt{0.53125} \approx 0.7288689$

Note: $\sqrt{\dfrac{17}{32}} \approx 0.7$ (nearest tenth), or 0.73 (nearest hundredth), or 0.729 (nearest thousandth), and so on, depending on what decimal place you round to.

4-10. Solving Word Problems Involving Decimals

To solve *addition* word problems containing decimals, you find how much two or more amounts are all together.

To solve *subtraction* word problems containing decimals, you find

(**a**) how much is left;

(**b**) how much more is needed;

(**c**) the difference between two amounts.

To solve *multiplication* word problems containing decimals, you find

(**a**) how much two or more equal amounts are all together;

(**b**) the product of one amount times another amount.

To solve *division* word problems containing decimals, you find

(**a**) how many equal amounts there are in all;

(**b**) how much is in each equal amount;

(**c**) how many times larger one amount is than another amount.

EXAMPLE 4-91   Solve the following word problems using addition or subtraction:

(**a**) Ginger had $25 in her wallet. When she cleaned out her purse, she found another 59¢. How much money did Ginger have then?

(**b**) Eric had $25 in his wallet. He spent 59¢ on candy. How much did Eric have then?

Solution

(**a**) (**1**) *Identify:*   Ginger had ($25). She found another (59¢). How much did Ginger have then?

   (**2**) *Understand:*   The question asks you to join amounts of $25 and 59¢ together.

   (**3**) *Decide:*   To join amounts together, you **add.**

   (**4**) *Compute:*
$$\begin{array}{r} \$2\,5.0\,0 \\ +\quad 0.5\,9 \\ \hline \$2\,5.5\,9 \end{array}$$

   (**5**) *Interpret:*   Ginger then had **$25.59.**

(**b**) (**1**) *Identify:*   Eric had ($25). He spent (59¢) of it. How much did Eric have then?

   (**2**) *Understand:*   The question asks you to take the amount of 59¢ away from $25.

   (**3**) *Decide:*   To take an amount away, you **subtract.**

   (**4**) *Compute:*
$$\begin{array}{r} \$2\,5.0\,0 \\ -\quad 0.5\,9 \\ \hline \$2\,4.4\,1 \end{array}$$

   (**5**) *Interpret:*   Eric then had **$24.41.**

EXAMPLE 4-92   Solve each word problem using multiplication or division:

(**a**) Morris earns $8.25 per hour typing papers. How much does Morris earn in a 40-hour work week?

(**b**) Rachel earns $390 for a 40-hour work week at the grocery store. How much does she earn per hour?

Solution

(**a**) (**1**) *Identify:*   Morris earns ($8.25) per hour. How much does Morris earn in a (40)-hour week?

   (**2**) *Understand:*   The question asks you to find 40 equal amounts of 8.25.

(**b**) (**1**) *Identify:*   Rachel earns ($390) for a (40)-hour work week. How much does she earn per hour?

   (**2**) *Understand:*   The question asks you to find how much is in each of 40 equal amounts.

	(3) *Decide:*	To join equal amounts together, you **multiply**.		(3) *Decide:*	To find how much is in each equal amount, you **divide**.
	(4) *Compute:*	$40 \times 8.25 = 330$		(4) *Compute:*	$390 \div 40 = 9.75$
	(5) *Interpret:*	330 means that in a 40-hour work week Morris earns **$330**.		(5) *Interpret:*	9.75 means that Rachel's hourly pay is **$9.75**.
	(6) *Check:*	Is $330 for 40 hours equal to $8.25 per hour? Yes. $330 \div 40 = 8.25$		(6) *Check:*	Is 40 hours at $9.75 per hour equal to $390? Yes: $40 \times 9.75 = 390$

SOLVED PROBLEMS

PROBLEM 4-1 Draw a line under the digit in each decimal that has the given value:

(a) hundredths: 5813.490267 (b) millionths: 4871.032965 (c) thousandths: 1026.589736

(d) tenths: 6105.478329 (e) hundred thousandths: 7580.613294 (f) hundreds: 4259.836107

Solution Recall that to help locate the correct digit in a decimal, you can use a place-value chart [see Example 4-3]:

(a) 5813.4<u>9</u>0267 (b) 4871.03296<u>5</u> (c) 1026.58<u>9</u>736

(d) 6105.4<u>7</u>8329 (e) 7580.6132<u>9</u>4 (f) 4<u>2</u>59.836107

PROBLEM 4-2 Write the place-value name for each underlined digit:

(a) 3587.019<u>2</u>46 (b) 5813.<u>6</u>49072 (c) 1705.8<u>9</u>2364 (d) 7258.39614<u>0</u>

(e) 6278.30<u>9</u>145 (f) 2340.785<u>1</u>69 (g) 486<u>5</u>.079132 (h) 9<u>0</u>75.481263

Solution Recall that to help identify the correct place-value name for a given digit, you can use a place-value chart [see Example 4-4]:

(a) hundred thousandths (b) tenths (c) hundredths (d) millionths

(e) thousandths (f) ten thousandths (g) ones (h) hundreds

PROBLEM 4-3 Write the value of 3 in each decimal:

(a) 5281.637094 (b) 8157.024693 (c) 2879.145306

(d) 7283.156940 (e) 4170.329568 (f) 1235.049876

Solution Recall that to write the value for a given digit, you first identify the place-value name of that digit [see Example 4-5]:

(a) 3 hundredths or 0.03 (b) 3 millionths or 0.000003 (c) 3 ten thousandths or 0.0003

(d) 3 ones or 3 (e) 3 tenths or 0.3 (f) 3 tens or 30

PROBLEM 4-4 Write the short decimal-word name for each decimal number:

(a) 1.5 (b) 3.25 (c) 25.075 (d) 187.0015 (e) 0.000001 (f) 0.00279

Solution Recall that to write the short decimal-word name for a given decimal number, you write *and* for the decimal point when the whole-number part is not zero and then name the whole number and the decimal-fraction part with the place-value name of the last digit on the right [see Examples 4-6 and 4-7]:

(a) 1 and 5 tenths   (b) 3 and 25 hundredths   (c) 25 and 75 thousandths

(d) 187 and 15 ten thousandths   (e) 1 millionth   (f) 279 hundred thousandths

PROBLEM 4-5   Write the decimal number for each short decimal-word name:

(a) 1 and 375 thousandths   (b) 2 and 3 tenths   (c) ninety-nine and 9 hundredths

(d) 3 and 2 ten thousandths   (e) 5 millionths   (f) 36 hundred thousandths

Solution   Recall that to write the correct decimal number for a short decimal-word name, you write the last digit of the decimal-fraction part so that it has the same value as the given short decimal-word name [see Example 4-8]:

(a) 1.375   (b) 2.3   (c) 99.09   (d) 3.0002   (e) 0.000005   (f) 0.00036

PROBLEM 4-6   Which is the smallest decimal in the following pairs of decimals?

(a) 0.9 and 1.0   (b) 0.05 and 0.5   (c) 1.03 and 0.4   (d) 25.0 and 25

(e) 49 and 48.9   (f) 0.625 and 0.75   (g) 3.14 and 3.141   (h) 2.25 and 3.99

Solution   Recall that to compare two decimals, you may find it helpful to write one decimal under the other while aligning the decimal points and like place values [see Example 4-11]:

(a) 0.9   (b) 0.05   (c) 0.4   (d) 25.0 = 25   (e) 48.9   (f) 0.625   (g) 3.14   (h) 2.25

PROBLEM 4-7   List each group of decimals from smallest to largest:

(a) 0.5, 0.8, 0.1   (b) 0.01, 0.1, 0.001   (c) 1.5, 0.5, 1.0

(d) 2.5, 0.75, 1.99   (e) 0.25, 0.125, 0.75, 0.375   (f) 5.49, 6.25, 4.99, 12.00

Solution   Recall that to order numbers from smallest to largest, you write the smallest number first, then the next smallest, and so on, until you have listed all the given numbers [see Example 4-12]:

(a) 0.1, 0.5, 0.8   (b) 0.001, 0.01, 0.1   (c) 0.5, 1.0, 1.5

(d) 0.75, 1.99, 2.5   (e) 0.125, 0.25, 0.375, 0.75   (f) 4.99, 5.49, 6.25, 12.00

PROBLEM 4-8   Round each decimal to the given place:

(a) 0.25 to the nearest tenth   (b) 0.138 to the nearest hundredth

(c) 1.82539 to the nearest thousandth   (d) 0.238504 to the nearest ten thousandth

(e) 0.851246 to the nearest hundred thousandth   (f) 0.834 to the nearest tenth

(g) 4582.1675 to the nearest hundred   (h) 3487.139 to the nearest whole number

(i) 8213.92 to the nearest thousand   (j) 8056.2 to the nearest ten

Solution   Recall that if the digit on the right of the digit to be rounded is 5 or more, you round up; if it is less than 5, you round down [see Examples 4-14 and 4-15]:

(a) 0.3   (b) 0.14   (c) 1.825   (d) 0.2385   (e) 0.85125

(f) 0.8   (g) 4600   (h) 3487   (i) 8000   (j) 8060

PROBLEM 4-9   Rename the whole numbers (a) 5, (b) 26, (c) 0, and (d) 1 as equal decimals.

Solution   Recall that to rename a whole number as an equal decimal, you write the decimal point following the whole number and then write one or more zeros following the decimal point [see Example 4-16]:

(a) $5 = 5.0 = 5.00 = 5.000 = \cdots$ (b) $26 = 26.0 = 26.00 = 26.000 = \cdots$

(c) $0 = 0.0 = 0.00 = 0.000 = \cdots$ (d) $1 = 1.0 = 1.00 = 1.000 = \cdots$

PROBLEM 4-10   Rename each decimal as an equal whole number:

(a) 2.0   **(b)** 5.000   **(c)** 1.00   **(d)**   0.0000   **(e)**   20.0   **(f)** 89.00   **(g)**   125.000   **(h)**   10.0000

Solution   Recall that when the decimal-fraction part of a decimal is zero, the decimal can always be renamed as an equal whole number [see Example 4-17]:

(a) 2.$\emptyset$ = 2. = 2   **(b)** 5.$\emptyset\emptyset\emptyset$ = 5. = 5   **(c)** 1.$\emptyset\emptyset$ = 1. = 1   **(d)** 0.$\emptyset\emptyset\emptyset\emptyset$ = 0. = 0

(e) 20.$\emptyset$ = 20. = 20   **(f)** 89.$\emptyset\emptyset$ = 89. = 89   **(g)** 125.$\emptyset\emptyset\emptyset$ = 125. = 125   **(h)** 10.$\emptyset\emptyset\emptyset\emptyset$ = 10. = 10

PROBLEM 4-11   Rename each decimal fraction as an equal proper fraction in simplest form:

(a) 0.1   **(b)** 0.7   **(c)** 0.4   **(d)** 0.8   **(e)** 0.75   **(f)** 0.01   **(g)** 0.125   **(h)** 0.025

Solution   Recall that to help rename a decimal fraction as an equal proper fraction, you first write the value of the decimal fraction, and then rename using tenths, hundredths, thousandths, ... [see Example 4-18]:

(a) $0.1 = 1 \text{ tenth} = \dfrac{1}{10}$ **(b)** $0.7 = 7 \text{ tenths} = \dfrac{7}{10}$

(c) $0.4 = 4 \text{ tenths} = \dfrac{4}{10} = \dfrac{\cancel{2} \times 2}{\cancel{2} \times 5} = \dfrac{2}{5}$ **(d)** $0.8 = 8 \text{ tenths} = \dfrac{8}{10} = \dfrac{\cancel{2} \times 4}{\cancel{2} \times 5} = \dfrac{4}{5}$

(e) $0.75 = 75 \text{ hundredths} = \dfrac{75}{100} = \dfrac{3 \times \cancel{25}}{4 \times \cancel{25}} = \dfrac{3}{4}$ **(f)** $0.01 = 1 \text{ hundredth} = \dfrac{1}{100}$

(g) $0.125 = 125 \text{ thousandths}$

$= \dfrac{125}{1000} = \dfrac{1 \times \cancel{125}}{8 \times \cancel{125}} = \dfrac{1}{8}$

(h) $0.025 = 25 \text{ thousandths}$

$= \dfrac{25}{1000} = \dfrac{1 \times \cancel{25}}{40 \times \cancel{25}} = \dfrac{1}{40}$

PROBLEM 4-12   Rename each fraction as an equal decimal:

(a) $\frac{3}{10}$ **(b)** $\frac{9}{10}$ **(c)** $\frac{11}{10}$ **(d)** $\frac{23}{10}$ **(e)** $\frac{9}{100}$ **(f)** $\frac{99}{100}$

Solution   Recall that when a fraction has a denominator of 10, 100, 1000, ..., you can rename that fraction as an equal decimal using tenths, hundredths, thousandths, ..., respectively [see Example 4-19]:

(a) $\frac{3}{10} = 3 \text{ tenths} = 0.3$ **(b)** $\frac{9}{10} = 9 \text{ tenths} = 0.9$ **(c)** $\frac{11}{10} = 11 \text{ tenths} = 1.1$

(d) $\frac{23}{10} = 23 \text{ tenths} = 2.3$ **(e)** $\frac{9}{100} = 9 \text{ hundredths} = 0.09$ **(f)** $\frac{99}{100} = 99 \text{ hundredths} = 0.99$

PROBLEM 4-13   Rename each decimal as an equal mixed number in simplest form:

(a) 5.3   **(b)** 2.9   **(c)** 1.5   **(d)** 3.2   **(e)** 4.25   **(f)** 8.03   **(g)** 6.375   **(h)** 10.075

Solution   Recall that when the whole-number part of a decimal is not zero, the decimal can be renamed as an equal mixed number using the short decimal-word name [see Example 4-20]:

(a) $5.3 = 5 \text{ and } 3 \text{ tenths} = 5 \text{ and } \frac{3}{10} = 5\frac{3}{10}$

(b) $2.9 = 2 \text{ and } 9 \text{ tenths} = 2 \text{ and } \frac{9}{10} = 2\frac{9}{10}$

(c) $1.5 = 1 \text{ and } 5 \text{ tenths} = 1 \text{ and } \frac{5}{10} = 1\frac{5}{10} = 1\frac{1}{2}$

(d) $3.2 = 3 \text{ and } 2 \text{ tenths} = 3 \text{ and } \frac{2}{10} = 3\frac{2}{10} = 3\frac{1}{5}$

(e) $4.25 = 4 \text{ and } 25 \text{ hundredths} = 4 \text{ and } \frac{25}{100} = 4\frac{1}{4}$

(f) $8.03 = 8 \text{ and } 3 \text{ hundredths} = 8 \text{ and } \frac{3}{100} = 8\frac{3}{100}$

(g) $6.375 = 6 \text{ and } 375 \text{ thousandths} = 6 \text{ and } \frac{375}{1000} = 6\frac{3}{8}$

(h) $10.075 = 10 \text{ and } 75 \text{ thousandths} = 10 \text{ and } \frac{75}{1000} = 10\frac{75}{1000} = 10\frac{3}{40}$

PROBLEM 4-14   Rename each decimal as an equal improper fraction in lowest terms:

(a) 1.6   **(b)** 2.5   **(c)** 3.625   **(d)** 4.875   **(e)** 1.75   **(f)** 2.03   **(g)** 3.25

Solution   Recall that when the whole-number part of a decimal is not zero, the decimal can be renamed as an equal improper fraction using place values [see Example 4-21]:

(a) $1.6 = 16 \text{ tenths} = \dfrac{16}{10} = \dfrac{2 \times 8}{2 \times 5} = \dfrac{8}{5}$

(b) $2.5 = 25 \text{ tenths} = \dfrac{25}{10} = \dfrac{5 \times 5}{2 \times 5} = \dfrac{5}{2}$

(c) $3.625 = 3625 \text{ thousandths} = \dfrac{3625}{1000} = \dfrac{29 \times 125}{8 \times 125} = \dfrac{29}{8}$

(d) $4.875 = 4875 \text{ thousandths} = \dfrac{4875}{1000} = \dfrac{39 \times 125}{8 \times 125} = \dfrac{39}{8}$

(e) $1.75 = 175 \text{ hundredths} = \dfrac{175}{100} = \dfrac{7 \times 25}{4 \times 25} = \dfrac{7}{4}$

(f) $2.03 = 203 \text{ hundredths} = \dfrac{203}{100}$

(g) $3.25 = 325 \text{ hundredths} = \dfrac{325}{100} = \dfrac{13 \times 25}{4 \times 25} = \dfrac{13}{4}$

PROBLEM 4-15   Write the number of decimal places in each decimal number:

(a) $3 + 0.1$ **(b)** $0.5 + 2$ **(c)** $8 + 1.2$ **(d)** $3.5 + 1$ **(e)** $4 + 1.25$ **(f)** $7.49 + 12$

Solution   Recall that in a decimal number, each digit in the decimal-fraction part counts as one decimal place [see Example 4-22]:

(a) 1.3 has 1 decimal place because there is 1 digit in the decimal-fraction part.

(b) 5.2 has 1 decimal place because there is 1 digit in the decimal-fraction part.

(c) 6.39 has 2 decimal places because there are 2 digits in the decimal-fraction part.

(d) 2.05 has 2 decimal places because there are 2 digits in the decimal-fraction part.

(e) 5.60 has 2 decimal places because there are 2 digits in the decimal-fraction part.

(f) 8.00 has 2 decimal places because there are 2 digits in the decimal-fraction part.

(g) 3.125 has 3 decimal places because there are 3 digits in the decimal-fraction part.

PROBLEM 4-16   Rename each number as an equal decimal with the given number of decimal places:

(a) 3 with 1 decimal place   **(b)** 5 with 2 decimal places   **(c)** 2 with 3 decimal places

(d) 9 with 4 decimal places   **(e)** 2.9 with 2 decimal places   **(f)** 0.1 with 3 decimal places

Solution   Recall that to rename a decimal as an equal decimal containing more decimal places, you can add the necessary number of zeros on the right-hand side of the decimal-fraction part [see Example 4-23]:

(a) $3 = 3. = 3.0$ **(b)** $5 = 5. = 5.00$ **(c)** $2 = 2. = 2.000$

(d) $9 = 9. = 9.0000$ **(e)** $2.9 = 2.90$ **(f)** $0.1 = 0.100$

PROBLEM 4-17   Add decimals:

(a) $1.5 + 0.8$ **(b)** $8.4 + 7.9$

(c) $12.49 + 5.89$ **(d)** $16.785 + 0.792$

(e) 0.85 + 10.6

(f) 51.6 + 8.925

(g) 15.49 + 8.79 + 8.49 + 5.99 + 25.69

(h) 0.812 + 0.05 + 0.1 + 0.725 + 0.3495 + 0.9

Solution   Recall that to add decimals, you first write in vertical form while aligning decimal points and like values, then rename to get an equal number of decimal places as needed. Add as you would for whole numbers, and then write the decimal point in the answer directly below the other decimal points [see Examples 4-24, 4-25, and 4-26]:

(a)
```
    1
   1.5
 +0.8
   1.3
```

(b)
```
    1
   8.4
 +7.9
 1 6.3
```

(c)
```
   1 1
 1 2.4 9
 +  5.8 9
 1 8.3 8
```

(d)
```
     1 1
 1 6.7 8 5
 +  0.7 9 2
 1 7.5 7 7
```

(e)
```
     1
   0.8 5
 +1 0.6 0
 1 1.4 5
```

(f)
```
    1 1
  5 1.6 0 0
 +  8.9 2 5
  6 0.5 2 5
```

(g)
```
    3 3 4
  1 5.4 9
    8.7 9
    8.4 9
    5.9 9
 +2 5.6 9
  6 4.4 5
```

(h)
```
    2 1 1
  0.8 1 2 0
  0.0 5 0 0
  0.1 0 0 0
  0.7 2 5 0
  0.3 4 9 5
 +0.9 0 0 0
  2.9 3 6 5
```

PROBLEM 4-18   Add decimals and whole numbers:

(a) 3 + 0.1   (b) 0.5 + 2   (c) 8 + 1.2   (d) 3.5 + 1   (e) 4 + 1.25   (f) 7.49 + 12

Solution   Recall that to subtract decimals, you first write in vertical form while aligning the decimal as an equal decimal with the necessary number of decimal places and then add as before [see Example 4-28]:

(a)
```
   3.0  ⟵ 3 = 3. = 3.0
 +0.1
   3.1
```

(b)
```
   0.5
 +2.0  ⟵ 2 = 2. = 2.0
   2.5
```

(c)
```
   8.0  ⟵ 8 = 8. = 8.0
 +1.2
   9.2
```

(d)
```
   3.5
 +1.0  ⟵ 1 = 1. = 1.0
   4.5
```

(e)
```
   4.0 0  ⟵ 4 = 4. = 4.00
 +1.2 5
   5.2 5
```

(f)
```
     7.4 9
 +1 2.0 0  ⟵ 12 = 12. = 12.00
  1 9.4 9
```

PROBLEM 4-19   Add with money:

(a) $15.85 + $4.19

(b) $6.37 + $25

(c) 87¢ + 37¢

(d) 79¢ + 95¢

(e) $8.75 + 27¢

(f) 85¢ + $15

(g) $8.21 + $4.75 + $3.99 + $45 + 27¢ + 32¢

(h) 15¢ + $5 ∔ $8.42 + $125.38 + 87¢ + $59

Solution   Recall that to add with money, you first rename each dollar-and-cent amount as an equal decimal with 2 decimal places and then add as before [see Example 4-29]:

(a)
```
    1 1 1
 $1 5.8 5
 +   4.1 9
 $2 0.0 4
```

(b)
```
     1
 $  6.3 7
 + 2 5.0 0  ⟵ $25 = $25.00
 $3 1.3 7
```

(c)
```
    1 1
 $0.8 7  ⟵ 87¢ = $0.87
 + 0.3 7  ⟵ 37¢ = $0.37
 $1.2 4
```

(d)
```
    1 1
 $0.7 9  ⟵ 79¢ = $0.79
 + 0.9 5  ⟵ 95¢ = $0.95
 $1.7 4
```

(e)
$$
\begin{array}{r}
{}^{1}\;{}^{1}\;\\
\$8.7\;5 \\
+\ 0.2\;7 \longleftarrow\;27\cancel{c}=\$0.27 \\
\hline
\$9.0\;2
\end{array}
$$

(f)
$$
\begin{array}{r}
\$\;\;\;0.8\;5 \longleftarrow\;85\cancel{c}=\$0.85 \\
+\ 1\;5.0\;0 \longleftarrow\;\$15=\$15.00 \\
\hline
\$1\;5.8\;5
\end{array}
$$

(g)
$$
\begin{array}{r}
{}^{2}\;{}^{2}\;{}^{2}\;\\
\$\;\;8.2\;1 \\
4.7\;5 \\
3.9\;9 \\
4\;5.0\;0 \longleftarrow\;\$45=\$45.00 \\
0.2\;7 \longleftarrow\;27\cancel{c}=\$0.27 \\
+\;\;.0.3\;2 \longleftarrow\;32\cancel{c}=\$0.32 \\
\hline
\$6\;2.5\;4
\end{array}
$$

(h)
$$
\begin{array}{r}
{}^{2}\;{}^{1}\;{}^{2}\;\\
\$\;\;\;\;0.1\;5 \longleftarrow\;15\cancel{c}=\$0.15 \\
5.0\;0 \longleftarrow\;\$5\;=\$5.00 \\
8.4\;2 \\
1\;2\;5.3\;8 \\
0.8\;7 \longleftarrow\;87\cancel{c}=\$0.87 \\
+\;\;\;5\;9.0\;0 \longleftarrow\;\$59=\$59.00 \\
\hline
\$1\;9\;8.8\;2
\end{array}
$$

PROBLEM 4-20 Subtract decimals:

(a) $0.9 - 0.1$ (b) $8.3 - 3.7$ (c) $258.79 - 168.85$ (d) $4.1857 - 2.3462$

(e) $4.15 - 0.9$ (f) $18.125 - 5.3$ (g) $2.1 - 1.89$ (h) $35.75 - 0.3652$

Solution Recall that to subtract decimals, you first write in vertical form while aligning the decimal points and like values, then rename to get an equal number of decimal places as needed. Subtract as you would for whole numbers, and then write the decimal point in the answer directly below the other decimal points [see Examples 4-30 through 4-33]:

(a)
$$
\begin{array}{r}
0.9 \\
-0.1 \\
\hline
0.8
\end{array}
$$

(b)
$$
\begin{array}{r}
{}^{7}{}_{1} \\
\cancel{8}.3 \\
-3.7 \\
\hline
4.6
\end{array}
$$

(c)
$$
\begin{array}{r}
{}^{1}\;{}^{14}\;{}^{17}{}_{1} \\
\cancel{2}\;\cancel{5}\;\cancel{8}.7\;9 \\
-1\;6\;8.8\;5 \\
\hline
8\;9.9\;4
\end{array}
$$

(d)
$$
\begin{array}{r}
{}^{3}{}_{1}\;\;\;{}^{7}{}_{1} \\
\cancel{4}.1\;\cancel{8}\;5\;7 \\
-2.3\;4\;6\;2 \\
\hline
1.8\;3\;9\;5
\end{array}
$$

(e)
$$
\begin{array}{r}
{}^{3}{}_{1} \\
\cancel{4}.1\;5 \\
-0.9\;0 \\
\hline
3.2\;5
\end{array}
$$

(f)
$$
\begin{array}{r}
{}^{7}{}_{1} \\
1\;\cancel{8}.1\;2\;5 \\
-\;\;5.3\;0\;0 \\
\hline
1\;2.8\;2\;5
\end{array}
$$

(g)
$$
\begin{array}{r}
{}^{1}\;{}^{10}{}_{1} \\
\cancel{2}.\cancel{1}\;0 \\
-1.8\;9 \\
\hline
0.2\;1
\end{array}
$$

(h)
$$
\begin{array}{r}
{}^{6}\;{}^{14}\;{}^{9}{}_{1} \\
3\;5.\cancel{7}\;\cancel{5}\;\cancel{0}\;0 \\
-\;\;0.3\;6\;5\;2 \\
\hline
3\;5.3\;8\;4\;8
\end{array}
$$

PROBLEM 4-21 Subtract decimals and whole numbers:

(a) $5.25 - 3$ (b) $9.5 - 6$ (c) $28.49 - 15$ (d) $629.029 - 587$

(e) $5 - 0.1$ (f) $3 - 1.85$ (g) $23 - 7.49$ (h) $500 - 8.205$

Solution Recall that to subtract decimals and whole numbers, you first rename the whole number as an equal decimal with the necessary number of decimal places and then subtract as before [see Examples 4-34 and 4-35]:

(a)
$$
\begin{array}{r}
5.2\;5 \\
-3.0\;0 \longleftarrow\;3=3.=3.00 \\
\hline
2.2\;5
\end{array}
$$

(b)
$$
\begin{array}{r}
9.5 \\
-6.0 \longleftarrow\;6=6.=6.0 \\
\hline
3.5
\end{array}
$$

(c)
$$
\begin{array}{r}
2\;8.4\;9 \\
-1\;5.0\;0 \longleftarrow\;15=15.=15.00 \\
\hline
1\;3.4\;9
\end{array}
$$

(d)
$$
\begin{array}{r}
{}^{5}{}_{1} \\
\cancel{6}\;2\;9.0\;2\;9 \\
-5\;8\;7.0\;0\;0 \longleftarrow\;587=587.=587.000 \\
\hline
4\;2.0\;2\;9
\end{array}
$$

(e)
$$
\begin{array}{r}
{}^{4}{}_{1} \\
\cancel{5}.0 \longleftarrow\;5=5.=5.0 \\
-0.1 \\
\hline
4.9
\end{array}
$$

(f)
$$
\begin{array}{r}
{}^{2}\;{}^{9}{}_{1} \\
\cancel{3}.\cancel{0}\;0 \longleftarrow\;3=3.=3.00 \\
-1.8\;5 \\
\hline
1.1\;5
\end{array}
$$

(g)
$$\begin{array}{r} \overset{1\ 12\ \ 9}{\cancel{2}\ \cancel{3}.\cancel{0}\ 0} \\ -\ \ \ 7.4\ 9 \\ \hline 1\ 5.5\ 1 \end{array}$$
$\longleftarrow 23 = 23. = 23.00$

(h)
$$\begin{array}{r} \overset{4\ 9\ 9\ 9\ 9}{\cancel{5}\ \cancel{0}\ \cancel{0}.\cancel{0}\ \cancel{0}\ 0} \\ -\ \ \ \ 8.2\ 0\ 5 \\ \hline 4\ 9\ 1.7\ 9\ 5 \end{array}$$
$\longleftarrow 500 = 500. = 500.000$

PROBLEM 4-22   Subtract with money:

(a) $25 − $17
(b) $108 − $57
(c) $6.89 − $3
(d) $27.45 − $19

(e) $45 − $8.45
(f) $82 − $37.99
(g) 85¢ − 32¢
(h) 91¢ − 17¢

(i) $5.25 − 63¢
(j) $14.23 − 97¢
(k) $5 − 6¢
(l) $15 − 63¢

Solution   Recall that to subtract with money, you first rename each dollar-and-cent amount as an equal decimal with 2 decimal places (when necessary) and then subtract as before [see Example 4-36]:

(a)
$$\begin{array}{r} \overset{1\ 1}{\$\cancel{2}\ 5} \\ -\ 1\ 7 \\ \hline \$\ \ 8 \end{array}$$

(b)
$$\begin{array}{r} \overset{0\ 1}{\$\cancel{1}\ 0\ 8} \\ -\ \ \ 5\ 7 \\ \hline \$\ \ \ 5\ 1 \end{array}$$

(c)
$$\begin{array}{r} \$6.8\ 9 \\ -\ 3.0\ 0 \\ \hline \$3.8\ 9 \end{array}$$

(d)
$$\begin{array}{r} \overset{1\ 1}{\$\cancel{2}\ 7.4\ 5} \\ -\ 1\ 9.0\ 0 \\ \hline \$\ \ 8.4\ 5 \end{array}$$

(e)
$$\begin{array}{r} \overset{3\ 14\ 9}{\$\cancel{4}\ \cancel{5}.\cancel{0}\ 0} \\ -\ \ \ \ 8.4\ 5 \\ \hline \$3\ 6.5\ 5 \end{array}$$

(f)
$$\begin{array}{r} \overset{7\ 11\ 9}{\$\cancel{8}\ \cancel{2}.\cancel{0}\ 0} \\ -\ \ \ 3\ 7.9\ 9 \\ \hline \$4\ 4.0\ 1 \end{array}$$

(g)
$$\begin{array}{r} 8\ 5¢ \\ -3\ 2¢ \\ \hline 5\ 3¢ \end{array}$$

(h)
$$\begin{array}{r} \overset{8\ 1}{\cancel{9}\ 1¢} \\ -1\ 7¢ \\ \hline 7\ 4¢ \end{array}$$

(i)
$$\begin{array}{r} \overset{4\ 1}{\$\cancel{5}.2\ 5} \\ -\ 0.6\ 3 \\ \hline \$4.6\ 2 \end{array}$$
$\longleftarrow 63¢ = \$0.63$

(j)
$$\begin{array}{r} \overset{3\ 11\ 1}{\$1\ \cancel{4}.\cancel{2}\ 3} \\ -\ \ \ 0.9\ 7 \\ \hline \$1\ 3.2\ 6 \end{array}$$
$\longleftarrow 97¢ = \$0.97$

(k)
$$\begin{array}{r} \overset{4\ 9\ 1}{\$\cancel{5}.\cancel{0}\ 0} \\ -\ 0.0\ 6 \\ \hline \$4.9\ 4 \end{array}$$
$\longleftarrow 6¢ = \$0.06$

(l)
$$\begin{array}{r} \overset{4\ 9\ 1}{\$1\ \cancel{5}.\cancel{0}\ 0} \\ -\ \ \ 0.6\ 3 \\ \hline \$1\ 4.3\ 7 \end{array}$$
$\longleftarrow 63¢ = \$0.63$

PROBLEM 4-23   Multiply decimals and whole numbers:

(a) 3 × 0.2
(b) 0.1 × 4
(c) 6 × 0.8
(d) 0.9 × 7

(e) 2 × 0.5
(f) 0.4 × 5
(g) 3.45 × 3
(h) 2 × 0.75

Solution   Recall that to multiply a decimal by a whole number, you first write in vertical form while aligning the right-hand digits, then multiply as you would for whole numbers, and then write the decimal point in the product so that it has the same number of decimal places as the decimal factor [see Examples 4-36 and 4-37]:

(a)
$$\begin{array}{r} 0.2 \\ \times\ \ \ 3 \\ \hline 0.6 \end{array}$$
$\longleftarrow$ 1 decimal place
$\longleftarrow$ 0 decimal places
$\longleftarrow$ 1 (= 1 + 0) decimal place

(b)
$$\begin{array}{r} 0.1 \\ \times\ \ \ 4 \\ \hline 0.4 \end{array}$$
1 decimal place

(c)
$$\begin{array}{r} 0.8 \\ \times\ \ \ 6 \\ \hline 4.8 \end{array}$$
1 decimal place

(d)
$$\begin{array}{r} 0.9 \\ \times\ \ \ 7 \\ \hline 6.3 \end{array}$$
1 decimal place

(e)
$$\begin{array}{r} 0.5 \\ \times\ \ \ 2 \\ \hline 1.0 = 1 \end{array}$$
1 decimal place

(f)
$$\begin{array}{r} 0.4 \\ \times\ \ \ 5 \\ \hline 2.0 = 2 \end{array}$$
1 decimal place

(g)
$$\begin{array}{r} 3.4\ 5 \\ \times\ \ \ \ \ 3 \\ \hline 1\ 0.3\ 5 \end{array}$$
2 decimal places

(h)
$$\begin{array}{r} 0.7\ 5 \\ \times\ \ \ \ 2 \\ \hline 1.5\ 0 = 1.5 \end{array}$$
2 decimal places

PROBLEM 4-24   Multiply two decimals:

(a) 0.3×0.6 (b) 0.4×0.7 (c) 3.49×0.9 (d) 0.1×1.25

(e) 5.25×0.35 (f) 2.15×8.03

Solution   Recall the Multiplication Rule for Two or More Decimals [see Example 4-40]:

(1) Write in vertical form while aligning the right-hand digits.

(2) Multiply as you would for two whole numbers.

(3) Put the same number of decimal places in the product as the total number of decimal places contained in both factors.

(a)
```
    0.6  ←—— 1 decimal place
  × 0.3  ←—— 1 decimal place
  ─────
  0.1 8  ←—— 2 (= 1 + 1) decimal places
```

(b)
```
    0.7
  × 0.4
  ─────
  0.2 8
```

(c)
```
    3.4 9  ←—— 2 decimal places
  ×   0.9  ←—— 1 decimal place
  ───────
  3.1 4 1  ←—— 3 (= 2 + 1) decimal places
```

(d)
```
    1.2 5
  ×   0.1
  ───────
  0.1 2 5
```

(e)
```
      5.2 5  ←—— 2 decimal places
    × 0.3 5  ←—— 2 decimal places
    ───────
    2 6 2 5
  1 5 7 5
  ─────────
  1.8 3 7 5  ←—— 4 (= 2 + 2) decimal places
```

(f)
```
        2.1 5
      × 8.0 3
      ───────
        6 4 5
  1 7 2 0 0
  ───────────
  1 7.2 6 4 5
```

PROBLEM 4-25   Multiply decimals when zeros are needed in the product:

(a) 0.3×0.1 (b) 0.1×0.1 (c) 0.4×0.02 (d) 0.04×0.5 (e) 0.06×0.01

(f) 0.05×0.04 (g) 0.003×0.03 (h) 0.25×0.038 (i) 0.002×0.002 (j) 0.014×0.215

Solution   Recall that when the number of decimal places that belong in the product is greater than the number of digits in the product, you must write zeros in the product until there are enough digits to write the decimal point in the correct place [see Example 4-41]:

(a)
```
    0.1  ←—— 1 decimal place
  × 0.3  ←—— 1 decimal place
  ─────
  0.0 3  ←—— 2 (= 1 + 1) decimal places
  write zeros
```

(b)
```
    0.1
  × 0.1
  ─────
  0.0 1
```

(c)
```
    0.0 2  ←—— 2 decimal places
  ×   0.4  ←—— 1 decimal places
  ───────
  0.0 0 8  ←—— 3 (= 2 + 1) decimal places
  write zeros
```

(d)
```
    0.0 4
  ×   0.5
  ───────
  0.0 2 0 = 0.02
```

(e)
```
      0.0 1  ←—— 2 decimal places
    × 0.0 6  ←—— 2 decimal places
    ───────
  0.0 0 0 6  ←—— 4 (= 2 + 2) decimal places
  write zeros
```

(f)
```
    0.0 4
  × 0.0 5
  ───────
  0.0 0 2 0 = 0.002
```

(g)
```
      0.0 0 3  ←—— 3 decimal places
  ×     0.0 3  ←—— 2 decimal places
  ───────────
  0.0 0 0 0 9  ←—— 5 (= 3 + 2) decimal places
  write zeros
```

(h)
```
      0.0 3 8
  ×     0.2 5
  ───────────
      1 9 0
    7 6
  ─────────────
  0.0 0 9 5 0 = 0.0095
```

(i)
```
        0.0 0 2  ←—— 3 decimal places
    ×   0.0 0 2  ←—— 3 decimal places
    ─────────────
  0.0 0 0 0 0 4  ←—— 6 (= 3 + 3) decimal places
  write zeros
```

(j)
```
        0.2 1 5
      × 0.0 1 4
      ───────────
        8 6 0
      2 1 5
  ─────────────────
  0.0 0 3 0 1 0 = 0.00301
```

PROBLEM 4-26 Multiply by 10:

(a) 10×4.25 (b) 10×13.125 (c) 10×1.5

(d) 10×423.8 (e) 10×0.75 (f) 10×9

(g) 10×0.02 (h) 10×8 (i) 10×50

Solution Recall that to multiply by 10, you move the decimal point 1 decimal place to the right [see Example 4-43]:

(a) $10 \times 4.25 = 42.5 = 42.5$ (b) $10 \times 13.125 = 131.25 = 131.25$ (c) $10 \times 1.5 = 15. = 15$

(d) $10 \times 423.8 = 4238. = 4238$ (e) $10 \times 0.75 = 07.5 = 7.5$ (f) $10 \times 9 = 90. = 90$

(g) $10 \times 0.02 = 00.2 = 0.2$ (h) $10 \times 8 = 80. = 80$ (i) $10 \times 50 = 500. = 500$

PROBLEM 4-27 Multiply by 100:

(a) 100×1.125 (b) 100×0.375 (c) 100×0.025 (d) 100×0.003 (e) 100×0.05

(f) 100×0.19 (g) 100×3.1 (h) 100×25.6 (i) 100×8 (j) 100×200

Solution Recall that to multiply by 100, you move the decimal point 2 places to the right [see Example 4-45]:

(a) $100 \times 1.125 = 112.5 = 112.5$ (b) $100 \times 0.375 = 037.5 = 37.5$

(c) $100 \times 0.025 = 002.5 = 2.5$ (d) $100 \times 0.003 = 000.3 = 0.3$

(e) $100 \times 0.05 = 005. = 5$ (f) $100 \times 0.19 = 019. = 19$

(g) $100 \times 3.1 = 100 \times 3.10 = 310. = 310$ (h) $100 \times 25.6 = 100 \times 25.60 = 2560$

(i) $100 \times 8 = 100 \times 8.00 = 800$ (j) $100 \times 200 = 100 \times 200.00 = 20,000$

PROBLEM 4-28 Multiply by 1000:

(a) 1000×5.8196 (b) 1000×0.2458 (c) 1000×0.0675 (d) 1000×0.0025 (e) 1000×0.0004

(f) 1000×0.009 (g) 1000×5.12 (h) 1000×3.4 (i) 1000×7 (j) 1000×800

Solution Recall that to multiply by 1000, you move the decimal point 3 places to the right [see Example 4-47]:

(a) $1000 \times 5.8196 = 5819.6 = 5819.6$ (b) $1000 \times 0.2458 = 0245.8 = 245.8$

(c) $1000 \times 0.0675 = 0067.5 = 67.5$ (d) $1000 \times 0.0025 = 0002.5 = 2.5$

(e) $1000 \times 0.0004 = 0000.4 = 0.4$ (f) $1000 \times 0.009 = 0009. = 9$

(g) $1000 \times 5.12 = 1000 \times 5.120 = 5120. = 5120$ (h) $1000 \times 3.4 = 1000 \times 3.400 = 3400$

(i) $1000 \times 7 = 1000 \times 7.000 = 7000$ (j) $1000 \times 800.000 = 800,000$

PROBLEM 4-29 Multiply by powers of 10:

(a) 10×2.5 (b) 100×2.5 (c) $10,000 \times 2.5$

(d) $100,000 \times 0.005$ (e) $1,000,000 \times 3$ (f) $10,000,000 \times 0.000002$

(g) $100,000,000 \times 800$ (h) $1,000,000,000 \times 1,000,000$

Solution Recall that to multiply by a power of 10, you move the decimal point 1 decimal place to the right for each zero in the power of 10 [See Example 4-48]:

1 zero   right 1 place

(a) $10 \times 2.5 = 25$

2 zeros   right 2 places

(b) $100 \times 2.5 = 250$

(c) $10,000 \times 2.5 = 25,000$

(d) $100,000 \times 0.005 = 500$

(e) $1,000,000 \times 3 = 3,000,000$

(f) $10,000,000 \times 0.000002 = 20$

8 zeros	2 zeros	10 (= 8 + 2) zeros

(g) $100,000,000 \times 800 = 80,000,000,000$

9 zeros	6 zeros	15 (= 9 + 6) zeros

(h) $1,000,000,000 \times 1,000,000 = 1,000,000,000,000,000$

PROBLEM 4-30 Divide decimals by whole numbers:

(a) $1.0 \div 5$ (b) $0.18 \div 3$ (c) $0.03 \div 2$ (d) $0.0504 \div 6$

(e) $22.42 \div 3$ (f) $11.4 \div 9$ (g) $77.5 \div 31$ (h) $40.0 \div 400$

Solution Recall that to divide a decimal by a whole number, you first write in division box form, divide as you would for whole numbers, then put the same number of decimal places in the quotient as in the decimal dividend [see Example 4-51]:

(a)
$$
\begin{array}{r}
0.2 \quad \leftarrow \text{1 decimal place} \\
5 \overline{)\ 1.0} \\
-1\,0 \\
\hline
0
\end{array}
$$

(b)
$$
\begin{array}{r}
0.0\,6 \quad \leftarrow \text{2 decimal places} \\
3 \overline{)\ 0.1\,8} \\
-\ 1\,8 \\
\hline
0
\end{array}
$$

(c)
$$
\begin{array}{r}
0.0\,1\,5 \quad \leftarrow \text{3 decimal places} \\
2 \overline{)\ 0.0\,3\,0} \\
-\ \ 2 \\
\hline
1\,0 \\
-1\,0 \\
\hline
0
\end{array}
$$

(d)
$$
\begin{array}{r}
0.0\,0\,8\,4 \quad \leftarrow \text{4 decimal places} \\
6 \overline{)\ 0.0\,5\,0\,4} \\
-\ \ 4\,8 \\
\hline
2\,4 \\
-2\,4 \\
\hline
0
\end{array}
$$

(e)
$$
\begin{array}{r}
7.4\,7\,3\,3\,3\ldots = 7.47\overline{3} \\
3 \overline{)\ 2\,2.4\,2\,0\,0\,0} \\
-2\,1 \\
\hline
1\,4 \\
-1\,2 \\
\hline
2\,2 \\
-2\,1 \\
\hline
1\,0 \\
-\ \ 9 \\
\hline
1\,0 \\
-\ \ 9 \\
\hline
1\,0 \\
-\ \ 9 \\
\hline
1
\end{array}
$$

(f)
$$
\begin{array}{r}
1.2\,6\,6\,6\ldots = 1.2\overline{6} \\
9 \overline{)\ 1\,1.4\,0\,0\,0} \\
-\ \ 9 \\
\hline
2\,4 \\
-1\,8 \\
\hline
6\,0 \\
-5\,4 \\
\hline
6\,0 \\
-5\,4 \\
\hline
6\,0 \\
-5\,4 \\
\hline
6
\end{array}
$$

(g)
$$
\begin{array}{r}
2.5 \\
31 \overline{)\ 7\,7.5} \\
-6\,2 \\
\hline
1\,5\,5 \\
-1\,5\,5 \\
\hline
0
\end{array}
$$

(h)
$$
\begin{array}{r}
0.1 \\
400 \overline{)\ 4\,0.0} \\
-4\,0\,0 \\
\hline
0
\end{array}
$$

PROBLEM 4-31 Divide decimals by decimals:

(a) $0.18 \div 0.2$ (b) $0.35 \div 0.5$ (c) $0.36 \div 0.3$ (d) $1.08 \div 0.4$

(e) $0.15 \div 1.2$ (f) $7.875 \div 2.1$ (g) $0.6603 \div 0.31$ (h) $1.1137 \div 0.43$

Solution Recall the Division Rule for Decimals [see Examples 4-54 and 4-55]:

(1) Write in division box form.

(2) If the division is a decimal, then make the divisor a whole number by multiplying both the divisor and the dividend by the same power of 10.

(3) Divide as you would for whole numbers.

(4) Put the same number of decimal places in the quotient as in the decimal dividend from Step 2.

(a)
$$0.2\overline{)0.18} = 2\overline{)\begin{array}{c}0.9 \\ 1.8 \\ -1\,8 \\ \hline 0\end{array}}$$

(b)
$$0.5\overline{)0.35} = 5\overline{)\begin{array}{c}0.7 \\ 3.5 \\ -3\,5 \\ \hline 0\end{array}}$$

(c)
$$0.3\overline{)0.36} = 3\overline{)\begin{array}{c}1.2 \\ 3.6 \\ -3 \\ \hline \varnothing\,6 \\ -\;6 \\ \hline 0\end{array}}$$

(d)
$$0.4\overline{)1.08} = 4\overline{)\begin{array}{c}2.7 \\ 1\,0.8 \\ -\;8 \\ \hline 2\,8 \\ -2\,8 \\ \hline 0\end{array}}$$

(e)
$$1.2\overline{)0.15} = 12\overline{)\begin{array}{c}0.1\,2\,5 \\ 1.5\,0\,0 \\ -1\,2 \\ \hline 3\,0 \\ -2\,4 \\ \hline 6\,0 \\ -6\,0 \\ \hline 0\end{array}}$$

(f)
$$2.1\overline{)7.875} = 21\overline{)\begin{array}{c}3.7\,5 \\ 7\,8.7\,5 \\ -6\,3 \\ \hline 1\,5\,7 \\ -1\,4\,7 \\ \hline 1\,0\,5 \\ -1\,0\,5 \\ \hline 0\end{array}}$$

(g)
$$0.31\overline{)0.6603} = 31\overline{)\begin{array}{c}2.1\,3 \\ 6\,6.0\,3 \\ -6\,2 \\ \hline 4\,0 \\ -3\,1 \\ \hline 9\,3 \\ -9\,3 \\ \hline 0\end{array}}$$

(h)
$$0.43\overline{)1.1137} = 43\overline{)\begin{array}{c}2.5\,9 \\ 1\,1\,1.3\,7 \\ -\;8\,6 \\ \hline 2\,5\,3 \\ -2\,1\,5 \\ \hline 3\,8\,7 \\ -3\,8\,7 \\ \hline 0\end{array}}$$

PROBLEM 4-32   Divide when zeros are needed in the quotient:

(a) 0.04 ÷ 2 **(b)** 0.015 ÷ 3 **(c)** 0.0008 ÷ 8 **(d)** 0.0001 ÷ 5

(e) 0.001 ÷ 0.4 **(f)** 0.002 ÷ 0.3 **(g)** 0.1 ÷ 2.5 **(h)** 0.011 ÷ 3.3

Solution   Recall that when the number of decimal places that belong in the quotient is greater than the number of digits in the quotient, you must write zeros in the quotient until there are enough digits to write the decimal point in the correct place [see Example 4-56]:

write zeros

(a)
$$2\overline{)\begin{array}{c}0.0\,2 \\ 0.0\,4 \\ -\;\;\;4 \\ \hline 0\end{array}}$$

(b)
$$3\overline{)\begin{array}{c}0.0\,0\,5 \\ 0.0\,1\,5 \\ -\;\;\;1\,5 \\ \hline 0\end{array}}$$

(c)
$$8\overline{)\begin{array}{c}0.0\,0\,0\,1 \\ 0.0\,0\,0\,8 \\ -\;\;\;\;\;8 \\ \hline 0\end{array}}$$

(d)
$$5\overline{)\begin{array}{c}0.0\,0\,0\,0\,2 \\ 0.0\,0\,0\,1\,0 \\ -\;\;\;\;\;1\,0 \\ \hline 0\end{array}}$$

(e)
$$0.4\overline{)0.001} = 4\overline{)\begin{array}{c}0.0\,0\,2\,5 \\ 0.0\,1\,0\,0 \\ -\;\;\;\;8 \\ \hline 2\,0 \\ -2\,0 \\ \hline 0\end{array}}$$

(f)
$$0.3\overline{)0.002} = 3\overline{)\begin{array}{c}0.0\,0\,6\,6\,6\ldots = 0.00\overline{6} \\ 0.0\,2\,0\,0\,0 \\ -\;\;\;\;1\,8 \\ \hline 2\,0 \\ -1\,8 \\ \hline 2\,0 \\ -1\,8 \\ \hline 2\end{array}}$$

(g) $2.5\overline{)0.1} = 25\overline{)\begin{array}{l}0.0\ 4\\1.0\ 0\\\underline{-1\ 0\ 0}\\0\end{array}}$

(h) $3.3\overline{)0.011} = 33\overline{)\begin{array}{l}0.0\ 0\ 3\ 3\ 3\dots = 0.00\overline{3}\\0.1\ 1\ 0\ 0\ 0\\\underline{-9\ 9}\\1\ 1\ 0\\\underline{-9\ 9}\\1\ 1\ 0\\\underline{-9\ 9}\\1\ 1\end{array}}$

PROBLEM 4-33 Divide and round the quotient to the given place value:

(a) $5 \div 0.3$ (nearest tenth)

(b) $0.1 \div 0.6$ (nearest tenth)

(c) $1.3 \div 0.07$ (nearest hundredth)

(d) $2 \div 0.09$ (nearest hundredth)

(e) $3.1 \div 1.2$ (nearest thousandth)

(f) $0.1 \div 0.15$ (nearest thousandth)

(g) $4 \div 0.021$ (nearest whole number)

(h) $35.9 \div 2.4$ (nearest whole number)

Solution Recall that to divide and round to a given place, you carry out the division to one place beyond the given place and then round the quotient [see Example 4-57]:

(a) $0.3\overline{)5} = 3\overline{)\begin{array}{l}1\ 6.6\ 6 \approx 16.7\\5\ 0.0\ 0\\\underline{-3}\\2\ 0\\\underline{-1\ 8}\\2\ 0\\\underline{-1\ 8}\\2\ 0\\\underline{-1\ 8}\\2\end{array}}$

(b) $0.6\overline{)0.1} = 6\overline{)\begin{array}{l}0.1\ 6 \approx 0.2\\1.0\ 0\\\underline{-6}\\4\ 0\\\underline{-3\ 6}\\4\end{array}}$

(c) $0.07\overline{)1.3} = 7\overline{)\begin{array}{l}1\ 8.5\ 7\ 1 \approx 18.57\\1\ 3\ 0.0\ 0\ 0\\\underline{-7}\\6\ 0\\\underline{-5\ 6}\\4\ 0\\\underline{-3\ 5}\\5\ 0\\\underline{-4\ 9}\\1\ 0\\\underline{-7}\\3\end{array}}$

(d) $0.09\overline{)2} = 9\overline{)\begin{array}{l}2\ 2.2\ 2\ 2 \approx 22.22\\2\ 0\ 0.0\ 0\ 0\\\underline{-1\ 8}\\2\ 0\\\underline{-1\ 8}\\2\ 0\\\underline{-1\ 8}\\2\ 0\\\underline{-1\ 8}\\2\ 0\\\underline{-1\ 8}\\2\end{array}}$

(e) $1.2\overline{)3.1} = 12\overline{)\begin{array}{l}2.5\ 8\ 3\ 3 \approx 2.583\\3\ 1.0\ 0\ 0\\\underline{-2\ 4}\\7\ 0\\\underline{-6\ 0}\\1\ 0\ 0\\\underline{-9\ 6}\\4\ 0\\\underline{-3\ 6}\\4\ 0\\\underline{-3\ 6}\\4\end{array}}$

(f) $0.15\overline{)0.1} = 15\overline{)\begin{array}{l}0.6\ 6\ 6\ 6 \approx 0.667\\1\ 0.0\ 0\ 0\ 0\\\underline{-9\ 0}\\1\ 0\ 0\\\underline{-9\ 0}\\1\ 0\ 0\\\underline{-9\ 0}\\1\ 0\ 0\\\underline{-9\ 0}\\1\ 0\end{array}}$

$$\begin{array}{r} 1\ 9\ 0.4 \approx 190 \\ \textbf{(g)}\ 0.021\overline{)4} = 21\overline{)\ 4\ 0\ 0\ 0.0} \\ -2\ 1 \\ \hline 1\ 9\ 0 \\ -1\ 8\ 9 \\ \hline 1\ 0\ 0 \\ -\ \ 8\ 4 \\ \hline 1\ 6 \end{array}$$

$$\begin{array}{r} 1\ 4.9 \approx 15 \\ \textbf{(h)}\ 2.4\overline{)35.9} = 24\overline{)\ 3\ 5\ 9.0} \\ -2\ 4 \\ \hline 1\ 1\ 9 \\ -\ \ 9\ 6 \\ \hline 2\ 3\ 0 \\ -2\ 1\ 6 \\ \hline 1\ 4 \end{array}$$

PROBLEM 4-34   Divide by 10:

(a) 25.3 ÷ 10   **(b)** 125.49 ÷ 10   **(c)** 1.6 ÷ 10   **(d)** 3.75 ÷ 10   **(e)** 43 ÷ 10

(f) 0.6 ÷ 10   **(g)** 20 ÷ 10   **(h)** 600 ÷ 10   **(i)** 7000 ÷ 10   **(j)** 0.0025 ÷ 10

Solution   Recall that to divide by 10, you move the decimal point 1 decimal place to the left [see Examples 4-58 and 4-59]:

(a) 25.3 ÷ 10 = 2.53   **(b)** 125.49 ÷ 10 = 12.549   **(c)** 1.6 ÷ 10 = .16 = 0.16

(d) 3.75 ÷ 10 = .375 = 0.375   **(e)** 43. ÷ 10 = 4.3   **(f)** 0.6 ÷ 10 = 0.06

(g) 20 ÷ 10 = 20. ÷ 10 = 2   or   20̸ ÷ 10̸ = 2 ÷ 1 = 2

(h) 60̸0̸ ÷ 10̸ = 60 ÷ 1 = 60   **(i)** 700̸0̸ ÷ 10̸ = 700 ÷ 1 = 700   **(j)** 0.0025 ÷ 10 = 0.00025

PROBLEM 4-35   Divide by 100:

(a) 525.3 ÷ 100   **(b)** 60035.1 ÷ 100   **(c)** 23.4 ÷ 100

(d) 1.5 ÷ 100   **(e)** 0.3 ÷ 100   **(f)** 0.05 ÷ 100

(g) 2 ÷ 100   **(h)** 30 ÷ 100   **(i)** 900 ÷ 100

Solution   Recall that to divide by 100, you move the decimal point 2 decimal places to the left [see Examples 4-60 and 4-61]:

(a) 525.3 ÷ 100 = 5.253   **(b)** 60035.1 ÷ 100 = 600.351

(c) 23.4 ÷ 100 = .234 = 0.234   **(d)** 1.5 ÷ 100 = .015 = 0.015

(e) 0.3 ÷ 100 = .003 = 0.003   **(f)** 0.05 ÷ 100 = .0005 = 0.0005

(g) 2 ÷ 100 = 2. ÷ 100 = .02 = 0.02   **(h)** 30 ÷ 100 = 30. ÷ 100 = .30 = 0.3

(i) 900 ÷ 100 = 900. ÷ 100 = 9   or   90̸0̸ ÷ 10̸0̸ = 9 ÷ 1 = 9

PROBLEM 4-36   Divide by 1000:

(a) 1257.5 ÷ 1000   **(b)** 382.4 ÷ 1000   **(c)** 25.6 ÷ 1000   **(d)** 1.85 ÷ 1000   **(e)** 0.3 ÷ 1000

(f) 0.025 ÷ 1000   **(g)** 8 ÷ 1000   **(h)** 900 ÷ 1000   **(i)** 7000 ÷ 1000   **(j)** 1,000,000 ÷ 1000

Solution   Recall that to divide by 1000, you move the decimal point 3 decimal places to the left [see Examples 4-62 and 4-63]:

(a) 1257.5 ÷ 1000 = 1.2575   **(b)** 382.4 ÷ 1000 = .3824 = 0.3824

(c) 25.6 ÷ 1000 = .0256 = 0.0256   **(d)** 1.85 ÷ 1000 = 0.00185

(e) 0.3 ÷ 1000 = 0.0003   **(f)** 0.025 ÷ 1000 = 0.000025

(g) 8 ÷ 1000 = 0.008   **(h)** 900 ÷ 1000 = 0.900 = 0.9

(i) 7000 ÷ 1000 = 7 or 700̸0̸ ÷ 100̸0̸ = 7 ÷ 1 = 7   **(j)** 1,000,00̸0̸ ÷ 100̸0̸ = 1000

PROBLEM 4-37   Divide by powers of 10:

(a) 2.8 ÷ 10   **(b)** 2.8 ÷ 100   **(c)** 2.8 ÷ 1000

(d) $2.8 \div 10,000$ **(e)** $5857 \div 100,000$ **(f)** $258,973 \div 1,000,000$

(g) $373,582,961 \div 10,000,000$ **(h)** $80 \div 100,000$ **(i)** $24.99 \div 10$

(j) $359.86 \div 1000$

Solution Recall that to divide by a power of 10, you move the decimal point 1 decimal place to the left for each zero in the power of 10 [see Example 4-64]:

(a) $2.8 \div 10 = 0.28$ **(b)** $2.8 \div 100 = 0.028$

(c) $2.8 \div 1000 = 0.0028$ **(d)** $2.8 \div 10,000 = 0.00028$

(e) $5857 \div 100,000 = 0.05857$ **(f)** $258,973 \div 1,000,000 = 0.258973$

(g) $373,582,961 \div 10,000,000 = 37.3582961$ **(h)** $80. \div 100,000 = 0.0008$

(i) $24.99 \div 10 = 2.499$ **(j)** $359.86 \div 1000 = 0.35986$

PROBLEM 4-38 Find the average (mean) of each group of numbers. Round to the nearest tenth when the average is a repeating decimal:

(a) 5, 2, 4, 9, 3, 7   **(b)** 18, 52, 34, 56   **(c)** 25, 36, 79

(d) 123, 142, 5, 83, 91, 125   **(e)** 0.3, 0.2, 0.9, 0.7   **(f)** 1.5, 1.8, 2.7, 2.8, 3.5, 1.6, 0.2, 3.1

(g) 15.82, 4.89, 5.83   **(h)** $\frac{1}{5}, \frac{3}{5}$

Solution Recall that to find the average (mean) of two or more numbers, you must add all the numbers together and then divide the sum by the number of given numbers [see Example 4-65]:

(a) $5 + 2 + 4 + 9 + 3 + 7 = 30$ and $30 \div 6 = 5$

(b) $18 + 52 + 34 + 56 = 160$ and $160 \div 4 = 40$

(c) $25 + 36 + 79 = 140$ and $140 \div 3 = 46.666 \ldots \approx 46.7$ (nearest tenth)

(d) $123 + 142 + 5 + 83 + 91 + 125 = 569$ and $569 \div 6 = 94.8333 \ldots \approx 94.8$

(e) $0.3 + 0.2 + 0.9 + 0.7 = 2.1$ and $2.1 \div 4 = 0.525$

(f) $1.5 + 1.8 + 2.7 + 2.8 + 3.5 + 1.6 + 0.2 + 3.1 = 17.2$ and $17.2 \div 8 = 2.15$

(g) $15.82 + 4.89 + 5.83 = 26.54$ and $26.54 \div 3 = 8.84666 \ldots \approx 8.8$

(h) $\frac{1}{5} + \frac{3}{5} = \frac{4}{5}$ and $\frac{4}{5} \div 2 = \frac{4}{5} \div \frac{2}{1} = \frac{4}{5} \times \frac{1}{2} = \frac{2 \times 2}{5} \times \frac{1}{2} = \frac{2}{5}$

PROBLEM 4-39 Find the average (mean) of each given number of addends using the given sum. Round to the nearest tenth when the average is a repeating decimal:

(a) 258, 4 addends   **(b)** 192, 2 addends   **(c)** 354, 3 addends   **(d)** 7890, 6 addends

(e) 25.5, 5 addends   **(f)** 135.2, 8 addends   **(g)** 51.25, 7 addends   **(h)** 3048.25, 9 addends

Solution Recall that to find the average (mean) of a given number of addends, you divide the sum by the number of addends [see Example 4-66]:

(a) $258 \div 4 = 64.5$ **(b)** $192 \div 2 = 96$ **(c)** $354 \div 3 = 118$ **(d)** $7890 \div 6 = 1315$

(e) $25.5 \div 5 = 5.1$ **(f)** $135.2 \div 8 = 16.9$ **(g)** $51.25 \div 7 = 7.3$ **(h)** $3048.25 \div 9 = 338.7$

PROBLEM 4-40 Identify each decimal as either terminating or repeating:

(a) 0.3   **(b)** 0.333   **(c)** $0.\overline{3}$ **(d)** 0.333 ...   **(e)** 5.1

(f) 5.16   **(g)** 5.1666   **(h)** 5.1666 ...   **(i)** $5.1\overline{6}$ **(j)** $63.\overline{63}$

Solution   Recall that a terminating decimal ends in a given place and a repeating decimal repeats one or more digits forever [see Example 4-69]:

(a) 0.3 is a terminating decimal because it ends in the tenths place.

(b) 0.333 is a terminating decimal because it ends in the thousandths place.

(c) 0.$\overline{3}$ is a repeating decimal because it repeats 3s forever.

(d) 0.333 . . . is a repeating decimal because it repeats 3s forever.

(e) 5.1 is a terminating decimal because it ends in the tenths place.

(f) 5.16 is a terminating decimal because it ends in the hundredths place.

(g) 5.1666 is a terminating decimal because it ends in the ten thousandths place.

(h) 5.1666 . . . is a repeating decimal because it repeats 6s forever.

(i) 5.1$\overline{6}$ is a repeating decimal because it repeats 6s forever.

(j) 63.$\overline{63}$ is a repeating decimal because it repeats 63s forever.

PROBLEM 4-41   Rename each repeating decimal using bar notation in simplest form:

(a) 0.333 . . .   **(b)** 0.666 . . .   **(c)** 33.333 . . .   **(d)** 66.666 . . .

(e) 0.1666 . . .   **(f)** 0.8333 . . .   **(g)** 0.181818 . . .   **(h)** 0.090909 . . .

Solution   Recall that to write a repeating decimal using bar notation, you write each digit in the decimal part that repeats under the bar [see Example 4-71]:

(a) 0.333 . . . = 0.$\overline{3}$ **(b)** 0.666 . . . = 0.$\overline{6}$ **(c)** 33.333 . . . = 33.$\overline{3}$ **(d)** 66.666 . . . = 66.$\overline{6}$

(e) 0.1666 . . . = 0.1$\overline{6}$ **(f)** 0.8333 . . . = 0.8$\overline{3}$ **(g)** 0.181818 . . . = 0.$\overline{18}$ **(h)** 0.090909 . . . = 0.$\overline{09}$

PROBLEM 4-42   Rename each fraction as an equal terminating decimal:

(a) $\frac{1}{2}$ **(b)** $\frac{1}{4}$ **(c)** $\frac{1}{5}$ **(d)** $\frac{4}{5}$ **(e)** $\frac{1}{8}$ **(f)** $\frac{5}{8}$ **(g)** $\frac{1}{10}$ **(h)** $\frac{7}{10}$

Solution   Recall that to rename a fraction as an equal terminating decimal, you divide the denominator into the numerator to get a terminating decimal quotient [see Example 4-72]:

$$\textbf{(a) } 2\overline{)\begin{array}{l}0.5 \leftarrow \frac{1}{2}\\ 1.0 \\ \underline{-1\ 0} \\ 0\end{array}} \qquad \textbf{(b) } 4\overline{)\begin{array}{l}0.2\ 5 \leftarrow \frac{1}{4}\\ 1.0\ 0 \\ \underline{-\ 8} \\ 2\ 0 \\ \underline{-2\ 0} \\ 0\end{array}} \qquad \textbf{(c) } 5\overline{)\begin{array}{l}0.2 \leftarrow \frac{1}{5}\\ 1.0 \\ \underline{-1\ 0} \\ 0\end{array}} \qquad \textbf{(d) } 5\overline{)\begin{array}{l}0.8 \leftarrow \frac{4}{5}\\ 4.0 \\ \underline{-4\ 0} \\ 0\end{array}}$$

$$\textbf{(e) } 8\overline{)\begin{array}{l}0.1\ 2\ 5 \leftarrow \frac{1}{8}\\ 1.0\ 0\ 0 \\ \underline{-\ 8} \\ 2\ 0 \\ \underline{-1\ 6} \\ 4\ 0 \\ \underline{-4\ 0} \\ 0\end{array}} \qquad \textbf{(f) } 8\overline{)\begin{array}{l}0.6\ 2\ 5 \leftarrow \frac{5}{8}\\ 5.0\ 0\ 0 \\ \underline{-4\ 8} \\ 2\ 0 \\ \underline{-1\ 6} \\ 4\ 0 \\ \underline{-4\ 0} \\ 0\end{array}} \qquad \textbf{(g) } 10\overline{)\begin{array}{l}0.1 \leftarrow \frac{1}{10}\\ 1.0 \\ \underline{-1\ 0} \\ 0\end{array}} \qquad \textbf{(h) } 10\overline{)\begin{array}{l}0.7 \leftarrow \frac{7}{10}\\ 7.0 \\ \underline{-7\ 0} \\ 0\end{array}}$$

PROBLEM 4-43   Rename each fraction as an equal repeating decimal in simplest bar notation:

(a) $\frac{1}{3}$ **(b)** $\frac{1}{6}$ **(c)** $\frac{1}{9}$ **(d)** $\frac{5}{9}$ **(e)** $\frac{1}{12}$ **(f)** $\frac{5}{12}$

Solution   Recall that to rename a fraction as an equal repeating decimal, you divide the denominator into the numerator to get a repeating decimal quotient [see Example 4-72]:

(a)
$$
\begin{array}{r}
0.3\ 3\ 3\ldots = 0.\overline{3} \longleftarrow \tfrac{1}{3} \\
3\overline{)\ 1.0\ 0\ 0} \\
-\ 9 \\
\hline
1\ 0 \\
-\ 9 \\
\hline
1\ 0 \\
-\ 9 \\
\hline
1
\end{array}
$$

(b)
$$
\begin{array}{r}
0.1\ 6\ 6\ 6\ldots = 0.1\overline{6} \longleftarrow \tfrac{1}{6} \\
6\overline{)\ 1.0\ 0\ 0\ 0} \\
-\ 6 \\
\hline
4\ 0 \\
-3\ 6 \\
\hline
4\ 0 \\
-3\ 6 \\
\hline
4\ 0 \\
-3\ 6 \\
\hline
4
\end{array}
$$

(c)
$$
\begin{array}{r}
0.1\ 1\ 1\ldots = 0.\overline{1} \longleftarrow \tfrac{1}{9} \\
9\overline{)\ 1.0\ 0\ 0} \\
-\ 9 \\
\hline
1\ 0 \\
-\ 9 \\
\hline
1\ 0 \\
-\ 9 \\
\hline
1
\end{array}
$$

(d)
$$
\begin{array}{r}
0.5\ 5\ 5\ldots = 0.\overline{5} \longleftarrow \tfrac{5}{9} \\
9\overline{)\ 5.0\ 0\ 0} \\
-4\ 5 \\
\hline
5\ 0 \\
-4\ 5 \\
\hline
5\ 0 \\
-4\ 5 \\
\hline
5
\end{array}
$$

(e)
$$
\begin{array}{r}
0.0\ 8\ 3\ 3\ 3\ldots = 0.08\overline{3} \longleftarrow \tfrac{1}{12} \\
12\overline{)\ 1.0\ 0\ 0\ 0\ 0} \\
-\ 9\ 6 \\
\hline
4\ 0 \\
-3\ 6 \\
\hline
4\ 0 \\
-3\ 6 \\
\hline
4\ 0 \\
-3\ 6 \\
\hline
4
\end{array}
$$

(f)
$$
\begin{array}{r}
0.4\ 1\ 6\ 6\ 6\ldots = 0.41\overline{6} \longleftarrow \tfrac{5}{12} \\
12\overline{)\ 5.0\ 0\ 0\ 0\ 0} \\
-4\ 8 \\
\hline
2\ 0 \\
-1\ 2 \\
\hline
8\ 0 \\
-7\ 2 \\
\hline
8\ 0 \\
-7\ 2 \\
\hline
8\ 0 \\
-7\ 2 \\
\hline
8
\end{array}
$$

PROBLEM 4-44   Write each product of repeated factors as power notation:

(a) 5×5

(b) 0.1×0.1

(c) $\dfrac{4}{5} \times \dfrac{4}{5}$

(d) $2 \times 2 \times 2$

(e) $2.5 \times 2.5 \times 2.5$

(f) $\dfrac{1}{2} \times \dfrac{1}{2} \times \dfrac{1}{2}$

(g) $6 \times 6 \times 6 \times 6$

(h) $1.5 \times 1.5 \times 1.5 \times 1.5$

(i) $\dfrac{2}{3} \times \dfrac{2}{3} \times \dfrac{2}{3} \times \dfrac{2}{3}$

(j) $\dfrac{8 \times 8}{3}$

(k) $\dfrac{7}{9 \times 9 \times 9}$

(l) $\dfrac{2 \times 2}{3 \times 3 \times 3}$

Solution   Recall that to write a product of repeated factors as power notation, you use the repeated factor as the base and the number of repeated factors as the power (exponent) [see Examples 4-74 through 4-76]:

repeated factor

number of repeated factors

(a) $5 \times 5 = 5^2$

(b) $0.1 \times 0.1 = 0.1^2$

use parentheses

(c) $\dfrac{4}{5} \times \dfrac{4}{5} = \left(\dfrac{4}{5}\right)^2$

(d) $2 \times 2 \times 2 = 2^3$

(e) $2.5 \times 2.5 \times 2.5 = 2.5^3$

(f) $\dfrac{1}{2} \times \dfrac{1}{2} \times \dfrac{1}{2} = \left(\dfrac{1}{2}\right)^3$

(g) $6 \times 6 \times 6 \times 6 = 6^4$ **(h)** $1.5 \times 1.5 \times 1.5 \times 1.5 = 1.5^4$ **(i)** $\dfrac{2}{3} \times \dfrac{2}{3} \times \dfrac{2}{3} \times \dfrac{2}{3} = \left(\dfrac{2}{3}\right)^4$

(j) $\dfrac{8 \times 8}{3} = \dfrac{8^2}{3}$ **(k)** $\dfrac{7}{9 \times 9 \times 9} = \dfrac{7}{9^3}$ **(l)** $\dfrac{2 \times 2}{3 \times 3 \times 3} = \dfrac{2^2}{3^3}$

PROBLEM 4-45   Evaluate each power notation:

(a) 4^2 **(b)** 1.5^2 **(c)** $\left(\dfrac{1}{3}\right)^2$ **(d)** 2^3 **(e)** 0.1^3 **(f)** $\left(\dfrac{3}{4}\right)^3$ **(g)** 10^2 **(h)** 10^3

(i) 10^5 **(j)** 10^{12} **(k)** 0^3 **(l)** 1^{10} **(m)** $\dfrac{5^2}{4}$ **(n)** $\left(\dfrac{5}{4}\right)^2$ **(o)** 3^1 **(p)** $\left(\dfrac{7}{10}\right)^1$

(q) 0^1 **(r)** 8.25^1 **(s)** 1^0 **(t)** $\left(\dfrac{9}{10}\right)^0$ **(u)** 0^0 **(v)** 198.756^0 **(w)** $\dfrac{2}{3^2}$ **(x)** $\dfrac{5^2}{2^3}$

Solution   Recall that to evaluate power notation, you first write the power notation as a product of repeated factors and then multiply to get a whole number, or fraction, or decimal. Also, $a^1 = a$ and $a^0 = 1$ if $a \neq 0$. [see Examples 4-77 through 4-80].

repeated factor
number of repeated factors

(a) $4^2 = 4 \times 4 = 16$ **(b)** $1.5^2 = 1.5 \times 1.5 = 2.25$ **(c)** $\left(\dfrac{1}{3}\right)^2 = \dfrac{1}{3} \times \dfrac{1}{3} = \dfrac{1}{9}$

3 zeros

(d) $2^3 = 2 \times 2 \times 2 = 8$ **(e)** $0.1^3 = 0.1 \times 0.1 \times 0.1 = 0.001$ **(f)** $\left(\dfrac{3}{4}\right)^3 = \dfrac{3}{4} \times \dfrac{3}{4} \times \dfrac{3}{4} = \dfrac{27}{64}$

2 zeros

(g) $10^2 = 100$

3 zeros

(h) $10^3 = 1000$

5 zeros

(i) $10^5 = 100{,}000$

12 zeros

(j) $10^{12} = 1{,}000{,}000{,}000{,}000$ **(k)** $0^3 = 0$ $(0^n = 0$ if $n \neq 0)$ **(l)** $1^{10} = 1$ $(1^n = 1)$

(m) $\dfrac{5^2}{4} = \dfrac{5 \times 5}{4} = \dfrac{25}{4}$ or $6\frac{1}{4}$ **(n)** $\left(\dfrac{5}{4}\right)^2 = \dfrac{5}{4} \times \dfrac{5}{4} = \dfrac{25}{16}$ or $1\frac{9}{16}$ **(o)** $3^1 = 3$ $(a^1 = a)$

(p) $\left(\dfrac{7}{10}\right)^1 = \dfrac{7}{10}$ **(q)** $0^1 = 0$ **(r)** $8.25^1 = 8.25$

(s) $1^0 = 1$ $(a^0 = 1$ if $a \neq 0)$ **(t)** $\left(\dfrac{9}{10}\right)^0 = 1$ **(u)** 0^0 is not defined

(v) $198.756^0 = 1$ **(w)** $\dfrac{2}{3^2} = \dfrac{2}{3 \times 3} = \dfrac{2}{9}$ **(x)** $\dfrac{5^2}{2^3} = \dfrac{5 \times 5}{2 \times 2 \times 2}$

$= \dfrac{25}{8}$ or $3\frac{1}{8}$

PROBLEM 4-46   Find each square using paper and pencil:

(a) 1^2 **(b)** 2^2 **(c)** 3^2 **(d)** 4^2 **(e)** 5^2 **(f)** 6^2 **(g)** 7^2 **(h)** 8^2 **(i)** 9^2 **(j)** 10^2

(k) $\left(\dfrac{1}{2}\right)^2$ **(l)** $\left(\dfrac{2}{3}\right)^2$ **(m)** $\left(\dfrac{3}{4}\right)^2$ **(n)** 0.2^2 **(o)** 0.03^2 **(p)** 0.004^2 **(q)** 11^2 **(r)** 12^2 **(s)** 20^2 **(t)** 300^2

Solution Recall that to find the square of a given number, you multiply the given number times it-self [see Examples 4-81 and 4-82]:

1 times itself

(a) $1^2 = \overbrace{1 \times 1}^{} = 1$ **(b)** $2^2 = 4$ **(c)** $3^2 = 9$ **(d)** $4^2 = 16$ **(e)** $5^2 = 25$

(f) $6^2 = 36$ **(g)** $7^2 = 49$ **(h)** $8^2 = 64$ **(i)** $9^2 = 81$ **(j)** $10^2 = 100$

(k) $\left(\dfrac{1}{2}\right)^2 = \dfrac{1}{4}$ **(l)** $\left(\dfrac{2}{3}\right)^2 = \dfrac{4}{9}$ **(m)** $\left(\dfrac{3}{4}\right)^2 = \dfrac{9}{16}$ **(n)** $0.2^2 = 0.04$ **(o)** $0.03^2 = 0.0009$

(p) $0.004^2 = 0.000016$ **(q)** $11^2 = 121$ **(r)** $12^2 = 144$ **(s)** $20^2 = 400$ **(t)** $300^2 = 90{,}000$

PROBLEM 4-47 Find each square root using paper and pencil:

(a) $\sqrt{1}$ **(b)** $\sqrt{100}$ **(c)** $\sqrt{49}$ **(d)** $\sqrt{4}$ **(e)** $\sqrt{16}$ **(f)** $\sqrt{81}$

(g) $\sqrt{25}$ **(h)** $\sqrt{9}$ **(i)** $\sqrt{36}$ **(j)** $\sqrt{64}$ **(k)** $\sqrt{\dfrac{4}{9}}$ **(l)** $\sqrt{\dfrac{25}{49}}$

(m) $\sqrt{\dfrac{1}{100}}$ **(n)** $\sqrt{\dfrac{1}{16}}$ **(o)** $\sqrt{121}$ **(p)** $\sqrt{225}$ **(q)** $\sqrt{400}$ **(r)** $\sqrt{144}$

Solution Recall that to find the square root of a given whole number square, you can divide the given whole number by various whole number divisors until the whole number divisor and the quotient are the the same [see Examples 4-84 and 4-85]:

(a) $1 \div 1 = 1$ means $\sqrt{1} = 1$ **(b)** $100 \div 10 = 10$ means $\sqrt{100} = 10$

(c) $49 \div 7 = 7$ means $\sqrt{49} = 7$ **(d)** $4 \div 2 = 2$ means $\sqrt{4} = 2$

(e) $16 \div 4 = 4$ means $\sqrt{16} = 4$ **(f)** $81 \div 9 = 9$ means $\sqrt{81} = 9$

(g) $25 \div 5 = 5$ means $\sqrt{25} = 5$ **(h)** $9 \div 3 = 3$ means $\sqrt{9} = 3$

(i) $36 \div 6 = 6$ means $\sqrt{36} = 6$ **(j)** $64 \div 8 = 8$ means $\sqrt{64} = 8$

(k) $\sqrt{\dfrac{4}{9}} = \dfrac{2}{3}$ **(l)** $\sqrt{\dfrac{25}{49}} = \dfrac{5}{7}$

(m) $\sqrt{\dfrac{1}{100}} = \dfrac{1}{10}$ **(n)** $\sqrt{\dfrac{1}{16}} = \dfrac{1}{4}$

(o) $121 \div 11 = 11$ means $\sqrt{121} = 11$ **(p)** $225 \div 15 = 15$ means $\sqrt{225} = 15$

(q) $400 \div 20 = 20$ means $\sqrt{400} = 20$ **(r)** $144 \div 12 = 12$ means $\sqrt{144} = 12$

PROBLEM 4-48 Find each square using Appendix Table 2:

(a) 17^2 **(b)** 23^2 **(c)** 39^2 **(d)** 41^2 **(e)** 58^2 **(f)** 60^2 **(g)** 72^2 **(h)** 84^2

Solution Recall that to find the square of any whole number from 0 to 100, you can use Appendix Table 2 by placing your left finger on the given base number in the "Number N" column and your right finger at the top of the "Number N^2" column and then moving your left finger straight across and your right finger straight down until they meet at the square of the given base number [see Example 4-86]:

(a) $17^2 = 289$ **(b)** $23^2 = 529$ **(c)** $39^2 = 1521$ **(d)** $41^2 = 1681$

(e) $58^2 = 3364$ **(f)** $60^2 = 3600$ **(g)** $72^2 = 5184$ **(h)** $84^2 = 7056$

PROBLEM 4-49 Find each square root using Appendix Table 2:

(a) $\sqrt{2}$ **(b)** $\sqrt{3}$ **(c)** $\sqrt{8}$ **(d)** $\sqrt{26}$ **(e)** $\sqrt{30}$ **(f)** $\sqrt{59}$ **(g)** $\sqrt{72}$ **(h)** $\sqrt{87}$

Solution Recall that to find the square root of any whole number from 0 to 100, you can use Appendix Table 2 by placing your left finger on the given radicand in the "Number N" column and your right finger at the top of the "Square Root $\sqrt{N}$" column and then moving your left finger straight across and your right finger straight down until they meet at the approximated square root of the given radicand [see Example 4-87]:

(a) $\sqrt{2} \approx 1.414$ (b) $\sqrt{3} \approx 1.732$ (c) $\sqrt{8} \approx 2.828$ (d) $\sqrt{26} \approx 5.099$

(e) $\sqrt{30} \approx 5.477$ (f) $\sqrt{59} \approx 7.681$ (g) $\sqrt{72} \approx 8.485$ (h) $\sqrt{87} \approx 9.327$

PROBLEM 4-50   Find each square using a calculator:

(a) 101^2 (b) 300^2 (c) 4.56^2 (d) 82.1^2 (e) 1000^2 (f) 9508^2 (g) 27.82^2 (h) 502.7^2

Solution   Recall that to find the square of any whole number or decimal with 4 digits or less, you can use a standard 8-digit display calculator. To find the square of any whole number or decimal with 5 digits or less, you can use a standard 10-digit display calculator [see Example 4-88].

(a) $101^2 = 10,201$ (b) $300^2 = 90,000$ (c) $4.56^2 = 20.7936$ (d) $82.1^2 = 6740.41$

(e) $1000^2 = 1,000,000$ (f) $9508^2 = 90,402,064$ (g) $27.82^2 = 773.9524$ (h) $502.7^2 = 252,707.29$

PROBLEM 4-51   Find or approximate each square root using a calculator:

(a) $\sqrt{101}$ (b) $\sqrt{1000}$ (c) $\sqrt{72,361}$

(d) $\sqrt{265,271}$ (e) $\sqrt{9,205,000}$ (f) $\sqrt{38,521.619}$

(g) $\sqrt{3.5}$ (h) $\sqrt{31.36}$ (i) $\sqrt{0.002589}$

Solution   Recall that to find or approximate the square root of any 8-digit whole number or decimal, you can use a standard 8-digit calculator. To find or approximate the square root of any 10-digit number or decimal, you can use a standard 10-digit display calculator [see Example 4-89].

(a) $\sqrt{101} \approx 10.049876$ (b) $\sqrt{1000} \approx 31.622777$ (c) $\sqrt{72,361} = 269$

(d) $\sqrt{265,271} \approx 515.04466$ (e) $\sqrt{9,205,000} \approx 3033.9743$ (f) $\sqrt{38,521.619} \approx 196.26925$

(g) $\sqrt{3.5} \approx 1.8708287$ (h) $\sqrt{31.36} = 5.6$ (i) $\sqrt{0.002589} \approx 0.0508822$

PROBLEM 4-52   Find or approximate each square root of a fraction using a calculator:

(a) $\sqrt{\dfrac{1}{2}}$ (b) $\sqrt{\dfrac{1}{3}}$ (c) $\sqrt{\dfrac{1}{4}}$ (d) $\sqrt{\dfrac{2}{3}}$ (e) $\sqrt{\dfrac{3}{4}}$ (f) $\sqrt{\dfrac{7}{10}}$ (g) $\sqrt{\dfrac{9}{16}}$ (h) $\sqrt{\dfrac{5}{32}}$

Solution   Recall that to find or approximate the square root of a fraction using a calculator, you first divide the numerator by the denominator and then press the square root key [see Example 4-90]:

(a) $\sqrt{\dfrac{1}{2}} = \sqrt{0.5} \approx 0.7071067$ (b) $\sqrt{\dfrac{1}{3}} \approx \sqrt{0.3333333} \approx 0.5773502$

(c) $\sqrt{\dfrac{1}{4}} = \sqrt{0.25} = 0.5$ (d) $\sqrt{\dfrac{2}{3}} \approx \sqrt{0.66666666} \approx 0.8164965$

(e) $\sqrt{\dfrac{3}{4}} = \sqrt{0.75} \approx 0.8660254$ (f) $\sqrt{\dfrac{7}{10}} = \sqrt{0.7} \approx 0.8366600$

(g) $\sqrt{\dfrac{9}{16}} = \sqrt{0.5625} = 0.75$ (h) $\sqrt{\dfrac{5}{32}} = \sqrt{0.15625} \approx 0.3952847$

PROBLEM 4-53   Decide what to do (add, subtract, multiply, and/or divide) and then solve the following word problems:

(a) A 6-pack of *Drink Me* soda costs $1.99. How many cents does each bottle cost? [*Hint:* Remember that stores always round *up* to the nearest cent.]

(b) Each bottle of *Drink Me* soda costs 25¢ on sale. How much does a 6-pack of *Drink Me* cost on sale if there is an additional discount of 1 cent when a full 6-pack is purchased?

(c) A can of corn costs $1.05. A can of beans costs 49¢. How much more does the corn cost than the beans?

(d) A large loaf of bread costs $1.49. A small loaf of bread costs 99¢. How much will it cost for both a large and small loaf of bread?

(e) A case of noodles costs $19.08. Each box of noodles in a case costs $1.59. How many boxes are in a case?

(f) Brand X oil costs $29.88 per case. Brand Y oil costs $9.96 for the same size case. How many times more expensive is brand X than brand Y?

(g) The road mileage between two cities is 125.6 km. How many kilometers must be driven to make the trip 20 times?

(h) Sam gives the clerk $20 for a $6.88 purchase amount. How much change should Sam receive?

Solution Recall that to solve addition word problems containing decimals, you find how much two or more amounts are all together. To solve subtraction word problems containing decimals, you find the difference between two amounts. To solve multiplication word problems with decimals, you find how much two or more equal amounts are all together. To solve division word problems with decimals, you find how many equal amounts there are in all [see Examples 4-91 and 4-92].

(a) divide and then round up to the next cent: $1.99 \div 6 = 0.331\overline{6} \approx 0.34$ or 34¢

(b) multiply and then subtract 1 cent: $6 \times 25 = 150$ and $150 - 1 = 149$ (cents) or $1.49

(c) subtract: $1.05 - 0.49 = 0.56$ (dollars) or 56 cents

(d) add: $1.49 + 0.99 = 2.48$ (dollars)

(e) divide: $19.08 \div 1.59 = 12$ (boxes)

(f) divide: $29.88 \div 9.96 = 3$ (times more)

(g) multiply: $20 \times 125.6 = 2512$ (km)

(h) subtract: $20 - 6.88 = 13.12$ (dollars)

Supplementary Exercises

PROBLEM 4-54 Draw a line under the digit in each decimal that has the given value:

(a) thousandths: 6582.197034

(b) tens: 3075.192864

(c) tenths: 4132.709865

(d) hundredths: 9827.651430

(e) hundreds: 7043.928165

(f) millionths: 5897.264013

(g) ten thousandths: 1087.649235

(h) hundred thousandths: 2540.738169

PROBLEM 4-55 Write the place-value name for each underlined digit:

(a) 6528.139704   (b) 5764.120839   (c) 4126.518093   (d) 3487.659102

(e) 9870.431625   (f) 8213.597604   (g) 2087.941365   (h) 7501.249863

PROBLEM 4-56 Write the value of 2 in each decimal:

(a) 4123.587609   (b) 8570.624139   (c) 2658.901743   (d) 6409.218753

(e) 1587.604932   (f) 9413.850287   (g) 7502.613489   (h) 3415.098726

PROBLEM 4-57 Write the short decimal-word name for each decimal number:

(a) 2.6   (b) 1.75   (c) 3.08   (d) 4.375   (e) 0.045   (f) 0.005   (g) 0.0015   (h) 0.0125

PROBLEM 4-58   Write the decimal number for each short decimal-word name:

(a) 5 and 5 tenths   (b) 3 and 4 hundredths   (c) 6 and 125 thousandths   (d) 12 and 2 thousandths

(e) 3 thousandths   (f) 75 millionths   (g) 815 tenths   (h) 7 and 8 tenths

PROBLEM 4-59   Write the largest decimal in the following pairs of decimals:

(a) 2.35 and 2.53   (b) 0.01 and 0.001   (c) 3.1 and 3.10   (d) 1.01 and 1.011

(e) 34.8l9 and 34.90   (f) 26.78 and 25.79   (g) 0.0101 and 0.01   (h) 7.601 and 7.0601

PROBLEM 4-60   List each group of decimals from largest to smallest:

(a) 1.1, 1.111, 1.11   (b) 0.9, 0.99, 0.909

(c) 3.1, 3.141, 3.14, 3.1415   (d) 3.025, 3.520, 3.052, 3.250, 3.502, 3.205

(e) 0.125, 12.5, 1.25, 125.0   (f) 1.414, 1, 1.732, 2.828, 2.646, 2, 2.449, 3, 2.236

PROBLEM 4-61   Round each decimal to the given place:

(a) 5.8361 to the nearest hundredth   (b) 3.14 to the nearest whole number

(c) 0.81539 to the nearest thousandth   (d) 2.7506 to the nearest tenth

(e) 0.815649 to the nearest ten thousandth   (f) 0.2185 to the nearest hundredth

PROBLEM 4-62   Rename each whole number as an equal decimal:

(a) 4   (b) 9   (c) 0   (d) 7   (e) 15   (f) 18   (g) 135   (h) 1

PROBLEM 4-63   Rename each decimal as an equal whole number:

(a) 5.0   (b) 1.0   (c) 0.00   (d) 4.00   (e) 15.00   (f) 7.000   (g) 8.0   (h) 3.0000

PROBLEM 4-64   Rename each decimal fraction as an equal proper fraction in simplest form:

(a) 0.3   (b) 0.9   (c) 0.2   (d) 0.6   (e) 0.25   (f) 0.05   (g) 0.375   (h) 0.1875

PROBLEM 4-65   Rename each fraction as an equal decimal:

(a) $\frac{1}{10}$ (b) $\frac{7}{10}$ (c) $\frac{29}{10}$ (d) $\frac{43}{10}$ (e) $\frac{1}{100}$ (f) $\frac{23}{100}$ (g) $\frac{859}{100}$ (h) $\frac{1}{1000}$

PROBLEM 4-66   Rename each decimal as an equal mixed number in simplest form and as an equal improper fraction in lowest terms:

(a) 2.9   (b) 1.3   (c) 3.75   (d) 6.50   (e) 9.125   (f) 7.375

PROBLEM 4-67   Rename each number as an equal decimal with the given number of decimal places:

(a) 2 with 1 decimal place   (b) 4 with 5 decimal places   (c) 8 with 3 decimal places

(d) 1 with 4 decimal places   (e) 1.5 with 2 decimal places   (f) 0.9 with 3 decimal places

PROBLEM 4-68   Add with decimals:

(a) 1.125 + 3.375   (b) 4 + 8.25   (c) 5.1 + 16   (d) 0.915 + 0.6

(e) 1.4 + 18.35   (f) 123 + 0.123   (g) 14.58 + 8 + 0.06 + 12.1 + 125

PROBLEM 4-69   Subtract with decimals:

(a) 8.451 − 0.079   (b) 6.85 − 3   (c) 5 − 2.84   (d) 6.3 − 5.859

(e) 7.285 − 6.9   (f) 18.49 − 9.99   (g) 8 − 0.1   (h) 17.5 − 9

PROBLEM 4-70   Add or subtract with money, as indicated:

(a) $15.49 + $8.37   (b) $19 + $8.50   (c) $21.75 − $16   (d) $43 − $29

(e) $5 − 97¢   (f) 82¢ − 57¢   (g) $20 − $8.43   (h) $8.52 + $19.75 + 87¢ + $25 + 83¢

PROBLEM 4-71   Multiply with decimals:

(a) 8 × 0.3   (b) 1.2 × 5   (c) 0.2 × 0.3   (d) 1.5 × 2.4   (e) 3.25 × 0.7

(f) 8 × 0.125   (g) 5.67 × 8.02   (h) 0.6 × 0.375   (i) 0.03 × 0.04   (j) 0.0005 × 0.001

PROBLEM 4-72   Multiply by powers of 10:

(a) 10 × 0.2   (b) 10 × 80   (c) 100 × 0.3   (d) 100 × 20   (e) 1000 × 0.005

(f) 1000 × 7000   (g) 10,000 × 0.15   (h) 10,000 × 2300   (i) 100,000 × 0.0025   (j) 1,000,000 × 1000

PROBLEM 4-73   Divide with decimals:

(a) 1.5 ÷ 3   (b) 1.25 ÷ 5   (c) 0.012 ÷ 0.2   (d) 0.492 ÷ 0.4   (e) 1.2 ÷ 1.5

(f) 10.419 ÷ 2.3   (g) 17.55 ÷ 0.75   (h) 0.0004 ÷ 0.08   (i) 1.5625 ÷ 1.25   (j) 0.000002 ÷ 0.001

PROBLEM 4-74   Divide and round each quotient to the nearest whole, tenth, hundredth, and thousandth:

(a) 1.87 ÷ 3   (b) 0.3 ÷ 7   (c) 0.285 ÷ 9   (d) 1 ÷ 64   (e) 0.283 ÷ 0.3

(f) 25 ÷ 0.6   (g) 31 ÷ 1.5   (h) 4.85 ÷ 1.2   (i) 800 ÷ 0.07   (j) 1 ÷ 1.05

PROBLEM 4-75   Divide by powers of 10:

(a) 0.2 ÷ 10   (b) 5 ÷ 10   (c) 1.3 ÷ 100   (d) 60 ÷ 100   (e) 25.8 ÷ 1000

(f) 100 ÷ 1000   (g) 2.5 ÷ 10,000   (h) 1,000,000 ÷ 10,000   (i) 0.1 ÷ 100,000   (j) 1 ÷ 1,000,000

PROBLEM 4-76   Find the average (mean) of:

(a) 82, 96, 87, 93   (b) 1.25, 1.46, 1.26, 1.75, 0.82   (c) 0.2, 0.1, 0.9, 1.1

(d) $\frac{3}{4}, \frac{7}{8}$ (e) 486 if there are 2 addends   (f) 1.08 if there are 4 addends

(g) 250.87 if there are 3 addends   (h) 0.91 if there are 6 addends   (i) $7\frac{1}{2}$ if there are 3 addends

PROBLEM 4-77   Rename each repeating decimal using bar notation in simplest form:

(a) 0.0333 . . .   (b) 0.00666 . . .   (c) 16.666 . . .

(d) 83.333 . . .   (e) 0.454545 . . .   (f) 0.636363 . . .

(g) 0.123123123 . . .   (h) 1.010101 . . .   (i) 0.285714285714285714 . . .

PROBLEM 4-78   Rename each fraction as an equal terminating or repeating decimal:

(a) $\frac{3}{4}$ (b) $\frac{2}{5}$ (c) $\frac{3}{5}$ (d) $\frac{3}{8}$ (e) $\frac{7}{8}$ (f) $\frac{3}{10}$ (g) $\frac{9}{10}$ (h) $\frac{3}{16}$ (i) $\frac{11}{32}$ (j) $\frac{1}{64}$

(k) $\frac{2}{3}$ (l) $\frac{5}{6}$ (m) $\frac{2}{9}$ (n) $\frac{7}{9}$ (o) $\frac{2}{11}$ (p) $\frac{10}{11}$ (q) $\frac{7}{12}$ (r) $\frac{11}{12}$ (s) $\frac{2}{7}$ (t) $\frac{5}{7}$

PROBLEM 4-79   Write each as power notation:

(a) 8 × 8   (b) 2.3 × 2.3   (c) $\frac{9}{10} \times \frac{9}{10}$ (d) 7 × 7 × 7

(e) 0.1 × 0.1 × 0.1   (f) $\frac{3}{4} \times \frac{3}{4} \times \frac{3}{4}$ (g) 10 × 10 × 10 × 10   (h) 24.75 × 24.75 × 24.75

(i) $\dfrac{2}{3} \times \dfrac{2}{3} \times \dfrac{2}{3} \times \dfrac{2}{3}$ **(j)** $\dfrac{6 \times 6}{5}$ **(k)** $\dfrac{2}{5 \times 5 \times 5}$ **(l)** $\dfrac{9 \times 9 \times 9}{3 \times 3}$

(m) 0   **(n)** 1   **(o)** 2   **(p)** 10

(q) 12.5   **(r)** $\dfrac{3}{4}$ **(s)** $5 \times 5 \times 4$ **(t)** $3 \times 6 \times 6 \times 6 \times 6$

(u) 0×0 **(v)** $1 \times 1 \times 1$ **(w)** 2×2 **(x)** $3 \times 3 \times 3$

PROBLEM 4-80   Evaluate each power notation:

(a) 9^2 **(b)** 0.2^2 **(c)** $\left(\dfrac{3}{4}\right)^2$ **(d)** 3^3 **(e)** 0.2^3 **(f)** $\left(\dfrac{1}{2}\right)^3$ **(g)** 10^1 **(h)** 10^4

(i) 10^6 **(j)** 10^{11} **(k)** 0^5 **(l)** 1^8 **(m)** $\dfrac{3^2}{2}$ **(n)** $\left(\dfrac{3}{2}\right)^2$ **(o)** 5^1 **(p)** $\left(\dfrac{2}{3}\right)^1$

(q) 0^1 **(r)** 25.3^1 **(s)** 10^0 **(t)** $\left(\dfrac{4}{5}\right)^0$ **(u)** 0^0 **(v)** 0.00856^0 **(w)** $\dfrac{18}{2^3}$ **(x)** $\left(\dfrac{3}{4}\right)^4$

PROBLEM 4-81   Find each square using paper and pencil and Appendix Table 2, or a calculator.

(a) 0^2 **(b)** 1^2 **(c)** 2^2 **(d)** 3^2 **(e)** 4^2 **(f)** 5^2 **(g)** 6^2 **(h)**7^2

(i) 8^2 **(j)** 9^2 **(k)** 10^2 **(l)** 11^2 **(m)** 12^2 **(n)** 13^2 **(o)** 14^2 **(p)** 15^2

(q) 225^2 **(r)** $\left(\dfrac{1}{3}\right)^2$ **(s)** $\left(\dfrac{1}{4}\right)^2$ **(t)** $\left(\dfrac{2}{5}\right)^2$ **(u)** $\left(\dfrac{9}{10}\right)^2$ **(v)** $\left(\dfrac{9}{100}\right)^2$ **(w)** 0.1^2 **(x)** 2.5^2

PEOBLEM 4-82   Find or approximate each square root using paper and pencil and Appendix Table 2, or a calculator. Round to the nearest thousandth when necessary:

(a) $\sqrt{0}$ **(b)** $\sqrt{1}$ **(c)** $\sqrt{2}$ **(d)** $\sqrt{3}$ **(e)** $\sqrt{4}$ **(f)** $\sqrt{5}$ **(g)** $\sqrt{6}$ **(h)** $\sqrt{7}$

(i) $\sqrt{8}$ **(j)** $\sqrt{9}$ **(k)** $\sqrt{10}$ **(l)** $\sqrt{11}$ **(m)** $\sqrt{12}$ **(n)** $\sqrt{13}$ **(o)** $\sqrt{14}$ **(p)** $\sqrt{15}$

(q) $\sqrt{225}$ **(r)** $\sqrt{\dfrac{1}{3}}$ **(s)** $\sqrt{\dfrac{1}{4}}$ **(t)** $\sqrt{\dfrac{2}{5}}$ **(u)** $\sqrt{\dfrac{9}{10}}$ **(v)** $\sqrt{\dfrac{9}{100}}$ **(w)** $\sqrt{0.1}$ **(x)** $\sqrt{2.5}$

PROBLEM 4-83   Solve each word problem containing decimals using addition, subtraction, multiplication, and/or division:

(a) Lyle received math scores for the semester of 96, 84, 98, 72, and 83. If an A is 96–100, and a B is 88–95, and a C is 77–87, then what grade will Lyle earn?

(b) What is the lowest score that Lyle, in problem (a), can get on the next test for a B average if each test is worth 100 points?

(c) Melody pays $650.75 rent each month. How much rent does she pay during a year (12 months)?

(d) Yale owes $10,269 for a new car he just purchased. The amount owed is to be paid back over a 36-month period. What are the monthly payments?

(e) There are 365 days in a normal year. How many 7-day weeks are in a normal year?

(f) There are 366 days in a leap year. How many days are in an average month during a leap year?

(g) How many more hours are in an average month during a leap year than a normal year?

(h) If Meg works 40 hours per week at $12.25 per hour, how much will she earn during an average 365-day year?

Answers to Supplementary Exercises

(4-54) **(a)** 6582.197034   **(b)** 3075.192864   **(c)** 4132.709865   **(d)** 9827.651430
    (e) 7043.928165   **(f)** 5897.264013   **(g)** 1087.649235   **(h)** 2540.738169

(4-55) **(a)** thousandths   **(b)** millionths   **(c)** hundredths   **(d)** hundred thousandths
    (e) tenths   **(f)** ten thousandths   **(g)** ones   **(h)** hundreds

(4-56) **(a)** 20   **(b)** 0.02   **(c)** 2000   **(d)** 0.2
    (e) 0.000002   **(f)** 0.0002   **(g)** 2   **(h)** 0.00002

(4-57) **(a)** 2 and 6 tenths   **(b)** 1 and 75 hundredths
    (c) 3 and 8 hundredths   **(d)** 4 and 375 thousandths
    (e) 45 thousandths   **(f)** 5 thousandths
    (g) 15 ten thousandths   **(h)** 125 ten thousandths

(4-58) **(a)** 5.5   **(b)** 3.04   **(c)** 6.125   **(d)** 12.002
    (e) 0.003   **(f)** 0.000075   **(g)** 81.5   **(h)** 7.8

(4-59) **(a)** 2.53   **(b)** 0.01   **(c)** $3.1 = 3.10$ **(d)** 1.011
    (e) 34.90   **(f)** 26.78   **(g)** 0.0101   **(h)** 7.601

(4-60) **(a)** 1.111, 1.11, 1.1   **(b)** 0.99, 0.909, 0.9
    (c) 3.1415, 3.141, 3.14, 3.1   **(d)** 3.520, 3.502, 3.250, 3.205, 3.052, 3.025
    (e) 125.0, 12.5, 1.25, 0.125   **(f)** 3, 2.828, 2.646, 2.449, 2.236, 2, 1.732, 1.414, 1

(4-61) **(a)** 5.84   **(b)** 3   **(c)** 0.815   **(d)** 2.8   **(e)** 0.8156   **(f)** 0.22

(4-62) **(a)** 4.0   **(b)** 9.0   **(c)** 0.0   **(d)** 7.0   **(e)** 15.0   **(f)** 18.0   **(g)** 135.0   **(h)** 1.0

(4-63) **(a)** 5   **(b)** 1   **(c)** 0   **(d)** 4   **(e)** 15   **(f)** 7   **(g)** 8   **(h)** 3

(4-64) **(a)** $\frac{3}{10}$ **(b)** $\frac{9}{10}$ **(c)** $\frac{1}{5}$ **(d)** $\frac{3}{5}$ **(e)** $\frac{1}{4}$ **(f)** $\frac{1}{20}$ **(g)** $\frac{3}{8}$ **(h)** $\frac{3}{16}$

(4-65) **(a)** 0.1   **(b)** 0.7   **(c)** 2.9   **(d)** 4.3   **(e)** 0.01   **(f)** 0.23   **(g)** 8.59   **(h)** 0.001

(4-66) **(a)** $2\frac{9}{10}, \frac{29}{10}$ **(b)** $1\frac{3}{10}, \frac{13}{10}$ **(c)** $3\frac{3}{4}, \frac{15}{4}$ **(d)** $6\frac{1}{2}, \frac{13}{2}$ **(e)** $9\frac{1}{8}, \frac{73}{8}$ **(f)** $7\frac{3}{8}, \frac{59}{8}$

(4-67) **(a)** 2.0   **(b)** 4.00000   **(c)** 8.000   **(d)** 1.0000   **(e)** 1.50   **(f)** 0.900

(4-68) **(a)** 4.5   **(b)** 12.25   **(c)** 21.1   **(d)** 1.515   **(e)** 19.75   **(f)** 123.123   **(g)** 159.74

(4-69) **(a)** 8.372   **(b)** 3.85   **(c)** 2.16   **(d)** 0.441   **(e)** 0.385   **(f)** 8.5   **(g)** 7.9   **(h)** 8.5

(4-70) **(a)** $23.86   **(b)** $27.50   **(c)** $5.75   **(d)** $14
    (e) $4.03   **(f)** $0.25 or 25¢   **(g)** $11.57   **(h)** $54.97

(4-71) **(a)** 2.4   **(b)** 6   **(c)** 0.06   **(d)** 3.6   **(e)** 2.275
    (f) 1   **(g)** 45.4734   **(h)** 0.225   **(i)** 0.0012   **(j)** 0.0000005

(4-72) **(a)** 2   **(b)** 800   **(c)** 30   **(d)** 2000   **(e)** 5
    (f) 7,000,000   **(g)** 1500   **(h)** 23,000,000   **(i)** 250   **(j)** 1,000,000,000

(4-73) **(a)** 0.5   **(b)** 0.25   **(c)** 0.06   **(d)** 1.23   **(e)** 0.8
    (f) 4.53   **(g)** 23.4   **(h)** 0.005   **(i)** 1.25   **(j)** 0.002

(4-74) **(a)** 1, 0.6, 0.62, 0.623   **(b)** 0, 0, 0.04, 0.043
(c) 0, 0, 0.03, 0.032   **(d)** 0, 0, 0.02, 0.016
(e) 1, 0.9, 0.94, 0.943   **(f)** 42, 41.7, 41.67, 41.667
(g) 21, 20.7, 20.67, 20.667   **(h)** 4, 4.0, 4.04, 4.042
(i) 11,429; 11,428.6; 11,428.57; 11,428.571   **(j)** 1, 1.0, 0.95, 0.952

(4-75) **(a)** 0.02   **(b)** 0.5   **(c)** 0.013   **(d)** 0.6   **(e)** 0.0258
(f) 0.1   **(g)** 0.00025   **(h)** 100   **(i)** 0.000001   **(j)** 0.000001

(4-76) **(a)** 89.5   **(b)** 1.308   **(c)** 0.575
(d) $\frac{13}{16}$ or 0.8125   **(e)** 243   **(f)** 0.27
(g) $83.62\overline{3}$ **(h)** $0.151\overline{6}$ **(i)** $2\frac{1}{2}$ or 2.5

(4-77) **(a)** $0.0\overline{3}$ **(b)** $0.00\overline{6}$ **(c)** $16.\overline{6}$
(d) $83.\overline{3}$ **(e)** $0.\overline{45}$ **(f)** $0.\overline{63}$
(g) $0.\overline{123}$ **(h)** $1.\overline{01}$ **(i)** $0.\overline{285714}$

(4-78) **(a)** 0.75   **(b)** 0.4   **(c)** 0.6   **(d)** 0.375   **(e)** 0.875
(f) 0.3   **(g)** 0.9   **(h)** 0.1875   **(i)** 0.34375   **(j)** 0.015625
(k) $0.\overline{6}$ **(l)** $0.8\overline{3}$ **(m)** $0.\overline{2}$ **(n)** $0.\overline{7}$ **(o)** $0.\overline{18}$
(p) $0.\overline{90}$ **(q)** $0.58\overline{3}$ **(r)** $0.91\overline{6}$ **(s)** $0.\overline{285714}$ **(t)** $0.\overline{714285}$

(4-79) **(a)** 8^2 **(b)** 2.3^2 **(c)** $\left(\dfrac{9}{10}\right)^2$ **(d)** 7^3

(e) 0.1^3 **(f)** $\left(\dfrac{3}{4}\right)^3$ **(g)** 10^4 **(h)** 24.75^3

(i) $\left(\dfrac{2}{3}\right)^4$ **(j)** $\dfrac{6^2}{5}$ **(k)** $\dfrac{2}{5^3}$ **(l)** $\dfrac{9^2}{3^2}$

(m) 0^1 **(n)** 1^1 or n^0 where $n \neq 0$ **(o)** 2^1 **(p)** 10^1

(q) 12.5^1 **(r)** $\left(\dfrac{3}{4}\right)^1$ **(s)** $5^2 \times 4$ **(t)** 3×6^4

(u) 0^2 or 0^n where $n \neq 0$ **(v)** 1^3 or 1^n **(w)** 2^2 **(x)** 3^3

(4-80) **(a)** 81   **(b)** 0.04   **(c)** $\frac{9}{16}$ **(d)** 27   **(e)** 0.008   **(f)** $\frac{1}{8}$
(g) 10   **(h)** 10,000   **(i)** 1,000,000   **(j)** 100,000,000,000   **(k)** 0   **(l)** 1
(m) $\frac{9}{2}$ **(n)** $\frac{9}{4}$ **(o)** 5   **(p)** $\frac{2}{3}$ **(q)** 0   **(r)** 25.3
(s) 1   **(t)** 1   **(u)** not defined   **(v)** 1   **(w)** $\frac{18}{8}$ or $\frac{9}{4}$ **(x)** $\frac{81}{256}$

(4-81) **(a)** 0   **(b)** 1   **(c)** 4   **(d)** 9   **(e)** 16   **(f)** 25
(g) 36   **(h)** 49   **(i)** 64   **(j)** 81   **(k)** 100   **(l)** 121
(m) 144   **(n)** 169   **(o)** 196   **(p)** 225   **(q)** 50,625   **(r)** $\frac{1}{9}$
(s) $\frac{1}{16}$ **(t)** $\frac{4}{25}$ **(u)** $\frac{81}{100}$ **(v)** $\frac{81}{10,000}$ **(w)** 0.01   **(x)** 6.25

(4-82) **(a)** 0   **(b)** 1   **(c)** 1.414   **(d)** 1.732   **(e)** 2   **(f)** 2.236
(g) 2.449   **(h)** 2.646   **(i)** 2.828   **(j)** 3   **(k)** 3.162   **(l)** 3.317
(m) 3.464   **(n)** 3.606   **(o)** 3.742   **(p)** 3.873   **(q)** 15   **(r)** 0.577
(s) 0.5   **(t)** 0.632   **(u)** 0.949   **(v)** 0.3   **(w)** 0.316   **(x)** 1.581

(4-83) **(a)** 86.6 = grade of C   **(b)** 92 points   **(c)** \$7809   **(d)** \$285.25
(e) $52.\overline{142857}$ 7-day weeks   **(f)** 30.5 days   **(g)** 2 hours   **(h)** \$25,550

5 PERCENTS

5-1. Renaming Percents as Decimals and Fractions

To rename part of a whole, you can use a number value, decimal, fraction, or **percent.**

The symbol % stands for the word *percent*. For example, read "49%" as "forty-nine percent." *Percent* can mean (**a**) hundredths, (**b**) out of one hundred, (**c**) per hundred, (**d**) $\times \frac{1}{100}$, or (**e**) $\div 100$.

EXAMPLE 5-1 Depict 49 hundredths as part of a whole and write it as a decimal, fraction, and percent.

Solution

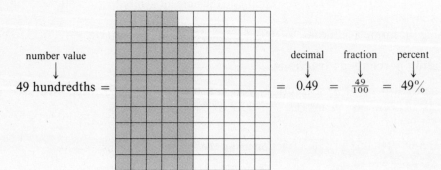

number value

49 hundredths =   =   0.49   = $\frac{49}{100}$ =   49%

decimal   fraction   percent

EXAMPLE 5-2 Rename 50% using (**a**) hundredths, (**b**) out of one hundred, (**c**) per hundred, (**d**) $\times \frac{1}{100}$, and (**e**) $\div 100$.

Solution

(**a**) 50% = 50 hundredths   (**b**) 50% = 50 out of one hundred   (**c**) 50% = 50 per hundred

(**d**) 50% = $50 \times \frac{1}{100}$ (**e**) 50% = $50 \div 100$

You can rename every percent as either a decimal or a fraction.

A. Renaming percents as decimals

To rename a percent as a decimal, you can use a formula: % means "÷ 100."

EXAMPLE 5-3   Rename 50% as a decimal using the formula % means "÷ 100."

Solution $\qquad\qquad\qquad 50\% = 50 \div 100 \qquad$ *Think:* % means "÷ 100."

$\qquad\qquad\qquad\qquad\qquad = 0.5 \qquad\qquad 50 \div 100 = 0.5$

To rename a percent as a decimal, you can also use Shortcut 5-1.

Shortcut 5-1   To rename a percent as a decimal:

(1) Move the decimal point two places to the *left* (divide by 100).
(2) Eliminate the percent symbol (%).

EXAMPLE 5-4   Rename 125% as a decimal using Shortcut 5-1.

Solution $\qquad\quad 125\% = 1.25\% \qquad$ Move the decimal point two places to the left.

$\qquad\qquad\qquad = 1.25 \qquad\quad$ Eliminate the percent symbol.

Check: Does 125% = 1.25 using the formula % means "÷ 100"? Yes: 125% = 125 ÷ 100 = 1.25

To rename a **fractional percent** (like $5\frac{3}{4}\%$) as a terminating decimal, you must first rename the fractional percent as a **decimal percent.**

EXAMPLE 5-5   Rename $5\frac{3}{4}\%$ as a decimal by first renaming as a decimal percent.

$\qquad\qquad$ fractional $\quad$ decimal
$\qquad\qquad$ percent $\qquad$ percent
$\qquad\qquad\qquad\downarrow\qquad\qquad\downarrow$

Solution $\qquad 5\frac{3}{4}\% \;=\; 5.75\% \qquad 5\frac{3}{4} = 5.75$

$\qquad\qquad\qquad\;\; = 0.0575\% \qquad$ Move the decimal point two places to the left.

$\qquad\qquad\qquad\;\; = 0.0575 \qquad\;\;$ Eliminate the percent symbol.

Check: Using the formula % means "÷ 100," $5\frac{3}{4}\% = 5.75 \div 100 = 0.0575$.

B. Renaming percents as fractions

To rename a percent as a fraction, you can use the same formula you used to rename percents as decimals: % means "÷ 100," or % means "× $\frac{1}{100}$."

EXAMPLE 5-6   Rename 50% as a fraction using the formula % means "÷ 100."

Solution $\qquad\qquad\quad 50\% = 50 \div 100 \qquad$ *Think:* % means "÷ 100."

$\qquad\qquad\qquad\qquad = \dfrac{50}{100} \qquad\qquad 50 \div 100 = \dfrac{50}{100}$

$\qquad\qquad\qquad\qquad = \dfrac{1}{2} \qquad\qquad \dfrac{50}{100} = \dfrac{1 \times 50}{2 \times 50} = \dfrac{1}{2}$

Note: 50%, 0.5, and $\frac{1}{2}$ all name the same value since $0.5 = \frac{5}{10} = \frac{1}{2}$.

EXAMPLE 5-7   Rename **(a)** $33\frac{1}{3}\%$ and **(b)** 62.5% as fractions using the formula % means "× $\frac{1}{100}$."

Solution

(a)
$$33\tfrac{1}{3}\% = 33\tfrac{1}{3} \times \frac{1}{100}$$

Think: % means "$\times \frac{1}{100}$."

$$= \frac{\cancel{100}}{3} \times \frac{1}{\cancel{100}}$$

$$33\tfrac{1}{3} = \frac{100}{3}$$

$$= \frac{1}{3}$$

$$\frac{\cancel{100}}{3} \times \frac{1}{\cancel{100}} = \frac{1}{3}$$

(b)
$$62.5\% = 62.5 \times \frac{1}{100}$$

Think: % means "$\times \frac{1}{100}$."

$$= 62\tfrac{1}{2} \times \frac{1}{100}$$

$$62.5 = 62\tfrac{1}{2}$$

$$= \frac{125}{2} \times \frac{1}{100}$$

$$62\tfrac{1}{2} = \frac{125}{2}$$

$$= \frac{5 \times \cancel{25}}{2} \times \frac{1}{4 \times \cancel{25}}$$

Multiply fractions.

$$= \frac{5}{2} \times \frac{1}{4} = \frac{5}{8}$$

5-2. Renaming Decimals and Fractions as Percents

You can rename every terminating decimal, repeating decimal, and fraction as a percent.

A. Renaming decimals as percents

Remember Shortcut 5-1 stated that to rename a percent as a decimal, you move the decimal point two places to the left, and then eliminate the percent symbol.

To rename a decimal as a percent, you do the opposite of each step in Shortcut 5-1. Shortcut 5-2 shows how to rename a decimal as a percent.

Shortcut 5-2   To rename a decimal as a percent:

(1) Move the decimal point two places to the *right* (multiply by 100).

(2) Add the percent symbol (%).

EXAMPLE 5-8   Rename 0.25 as a percent using Shortcut 5-2.

Solution $0.25 = 025.\%$ Move the decimal point two places to the right.
Add the percent symbol (%).

$= 25\%$ *Think:* 025. = 25

Check: Does $25\% = 0.25$ using the formula % means "$\div 100$"? Yes: $25\% = 25 \div 100 = 0.25$

B. Renaming fractions as percents

To rename a fraction as a percent, you can first rename as a decimal and then as a percent.

EXAMPLE 5-9   Rename $\tfrac{5}{8}$ as a percent by first renaming as a decimal.

Solution $\dfrac{5}{8} = 0.625 \longleftarrow 8\overline{)5.000}^{\,0.625}$

$= 62.5\% \longleftarrow 0.625 = 62.5\%$

Check: Does $62.5\% = \frac{5}{8}$ using the formula % means "÷100"? Yes:

$$62.5\% = 62.5 \div 100 = 0.625 = \frac{625}{1000} = \frac{5 \times \cancel{125}}{8 \times \cancel{125}} = \frac{5}{8}$$

Note: To check $62.5\% = \frac{5}{8}$, you divide by 100 and eliminate the percent symbol (%):
$62.5\% = 62.5 \div 100 = \frac{5}{8}$

You can also rename a fraction as a percent by using Shortcut 5-3.

Shortcut 5-3   To rename a fraction as a percent:

(1) Multiply by 100.

(2) Write the percent symbol (%).

EXAMPLE 5-10   Rename $\frac{2}{3}$ as a percent using Shortcut 5-3:

Solution $\dfrac{2}{3} = \left(\dfrac{2}{3} \times 100\right)\%$ Multiply by 100. Write the percent symbol (%).

$= \dfrac{200}{3}\%$ $\dfrac{2}{3} \times 100 = \dfrac{2}{3} \times \dfrac{100}{1} = \dfrac{200}{3}$

$= 66\frac{2}{3}\%$ $3\overline{)200}^{\,66\frac{2}{3}}$

Note: When renaming a number like $\frac{2}{3}$ as a percent, it is easier to use Shortcut 5-3. It would be more difficult to rename $\frac{2}{3}$ as a percent by first renaming as a decimal because $\frac{2}{3}$ equals the repeating decimal of $0.\overline{6}$.

5-3. Solving Percent Problems

A. Finding the amount in a percent problem

To solve percent problems, you can use a **Percent Diagram:**

Percent Diagram: 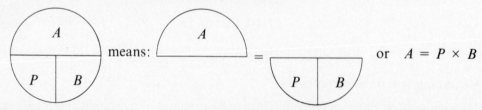   where $\begin{cases} A \text{ is the amount.} \\ B \text{ is the base.} \\ P \text{ is the percent.} \end{cases}$

When both the percent P and base B are known, you can use the Percent Diagram to write a formula for finding the amount A as follows:

means: =   or $A = P \times B$

The formula $A = P \times B$ is called the **Amount Formula.** The Amount Formula tells you that to find the amount A when both the percent P and base B are known, you multiply the base by the percent.

Caution: To multiply by a percent, you must first rename the percent as either a terminating decimal or a fraction.

When a percent can be renamed as a terminating decimal, you can multiply to find the amount.

EXAMPLE 5-11 What number is 75% of 140?

Solution

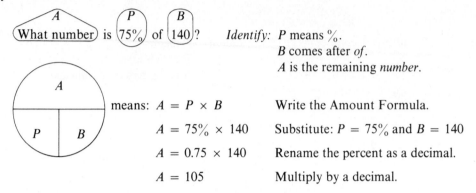

means: $A = P \times B$	Write the Amount Formula.
$A = 75\% \times 140$	Substitute: $P = 75\%$ and $B = 140$
$A = 0.75 \times 140$	Rename the percent as a decimal.
$A = 105$	Multiply by a decimal.

Note: Because 75% can be renamed as either a terminating decimal (0.75) or a fraction ($\frac{3}{4}$), you can use either form to multiply.

When a percent is equal to a repeating decimal, you must rename the percent as a fraction and then multiply to find the amount.

EXAMPLE 5-12 What number is $66\frac{2}{3}\%$ of 24?

Solution

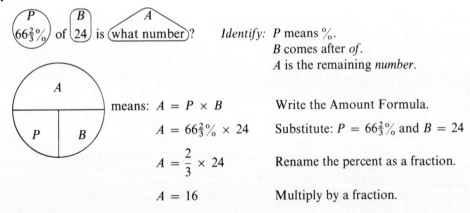

means: $A = P \times B$	Write the Amount Formula.
$A = 66\frac{2}{3}\% \times 24$	Substitute: $P = 66\frac{2}{3}\%$ and $B = 24$
$A = \dfrac{2}{3} \times 24$	Rename the percent as a fraction.
$A = 16$	Multiply by a fraction.

Note: Because $66\frac{2}{3}\%$ is equal to the repeating decimal $0.\overline{6}$, you must rename $66\frac{2}{3}\%$ as the fraction $\frac{2}{3}$ in order to multiply.

B. Finding the base in a percent problem

When both the percent P and the amount A are known, you can use the Percent Diagram to write a formula for finding the base B as follows:

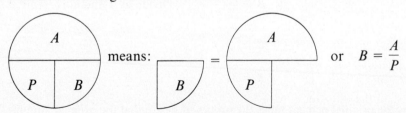

The formula $B = \frac{A}{P}$ is called the **Base Formula.** The Base Formula tells you that to find the base B when both the percent P and amount A are known, you divide the amount by the percent.

To divide by a percent, you must first rename the percent as either a terminating decimal or a fraction. When the percent can be renamed as a terminating decimal, you can divide to find the amount.

EXAMPLE 5-13 25% of what number is 10?

Solution

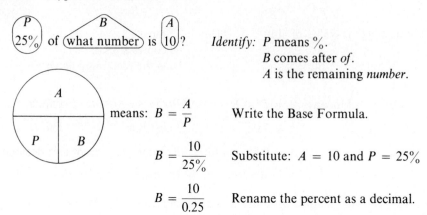

Identify: P means %.
$\quad\quad\quad\;$ B comes after *of*.
$\quad\quad\quad\;$ A is the remaining *number*.

means: $B = \dfrac{A}{P}$ Write the Base Formula.

$B = \dfrac{10}{25\%}$ Substitute: $A = 10$ and $P = 25\%$

$B = \dfrac{10}{0.25}$ Rename the percent as a decimal.

$B = 40$ Divide by a decimal.

Check: Does the original percent times the proposed base equal the original amount?
$\quad\quad$ Yes: $25\% \times 40 = 0.25 \times 40 = 10$

Note: Because 25% can be renamed as either a terminating decimal (0.25) or a fraction ($\frac{1}{4}$), you can use either form to divide.

When the percent is equal to a repeating decimal, you must rename the percent as a fraction and then divide to find the base.

EXAMPLE 5-14 $5\frac{1}{2}$ is $16\frac{2}{3}\%$ of what number?

Solution

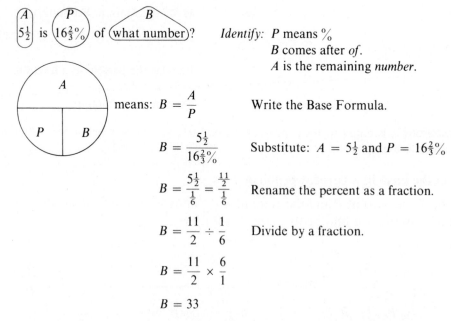

Identify: P means %
$\quad\quad\quad\;$ B comes after *of*.
$\quad\quad\quad\;$ A is the remaining *number*.

means: $B = \dfrac{A}{P}$ Write the Base Formula.

$B = \dfrac{5\frac{1}{2}}{16\frac{2}{3}\%}$ Substitute: $A = 5\frac{1}{2}$ and $P = 16\frac{2}{3}\%$

$B = \dfrac{5\frac{1}{2}}{\frac{1}{6}} = \dfrac{\frac{11}{2}}{\frac{1}{6}}$ Rename the percent as a fraction.

$B = \dfrac{11}{2} \div \dfrac{1}{6}$ Divide by a fraction.

$B = \dfrac{11}{2} \times \dfrac{6}{1}$

$B = 33$

Check: Does the original percent times the proposed base equal the original amount?
$\quad\quad$ Yes: $16\frac{2}{3}\% \times 33 = \frac{1}{6} \times 33 = \frac{11}{2} = 5\frac{1}{2}$

Note: Because $16\frac{2}{3}\%$ is equal to the repeating decimal $0.1\overline{6}$, you must rename $16\frac{2}{3}\%$ as the fraction $\frac{1}{6}$ in order to divide.

C. Finding the percent in a percent problem

When both the amount A and base B are known, you can use the Percent Diagram to write a formula for finding the percent P as follows:

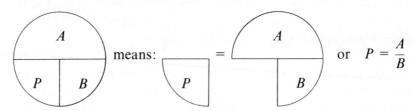

The formula $P = \frac{A}{B}$ is called the **Percent Formula.** The Percent Formula tells you that to find the percent P when both the amount A and base B are known, divide the amount by the base.

After dividing the amount by the base, you must rename the decimal or fractional quotient as a percent.

EXAMPLE 5-15 What percent of 16 is 5?

Solution

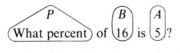

Identify: P means *percent.*
B comes after *of.*
A is the remaining *number.*

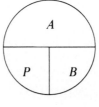 means: $P = \dfrac{A}{B}$ Write the Percent Formula.

$P = \dfrac{5}{16}$ Substitute: $A = 5$ and $B = 16$

$P = 0.3125$ Divide to get a decimal quotient.

$P = 31.25\%$ Rename as a percent.

Check: Does the proposed percent times the original base equal the original amount?
Yes: $31.25\% \times 16 = 0.3125 \times 16 = 5$

Note: Because 0.3125 is a terminating decimal, you can rename 0.3125 as either a decimal percent (31.25%) or a fractional percent $(31\frac{1}{4}\%)$.

EXAMPLE 5-16 4 is what percent of 3?

Solution

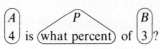

Identify: P means *percent.*
B comes after *of.*
A is the remaining *number.*

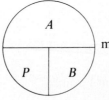 means: $P = \dfrac{A}{B}$ Write the Percent Formula.

$P = \dfrac{4}{3}$ Substitute: $A = 4$ and $B = 3$

$P = 1\frac{1}{3}$ Divide to get a fractional quotient.

$P = (100 \times 1\frac{1}{3})\%$ Rename as a percent.

$P = 133\frac{1}{3}\%$

Check: Does the proposed percent times the original base equal the original amount?

Yes: $133\frac{1}{3}\% \times 3 = \frac{400}{3}\% \times 3 = \frac{4}{3} \times 3 = 4$

Note: $1\frac{1}{3}$ is equal to the repeating decimal $1.\overline{3}$. You should rename $1\frac{1}{3}$ as the fractional percent $133\frac{1}{3}\%$ because percents are not usually written in repeating decimal form, such as $133.\overline{3}\%$.

5-4. Finding Percent Increase and Decrease

When the *original amount* of something increases to form a *new amount*, the *difference* between the two amounts is called the **amount of increase.** The percent found by dividing the amount of increase by the original amount is called the **percent increase.**

Similarly, when the original amount of something decreases to form a new amount, the difference between the two amounts is called the **amount of decrease.** The percent found by dividing the amount of decrease by the original amount is called the **percent decrease.**

A. Finding the percent increase

Recall that to find the percent P in a percent problem, you must know the base B and the amount A.

In percent increase problems:

(1) The given original amount (the smaller number) is used as the base B.

(2) The amount of increase (the difference between the original amount and the new amount) is used as the amount A.

EXAMPLE 5-17   What is the percent increase from 2 to 3?

Solution   The given original amount is 2. ←——————— base B
The amount of increase from 2 to 3 is 1. ←—— amount A

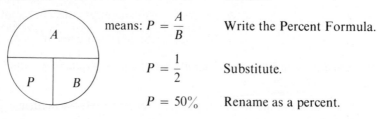

means: $P = \dfrac{A}{B}$ Write the Percent Formula.

$P = \dfrac{1}{2}$ Substitute.

$P = 50\%$ Rename as a percent.

B. Finding the percent decrease

In percent decrease problems:

(1) The given original amount (the larger number) is used as the base B.

(2) The amount of decrease (the difference between the original amount and the new amount) is used as the amount A.

EXAMPLE 5-18   What is the percent decrease from 3 to 2?

Solution   The given original amount is 3. ←——————— base B
The amount of decrease from 3 to 2 is 1. ←—— amount A

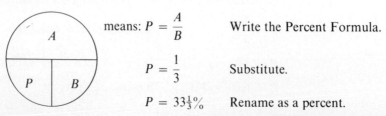

means: $P = \dfrac{A}{B}$ Write the Percent Formula.

$P = \dfrac{1}{3}$ Substitute.

$P = 33\frac{1}{3}\%$ Rename as a percent.

Note: The percent decrease from 3 to 2 is $33\frac{1}{3}\%$, while the percent increase from 2 to 3 is 50%.

To help you understand percent increase and decrease better, think of a $3 necktie. If a $3 tie is marked down $33\frac{1}{3}\%$, its price will be $2. Then, if the $2 tie is marked back up 50%, its price will be $3—the original price.

5-5. Solving Problems Involving Sales Tax

The price marked on an item in a retail store is called the **purchase price.** On most items sold in retail stores, you must also pay **sales tax.** The **total price** for an item is the purchase price plus any sales tax that is applicable.

A. Finding the total price

To find the total price for an item, you add the purchase price and applicable sales tax (if any) together.

Total Price Formula: Purchase Price + Sales Tax = Total Price

EXAMPLE 5-19 A certain coat has a purchase price of $89.99 with a sales tax of $5.40. What is the total price of the coat?

Solution

$$\underbrace{\$89.99}_{\text{purchase price}} + \underbrace{\$5.40}_{\text{sales tax}} = \underbrace{\$95.39}_{\text{total price}} \quad \text{Add.}$$

B. Finding the sales tax

To find the sales tax on an item, you multiply the purchase price by the **sales-tax rate.**

Sales Tax Formula: Sales-Tax Rate × Purchase Price = Sales Tax

Caution: When the sales-tax rate is given as a percent, you must first rename the sales-tax rate as either a terminating decimal or a fraction in order to multiply.

EXAMPLE 5-20 The sales-tax rate in Atlanta, Georgia, is 4%. What is the sales tax on a dress that has a purchase price of $45 in Atlanta, Georgia?

Solution

$$\underbrace{4\%}_{\substack{\text{sales-tax}\\\text{rate}}} \times \underbrace{\$45}_{\substack{\text{purchase}\\\text{price}}} = 0.04 \times \$45 \qquad \text{Rename the percent.}$$

$$= \$1.80 \longleftarrow \text{sales tax} \qquad \text{Multiply.}$$

Note: The total price for this dress is

$$\$45 + \$1.80 = \$46.80 \longleftarrow \text{total price}$$

5-6. Solving Problems Involving Discounts

The price at which an item normally sells is called the **regular price.** When the regular price of an item is reduced to a lower price, the new lower price is called the **sale price.** The dollar amount by which the regular price is reduced to get the sale price is called the **discount.**

A. Finding the sale price

To find the sale price for an item, you subtract the discount from the regular price.

Sale Price Formula: Regular Price − Discount = Sale Price

EXAMPLE 5-21 A certain radio has a regular price of $49.99. What is the sale price if the radio is discounted $25?

	regular price	discount	sale price	
Solution	$49.99	− $25	= $24.99	Subtract.

B. Finding the discount

To find the discount on a given sale item, you multiply the regular price by the **discount rate** or **rate of discount.**

Discount Formula: Discount Rate × Regular Price = Discount

Caution: When the discount rate is given as a percent, you must rename the discount rate as either a terminating decimal or a fraction in order to multiply.

EXAMPLE 5-22 A $19.99 calculator is marked down 25%. How much is the discount?

	discount rate	regular price		
Solution	25%	× $19.99	= 0.25 × $19.99	Rename the percent.
			= $4.9975	Multiply.
			≈ $5.00	Round to the nearest cent.
			= $5 ⟵ discount	

Note: The sale price for the calculator is:

$$\$19.99 - \$5 = \$14.99 \longleftarrow \text{sale price}$$

5-7. Solving Problems Involving Commissions

The amount paid for working a given number of hours is called a **salary.** The amount paid for selling a given dollar amount is called a **commission.** The **total pay** is the salary plus any commission that is applicable.

A. Finding the total pay

To find the total pay for salary plus commission work, you add the salary and the applicable commission (if any) together.

Total Pay Formula: Salary + Commission = Total Pay

EXAMPLE 5-23 A used-car salesperson earns $200 salary per week. This week the salesperson earned $450 in commissions. What is the salesperson's total pay this week?

	salary	commission	total pay	
Solution	$200 +	$450	= $650	Add.

B. Finding the commission

To find the commission on sales, you multiply the sales by the commission rate.

Commission Formula: Commission Rate × Sales = Commission

Caution: When the commission rate is given as a percent, you must rename the commission rate as either a terminating decimal or a fraction in order to multiply.

EXAMPLE 5-24 What is the commission on sales of $500 if the commission rate is 15%?

	commission rate	sales		
Solution	15%	× $500	= 0.15 × $500	Rename the percent.
			= $75 ⟵ commission	Multiply.

5-8. Solving Problems Involving Simple Interest

A. Understanding interest rates and time periods

Money that is paid for the use of money is called **interest**. The money on which interest is paid is called the **principal**. The rate that interest is charged as a percent of the principal over a period of time is called the **interest rate**. The period of time that interest is paid on the principal is called the **time period**. The three most popular time periods for interest rates are the day, the month, and the year.

EXAMPLE 5-25 Explain the meaning of (a) 15% per year, (b) 1%/month, and (c) $\dfrac{0.04\%}{\text{day}}$.

Solution

(a) 15% per year means that interest is paid on the principal at a rate of 15% each year.

(b) 1%/month or 1% per month means that interest is paid on the principal at a rate of 1% each month.

(c) $\dfrac{0.04\%}{\text{day}}$ or 0.04% per day means that interest is paid on the principal at a rate of 0.04% each day.

Interest rates can be written in several different ways.

EXAMPLE 5-26 Write the interest rate 12% per year in three different ways.

Solution

$$\overbrace{12\% \text{ per year}}^{\text{word form}} = 12\%/\text{year} \longleftarrow \text{slash form}$$

$$= \frac{12\%}{\text{year}} \longleftarrow \text{fraction form}$$

Note: When computing interest in this book, we'll use the following **business time units:**

$$1 \text{ business year} = 360 \text{ days} \qquad 1 \text{ business month} = 30 \text{ days}$$

For the remainder of this chapter, all years and months are assumed to be business years and business months.

When computing interest, it is often necessary to rename a given time period in terms of a different time unit.

EXAMPLE 5-27 Rename 8 months as (a) days and (b) years.

Solution

(a) To rename a time period (like 8 months) in terms of a shorter time unit (like days), you multiply.

$$8 \text{ months} = 8 \times 30 \text{ days} = 240 \text{ days} \qquad \textit{Think: } 1 \text{ month} = 30 \text{ days}$$

(b) To rename a time period (like 8 months) in terms of a longer time unit (like years), you divide.

$$8 \text{ months} = \frac{8}{12} \text{ year} = \frac{2}{3} \text{ year} \qquad \textit{Think: } 12 \text{ months} = 1 \text{ year}$$

B. Finding simple interest

Interest that is paid on principal only is called **simple interest**. To find the simple interest, you multiply the principal times the interest rate times the time period.

Simple Interest Formula: Simple Interest = Principal × Interest Rate × Time Period

$$\text{or} \quad I = P \times r \times t$$

Caution: Before you can evaluate the simple interest formula, you need to know that

(a) both the interest rate r and the time period t must be expressed with a common time unit

(b) the interest rate r must be renamed as either a terminating decimal or a fraction

EXAMPLE 5-28   What is the simple interest paid on \$500 at 8% per year simple interest for 90 days?

Solution $I = P \times r \times t$ Write the simple interest formula.

$ = \$500 \times \dfrac{8\%}{\text{year}} \times 90 \text{ days}$ Substitute.

$ = \$500 \times \dfrac{8\%}{\text{year}} \times \dfrac{90}{360} \text{ year}$ Rename to get a common time unit: $90 \text{ days} = \dfrac{90}{360} \text{ year.}$

$ = \$500 \times \dfrac{8\%}{\cancel{\text{year}}} \times \dfrac{90}{360} \cancel{\text{year}}$ Eliminate the common time unit.

$ = \$500 \times 8\% \times \dfrac{1}{4}$ Simplify the fraction.

$ = \$500 \times 0.08 \times \dfrac{1}{4}$ Rename the percent.

$ = \$40 \times \dfrac{1}{4}$ Multiply.

$ = \$10$ ⟵ simple interest on \$500 at 8% for 90 days

Note: If you borrow \$500 at 8% per year simple interest for 90 days, you will need to repay the \$500 principal plus an additional \$10 in interest by the end of the 90-day time period.

C. Finding the total amount

To find the **total amount,** you add the principal and the interest together.

Total Amount Formula: Total Amount = Principal + Interest

or $A = P + I$

Caution: Before you can evaluate the total amount formula, you must find the interest.

EXAMPLE 5-29   What is the total amount to be repaid for a loan of \$2000 at $1\frac{1}{2}\%$ per month simple interest for 540 days?

Solution $I = P \times r \times t$ Find the simple interest.

$ = \$2000 \times \dfrac{1\frac{1}{2}\%}{\text{month}} \times 540 \text{ days}$ Substitute.

$ = \$2000 \times \dfrac{1\frac{1}{2}\%}{\text{month}} \times 18 \text{ months}$ Rename to get a common time unit.

$ = \$2000 \times 1\frac{1}{2}\% \times 18$ Eliminate the common time unit.

$ = \$2000 \times 0.015 \times 18$ Rename the percent: $1\frac{1}{2}\% = 1.5\%$

$ = \$30 \times 18$ Multiply.

$ = \$540$ ⟵ simple interest

$A = P + I$ Find the total amount to be repaid.

$= \$2000 + \$540$

$= \$2540 \longleftarrow$ total amount to be repaid

SOLVED PROBLEMS

PROBLEM 5-1 Rename each percent using both "$\times \frac{1}{100}$" and "$\div 100$":

(a) 25% **(b)** 75% **(c)** 10% **(d)** 1% **(e)** 90%

(f) 100% **(g)** 150% **(h)** 5% **(i)** $\frac{1}{2}$% **(j)** 0.25%

(k) $133\frac{1}{3}$% **(l)** 0.01% **(m)** 1000% **(n)** 12.5% **(o)** $66\frac{2}{3}$%

Solution Recall that % means "hundredths," "out of one hundred," "per hundred," "$\times \frac{1}{100}$," or "$\div 100$" [see Example 5-2]:

(a) $25 \times \frac{1}{100}$, $25 \div 100$ **(b)** $75 \times \frac{1}{100}$, $75 \div 100$ **(c)** $10 \times \frac{1}{100}$, $10 \div 100$

(d) $1 \times \frac{1}{100}$, $1 \div 100$ **(e)** $90 \times \frac{1}{100}$, $90 \div 100$ **(f)** $100 \times \frac{1}{100}$, $100 \div 100$

(g) $150 \times \frac{1}{100}$, $150 \div 100$ **(h)** $5 \times \frac{1}{100}$, $5 \div 100$ **(i)** $\frac{1}{2} \times \frac{1}{100}$, $\frac{1}{2} \div 100$

(j) $0.25 \times \frac{1}{100}$, $0.25 \div 100$ **(k)** $133\frac{1}{3} \times \frac{1}{100}$, $133\frac{1}{3} \div 100$ **(l)** $0.01 \times \frac{1}{100}$, $0.01 \div 100$

(m) $1000 \times \frac{1}{100}$, $1000 \div 100$ **(n)** $12.5 \times \frac{1}{100}$, $12.5 \div 100$ **(o)** $66\frac{2}{3} \times \frac{1}{100}$, $66\frac{2}{3} \div 100$

PROBLEM 5-2 Rename each percent as a decimal:

(a) 37.5% **(b)** 12.5% **(c)** 7.5% **(d)** 6.25% **(e)** 0.5%

(f) 0.125% **(g)** 10% **(h)** 5% **(i)** 150% **(j)** 100%

(k) $\frac{1}{4}$% **(l)** $\frac{7}{10}$% **(m)** $4\frac{3}{4}$% **(n)** $12\frac{5}{8}$% **(o)** $215\frac{1}{2}$%

Solution To rename a percent as a decimal, you move the decimal point two places to the left and then eliminate the percent symbol [see Example 5-4]:

(a) $37.5\% = 0.375$ **(b)** $12.5\% = 0.125$ **(c)** $7.5\% = 0.075$

(d) $6.25\% = 0.0625$ **(e)** $0.5\% = 0.005$ **(f)** $0.125\% = 0.00125$

(g) $10\% = 0.10 = 0.1$ **(h)** $5\% = 0.05$ **(i)** $150\% = 1.50 = 1.5$

(j) $100\% = 1.00 = 1$ **(k)** $\frac{1}{4}\% = 0.25\% = 0.0025$ **(l)** $\frac{7}{10}\% = 0.7\% = 0.007$

(m) $4\frac{3}{4}\% = 4.75\% = 0.0475$ **(n)** $12\frac{5}{8}\% = 12.625\% = 0.12625$ **(o)** $215\frac{1}{2}\% = 215.5\% = 2.155$

PROBLEM 5-3 Rename each percent as a fraction:

(a) 37.5% **(b)** 12.5% **(c)** 0.25% **(d)** 0.5% **(e)** 0.1%

(f) 0.01% **(g)** 50% **(h)** 25% **(i)** 175% **(j)** 100%

(k) $\frac{2}{3}$% **(l)** $\frac{5}{6}$% **(m)** $\frac{5}{12}$% **(n)** $16\frac{2}{3}$% **(o)** $83\frac{1}{3}$%

Solution   To rename a percent as a fraction, you multiply by $\frac{1}{100}$ and then eliminate the percent symbol [see Example 5-7]:

(a) $37.5\% = 37.5 \times \dfrac{1}{100} = 37\frac{1}{2} \times \dfrac{1}{100} = \dfrac{75}{2} \times \dfrac{1}{100} = \dfrac{3}{2} \times \dfrac{1}{4} = \dfrac{3}{8}$

(b) $12.5\% = 12.5 \times \dfrac{1}{100} = 12\frac{1}{2} \times \dfrac{1}{100} = \dfrac{25}{2} \times \dfrac{1}{100} = \dfrac{1}{2} \times \dfrac{1}{4} = \dfrac{1}{8}$

(c) $0.25\% = 0.25 \times \dfrac{1}{100} = \dfrac{1}{4} \times \dfrac{1}{100} = \dfrac{1}{400}$

(d) $0.5\% = 0.5 \times \dfrac{1}{100} = \dfrac{1}{2} \times \dfrac{1}{100} = \dfrac{1}{200}$

(e) $0.1\% = 0.1 \times \dfrac{1}{100} = \dfrac{1}{10} \times \dfrac{1}{100} = \dfrac{1}{1000}$

(f) $0.01\% = 0.01 \times \dfrac{1}{100} = \dfrac{1}{100} \times \dfrac{1}{100} = \dfrac{1}{10,000}$

(g) $50\% = 50 \times \dfrac{1}{100} = \dfrac{50}{1} \times \dfrac{1}{100} = \dfrac{1}{1} \times \dfrac{1}{2} = \dfrac{1}{2}$

(h) $25\% = 25 \times \dfrac{1}{100} = \dfrac{25}{1} \times \dfrac{1}{100} = \dfrac{1}{1} \times \dfrac{1}{4} = \dfrac{1}{4}$

(i) $175\% = 175 \times \dfrac{1}{100} = \dfrac{175}{1} \times \dfrac{1}{100} = \dfrac{7}{1} \times \dfrac{1}{4} = \dfrac{7}{4}$ or $1\frac{3}{4}$

(j) $100\% = 100 \times \dfrac{1}{100} = \dfrac{100}{1} \times \dfrac{1}{100} = \dfrac{1}{1} = 1$

(k) $\dfrac{2}{3}\% = \dfrac{2}{3} \times \dfrac{1}{100} = \dfrac{1}{3} \times \dfrac{1}{50} = \dfrac{1}{150}$

(l) $\dfrac{5}{6}\% = \dfrac{5}{6} \times \dfrac{1}{100} = \dfrac{1}{6} \times \dfrac{1}{20} = \dfrac{1}{120}$

(m) $\dfrac{5}{12}\% = \dfrac{5}{12} \times \dfrac{1}{100} = \dfrac{1}{12} \times \dfrac{1}{20} = \dfrac{1}{240}$

(n) $16\frac{2}{3}\% = 16\frac{2}{3} \times \dfrac{1}{100} = \dfrac{50}{3} \times \dfrac{1}{100} = \dfrac{1}{3} \times \dfrac{1}{2} = \dfrac{1}{6}$

(o) $83\frac{1}{3}\% = 83\frac{1}{3} \times \dfrac{1}{100} = \dfrac{250}{3} \times \dfrac{1}{100} = \dfrac{5}{3} \times \dfrac{1}{2} = \dfrac{5}{6}$

PROBLEM 5-4   Rename each decimal as a percent:

(a) 0.375   (b) 0.125   (c) 0.75   (d) 0.25   (e) 0.4   (f) 0.2   (g) 0.01   (h) 0.05   (i) 1

Solution   To rename a decimal as a percent, you move the decimal point two places to the right and then write the percent symbol [see Example 5-8]:

(a) $0.375 = 037.5\% = 37.5\%$ (b) $0.125 = 012.5\% = 12.5\%$ (c) $0.75 = 075.\% = 75\%$

(d) $0.25 = 025.\% = 25\%$ (e) $0.4 = 040.\% = 40\%$ (f) $0.2 = 020.\% = 20\%$

(g) $0.01 = 001.\% = 1\%$ (h) $0.05 = 005.\% = 5\%$ (i) $1 = 1. = 100.\% = 100\%$

PROBLEM 5-5   Rename each fraction or mixed number as a percent:

(a) $\frac{1}{2}$ (b) $\frac{1}{4}$ (c) $\frac{3}{4}$ (d) $\frac{1}{3}$ (e) $\frac{2}{3}$

(f) $\frac{1}{5}$ **(g)** $\frac{1}{8}$ **(h)** $\frac{5}{12}$ **(i)** $\frac{9}{10}$ **(j)** $\frac{1}{50}$

(k) $\frac{1}{100}$ **(l)** $\frac{1}{300}$ **(m)** $\frac{1}{500}$ **(n)** $1\frac{3}{8}$ **(o)** $2\frac{3}{5}$

Solution To rename a fraction or mixed number as a percent, you multiply by 100 and then write the percent symbol [see Example 5-10]:

(a) $\frac{1}{2} = (100 \times \frac{1}{2})\% = 50\%$ **(b)** $\frac{1}{4} = (100 \times \frac{1}{4})\% = 25\%$ **(c)** $\frac{3}{4} = (100 \times \frac{3}{4})\% = 75\%$

(d) $\frac{1}{3} = (100 \times \frac{1}{3})\% = 33\frac{1}{3}\%$ **(e)** $\frac{2}{3} = (100 \times \frac{2}{3})\% = 66\frac{2}{3}\%$ **(f)** $\frac{1}{5} = (100 \times \frac{1}{5})\% = 20\%$

(g) $\frac{1}{8} = (100 \times \frac{1}{8})\% = 12\frac{1}{2}\%$ **(h)** $\frac{5}{12} = (100 \times \frac{5}{12})\% = 41\frac{2}{3}\%$ **(i)** $\frac{9}{10} = (100 \times \frac{9}{10})\% = 90\%$

(j) $\frac{1}{50} = (100 \times \frac{1}{50})\% = 2\%$ **(k)** $\frac{1}{100} = (100 \times \frac{1}{100})\% = 1\%$ **(l)** $\frac{1}{300} = (100 \times \frac{1}{300})\% = \frac{1}{3}\%$

(m) $\frac{1}{500} = (100 \times \frac{1}{500})\% = \frac{1}{5}\%$ **(n)** $1\frac{3}{8}\% = (100 \times 1\frac{3}{8})\% = 137\frac{1}{2}\%$ **(o)** $2\frac{3}{5} = (100 \times 2\frac{3}{5})\% = 260\%$

PROBLEM 5-6 Find the following amounts by renaming the percent as a decimal.

(a) What number is 20% of 55? **(b)** 40% of 25 is what number?

(c) What number is 25% of 16? **(d)** 5% of 20 is what number?

(e) What number is $\frac{1}{2}\%$ of 1600? **(f)** 200% of 4.5 is what number?

(g) What number is 30% of 50? **(h)** 10% of 400 is what number?

Solution To find a decimal amount, you write the Amount Formula, substitute for P and B, rename the percent as a decimal, and then multiply [see Example 5-11]:

(a) $A = P \times B = 20\% \times 55 = 0.2 \times 55 = 11$

(b) $A = P \times B = 40\% \times 25 = 0.4 \times 25 = 10$

(c) $A = P \times B = 25\% \times 16 = 0.25 \times 16 = 4$

(d) $A = P \times B = 5\% \times 20 = 0.05 \times 20 = 1$

(e) $A = P \times B = \frac{1}{2}\% \times 1600 = 0.5\% \times 1600 = 0.005 \times 1600 = 8$

(f) $A = P \times B = 200\% \times 4.5 = 2 \times 4.5 = 9$

(g) $A = P \times B = 30\% \times 50 = 0.3 \times 50 = 15$

(h) $A = P \times B = 10\% \times 400 = 0.1 \times 400 = 40$

PROBLEM 5-7 Find the following amounts by renaming the percent as a fraction:

(a) What number is $66\frac{2}{3}\%$ of 90? **(b)** $8\frac{1}{3}\%$ of 36 is what number?

(c) What number is 25% of $5\frac{3}{5}$? **(d)** 75% of 4.8 is what number?

(e) What number is $\frac{5}{6}\%$ of 24? **(f)** What number is $33\frac{1}{3}\%$ of 45?

(g) $16\frac{2}{3}\%$ of 24 is what number? **(h)** 50% of 0.8 is what number?

Solution To find a fractional amount, you write the Amount Formula, substitute for P and B, rename the percent as a fraction, and then multiply [see Example 5-12]:

(a) $A = P \times B = 66\frac{2}{3}\% \times 90 = \dfrac{2}{3} \times 90 = 60$

(b) $A = P \times B = 8\frac{1}{3}\% \times 36 = \dfrac{1}{12} \times 36 = 3$

(c) $A = P \times B = 25\% \times 5\frac{3}{5} = \dfrac{1}{4} \times \dfrac{28}{5} = \dfrac{7}{5}$ or $1\frac{2}{5}$

(d) $A = P \times B = 75\% \times 4.8 = \dfrac{3}{4} \times 4\frac{4}{5} = \dfrac{3}{4} \times \dfrac{24}{5} = \dfrac{18}{5}$ or $3\frac{3}{5}$

(e) $A = P \times B = \dfrac{5}{6}\% \times 24 = \dfrac{1}{120} \times 24 = \dfrac{1}{5}$

(f) $A = P \times B = 33\tfrac{1}{3}\% \times 45 = \dfrac{1}{3} \times 45 = 15$

(g) $A = P \times B = 16\tfrac{2}{3}\% \times 24 = \dfrac{1}{6} \times 24 = 4$

(h) $A = P \times B = 50\% \times 0.8 = \dfrac{1}{2} \times \dfrac{8}{10} = \dfrac{4}{10} = \dfrac{2}{5}$

PROBLEM 5-8   Find the base by renaming the percent as a decimal:

(a) 80% of what number is 5?   (b) 20 is 50% of what number?   (c) 65% of what number is 39?

(d) 2.25 is 9% of what number?   (e) 35 is 20% of what number?   (f) 45% of what number is 9?

Solution   To find the base in decimal form, you write the Base Formula, substitute for A and P, rename the percent as a decimal, and then divide [see Example 5-13]:

(a) $B = \dfrac{A}{P} = \dfrac{5}{80\%} = \dfrac{5}{0.8} = 6.25$ (b) $B = \dfrac{A}{P} = \dfrac{20}{50\%} = \dfrac{20}{0.5} = 40$

(c) $B = \dfrac{A}{P} = \dfrac{39}{65\%} = \dfrac{39}{0.65} = 60$ (d) $B = \dfrac{A}{P} = \dfrac{2.25}{9\%} = \dfrac{2.25}{0.09} = 25$

(e) $B = \dfrac{A}{P} = \dfrac{35}{20\%} = \dfrac{35}{0.2} = 175$ (f) $B = \dfrac{A}{P} = \dfrac{9}{45\%} = \dfrac{9}{0.45} = 20$

PROBLEM 5-9   Find each base by renaming the percent as a fraction:

(a) $83\tfrac{1}{3}\%$ of what number is 25?

(b) 5 is $8\tfrac{1}{3}\%$ of what number?

(c) $91\tfrac{2}{3}\%$ of what number is $1\tfrac{5}{6}$?

(d) 70 is $16\tfrac{2}{3}\%$ of what number?

(e) $66\tfrac{2}{3}\%$ of what number is 5?

(f) $\tfrac{1}{8}$ is $16\tfrac{2}{3}\%$ of what number?

(g) 60% of what number is 1.5?

(h) $2\tfrac{3}{4}$ is $33\tfrac{1}{3}\%$ of what number?

(i) $41\tfrac{2}{3}\%$ of what number is 250?

Solution   To find the base in fraction form, you write the Base Formula, substitute for A and P, rename the percent as a fraction, and then divide [see Example 5-14]:

(a) $B = \dfrac{A}{P} = \dfrac{25}{83\tfrac{1}{3}\%} = \dfrac{25}{\frac{5}{6}} = 25 \div \dfrac{5}{6} = \dfrac{25}{1} \times \dfrac{6}{5} = 30$

(b) $B = \dfrac{A}{P} = \dfrac{5}{8\tfrac{1}{3}\%} = \dfrac{5}{\frac{1}{12}} = 5 \div \dfrac{1}{12} = \dfrac{5}{1} \times \dfrac{12}{1} = 60$

(c) $B = \dfrac{A}{P} = \dfrac{1\tfrac{5}{6}}{91\tfrac{2}{3}\%} = \dfrac{1\tfrac{5}{6}}{\frac{11}{12}} = 1\tfrac{5}{6} \div \dfrac{11}{12} = \dfrac{11}{6} \times \dfrac{12}{11} = 2$

(d) $B = \dfrac{A}{P} = \dfrac{70}{16\tfrac{2}{3}\%} = \dfrac{70}{\frac{1}{6}} = 70 \div \dfrac{1}{6} = \dfrac{70}{1} \times \dfrac{6}{1} = 420$

(e) $B = \dfrac{A}{P} = \dfrac{5}{66\tfrac{2}{3}\%} = \dfrac{5}{\frac{2}{3}} = 5 \div \dfrac{2}{3} = \dfrac{5}{1} \times \dfrac{3}{2} = \dfrac{15}{2}$ or $7\tfrac{1}{2}$

(f) $B = \dfrac{A}{P} = \dfrac{\frac{1}{8}}{16\tfrac{2}{3}\%} = \dfrac{\frac{1}{8}}{\frac{1}{6}} = \dfrac{1}{8} \div \dfrac{1}{6} = \dfrac{1}{8} \times \dfrac{6}{1} = \dfrac{3}{4}$

(g) $B = \dfrac{A}{P} = \dfrac{1.5}{60\%} = \dfrac{1.5}{\frac{3}{5}} = 1\tfrac{1}{2} \div \dfrac{3}{5} = \dfrac{3}{2} \times \dfrac{5}{3} = \dfrac{5}{2}$ or $2\tfrac{1}{2}$

(h) $B = \dfrac{A}{P} = \dfrac{2\frac{3}{4}}{33\frac{1}{3}\%} = \dfrac{2\frac{3}{4}}{\frac{1}{3}} = 2\frac{3}{4} \div \dfrac{1}{3} = \dfrac{11}{4} \times \dfrac{3}{1} = \dfrac{33}{4}$ or $8\frac{1}{4}$

(i) $B = \dfrac{A}{P} = \dfrac{250}{41\frac{2}{3}\%} = \dfrac{250}{\frac{5}{12}} = 250 \div \dfrac{5}{12} = \dfrac{250}{1} \times \dfrac{12}{5} = 600$

PROBLEM 5-10   Find each percent by dividing to get a decimal quotient:

(a) What percent of 125 is 25?   **(b)** 7 is what percent of 10?   **(c)** What percent of 200 is 130?

(d) 110 is what percent of 160?   **(e)** What percent of 84 is 0.63?   **(f)** 3 is what percent of 5?

Solution   To find a percent in decimal quotient form, you write the Percent Formula, substitute for *A* and *B*, divide to get a decimal quotient, and then rename as a percent [see Example 5-15]:

(a) $P = \dfrac{A}{B} = \dfrac{25}{125} = 0.2 = 20\%$ **(b)** $P = \dfrac{A}{B} = \dfrac{7}{10} = 0.7 = 70\%$

(c) $P = \dfrac{A}{B} = \dfrac{130}{200} = 0.65 = 65\%$ **(d)** $P = \dfrac{A}{B} = \dfrac{110}{160} = 0.6875 = 68.75\%$

(e) $P = \dfrac{A}{B} = \dfrac{0.63}{84} = 0.0075 = 0.75\%$ **(f)** $P = \dfrac{A}{B} = \dfrac{3}{5} = 0.6 = 60\%$

PROBLEM 5-11   Find each percent by dividing to get a fractional quotient:

(a) What percent of $\frac{3}{5}$ is $\frac{1}{5}$?   **(b)** $\frac{1}{6}$ is what percent of $\frac{2}{3}$?

(c) What percent of $7\frac{1}{2}$ is 3?   **(d)** $\frac{1}{4}$ is what percent of $1\frac{1}{2}$?

(e) What percent of 120 is 10?   **(f)** $\frac{1}{5}$ is what percent of $\frac{3}{10}$?

(g) What percent of $5\frac{1}{3}$ is 4?   **(h)** $\frac{2}{3}$ is what percent of $3\frac{1}{3}$?

Solution   To find a percent in fractional quotient form, you write the Percent Formula, substitute for *A* and *B*, divide to get a fractional quotient, and then rename as a percent [see Example 5-16]:

(a) $P = \dfrac{A}{B} = \dfrac{\frac{1}{5}}{\frac{3}{5}} = \dfrac{1}{5} \div \dfrac{3}{5} = \dfrac{1}{5} \times \dfrac{5}{3} = \dfrac{1}{3} = 33\frac{1}{3}\%$ **(b)** $P = \dfrac{A}{B} = \dfrac{\frac{1}{6}}{\frac{2}{3}} = \dfrac{1}{6} \div \dfrac{2}{3} = \dfrac{1}{6} \times \dfrac{3}{2} = \dfrac{1}{4} = 25\%$

(c) $P = \dfrac{A}{B} = \dfrac{3}{7\frac{1}{2}} = 3 \div \dfrac{15}{2} = \dfrac{3}{1} \times \dfrac{2}{15} = \dfrac{2}{5} = 40\%$ **(d)** $P = \dfrac{A}{B} = \dfrac{\frac{1}{4}}{1\frac{1}{2}} = \dfrac{1}{4} \div \dfrac{3}{2} = \dfrac{1}{4} \times \dfrac{2}{3} = \dfrac{1}{6} = 16\frac{2}{3}\%$

(e) $P = \dfrac{A}{B} = \dfrac{10}{120} = \dfrac{1}{12} = 8\frac{1}{3}\%$ **(f)** $P = \dfrac{A}{B} = \dfrac{\frac{1}{5}}{\frac{3}{10}} = \dfrac{1}{5} \div \dfrac{3}{10} = \dfrac{1}{5} \times \dfrac{10}{3} = \dfrac{2}{3} = 66\frac{2}{3}\%$

(g) $P = \dfrac{A}{B} = \dfrac{4}{5\frac{1}{3}} = 4 \div \dfrac{16}{3} = \dfrac{4}{1} \times \dfrac{3}{16} = \dfrac{3}{4} = 75\%$ **(h)** $P = \dfrac{A}{B} = \dfrac{\frac{2}{3}}{3\frac{1}{3}} = \dfrac{2}{3} \div \dfrac{10}{3} = \dfrac{2}{3} \times \dfrac{3}{10} = \dfrac{1}{5} = 20\%$

PROBLEM 5-12   Find each percent increase from:

(a) 50 to 75   **(b)** 8.24 to 14.42   **(c)** $18\frac{3}{4}$ to $31\frac{1}{4}$ **(d)** 80 to 112

(e) 5.1 to 9.18   **(f)** $21\frac{1}{3}$ to $29\frac{1}{3}$ **(g)** 60 to 75   **(h)** 3.87 to 5.16

Solution   To find a percent increase, you identify the given original amount (the smaller number) as the base *B*, the amount of increase (the difference between the two given amounts) as the amount *A*, and then find the percent *P* using the Percent Formula [see Example 5-17]:

(a) $B = 50$, $A = 75 - 50 = 25$, $P = \dfrac{A}{B} = \dfrac{25}{50} = \dfrac{1}{2} = 50\%$

(b) $B = 8.24$, $A = 14.42 - 8.24 = 6.18$, $P = \dfrac{A}{B} = \dfrac{6.18}{8.24} = 0.75 = 75\%$

(c) $B = 18\frac{3}{4},$ $A = 31\frac{1}{4} - 18\frac{3}{4} = 12\frac{1}{2},$ $P = \dfrac{A}{B} = \dfrac{12\frac{1}{2}}{18\frac{3}{4}} = \dfrac{25}{4} \div \dfrac{75}{4} = \dfrac{25}{2} \times \dfrac{4}{75} = \dfrac{2}{3} = 66\frac{2}{3}\%$

(d) $B = 80,$ $A = 112 - 80 = 32,$ $P = \dfrac{A}{B} = \dfrac{32}{80} = \dfrac{2}{5} = 40\%$

(e) $B = 5.1,$ $A = 9.18 - 5.1 = 4.08,$ $P = \dfrac{A}{B} = \dfrac{4.08}{5.1} = 0.8 = 80\%$

(f) $B = 21\frac{1}{3},$ $A = 29\frac{1}{3} - 21\frac{1}{3} = 8,$ $P = \dfrac{A}{B} = \dfrac{8}{21\frac{1}{3}} = \dfrac{8}{1} \div \dfrac{64}{3} = \dfrac{8}{1} \times \dfrac{3}{64} = \dfrac{3}{8} = 37\frac{1}{2}\%$

(g) $B = 60,$ $A = 75 - 60 = 15,$ $P = \dfrac{A}{B} = \dfrac{15}{60} = \dfrac{1}{4} = 25\%$

(h) $B = 3.87,$ $A = 5.16 - 3.87 = 1.29,$ $P = \dfrac{A}{B} = \dfrac{1.29}{3.87} = 0.\overline{3} = \dfrac{1}{3} = 33\frac{1}{3}\%$

PROBLEM 5-13   Find each percent decrease from:

(a) 120 to 42   **(b)** 36.96 to 21.56   **(c)** $28\frac{4}{5}$ to $2\frac{2}{5}$ **(d)** 500 to 480

(e) 35.6 to 33.82   **(f)** $52\frac{1}{2}$ to $34\frac{1}{8}$ **(g)** 80 to 44   **(h)** 30.36 to 27.83

Solution   To find percent decrease, you identify the given original amount (the larger number) as the base B, the amount of decrease (the difference between the two given amounts) as the amount A, and then find the percent P using the Percent Formula [see Example 5-18]:

(a) $B = 120,$ $A = 120 - 42 = 78,$ $P = \dfrac{A}{B} = \dfrac{78}{120} = 0.65 = 65\%$

(b) $B = 36.96,$ $A = 36.96 - 21.56 = 15.4,$ $P = \dfrac{A}{B} = \dfrac{15.4}{36.96} = 0.41\overline{6} = 41\frac{2}{3}\%$

(c) $B = 28\frac{4}{5},$ $A = 28\frac{4}{5} - 2\frac{2}{5} = 26\frac{2}{5},$ $P = \dfrac{A}{B} = \dfrac{26\frac{2}{5}}{28\frac{4}{5}} = \dfrac{132}{5} \div \dfrac{144}{5} = \dfrac{132}{5} \times \dfrac{5}{144} = \dfrac{11}{12} = 91\frac{2}{3}\%$

(d) $B = 500,$ $A = 500 - 480 = 20,$ $P = \dfrac{A}{B} = \dfrac{20}{500} = \dfrac{1}{25} = 4\%$

(e) $B = 35.6,$ $A = 35.6 - 33.82 = 1.78,$ $P = \dfrac{A}{B} = \dfrac{1.78}{35.6} = 0.05 = 5\%$

(f) $B = 52\frac{1}{2},$ $A = 52\frac{1}{2} - 34\frac{1}{8} = 18\frac{3}{8},$ $P = \dfrac{A}{B} = \dfrac{18\frac{3}{8}}{52\frac{1}{2}} = \dfrac{147}{8} \div \dfrac{105}{2} = \dfrac{147}{8} \times \dfrac{2}{105} = \dfrac{7}{20} = 35\%$

(g) $B = 80,$ $A = 80 - 44 = 36,$ $P = \dfrac{A}{B} = \dfrac{36}{80} = \dfrac{9}{20} = 45\%$

(h) $B = 30.36,$ $A = 30.36 - 27.83 = 2.53,$ $P = \dfrac{A}{B} = \dfrac{2.53}{30.36} = 0.08\overline{3} = 8\frac{1}{3}\%$

PROBLEM 5-14   Find the sales tax and total price given the purchase price and sales-tax rate. Round the sales tax to the nearest cent when necessary.

	Purchase Price	Sales-Tax Rate	Sales Tax	Total Price
(a)	$45	5% (Chicago, IL)	?	?
(b)	$89	6% (Denver, CO)	?	?
(c)	$25.50	7% (Buffalo, NY)	?	?
(d)	$18.39	8% (New York, NY)	?	?
(e)	$50	$4\frac{1}{2}\%$ (Cincinnati, OH)	?	?

(f)	$39	$6\frac{1}{2}\%$ (San Francisco, CA)	?	?
(g)	$15.80	$4\frac{1}{8}\%$ (St. Louis, MO)	?	?
(h)	$62.99	$3\frac{5}{8}\%$ (Kansas City, MO)	?	?

Solution   Recall that to find the sales tax and total price on an item, you first multiply the purchase price by the sales-tax rate to find the sales tax and then you add the purchase price and sales tax together to find the total price [see Examples 5-19 and 5-20]:

(a) $5\% \times \$45 = 0.05 \times \$45 = \$2.25$ and $\$45 + \$2.25 = \$47.25$

(b) $6\% \times \$89 = 0.06 \times \$89 = \$5.34$ and $\$89 + \$5.34 = \$94.34$

(c) $7\% \times \$25.50 = 0.07 \times \$25.50 = \$1.785 \approx \$1.79$ and $\$25.50 + \$1.79 = \$27.29$

(d) $8\% \times \$18.39 = 0.08 \times \$18.39 = \$1.4712 \approx \$1.47$ and $\$18.39 + \$1.47 = \$19.86$

(e) $4\frac{1}{2}\% \times \$50 = 0.045 \times \$50 = \$2.25$ and $\$50 + \$2.25 = \$52.25$

(f) $6\frac{1}{2}\% \times \$39 = 0.065 \times \$39 = \$2.535 \approx \$2.54$ and $\$39 + \$2.54 = \$41.54$

(g) $4\frac{1}{8}\% \times \$15.80 = 0.04125 \times \$15.80 = \$0.65175 \approx \$0.65$ and $\$15.80 + \$0.65 = \$16.45$

(h) $3\frac{5}{8}\% \times \$62.99 = 0.03625 \times \$62.99 = \$2.2833875 \approx \$2.28$ and $\$62.99 + \$2.28 = \$65.27$

PROBLEM 5-15   Find the discount and sale price given the regular price and the discount rate. (Round the discount to the nearest cent when necessary.)

	Regular Price	Discount Rate	Discount	Sale Price
(a)	$25	$\frac{1}{2}$	?	?
(b)	$10	$\frac{1}{3}$	?	?
(c)	$32	30%	?	?
(d)	$15	40%	?	?
(e)	$59.30	15%	?	?
(f)	$75.85	25%	?	?

Solution   Recall that to find the discount and sale price on a given item, you first multiply the regular price by the discount rate to find the discount and then you subtract the discount from the regular price to find the sale price [see Examples 5-21 and 5-22]:

(a) $\frac{1}{2} \times \$25 = \dfrac{\$25}{2} = \$12.50$ and $\$25 - \$12.50 = \$12.50$

(b) $\frac{1}{3} \times \$10 = \dfrac{\$10}{3} = \$3.33\overline{3} \approx \$3.33$ and $\$10 - \$3.33 = \$6.67$

(c) $30\% \times \$32 = 0.3 \times \$32 = \$9.60$ and $\$32 - \$9.60 = \$22.40$

(d) $40\% \times \$15 = 0.4 \times \$15 = \$6$ and $\$15 - \$6 = \$9$

(e) $15\% \times \$59.30 = 0.15 \times \$59.30 = \$8.895 \approx \$8.90$ and $\$59.30 - \$8.90 = \$50.40$

(f) $25\% \times \$75.85 = 0.25 \times \$75.85 = \$18.9625 \approx \$18.96$ and $\$75.85 - \$18.96 = \$56.89$

PROBLEM 5-16   Find the commission and total pay given the sales, commission rate, and salary. (Round the commission to the nearest cent when necessary.)

	Sales	Commission Rate	Commission	Salary	Total Pay
(a)	$100	5%	?	$ 50	?
(b)	$500	10%	?	$100	?
(c)	$250	20%	?	$ 75	?

(d)	$725	25%	?	$200	?
(e)	$ 85.98	$12\frac{1}{2}\%$	?	$ 25	?
(f)	$112.53	$6\frac{1}{4}\%$	?	$ 52	?

Solution   Recall that to find the commission and total pay, you first multiply the sales by the commission rate to find the commission and then you add the salary to the commission to find the total pay [see Examples 5-23 and 5-24]:

	Commission Rate	Sales		Commission	Salary	Total Pay
(a)	5%	× $100	= 0.05 × $100	= $ 5	$ 50 + $ 5	= $ 55
(b)	10%	× $500	= 0.1 × $500	= $ 50	$100 + $ 50	= $150
(c)	20%	× $250	= 0.2 × $250	= $ 50	$ 75 + $ 50	= $125
(d)	25%	× $725	= 0.25 × $725	= $181.25	$200 + $181.25	= $381.25
(e)	$12\frac{1}{2}\%$	× $ 85.98	= 0.125 × $ 85.98	= $ 10.7475 ≈ $10.75	$ 25 + $ 10.75	= $ 35.75
(f)	$6\frac{1}{4}\%$	× $112.53	= 0.0625 × $112.53	= $ 7.033125 ≈ $7.03	$ 52 + $ 7.03	= $ 59.03

PROBLEM 5-17   Rename each time period in terms of the given shorter time unit:

(a) 3 years = ? months   **(b)** $2\frac{1}{2}$ years = ? months   **(c)** 2 years = ? days

(d) $1\frac{1}{2}$ years = ? days   **(e)** 6 months = ? days   **(f)** $5\frac{1}{2}$ months = ? days

Solution   Recall that to rename a time period in terms of a shorter time unit, you multiply [see Example 5-27a]:

(a) 3 years = 3 × 12 months = 36 months   **(b)** $2\frac{1}{2}$ years = $2\frac{1}{2}$ × 12 months = 30 months

(c) 2 years = 2 × 360 days = 720 days   **(d)** $1\frac{1}{2}$ years = $1\frac{1}{2}$ × 360 days = 540 days

(e) 6 months = 6 × 30 days = 180 days   **(f)** $5\frac{1}{2}$ months = $5\frac{1}{2}$ × 30 days = 165 days

PROBLEM 5-18   Rename each time period in terms of the given larger time unit:

(a) 48 months = ? years   **(b)** 42 months = ? years   **(c)** 270 days = ? months

(d) 135 days = ? months   **(e)** 1440 days = ? years   **(f)** 900 days = ? years

Solution   Recall that to rename a time period in terms of a longer time unit, you divide [see Example 5-27b]:

(a) 48 months = $\frac{48}{12}$ years = 4 years   **(b)** 42 months = $\frac{42}{12}$ years = $3\frac{1}{2}$ years

(c) 270 days = $\frac{270}{30}$ months = 9 months   **(d)** 135 days = $\frac{135}{30}$ months = $4\frac{1}{2}$ months

(e) 1440 days = $\frac{1440}{360}$ years = 4 years   **(f)** 900 days = $\frac{900}{360}$ years = $2\frac{1}{2}$ years

PROBLEM 5-19   Find the simple interest and total amount:

	Principal (P)	Interest Rate (r)	Time Period (t)	Simple Interest (I)	Total Amount (A)
(a)	$100	10% per year	5 years	?	?
(b)	$250	$6\frac{1}{2}\%$ per year	18 months	?	?
(c)	$425	1% per month	9 months	?	?
(d)	$ 80.50	$1\frac{1}{2}\%$ per month	90 days	?	?
(e)	$525.10	$\frac{1}{2}\%$ per day	30 days	?	?
(f)	$135.68	0.45% per day	3 months	?	?

Solution Recall that to find the simple interest and total amount, you first multiply the principal, interest rate, and time period together to find the simple interest and then add the principal to the simple interest to find the total amount [See Examples 5-28 and 5-29]:

(a) $I = P \times r \times t$

$\quad = \$100 \times \dfrac{10\%}{\text{year}} \times 5 \text{ years}$

$\quad = \$100 \times 0.1 \times 5$

$\quad = \$10 \times 5$

$\quad = \$50 \longleftarrow \text{ simple interest}$

$A = P + I$

$\quad = \$100 + \$50$

$\quad = \$150 \longleftarrow \text{ total amount}$

(b) $I = P \times r \times t$

$\quad = \$250 \times \dfrac{6\frac{1}{2}\%}{\text{year}} \times 18 \text{ months}$

$\quad = \$250 \times \dfrac{0.065}{\text{year}} \times 1\frac{1}{2} \text{ years}$

$\quad = \$16.25 \times 1.5$

$\quad = \$24.375$

$\quad \approx \$24.38 \longleftarrow \text{ simple interest}$

$A = P + I$

$\quad \approx \$250 + \$24.38$

$\quad = \$274.38 \longleftarrow \text{ total amount}$

(c) $I = P \times r \times t$

$\quad = \$425 \times \dfrac{1\%}{\text{month}} \times 9 \text{ months}$

$\quad = \$425 \times 0.01 \times 9$

$\quad = \$4.25 \times 9$

$\quad = \$38.25 \longleftarrow \text{ simple interest}$

$A = P + I$

$\quad = \$425 + \$38.25$

$\quad = \$463.25 \longleftarrow \text{ total amount}$

(d) $I = P \times r \times t$

$\quad = \$80.50 \times \dfrac{1\frac{1}{2}\%}{\text{month}} \times 90 \text{ days}$

$\quad = \$80.50 \times \dfrac{0.015}{\text{month}} \times 3 \text{ months}$

$\quad = \$1.2075 \times 3$

$\quad = \$3.6225$

$\quad \approx \$3.62 \longleftarrow \text{ simple interest}$

$A = P + I$

$\quad \approx \$80.50 + \$3.62$

$\quad = \$84.12 \longleftarrow \text{ total amount}$

(e) $I = P \times r \times t$

$\quad = \$525.10 \times \dfrac{\frac{1}{2}\%}{\text{day}} \times 30 \text{ days}$

$\quad = \$525.10 \times 0.005 \times 30$

$\quad = \$2.6255 \times 30$

$\quad = \$78.765$

$\quad \approx \$78.77$

$A = P + I \longleftarrow \text{ simple interest}$

$\quad \approx \$525.10 + \$78.77$

$\quad = \$603.87 \longleftarrow \text{ total amount}$

(f) $I = P \times r \times t$

$\quad = \$135.68 \times \dfrac{0.45\%}{\text{day}} \times 3 \text{ months}$

$\quad = \$135.68 \times \dfrac{0.0045}{\text{day}} \times 90 \text{ days}$

$\quad = \$0.61056 \times 90$

$\quad = \$54.9504$

$\quad \approx \$54.95 \longleftarrow \text{ simple interest}$

$A = P + I$

$\quad \approx \$135.68 + \$54.95$

$\quad = \$190.63 \longleftarrow \text{ total amount}$

Supplementary Exercises

PROBLEM 5-20   Rename each percent using both "$\times \frac{1}{100}$" and "$\div 100$":

(a) 30%   **(b)** 5%   **(c)** 90%   **(d)** $33\frac{1}{3}\%$ **(e)** $166\frac{2}{3}\%$ **(f)** 150%

(g) 37.5%   **(h)** 87.5%   **(i)** $16\frac{2}{3}\%$ **(j)** $83\frac{1}{3}\%$ **(k)** 62.5%   **(l)** 20%

(m) 80%   **(n)** $\frac{1}{3}\%$ **(o)** $\frac{3}{4}\%$ **(p)** $6\frac{1}{4}\%$ **(q)** $\frac{2}{3}\%$ **(r)** $8\frac{1}{2}\%$

PROBLEM 5-21   Rename each percent as a decimal:

(a) 25%   **(b)** 50%   **(c)** 75%   **(d)** 90%   **(e)** 10%

(f) 20%   **(g)** 12.5%   **(h)** 37.5%   **(i)** 30%   **(j)** 40%

(k) $62\frac{1}{2}\%$ **(l)** 87.5%   **(m)** 60%   **(n)** 80%   **(o)** $6\frac{1}{4}\%$

(p) 18.75%   **(q)** 100%   **(r)** 125%   **(s)** 31.25%   **(t)** $43\frac{3}{4}\%$

PROBLEM 5-22   Rename each percent as a fraction:

(a) $16\frac{2}{3}\%$ **(b)** $83\frac{1}{3}\%$ **(c)** $33\frac{1}{3}\%$ **(d)** $66\frac{2}{3}\%$ **(e)** $8\frac{1}{3}\%$

(f) $41\frac{2}{3}\%$ **(g)** $58\frac{1}{3}\%$ **(h)** $91\frac{2}{3}\%$ **(i)** 1%   **(j)** 25%

(k) 50%   **(l)** 75%   **(m)** 10%   **(n)** 20%   **(o)** 30%

(p) 40%   **(q)** 60%   **(r)** 70%   **(s)** 80%   **(t)** 90%

PROBLEM 5-23   Rename each decimal as a percent:

(a) 0.1   **(b)** 0.2   **(c)** 0.625   **(d)** 0.875   **(e)** 0.5

(f) 0.6   **(g)** 0.3125   **(h)** 0.4375   **(i)** 0.9   **(j)** 0.25

(k) 0.75   **(l)** 0.05   **(m)** 0.125   **(n)** 0.375   **(o)** 0.3

(p) 0.4   **(q)** 0.0625   **(r)** 0.1875   **(s)** 0.7   **(t)** 0.8

PROBLEM 5-24   Rename each fraction as a percent:

(a) $\frac{1}{2}$ **(b)** $\frac{1}{4}$ **(c)** $\frac{1}{5}$ **(d)** $\frac{1}{8}$ **(e)** $\frac{1}{10}$ **(f)** $\frac{1}{12}$ **(g)** $\frac{1}{16}$ **(h)** $\frac{3}{4}$ **(i)** $\frac{2}{5}$ **(j)** $\frac{3}{8}$

(k) $\frac{3}{10}$ **(l)** $\frac{5}{12}$ **(m)** $\frac{3}{16}$ **(n)** $\frac{3}{5}$ **(o)** $\frac{5}{8}$ **(p)** $\frac{7}{10}$ **(q)** $\frac{7}{12}$ **(r)** $\frac{5}{16}$ **(s)** $\frac{4}{5}$ **(t)** $\frac{7}{8}$

PROBLEM 5-25   Find each amount using the Percent Diagram to write the Amount Formula:

(a) What number is 50% of 30?   **(b)** $33\frac{1}{3}\%$ of 60 is what number?

(c) What number is $66\frac{2}{3}\%$ of 15?   **(d)** 25% of 100 is what number?

(e) What number is 75% of 84?   **(f)** 20% of 56 is what number?

(g) What number is 40% of 102?   **(h)** 60% of 1 is what number?

(i) What number is 80% of 2.5?   **(j)** $16\frac{2}{3}\%$ of 3.6 is what number?

(k) What number is $83\frac{1}{3}\%$ of 6?   **(l)** 12.5% of 0.1 is what number?

(m) What number is $37\frac{1}{2}\%$ of 16.50?   **(n)** 62.5% of 800 is what number?

(o) What number is $87\frac{1}{2}\%$ of 160?   **(p)** 10% of $\frac{1}{2}$ is what number?

(q) What number is 20% of $3\frac{1}{4}$?   **(r)** $8\frac{1}{3}\%$ of 12 is what number?

(s) What number is $41\frac{2}{3}\%$ of 4.8?   **(t)** 5% of 5 is what number?

PROBLEM 5-26   Find each base using the Percent Diagram to write the Base Formula:

(a) 50% of what number is 8?   **(b)** 28 is 70% of what number?

(c) $33\frac{1}{3}\%$ of what number is 100?

(d) 22.5 is 90% of what number?

(e) $66\frac{2}{3}\%$ of what number is 2.1?

(f) 2.1 is $58\frac{1}{3}\%$ of what number?

(g) 25% of what number is 5.2?

(h) 8.8 is $91\frac{2}{3}\%$ of what number?

(i) 75% of what number is $4\frac{4}{5}$?

(j) $2\frac{1}{2}$ is 31.25% of what number?

(k) 20% of what number is $5\frac{2}{3}$?

(l) $1\frac{3}{4}$ is 43.75% of what number?

(m) 40% of what number is 90?

(n) 55 is $68\frac{3}{4}\%$ of what number?

(o) 60% of what number is 82?

(p) 42.25 is $81\frac{1}{4}\%$ of what number?

(q) 80% of what number is 80?

(r) 0.6 is 93.75% of what number?

(s) $16\frac{2}{3}\%$ of what number is $\frac{1}{6}$?

(t) 9.2 is 4% of what number?

PROBLEM 5-27 Find each percent using the Percent Diagram to write the Percent Formula:

(a) What percent of 2 is 1?

(b) 1 is what percent of 8?

(c) What percent of 30 is 10?

(d) 30 is what percent of 80?

(e) What percent of 4.5 is 3?

(f) 1.25 is what percent of 2?

(g) What percent of 1 is $\frac{1}{4}$?

(h) $\frac{7}{16}$ is what percent of $\frac{1}{2}$?

(i) What percent of 4 is 3?

(j) 1 is what percent of 10?

(k) What percent of 50 is 10?

(l) 30 is what percent of 100?

(m) What percent of 32.5 is 13?

(n) 8.75 is what percent of 12.5?

(o) What percent of $\frac{1}{2}$ is $\frac{3}{10}$?

(p) $\frac{9}{10}$ is what percent of 1?

(q) What percent of 39.25 is 31.4?

(r) 1000 is what percent of 12,000?

(s) What percent of 1 is 6?

(t) 12 is what percent of 5?

PROBLEM 5-28 Find the amount, base, or percent using the Percent Diagram to write the correct formula:

(a) What number is $16\frac{2}{3}\%$ of 210?

(b) $83\frac{1}{3}\%$ of 21 is what number?

(c) $41\frac{2}{3}\%$ of what number is 144?

(d) 33 is $91\frac{2}{3}\%$ of what number?

(e) What percent of 16 is 1?

(f) 1 is what percent of 20?

(g) $18\frac{3}{4}\%$ of 20 is what number?

(h) 4% of what number is 9.2?

(i) 16 is what percent of 5?

(j) What number is $43\frac{3}{4}\%$ of 100?

(k) $8\frac{1}{2}$ is 56.25% of what number?

(l) What percent of 3 is 8?

(m) What number is 150% of 6?

(n) $56\frac{3}{4}\%$ of 25 is what number?

(o) 68.75% of what number is $5\frac{1}{2}$?

(p) 11.5 is $5\frac{3}{4}\%$ of what number?

(q) What percent of 32 is 26?

(r) 2 is what percent of 50?

(s) $93\frac{3}{4}\%$ of 18 is what number?

(t) 1% of what number is $1\frac{1}{2}$?

PROBLEM 5-29 Find each percent increase from:

(a) 8 to 13

(b) 1 to 2

(c) 20 to 30

(d) 45 to 60

(e) 4 to 5

(f) 50 to 60

(g) 72 to 84

(h) 8 to 9

(i) 100 to 110

(j) 1.2 to 1.3

(k) 16 to 17

(l) 20 to 21

(m) 25 to 26

(n) 50 to 51

(o) 1 to 1.01

(p) 40 to 46

(q) $\frac{4}{5}$ to $1\frac{2}{5}$

(r) 1 to $1\frac{2}{3}$

(s) $\frac{5}{8}$ to $\frac{7}{8}$

(t) 8 to 11

PROBLEM 5-30   Find each percent decrease from:

(a) 2 to 1   (b) $\frac{3}{4}$ to $\frac{1}{4}$ (c) 40 to 10   (d) 1.25 to 0.75   (e) 5 to 1

(f) $\frac{3}{4}$ to $\frac{1}{8}$ (g) 80 to 50   (h) 1 to 0.125   (i) 12 to 0   (j) 100 to 70

(k) $\frac{1}{2}$ to $\frac{1}{20}$ (l) 6 to 2.5   (m) 120 to 70   (n) 4 to $3\frac{2}{3}$ (o) 160 to 150

(p) 4.5 to 3.75   (q) $\frac{5}{8}$ to $\frac{1}{4}$ (r) 2 to 1.9   (s) $1\frac{2}{3}$ to $1\frac{1}{3}$ (t) 4 to 3

PROBLEM 5-31   Find the missing total price and sales tax.

	Purchase Price	Sales-Tax Rate	Sales Tax	Total Price
(a)	$ 59	4%	?	?
(b)	$145	5%	?	?
(c)	$349.99	$6\frac{1}{2}\%$	?	?

PROBLEM 5-32   Find the missing sale price and discount.

	Regular Price	Discount Rate	Discount	Sale Price
(a)	$ 85	20%	?	?
(b)	$106	10%	?	?
(c)	$ 7.99	$\frac{1}{4}$ off	?	?

PROBLEM 5-33   Find the missing total pay and commission.

	Sales	Commission Rate	Commission	Salary	Total Pay
(a)	$100	2%	?	$ 75	?
(b)	$ 50	3%	?	$ 10	?
(c)	$200	5%	?	$100	?

PROBLEM 5-34   Rename each time period in terms of each given time unit:

	Years (yr)	Quarterly (q)	Months (mo)	Days (d)
(a)	5	?	?	?
(b)	$4\frac{1}{2}$	?	?	?
(c)	4	?	?	?
(d)	$3\frac{1}{4}$	?	?	?
(e)	3	?	?	?
(f)	$2\frac{3}{4}$	?	?	?
(g)	2	?	?	?
(h)	$1\frac{1}{2}$	?	?	?
(i)	1	?	?	?
(j)	$\frac{1}{2}$	?	?	?
(k)	?	6	?	?
(l)	?	10	?	?
(m)	?	3	?	?
(n)	?	?	48	?
(o)	?	?	30	?

(p)	?	?	9	?
(q)	?	?	?	360
(r)	?	?	?	990

PROBLEM 5-35 Find the simple interest and total amount for each simple interest problem:

	Principal P	Interest Rate r	Time Period t	Simple Interest	Total Amount
(a)	\$ 100	10% per year	5 years	?	?
(b)	\$ 200	$1\frac{1}{2}$% per month	$2\frac{1}{2}$ years	?	?
(c)	\$ 350	0.04% per day	$\frac{1}{4}$ year	?	?
(d)	\$1000	5% per year	12 quarters	?	?
(e)	\$3000	$1\frac{1}{4}$% per month	2 quarters	?	?
(f)	\$4500	0.05% per day	1 quarter	?	?
(g)	\$ 300	6% per year	48 months	?	?
(h)	\$2000	$1\frac{1}{3}$% per month	30 months	?	?
(i)	\$ 250	0.0475% per day	3 months	?	?
(j)	\$1500	12% per year	720 days	?	?
(k)	\$ 500	$1\frac{2}{3}$% per month	120 days	?	?
(l)	\$5000	0.0525% per day	180 days	?	?

Answers to Supplementary Exercises

(5-20) (a) $30 \times \frac{1}{100}$, $30 \div 100$ (b) $15 \times \frac{1}{100}$, $5 \div 100$ (c) $90 \times \frac{1}{100}$, $90 \div 100$
(d) $33\frac{1}{3} \times \frac{1}{100}$, $33\frac{1}{3} \div 100$ (e) $166\frac{2}{3} \times \frac{1}{100}$, $166\frac{2}{3} \div 100$ (f) $150 \times \frac{1}{100}$, $150 \div 100$
(g) $37.5 \times \frac{1}{100}$, $37.5 \div 100$ (h) $87.5 \times \frac{1}{100}$, $87.5 \div 100$ (i) $16\frac{2}{3} \times \frac{1}{100}$, $16\frac{2}{3} \div 100$
(j) $83\frac{1}{3} \times \frac{1}{100}$, $83\frac{1}{3} \div 100$ (k) $62.5 \times \frac{1}{100}$, $62.5 \div 100$ (l) $20 \times \frac{1}{100}$, $20 \div 100$
(m) $80 \times \frac{1}{100}$, $80 \div 100$ (n) $\frac{1}{3} \times \frac{1}{100}$, $\frac{1}{3} \div 100$ (o) $\frac{3}{4} \times \frac{1}{100}$, $\frac{3}{4} \div 100$
(p) $6\frac{1}{4} \times \frac{1}{100}$, $6\frac{1}{4} \div 100$ (q) $\frac{2}{3} \times \frac{1}{100}$, $\frac{2}{3} \div 100$ (r) $8\frac{1}{2} \times \frac{1}{100}$, $8\frac{1}{2} \div 100$

(5-21) (a) 0.25   (b) 0.5   (c) 0.75   (d) 0.9   (e) 0.1
(f) 0.2   (g) 0.125   (h) 0.375   (i) 0.3   (j) 0.4
(k) 0.625   (l) 0.875   (m) 0.6   (n) 0.8   (o) 0.0625
(p) 0.1875   (q) 1   (r) 1.25   (s) 0.3125   (t) 0.4375

(5-22) (a) $\frac{1}{6}$ (b) $\frac{5}{6}$ (c) $\frac{1}{3}$ (d) $\frac{2}{3}$ (e) $\frac{1}{12}$
(f) $\frac{5}{12}$ (g) $\frac{7}{12}$ (h) $\frac{11}{12}$ (i) $\frac{1}{100}$ (j) $\frac{1}{4}$
(k) $\frac{1}{2}$ (l) $\frac{3}{4}$ (m) $\frac{1}{10}$ (n) $\frac{1}{5}$ (o) $\frac{3}{10}$
(p) $\frac{2}{5}$ (q) $\frac{3}{5}$ (r) $\frac{7}{10}$ (s) $\frac{4}{5}$ (t) $\frac{9}{10}$

(5-23) (a) 10%   (b) 20%   (c) 62.5% or $62\frac{1}{2}$%   (d) 87.5% or $87\frac{1}{2}$%
(e) 50%   (f) 60%   (g) 31.25% or $31\frac{1}{4}$%   (h) 43.75% or $43\frac{3}{4}$%
(i) 90%   (j) 25%   (k) 75%   (l) 5%
(m) 12.5% or $12\frac{1}{2}$%   (n) 37.5% or $37\frac{1}{2}$%   (o) 30%   (p) 40%
(q) 6.25% or $6\frac{1}{4}$%   (r) 18.75% or $18\frac{3}{4}$%   (s) 70%   (t) 80%

(5-24) (a) 50%   (b) 25%   (c) 20%   (d) $12\frac{1}{2}$% or 12.5%
(e) 10%   (f) $8\frac{1}{3}$%   (g) $6\frac{1}{4}$% or 6.25%   (h) 75%

(i) 40% (j) $37\frac{1}{2}$% or 37.5% (k) 30% (l) $41\frac{2}{3}$%
(m) $18\frac{3}{4}$% or 18.75% (n) 60% (o) $62\frac{1}{2}$% or 62.5% (p) 70%
(q) $58\frac{1}{3}$% (r) $31\frac{1}{4}$% or 31.25% (s) 80% (t) $87\frac{1}{2}$% or 87.5%

(5-25) (a) 15 (b) 20 (c) 10 (d) 25 (e) 63
(f) 11.2 (g) 40.8 (h) 0.6 (i) 2 (j) $\frac{3}{5}$ or 0.6
(k) 5 (l) 0.0125 (m) $6\frac{3}{16}$ or 6.1875 (n) 500 (o) 140
(p) $\frac{1}{20}$ or 0.05 (q) $\frac{13}{20}$ or 0.65 (r) 1 (s) 2 (t) 0.25

(5-26) (a) 16 (b) 40 (c) 300 (d) 25 (e) 3.15 or $3\frac{3}{20}$
(f) 3.6 (g) 20.8 (h) 9.6 or $9\frac{3}{5}$ (i) $6\frac{2}{5}$ or 6.4 (j) 8
(k) $28\frac{1}{3}$ (l) 4 (m) 225 (n) 80 (o) $136\frac{2}{3}$
(p) 52 (q) 100 (r) 0.64 (s) 1 (t) 230

(5-27) (a) 50% (b) 12.5% or $12\frac{1}{2}$% (c) $33\frac{1}{3}$% (d) 37.5% or $37\frac{1}{2}$%
(e) $66\frac{2}{3}$% (f) 62.5% or $62\frac{1}{2}$% (g) 25% (h) 87.5% or $87\frac{1}{2}$%
(i) 75% (j) 10% (k) 20% (l) 30%
(m) 40% (n) 70% (o) 60% (p) 90%
(q) 80% (r) $8\frac{1}{3}$% (s) 600% (t) 240%

(5-28) (a) 35 (b) 17.5 (c) $345\frac{3}{5}$ or 345.6 (d) 36
(e) 6.25% or $6\frac{1}{4}$% (f) 5% (g) 3.75 or $3\frac{3}{4}$ (h) 230
(i) 320% (j) 43.75 (k) $15\frac{1}{9}$ (l) $266\frac{2}{3}$%
(m) 9 (n) 14.1875 or $14\frac{3}{16}$ (o) 8 (p) 200
(q) 81.25% or $81\frac{1}{4}$% (r) 4% (s) 16.875 or $16\frac{7}{8}$ (t) 150

(5-29) (a) 62.5% or $62\frac{1}{2}$% (b) 100% (c) 50% (d) $33\frac{1}{3}$%
(e) 25% (f) 20% (g) $16\frac{2}{3}$% (h) 12.5% or $12\frac{1}{2}$%
(i) 10% (j) $8\frac{1}{3}$% (k) 6.25% or $6\frac{1}{4}$% (l) 5%
(m) 4% (n) 2% (o) 1% (p) 15%
(q) 75% (r) $66\frac{2}{3}$% (s) 40% (t) 37.5% or $37\frac{1}{2}$%

(5-30) (a) 50% (b) $66\frac{2}{3}$% (c) 75% (d) 40% (e) 80%
(f) $83\frac{1}{3}$% (g) 37.5% or $37\frac{1}{2}$% (h) 87.5% or $87\frac{1}{2}$% (i) 100% (j) 30%
(k) 90% (l) $58\frac{1}{3}$% (m) $41\frac{2}{3}$% (n) $8\frac{1}{3}$% (o) 6.25% or $6\frac{1}{4}$%
(p) $16\frac{2}{3}$% (q) 60% (r) 5% (s) 20% (t) 25%

(5-31) (a) $2.36, $61.36 (b) $7.25, $152.25 (c) $22.75, $372.74

(5-32) (a) $17, $68 (b) $10.60, $95.40 (c) $2, $5.99

(5-33) (a) $2, $77 (b) $1.50, $11.50 (c) $10, $110

(5-34) (a) 20 q, 60 mo, 1800 d (b) 18 q, 54 mo, 1620 d (c) 16 q, 48 mo, 1440 d
(d) 13 q, 39 mo, 1170 d (e) 12 q, 36 mo, 1080 d (f) 11 q, 33 mo, 990 d
(g) 8 q, 24 mo, 720 d (h) 6 q, 18 mo, 540 d (i) 4 q, 12 mo, 360 d
(j) 2 q, 6 mo, 180 d (k) $1\frac{1}{2}$ yr, 18 mo, 540 d (l) $2\frac{1}{2}$ yr, 30 mo, 900 d
(m) $\frac{3}{4}$ yr, 9 mo, 270 d (n) 4 yr, 16 q, 1440 d (o) $2\frac{1}{2}$ yr, 10 q, 900 d
(p) $\frac{3}{4}$ yr, 3 q, 270 d (q) 1 yr, 4 q, 12 mo (r) $2\frac{3}{4}$ yr, 11 q, 33 mo

(5-35) (a) $50, $150 (b) $90, $290 (c) $12.60, $362.60 (d) $150, $1150
(e) $225, $3225 (f) $202.50, $4702.50 (g) $72, $372 (h) $800, $2800
(i) $10.69, $260.69 (j) $360, $1860 (k) $33.33, $533.33 (l) $472.50, $5472.50

MIDTERM EXAM

Chapters 1-5

Part 1: Skills and Concepts (108 questions)

1. The place-value name of the underlined digit in 718,054,693 is

   (a) hundred millions   (b) ten thousands   (c) zero   (d) hundred thousands

   (e) none of these

2. What is the value of 2 in 619,082,583?

   (a) thousands   (b) millions   (c) ten thousands   (d) hundreds

   (e) none of these

3. Six thousand twenty-five written as a whole number is

   (a) 625   (b) 6025   (c) 6250   (d) 60,025   (e) none of these

4. 705 written as a whole-number-word name is

   (a) seventy-five   (b) seven hundred and five   (c) seven hundred five

   (d) seven hundred fifty   (e) none of these

5. The correct symbol for ? in 17,843 ? 17,834 is

   (a) <   (b) >   (c) =   (d) all of these   (e) none of these

6. 546, 456, 465, 645, 654, 564 listed from smallest to largest is

   (a) 654, 645, 564, 546, 456, 465   (b) 456, 465, 564, 546, 645, 654

   (c) 465, 456, 546, 564, 645, 654   (d) 654, 645, 564, 546, 465, 456

   (e) none of these

7. 6,375,418 rounded to the nearest ten thousand is

   (a) 6,400,000   (b) 6,370,000   (c) 6,380,000   (d) 6,375,000   (e) none of these

8. The sum of $83 + 585 + 109$ is

   (a) 677   (b) 667   (c) 767   (d) 777   (e) none of these

9. $642 - 285$ equals

   (a) 357   (b) 457   (c) 367   (d) 467   (e) none of these

10. $803 - 504$ equals

    (a) 399   (b) 299   (c) 309   (d) 209   (e) none of these

11. The product of 4×368 is

    (a) 1272   (b) 1442   (c) 1242   (d) 1482   (e) none of these

12. The product of 25×846 is

    (a) 21,050   (b) 22,150   (c) 21,150   (d) 5859   (e) none of these

13. The product of 235×346 is

    (a) 91,310   (b) 89,310   (c) 81,210   (d) 81,310   (e) none of these

14. The product of 800×3000 is

  (a) 2400   (b) 240,000   (c) 24,000   (d) 2,400,000   (e) none of these

15. The product of 203×324 is

  (a) 65,772   (b) 7452   (c) 65,762   (d) 648,972   (e) none of these

16. The quotient of $58 \div 3$ is

  (a) 19   (b) 18 R1   (c) 18 R2   (d) 19 R2   (e) none of these

17. The quotient of $108 \div 8$ is

  (a) 13 R1   (b) 13 R2   (c) 13 R3   (d) 13 R4   (e) none of these

18. The quotient of $816 \div 4$ is

  (a) 204   (b) 24   (c) 240   (d) 200 R16   (e) none of these

19. The quotient of $385 \div 20$ is

  (a) 18 R5   (b) 19   (c) 14 R5   (d) 14   (e) none of these

20. The quotient of $542 \div 29$ is

  (a) 19   (b) 18 R20   (c) 18   (d) 19 R20   (e) none of these

21. The quotient of $2654 \div 18$ is

  (a) 137 R8   (b) 147   (c) 147 R8   (d) 14 R13   (e) none of these

22. The quotient of $8400 \div 200$ is

  (a) 420   (b) 4200   (c) 42   (d) 42,000   (e) none of these

23. The whole-number factors of 18 are

  (a) 2, 3, 6, 9   (b) 1, 18   (c) 1, 2, 3, 6, 9   (d) 1, 2, 3, 6, 9, 18   (e) none of these

24. 18 factored as a product of primes is

  (a) $2 \times 2 \times 3$ (b) 2×9 (c) $2 \times 3 \times 3$ (d) 3×6 (e) none of these

25. The opposite of -5 is

  (a) 5   (b) -5 (c) $\frac{1}{5}$ (d) $-\frac{1}{5}$ (e) none of these

26. The absolute value of -5 is

  (a) 5   (b) -5 (c) $\frac{1}{5}$ (d) $-\frac{1}{5}$ (e) none of these

27. The correct symbol for ? in -5 ? -4 is

  (a) $<$ (b) $>$ (c) $=$ (d) all of these   (e) none of these

28. The sum of $-5 + 4$ is

  (a) 1   (b) -1 (c) 9   (d) -9 (e) none of these

29. The difference of $2 - (-7)$ is

  (a) 5   (b) -5 (c) 9   (d) -9 (e) none of these

30. The product of $-3(-2)$ is

(a) -6 (b) -5 (c) 5   (d) 6   (e) none of these

31. The quotient of $\frac{-12}{3}$ is

(a) -4 (b) 4   (c) $\frac{1}{4}$ (d) $-\frac{1}{4}$ (e) none of these

32. Which of the following fractions is not defined?

(a) $\frac{0}{3}$ (b) $\frac{3}{0}$ (c) $\frac{3}{3}$ (d) $\frac{0}{1}$ (e) none of these

33. $\frac{12}{18}$ is equal to

(a) $\frac{4}{6}$ (b) $\frac{3}{2}$ (c) $\frac{3}{4}$ (d) $\frac{6}{8}$ (e) none of these

34. $2 \div 5$ renamed as a fraction is

(a) $\frac{2}{5}$ (b) $\frac{5}{2}$ (c) $2\frac{1}{2}$ (d) $2\frac{1}{5}$ (e) none of these

35. $\frac{4}{3}$ renamed as a division problem is

(a) $3 \div 4$ (b) $4 \div 3$ (c) $1\frac{1}{3}$ (d) $4\overline{)3}$ (e) none of these

36. 4 renamed as an equal fraction is

(a) $\frac{4}{1}$ (b) $\frac{12}{3}$ (c) $\frac{8}{2}$ (d) $\frac{16}{4}$ (e) all of these

37. $4\frac{2}{3}$ renamed as an equal fraction is

(a) $\frac{2}{3}$ (b) $\frac{12}{3}$ (c) $\frac{24}{3}$ (d) $\frac{4}{1}$ (e) none of these

38. $\frac{15}{4}$ renamed as an equal mixed number is

(a) $3\frac{1}{4}$ (b) $4\frac{1}{4}$ (c) $2\frac{3}{4}$ (d) $3\frac{3}{4}$ (e) none of these

39. Which of the following fractions is in lowest terms?

(a) $\frac{4}{6}$ (b) $\frac{8}{18}$ (c) $\frac{15}{9}$ (d) $\frac{21}{14}$ (e) none of these

40. $\frac{18}{12}$ in simplest form is

(a) $\frac{4}{6}$ (b) $\frac{1}{2}$ (c) $\frac{2}{3}$ (d) $\frac{8}{12}$ (e) none of these

41. $\frac{15}{9}$ in simplest form is

(a) $\frac{15}{9}$ (b) $\frac{5}{3}$ (c) $1\frac{6}{9}$ (d) $1\frac{2}{3}$ (e) none of these

42. $2\frac{9}{3}$ in simplest form is

(a) $2\frac{3}{1}$ (b) 3   (c) $2\frac{9}{3}$ (d) $2 + \frac{9}{3}$ (e) none of these

43. The product of $\frac{2}{3} \times \frac{3}{4}$ in simplest form is

(a) $\frac{1}{2}$ (b) $\frac{6}{16}$ (c) $\frac{2}{4}$ (d) $\frac{3}{6}$ (e) all of these

44. The product of $5 \times \frac{7}{10}$ in simplest form is

(a) $\frac{35}{10}$ (b) $\frac{7}{50}$ (c) $5\frac{7}{10}$ (d) $\frac{57}{10}$ (e) none of these

45. The product of $2\frac{2}{3} \times 2\frac{1}{4}$ in simplest form is

(a) $4\frac{1}{6}$ (b) $\frac{6}{1}$ (c) $\frac{72}{12}$ (d) 6   (e) none of these

46. The reciprocal of $\frac{8}{6}$ is

(a) $\frac{8}{6}$ (b) $\frac{6}{8}$ (c) $\frac{4}{3}$ (d) $\frac{3}{4}$ (e) none of these

47. The quotient of $\frac{3}{4} \div \frac{8}{9}$ in simplest form is

(a) $\frac{2}{3}$ (b) $\frac{3}{2}$ (c) $\frac{27}{32}$ (d) $\frac{32}{27}$ (e) none of these

48. The quotient of $\frac{5}{6} \div 10$ in simplest form is

(a) $\frac{1}{12}$ (b) $\frac{25}{3}$ (c) $8\frac{1}{4}$ (d) $\frac{5}{60}$ (e) none of these

49. The quotient of $12 \div \frac{3}{4}$ in simplest form is

(a) $\frac{9}{1}$ (b) 9   (c) $\frac{48}{3}$ (d) $\frac{16}{1}$ (e) none of these

50. The quotient of $2\frac{1}{2} \div 1\frac{1}{4}$ in simplest form is

(a) $\frac{25}{8}$ (b) $3\frac{1}{8}$ (c) $\frac{4}{2}$ (d) 2   (e) none of these

51. Which of the following fraction pairs are like fractions?

(a) $\frac{2}{3}$ and $\frac{3}{2}$ (b) $\frac{2}{4}$ and $\frac{4}{6}$ (c) $\frac{2}{3}$ and $\frac{2}{5}$ (d) $\frac{5}{8}$ and $\frac{2}{8}$ (e) none of these

52. The sum of $\frac{3}{4} + \frac{3}{4}$ in simplest form is

(a) $\frac{9}{16}$ (b) $\frac{6}{4}$ (c) $\frac{3}{2}$ (d) $1\frac{1}{2}$ (e) none of these

53. The LCD for $\frac{5}{6}$ and $\frac{7}{8}$ is

(a) 48   (b) 8   (c) 24   (d) 14   (e) none of these

54. $\frac{3}{4}$ built up as an equal fraction with a denominator of 20 is

(a) $\frac{3}{20}$ (b) $\frac{7}{20}$ (c) $\frac{15}{20}$ (d) $\frac{16}{20}$ (e) none of these

55. $\frac{2}{3}, \frac{3}{4},$ and $\frac{5}{6}$ renamed as equivalent like fractions using the LCD are

(a) $\frac{16}{24}, \frac{18}{24}, \frac{20}{24}$ (b) $\frac{8}{12}, \frac{9}{12}, \frac{10}{12}$ (c) $\frac{2}{12}, \frac{3}{12}, \frac{5}{12}$ (d) $\frac{2}{24}, \frac{3}{24}, \frac{5}{24}$ (e) none of these

56. The sum of $\frac{2}{3} + \frac{3}{4}$ in simplest form is

(a) $\frac{17}{12}$ (b) $\frac{5}{12}$ (c) $\frac{6}{12}$ (d) $\frac{1}{2}$ (e) none of these

57. The sum of $\frac{1}{2}, \frac{3}{4},$ and $\frac{5}{8}$ in simplest form is

(a) $1\frac{7}{8}$ (b) $\frac{15}{8}$ (c) $\frac{15}{64}$ (d) 1   (e) none of these

58. The sum of $3\frac{1}{2} + 4$ in simplest form is

(a) $7\frac{1}{2}$ (b) $12\frac{1}{2}$ (c) $1\frac{1}{2}$ (d) 14   (e) none of these

59. The sum of $\frac{3}{4} + 2\frac{1}{6}$ in simplest form is

(a) $2\frac{3}{12}$ (b) $2\frac{1}{4}$ (c) $\frac{35}{12}$ (d) $2\frac{11}{12}$ (e) none of these

60. The sum of $2\frac{1}{2} + 2\frac{1}{2} + 3$ in simplest form is

(a) $12\frac{1}{4}$ (b) $3\frac{10}{2}$ (c) $3\frac{25}{4}$ (d) $7\frac{1}{2}$ (e) none of these

61. $\frac{5}{8} - \frac{3}{8}$ in simplest form equals

(a) $\frac{2}{8}$ (b) $\frac{8}{8}$ (c) $\frac{1}{4}$ (d) 1   (e) none of these

62. $\frac{7}{10} - \frac{1}{5}$ in simplest form equals

    (a) $\frac{6}{5}$ **(b)** $\frac{1}{2}$ **(c)** $\frac{5}{10}$ **(d)** $\frac{9}{10}$ **(e)** none of these

63. $5\frac{1}{2} - 4$ in simplest form equals

    (a) $1\frac{1}{2}$ **(b)** $9\frac{1}{2}$ **(c)** $20\frac{1}{2}$ **(d)** $4\frac{11}{2}$ **(e)** none of these

64. $2\frac{5}{8} - \frac{1}{3}$ in simplest form equals

    (a) $2\frac{23}{24}$ **(b)** $\frac{21}{5}$ **(c)** $4\frac{1}{5}$ **(d)** $2\frac{1}{4}$ **(e)** none of these

65. $8\frac{1}{10} - 3\frac{1}{12}$ in simplest form equals

    (a) $5\frac{1}{60}$ **(b)** $11\frac{11}{60}$ **(c)** $24\frac{1}{120}$ **(d)** $5\frac{1}{2}$ **(e)** none of these

66. $6 - 1\frac{3}{4}$ in simplest form equals

    (a) $5\frac{3}{4}$ **(b)** $5\frac{1}{4}$ **(c)** $4\frac{1}{4}$ **(d)** $4\frac{3}{4}$ **(e)** none of these

67. $3\frac{1}{3} - \frac{1}{2}$ in simplest form equals

    (a) $3\frac{1}{6}$ **(b)** $2\frac{5}{6}$ **(c)** $2\frac{11}{6}$ **(d)** $3\frac{5}{6}$ **(e)** none of these

68. Which digit is in the thousandths place in 1325.416908?

    (a) 1   **(b)** 8   **(c)** 0   **(d)** 9   **(e)** none of these

69. The decimal-word name for 1.03 is

    (a) thirteen   **(b)** 13 tenths   **(c)** 1 and 3 tenths   **(d)** 1 and 3 hundredths   **(e)** none of these

70. The decimal number for 25 and 8 thousandths is

    (a) 25.8   **(b)** 258   **(c)** 25.08   **(d)** 25.008   **(e)** none of these

71. The correct symbol for ? in 365.24717 ? 365.24709 is

    (a) <   **(b)** >   **(c)** =   **(d)** all of these   **(e)** none of these

72. The correct order from largest to smallest of 0.3, 0.03, 3, 3.03 is

    (a) 0.03, 0.3, 3, 3.03   **(b)** 3, 3.03, 0.3, 0.03   **(c)** 3.03, 3, 0.03, 0.3   **(d)** 3, 3.03, 0.03, 0.3

    (e) none of these

73. 325.963 rounded to the nearest tenth is

    (a) 325.9   **(b)** 325   **(c)** 326.0   **(d)** 326   **(e)** none of these

74. 5 renamed as an equal decimal is

    (a) 5.0   **(b)** 5.00   **(c)** 5.000   **(d)** 5.0000   **(e)** all of these

75. 245.00 renamed as an equal whole number is

    (a) 2.45   **(b)** 24.5   **(c)** 245   **(d)** 245.0   **(e)** none of these

76. $\frac{9}{100}$ renamed as an equal decimal is

    (a) 0.9   **(b)** 0.09   **(c)** 0.90   **(d)** 0.009   **(e)** none of these

77. 2.3 renamed as an equal decimal with 3 decimal places is

    (a) 2.300   **(b)** 2.30   **(c)** 2.3   **(d)** 0.023   **(e)** none of these

78. The sum of 8.25 + 4.63 + 0.25 + 0.125 + 128 is

   (a) 1566   (b) 141.255   (c) 141.155   (d) 140.255   (e) none of these

79. The sum of $14.89 + 99¢ is

   (a) $15.88   (b) $14.88   (c) $113.89   (d) $15.87   (e) none of these

80. 8.005 − 0.579 equals

   (a) 7.426   (b) 8.426   (c) 7.996   (d) 8.584   (e) none of these

81. 1.5 − 0.125 equals

   (a) 2.5   (b) 1.425   (c) 1.375   (d) 1.625   (e) none of these

82. 3.75 − 2 equals

   (a) 5.75   (b) 2.25   (c) 5.25   (d) 1.75   (e) none of these

83. $8 − 67¢ equals

   (a) $75   (b) 75¢   (c) $8.67   (d) $7.33   (e) none of these

84. The product of 5 × 0.2 is

   (a) 10   (b) 1   (c) 0.1   (d) 0.01   (e) none of these

85. The product of 3.4 × 1.02 is

   (a) 3.468   (b) 34.68   (c) 346.8   (d) 3468   (e) none of these

86. The product of 0.002 × 0.003 is

   (a) 0.06   (b) 0.006   (c) 0.0006   (d) 0.00006   (e) none of these

87. The product of 1000 × 0.05 is

   (a) 5   (b) 50   (c) 500   (d) 5000   (e) none of these

88. The quotient of 0.2 ÷ 5 is

   (a) 4   (b) 0.4   (c) 0.04   (d) 0.004   (e) none of these

89. The quotient of 25.2 ÷ 0.02 is

   (a) 126   (b) 12.6   (c) 1260   (d) 12,600   (e) none of these

90. The quotient of 0.001 ÷ 4 is

   (a) 0.25   (b) 0.025   (c) 0.0025   (d) 0.00025   (e) none of these

91. The quotient of 25,000 ÷ 1000 is

   (a) 25   (b) 250   (c) 2.5   (d) 25,000,000   (e) none of these

92. The average of 86, 87, 92, 53, 69 is

   (a) 77   (b) 774   (c) 7.74   (d) 77.4   (e) none of these

93. $\frac{5}{8}$ renamed as an equal terminating decimal is

   (a) 6.25   (b) 0.75   (c) 1.6   (d) 0.625   (e) none of these

94. $\frac{2}{3}$ renamed as an equal repeating decimal is

(a) $0.\overline{6}$ (b) $66.\overline{6}$ (c) $0.\overline{3}$ (d) 1.5 (e) none of these

95. $\frac{3}{10}$ renamed as an equal decimal is

(a) 3 (b) 0.3 (c) 0.03 (d) $3.\overline{3}$ (e) none of these

96. $\frac{2}{5} \times \frac{2}{5} \times \frac{2}{5}$ in power notation is

(a) $\frac{2^3}{5}$ (b) $\frac{8}{5}$ (c) $\frac{8}{125}$ (d) all of these   (e) none of these

97. $\frac{2^3}{5}$ evaluated is

(a) $\frac{8}{125}$ (b) $\frac{8}{5}$ (c) $\frac{6}{5}$ (d) all of these   (e) none of these

98. 2^0 is just another way of writing

(a) 0 (b) 2 (c) 1 (d) all of these   (e) none of these

99. The square root of $\frac{49}{100}$ is

(a) $\frac{7}{10}$ (b) 0.7 (c) 70% (d) all of these   (e) none of these

100. 12% renamed as an equal decimal is

(a) 12.0 (b) 1.2 (c) 0.12 (d) 120 (e) none of these

101. $66\frac{2}{3}\%$ renamed as an equal fraction is

(a) $\frac{2}{3}$ (b) $\frac{2}{300}$ (c) $\frac{1}{3}$ (d) $\frac{200}{3}$ (e) none of these

102. 0.08 renamed as an equal percent is

(a) 80% (b) 8% (c) 800% (d) 0.8% (e) none of these

103. $\frac{1}{3}$ renamed as an equal percent is

(a) $\frac{1}{3}\%$ (b) $\frac{100}{3}$ (c) $\frac{3}{100}\%$ (d) 33% (e) none of these

104. 25% of 40 is

(a) 0.625 (b) 100 (c) 1.6 (d) 10 (e) none of these

105. 12 is 75% of

(a) 16 (b) 9 (c) 0.0625 (d) 0.16 (e) none of these

106. The percent that 1 is of 4 is

(a) 400% (b) 250% (c) 4% (d) 25% (e) none of these

107. The percent increase from 3 to 4 is

(a) $133\frac{1}{3}\%$ (b) $33\frac{1}{3}\%$ (c) 25% (d) 75% (e) none of these

108. The percent decrease from 4 to 3 is

(a) $133\frac{1}{3}\%$ (b) $33\frac{1}{3}\%$ (c) 25% (d) 75% (e) none of these

Part 2: Problem Solving (18 questions)

109. There are 24 cans in one box and 8 cans in another box. How many cans are there in all?

110. There are 24 cans in each box. There are 8 boxes in all. How many cans are there in all?

111. There are 24 cans in one box and 8 cans in another box. What is the difference between the number of cans in the two boxes?

112. There are 24 cans in 8 boxes. If each box contains the same number of cans, how many cans are in each box?

113. A certain fence is 100 feet long. On Monday $\frac{3}{4}$ of the fence was painted. How many feet long was the painted section of the fence?

114. 40 feet of a certain fence was painted on Tuesday. This was $\frac{1}{5}$ of its total length. How long is the fence?

115. On Monday $\frac{1}{2}$ of a certain fence was painted. Tuesday another $\frac{1}{4}$ of the fence was painted. How much of the total fence was painted by the end of the day on Tuesday?

116. $\frac{1}{3}$ of a certain fence was painted on Monday. What fractional part of the fence remained to be painted on Tuesday?

117. Handkerchiefs are 3 for $1. How much will the store charge for 1 handkerchief?

118. Workshirts are $12.89 each. How much will 5 workshirts cost if the sales tax is $3.87?

119. Dress pants are $43.89 each and blouses are $25.99 each. How much more will 3 pairs of dress pants cost than 2 blouses, not including sales tax?

120. Jan earns $18.25 per hour. How much does Jan earn in a year if she works 40 hours per week, 52 weeks per year?

121. The purchase price of a radio is $84. The sales tax is $4. What is the total price?

122. The purchase price of a radio is $84. The sales-tax rate is 4%. What is the sales tax?

123. The regular price of a suit is $120. If the suit is marked down 40%, how much is the cash discount?

124. If a person earns $500 salary and $300 commission, how much is the total pay?

125. If a person has sales of $500 based on a commission rate of 30%, how much is the commission?

126. How much interest is paid on $100 at 5% per year simple interest for 120 days?

Midterm Exam Answers

Part 1

1. d	**2.** a	**3.** b	**4.** c	**5.** b	**6.** e
7. c	**8.** d	**9.** a	**10.** b	**11.** e	**12.** c
13. d	**14.** d	**15.** a	**16.** e	**17.** d	**18.** a
19. e	**20.** b	**21.** c	**22.** c	**23.** d	**24.** c
25. a	**26.** a	**27.** a	**28.** b	**29.** c	**30.** d
31. a	**32.** b	**33.** a	**34.** a	**35.** b	**36.** e
37. e	**38.** d	**39.** e	**40.** e	**41.** d	**42.** e
43. a	**44.** e	**45.** d	**46.** b	**47.** c	**48.** a
49. e	**50.** d	**51.** e	**52.** d	**53.** c	**54.** c
55. b	**56.** e	**57.** a	**58.** a	**59.** d	**60.** e
61. c	**62.** b	**63.** a	**64.** e	**65.** a	**66.** c
67. b	**68.** e	**69.** d	**70.** d	**71.** b	**72.** e
73. c	**74.** e	**75.** c	**76.** b	**77.** a	**78.** b

79. a	**80.** a	**81.** c	**82.** d	**83.** d	**84.** b
85. a	**86.** e	**87.** b	**88.** c	**89.** c	**90.** d
91. a	**92.** d	**93.** d	**94.** a	**95.** b	**96.** e
97. b	**98.** c	**99.** d	**100.** c	**101.** a	**102.** b
103. e	**104.** d	**105.** a	**106.** d	**107.** b	**108.** c

Part 2

109. 32 cans	**110.** 192 cans	**111.** 16 cans	**112.** 3 cans
113. 75 feet	**114.** 200 feet	**115.** $\frac{3}{4}$ of the fence	**116.** $\frac{2}{3}$ of the fence
117. 34¢	**118.** $68.32	**119.** $79.69	**120.** $37,960 per year
121. $88	**122.** $3.36	**123.** $48	**124.** $800
125. $150	**126.** $1.67		

6 REAL NUMBERS: PROPERTIES AND EXPRESSIONS

THIS CHAPTER IS ABOUT

- ☑ Identifying Real Numbers
- ☑ Computing with Rational Numbers
- ☑ Identifying Properties
- ☑ Using the Distributive Properties
- ☑ Clearing Parentheses
- ☑ Combining Like Terms
- ☑ Evaluating Expressions
- ☑ Evaluating Formulas

6-1. Identifying Real Numbers

Recall: The collection of all positive integers (1, 2, 3, . . .), negative integers
(. . . , −3, −2, −1), and zero (0) is called the set of **integers** [see Example 2-5].

EXAMPLE 6-1 List the set of integers.

Solution   . . . , −3, −2, −1, 0, 1, 2, 3, . . .

If a and b are integers, where $b \neq 0$, then the collection of all numbers that can be written in the fractional form $\frac{a}{b}$ is called the set of **rational numbers.**

EXAMPLE 6-2 Identify which of the following numbers are rational numbers:

(a) $\dfrac{1}{2}$ (b) $\dfrac{-3}{4}$ (c) $\dfrac{2}{-5}$ (d) $1\frac{1}{2}$ (e) 5   (f) −2

(g) 0.1   (h) −1.3   (i) $0.\overline{3}$ (j) $\sqrt{2}$ (k) π

Solution

(a) $\frac{1}{2}$ is a rational number because 1 and 2 are integers.

(b) $\dfrac{-3}{4}$ is a rational number because −3 and 4 are integers.

(c) $\dfrac{2}{-5}$ is a rational number because 2 and −5 are integers.

(d) $1\frac{1}{2}$ is a rational number because $1\frac{1}{2}$ can be written as $\frac{3}{2}$, and 3 and 2 are integers.

(e) 5 is a rational number because 5 can be written as $\frac{5}{1}$, and 5 and 1 are integers.

(f) −2 is a rational number because −2 can be written as $\dfrac{-2}{1}$, and −2 and 1 are integers.

(g) 0.1 is a rational number because 0.1 can be written as $\frac{1}{10}$, and 1 and 10 are integers.

(h) −1.3 is a rational number because −1.3 can be written as $\dfrac{-13}{10}$, and −13 and 10 are integers.

(i) $0.\overline{3}$ is a rational number because $0.\overline{3}$ can be written as $\frac{1}{3}$, and 1 and 3 are integers.

(j) $\sqrt{2}$ is not a rational number because $\sqrt{2}$ cannot be written in the form $\frac{a}{b}$ where a and b are integers: $\sqrt{2} = 1.41421356237\ldots$ and is not a terminating or repeating decimal.

(k) π is not a rational number because π cannot be written in the form $\frac{a}{b}$ where a and b are integers: $\pi = 3.1415926536\ldots$ and is not a terminating or repeating decimal.

All integers and whole numbers are rational numbers.

Some decimals are rational numbers. If the decimal form of a number terminates or repeats, the number is a rational number. But if the decimal form of a number does not terminate or repeat, the number is *not* a rational number.

A. Identifying irrational numbers

Numbers whose decimal forms do not terminate and do not repeat are called **irrational numbers.**

EXAMPLE 6-3 List some irrational numbers.

Solution $\sqrt{2}, -\sqrt{3}, \sqrt{5}, -\sqrt{6}, \pi, -\pi$ ⟵── irrational numbers

The positive or negative square root of any whole number that is not a square is an irrational number ($\ldots, -\sqrt{5}, -\sqrt{3}, -\sqrt{2}, \sqrt{2}, \sqrt{3}, \sqrt{5}, \ldots$).

B. Identifying real numbers

The collection of all rational and irrational numbers is called the set of **real numbers.**

Note 1: The real numbers include whole numbers, integers, fractions, and decimals.

Note 2: The only numbers that will be used in this text are real numbers.

Because every real number can be represented by a point on a number line, a number line is sometimes called a **real number line.** Every point on a number line represents a real number; that is, a number line is a pictorial representation of the collection of all real numbers.

EXAMPLE 6-4 Graph some positive and negative real numbers on the number line.

Solution

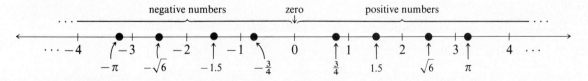

EXAMPLE 6-5 Draw a diagram that shows the relationships between real numbers, irrational numbers, rational numbers, integers, and whole numbers.

Solution

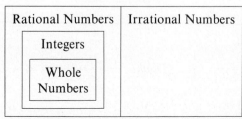

6-2. Computing with Rational Numbers

A. Renaming rational numbers

Recall: If a and b are integers ($b \neq 0$), then any number that can be written in the form $\frac{a}{b}$ is called a rational number.

Every rational number in fraction form has three signs: the sign of the fraction, the sign of the numerator, and the sign of the denominator.

EXAMPLE 6-6   Identify the three signs for the rational numbers **(a)** $\frac{3}{4}$ and **(b)** $-\frac{-3}{4}$.

Solution

$$\text{(a)} \quad \frac{3}{4} = +\frac{+3}{+4} \qquad\qquad \text{(b)} \quad -\frac{-3}{4} = -\frac{-3}{+4}$$

with labels: sign of the fraction, sign of the numerator, sign of the denominator.

If exactly two of a fraction's three signs are changed, the result is always the same—you get an equal fraction. That is, the following are all equivalent ways to write a fraction with form $\frac{a}{b}$:

$$\frac{a}{b} = -\frac{-a}{b} = -\frac{a}{-b} = \frac{-a}{-b}$$

EXAMPLE 6-7   Rename $\frac{3}{4}$ in three equivalent ways using sign changes.

Solution

$$\frac{3}{4} = -\frac{-3}{4} \qquad \textit{Think:} \quad +\frac{+3}{+4} = -\frac{-3}{+4} \quad \longleftarrow \text{ two sign changes}$$

$$\frac{3}{4} = -\frac{3}{-4} \qquad \textit{Think:} \quad +\frac{+3}{+4} = -\frac{+3}{-4} \quad \longleftarrow \text{ two sign changes}$$

$$\frac{3}{4} = \frac{-3}{-4} \qquad \textit{Think:} \quad +\frac{+3}{+4} = +\frac{-3}{-4} \quad \longleftarrow \text{ two sign changes}$$

Note: The fractions $\frac{3}{4}$, $-\frac{-3}{4}$, $-\frac{3}{-4}$, and $\frac{-3}{-4}$ are equal fractions.

The following are all equivalent ways to write a fraction with form $-\frac{a}{b}$:

$$-\frac{a}{b} = \frac{-a}{b} = \frac{a}{-b} = -\frac{-a}{-b}$$

EXAMPLE 6-8   Rename $-\frac{3}{4}$ in three equivalent ways using sign changes.

Solution

$$-\frac{3}{4} = \frac{-3}{4} \qquad \textit{Think:} \quad -\frac{+3}{+4} = +\frac{-3}{+4} \quad \longleftarrow \text{ two sign changes}$$

$$-\frac{3}{4} = \frac{3}{-4} \qquad \textit{Think:} \quad -\frac{+3}{+4} = +\frac{+3}{-4} \quad \longleftarrow \text{ two sign changes}$$

$$-\frac{3}{4} = -\frac{-3}{-4} \qquad \textit{Think:} \quad -\frac{+3}{+4} = -\frac{-3}{-4} \quad \longleftarrow \text{ two sign changes}$$

Note: The fractions $-\frac{3}{4}$, $\frac{-3}{4}$, $\frac{3}{-4}$, and $-\frac{-3}{-4}$ are equal fractions.

B. Simplifying rational numbers

A rational number in fraction form for which the numerator and denominator do not share a common integer factor other than 1 or -1 is called a **rational number in lowest terms.**

EXAMPLE 6-9 Which of the following rational numbers are in lowest terms?

$$\text{(a) } \frac{3}{4} \quad \text{(b) } \frac{-5}{2} \quad \text{(c) } \frac{-6}{-10}$$

Solution

(a) $\dfrac{3}{4}$ is a rational number in lowest terms because the only common integer factors of 3 and 4 are 1

and -1: $\dfrac{3}{4} = \dfrac{1(3)}{1(4)}$ and $\dfrac{3}{4} = \dfrac{-3}{-4} = \dfrac{-1(3)}{-1(4)}$.

(b) $\dfrac{-5}{2}$ is a rational number in lowest terms because the only common integer factors of -5 and 2

are 1 and -1: $\dfrac{-5}{2} = \dfrac{1(-5)}{1(2)}$ and $\dfrac{-5}{2} = \dfrac{5}{-2} = \dfrac{-1(-5)}{-1(2)}$.

(c) $\dfrac{-6}{-10}$ is not a rational number in lowest terms because 6 and 10 share a common integer factor

of 2 or -2: $\dfrac{-6}{-10} = \dfrac{2(-3)}{2(-5)}$ or $\dfrac{-6}{-10} = \dfrac{-2(3)}{-2(5)}$.

To simplify a rational number in fraction form, you write the rational number using as few negative signs as possible and then reduce the rational number in fraction form to lowest terms using the following rule:

Fundamental Rule for Rational Numbers

$$\text{If } a, b, \text{ and } c \text{ are integers } (b \neq 0 \text{ and } c \neq 0), \text{ then } \frac{a \cdot c}{b \cdot c} = \frac{a}{b}.$$

Note: By the Fundamental Rule for Rational Numbers, the value of a rational number in fraction form will not change when you divide both the numerator and denominator by the same nonzero number: $\dfrac{a \cdot \ell}{b \cdot \ell} = \dfrac{a}{b}$

EXAMPLE 6-10 Simplify (a) $\dfrac{6}{10}$, (b) $-\dfrac{6}{3}$, (c) $\dfrac{-12}{8}$, and (d) $-\dfrac{-10}{-25}$.

Solution

(a) $\dfrac{6}{10} = \dfrac{2 \cdot 3}{2 \cdot 5}$ Factor both the numerator and denominator.

$= \dfrac{\not{2} \cdot 3}{\not{2} \cdot 5}$ Use the Fundamental Rule for Rational Numbers.

$= \dfrac{3}{5}$ ⟵ simplest form

(b) $-\dfrac{6}{3} = -\dfrac{2 \cdot \not{3}}{1 \cdot \not{3}}$ Use the Fundamental Rule for Rational Numbers.

$= -\dfrac{2}{1}$

$= -2$ ⟵ simplest form

(c) $\dfrac{-12}{8} = -\dfrac{12}{8}$

$= -\dfrac{\cancel{2} \cdot \cancel{2} \cdot 3}{\cancel{2} \cdot \cancel{2} \cdot 2}$ or $-\dfrac{\cancel{4} \cdot 3}{\cancel{4} \cdot 2}$ Use the Fundamental Rule for Rational Numbers.

$= -\dfrac{3}{2}$ ⟵ simplest form

or $-1\dfrac{1}{2}$

(d) $-\dfrac{-10}{-25} = -\dfrac{10}{25}$ Write as few negative signs as possible: $-\dfrac{-10}{-25} = -\dfrac{+10}{+25}$

$= -\dfrac{2 \cdot \cancel{5}}{5 \cdot \cancel{5}}$ Use the Fundamental Rule for Rational Numbers.

$= -\dfrac{2}{5}$ ⟵ simplest form

C. Adding rational numbers

To **add rational numbers,** use the Addition Rules for Numbers with Like or Unlike Signs given in Section 2-4A.

EXAMPLE 6-11   Add (a) $\frac{1}{4} + 0.5$, (b) $-2.5 + 1.75$, (c) $\frac{3}{4} + (-\frac{1}{2})$, and (d) $-3.2 + (-1\frac{1}{3})$.

Solution

(a) <u>Rename to get fractions</u>   or   <u>Rename to get decimals</u>

$\dfrac{1}{4} + 0.5 = \dfrac{1}{4} + \dfrac{5}{10}$ $\dfrac{1}{4} + 0.5 = 0.25 + 0.5$

$= \dfrac{1}{4} + \dfrac{1}{2}$ $= 0.75$

$= \dfrac{1}{4} + \dfrac{2}{4}$

$= \dfrac{3}{4}$

(b) $-2.5 + 1.75 = ?\,0.75$ *Think:* $\overset{\overset{\text{larger absolute value}}{\downarrow}}{2.5} - 1.75 = 0.75$ ⟵ difference between absolute values

$= -0.75$ *Think:* Because -2.5 has the larger absolute value, the sum is negative.

(c) $\dfrac{3}{4} + \left(-\dfrac{1}{2}\right) = ?\,\dfrac{1}{4}$ *Think:* $\overset{\overset{\text{larger absolute value}}{\downarrow}}{\dfrac{3}{4}} - \dfrac{1}{2} = \dfrac{1}{4}$ ⟵ difference between the absolute values

$= +\dfrac{1}{4}$ or $\dfrac{1}{4}$ *Think:* Because $\frac{3}{4}$ has the larger absolute value, the sum is positive.

(d) $-3.2 + (-1\frac{1}{3}) = -3\frac{2}{10} + (-1\frac{1}{3})$ Rename -3.2 as a mixed number $(-3\frac{2}{10})$.

$\qquad\qquad\qquad = -3\frac{1}{5} + (-1\frac{1}{3})$ Simplify $-3\frac{2}{10}$ to $-3\frac{1}{5}$.

$\qquad\qquad\qquad = -\dfrac{16}{5} + \left(-\dfrac{4}{3}\right)$ Rename as fractions.

$\qquad\qquad\qquad = -\dfrac{48}{15} + \left(-\dfrac{20}{15}\right)$ Use the LCD 15 to get like fractions.

$\qquad\qquad\qquad = ?\,\dfrac{68}{15}$ *Think:* $\dfrac{48}{15} + \dfrac{20}{15} = \dfrac{68}{15}$

$\qquad\qquad\qquad = -\dfrac{68}{15}$ *Think:* Because both addends are negative, the sum is negative.

D. Subtracting rational numbers

To **subtract rational numbers,** use the Subtraction Rules for Numbers given in Section 2-4B.

EXAMPLE 6-12   Subtract $-4.5 - 2\frac{3}{4}$.

Solution   You can subtract by first renaming as decimals, or first renaming as fractions.

Rename to get decimals

$-4.5 - 2\frac{3}{4} = -4.5 - 2.75$ Rename $2\frac{3}{4}$ as 2.75.

$\qquad\qquad\quad = -4.5 + (-2.75)$ Add the opposite of the subtrahend.

$\qquad\qquad\quad = -7.25$

Rename to get fractions

$-4.5 - 2\frac{3}{4} = -4\frac{5}{10} - 2\frac{3}{4}$ Rename -4.5 as $-4\frac{5}{10}$.

$\qquad\qquad\quad = -4\frac{5}{10} + (-2\frac{3}{4})$ Add the opposite of the subtrahend.

$\qquad\qquad\quad = -4\frac{1}{2} + (-2\frac{3}{4})$ Rename $-4\frac{5}{10}$ as $-4\frac{1}{2}$.

$\qquad\qquad\quad = -4\frac{2}{4} + (-2\frac{3}{4})$ Use the LCD 4 to get like fractions.

$\qquad\qquad\quad = -6\frac{5}{4}$ Add rational numbers.

$\qquad\qquad\quad = -7\frac{1}{4}$ Rename $\frac{5}{4}$ as $1\frac{1}{4}$.

E. Multiplying rational numbers

To **multiply rational numbers,** use the Multiplication Rules for Numbers given in Section 2-4C.

EXAMPLE 6-13   Multiply the following rational numbers:

$$\textbf{(a)}\ \frac{1}{2}(0.25) \qquad \textbf{(b)}\ -\frac{1}{4}\cdot\frac{2}{3} \qquad \textbf{(c)}\ 1.5(-2.3) \qquad \textbf{(d)}\ -0.5\left(-\frac{2}{3}\right)$$

Solution

(a) Rename to get fractions   or   Rename to get decimals

$\dfrac{1}{2}(0.25) = \dfrac{1}{2}\cdot\dfrac{1}{4}$ $\qquad$ $\dfrac{1}{2}(0.25) = 0.5(0.25)$

$\qquad\quad = \dfrac{1}{8}$ $\qquad\qquad\qquad = 0.125$

(b) $-\dfrac{1}{4} \cdot \dfrac{2}{3} = -\dfrac{1}{2 \cdot \cancel{2}} \cdot \dfrac{\cancel{2}}{3}$ Eliminate common factors.

$\qquad = ?\dfrac{1}{6}$ *Think:* $\dfrac{1}{2} \cdot \dfrac{1}{3} = \dfrac{1}{6}$ ⟵ product of the absolute values

$\qquad = -\dfrac{1}{6}$ *Think:* Because the factors have unlike signs, the product is negative.

(c) $1.5(-2.3) = ?\,3.45$ *Think:* $1.5(2.3) = 3.45$ ⟵ product of the absolute values

$\qquad = -3.45$ *Think:* Because the fractions have unlike signs, the product is negative.

(d) $-0.5\left(-\dfrac{2}{3}\right) = -\dfrac{1}{2}\left(-\dfrac{2}{3}\right)$ Rename 0.5 as $\frac{1}{2}$

$\qquad = -\dfrac{1}{\cancel{2}}\left(-\dfrac{\cancel{2}}{3}\right)$ Multiply fractions.

$\qquad = ?\dfrac{1}{3}$ *Think:* $\dfrac{1}{\cancel{2}} \cdot \dfrac{\cancel{2}}{3} = \dfrac{1}{3}$ ⟵ product of the absolute values

$\qquad = +\dfrac{1}{3}$ or $\dfrac{1}{3}$ *Think:* Because both factors are negative, the product is positive.

F. Dividing rational numbers

To **divide rational numbers,** use the Division Rules for Numbers in Section 2-4D.

EXAMPLE 6-14   Divide $\frac{3}{4} \div (-0.25)$.

Solution   You can divide by first renaming as decimals, or first renaming as fractions.

Rename to get decimals

$\dfrac{3}{4} \div (-0.25) = 0.75 \div (-0.25)$ Rename $\frac{3}{4}$ as 0.75.

$\qquad = ?\,3$ Divide decimals.

$\qquad = -3$ *Think:* Because the decimals have unlike signs, the quotient is negative.

Rename to get fractions

$\dfrac{3}{4} \div (-0.25) = \dfrac{3}{4} \div \left(-\dfrac{1}{4}\right)$ Rename (-0.25) as $\left(-\dfrac{1}{4}\right)$.

$\qquad = \dfrac{3}{4}\left(-\dfrac{4}{1}\right)$ Multiply by the reciprocal of the divisor.

$\qquad = -\dfrac{3}{1}$ Multiply fractions and simplify.

$\qquad = -3$

Check: $-3(-0.25) = +0.75$ or 0.75 ⟵ -3 checks

6-3. Identifying Properties

A. Writing indicated products and quotients that involve variables

It is often necessary in algebra to write **indicated products** of real numbers and/or **variables** (letters that stand for changeable numbers).

Recall: There are several correct ways to write the indicated product of two real numbers. For example, the real number expression "2 times 5" can be written as

$$2 \times 5 \qquad 2 \cdot 5 \qquad 2(5) \qquad (2)5 \qquad (2)(5)$$

There are five correct ways to write the indicated product of variables and real numbers, or the indicated product of variables and variables.

EXAMPLE 6-15 Write five correct ways to indicate the products of the expressions **(a)** 2 times x, **(b)** -3 times y, and **(c)** m times n.

Solution

(a) 2 times $x \quad = 2 \cdot x \quad = 2(x) \quad = (2)x \quad = (2)(x) \quad = 2x$

(b) -3 times $y = -3 \cdot y = -3(y) = (-3)y = (-3)(y) = -3y \quad \longrightarrow$ preferred form

(c) m times $n \quad = m \cdot n \quad = m(n) \quad = (m)n \quad = (m)(n) \quad = mn$

Note: When a variable is used as a factor, the traditional "times" symbol $\times$ is never used because it looks too much like the letter x.

It is often necessary in algebra to write **indicated quotients** of real numbers and/or variables.

Recall: There are several correct ways to write the indicated quotient of two real numbers. For example, the real number expression "8 divided by -3" can be written as

$$8 \div (-3) \qquad 8/(-3) \qquad 8 \cdot \frac{1}{-3} \qquad \frac{1}{-3} \cdot 8 \qquad 8\left(\frac{1}{-3}\right) \qquad \left(\frac{1}{-3}\right)8 \qquad \frac{8}{-3}$$

There are five correct ways to write the indicated quotient of variables and real numbers, or the indicated quotient of variables and variables.

EXAMPLE 6-16 Write five correct ways to indicate the quotients of the expressions **(a)** x divided by -3, **(b)** 8 divided by y, **(c)** h divided by k.

Solution

(a) x divided by $-3 = x \div (-3) = x/(-3) = x \cdot \dfrac{1}{-3} = \dfrac{1}{-3} \cdot x = \dfrac{x}{-3}$

(b) 8 divided by $y \quad = 8 \div y \quad = 8/y \quad = 8 \cdot \dfrac{1}{y} \quad = \dfrac{1}{y} \cdot 8 \quad = \dfrac{8}{y} \quad \longleftrightarrow$ preferred form

(c) h divided by $k \quad = h \div k \quad = h/k \quad = h \cdot \dfrac{1}{k} \quad = \dfrac{1}{k} \cdot h \quad = \dfrac{h}{k}$

Note: When variables are used in algebra, the traditional division symbol $\div$ and the slash division symbol $/$, although acceptable, are seldom used to write an indicated quotient. The preferred form of division notation is $\frac{a}{b}$.

EXAMPLE 6-17 Write the following expressions using the preferred form of division notation:

(a) x divided by $(x + 3)$ **(b)** $(y - 2)$ divided by 6   **(c)** $(m + n)$ divided by $(m - n)$

Solution

(a) x divided by $(x + 3) \qquad = \dfrac{x}{x + 3}$

(b) $(y - 2)$ divided by 6 $\qquad = \dfrac{y - 2}{6} \qquad \longleftrightarrow$ preferred division form

(c) $(m + n)$ divided by $(m - n) = \dfrac{m + n}{m - n}$

If you choose to use a form other than the preferred form of division notation, you must use parentheses around the divisor to avoid any errors in computation, as in the expressions $x \div (x + 3)$ or $x/(x + 3)$.

Caution: Never write an expression like x divided by $(x + 3)$ as $x \div x + 3$ or $x/x + 3$. Omitting the parentheses will cause you to get a wrong answer when you solve the expression.

B. The commutative properties

The set of real numbers follows certain rules called **properties.** Among these are the *commutative properties,* which have to do with *order.*

- When you change the *order* of real number addends in an addition problem, the sum does not change.

EXAMPLE 6-18   Show that the sum does not change when the order of the addends is changed in $5 + 2$.

Solution

different order $\rightarrow$ $5 + 2 = 7$ $\leftarrow$ same sum
$\rightarrow$ $2 + 5 = 7$ $\leftarrow$

Stated algebraically, the property illustrated by Example 6-18 is:

Commutative Property of Addition

If a and b are real numbers, then $a + b = b + a$.

- When you change the the order of real number factors in a multiplication problem, the product does not change.

EXAMPLE 6-19   Show that the product does not change when the order of the factors is changed in $2 \cdot 3$.

Solution

different order $\rightarrow$ $2 \cdot 3 = 6$ $\leftarrow$ same product
$\rightarrow$ $3 \cdot 2 = 6$ $\leftarrow$

Stated algebraically, the property illustrated by Example 6-19 is:

Commutative Property of Multiplication

If a and b are real numbers, then $ab = ba$.

The Commutative Property of Addition and the Commutative Property of Multiplication are jointly called the **commutative properties.**

Caution: Subtraction and division of real numbers are *not* commutative.

EXAMPLE 6-20   Show that the difference changes when the order of the minuend and subtrahend is changed in $5 - 2$.

Solution

different order $\rightarrow$ $5 - 2 = 3$ $\leftarrow$ different differences
$\rightarrow$ $2 - 5 = -3$ $\leftarrow$

EXAMPLE 6-21   Show that the quotient changes when the order of the dividend and divisor is changed in $2 \div 1$.

Solution

different order $\rightarrow$ $2 \div 1 = 2$ $\leftarrow$ different quotients
$\rightarrow$ $1 \div 2 = \frac{1}{2}$ $\leftarrow$

C. The associative properties

The associative properties have to do with *grouping*.

- When you change the grouping of real number addends in an addition problem, the sum does not change.

EXAMPLE 6-22   Show that the sum does not change when the grouping of the addends is changed in $(2 + 3) + 4$.

Solution

different groupings $\begin{cases} \rightarrow (2 + 3) + 4 = 5 + 4 = 9 \\ \rightarrow 2 + (3 + 4) = 2 + 7 = 9 \end{cases}$ same sum

Note: Use the Order of Operations first to clear grouping symbols.

Stated algebraically, the property illustrated by Example 6-22 is:

Associative Property of Addition
If a, b, and c are real numbers, then $(a + b) + c = a + (b + c)$.

- When you change the grouping of real number factors in a multiplication problem, the product does not change.

EXAMPLE 6-23   Show that the product does not change when the grouping of the factors is changed in $(2 \cdot 3)4$.

Solution

different groupings $\begin{cases} \rightarrow (2 \cdot 3)4 = 6 \cdot 4 = 24 \\ \rightarrow 2(3 \cdot 4) = 2 \cdot 12 = 24 \end{cases}$ same product

Note: Use the Order of Operations first to clear grouping symbols.

Stated algebraically, the property illustrated by Example 6-23 is:

Associative Property of Multiplication
If a, b, and c are real numbers, then $(ab)c = a(bc)$.

The Associative Property of Addition and the Associative Property of Multiplication are jointly called the **associative properties.**

Caution: Subtraction and division of real numbers are *not* associative.

EXAMPLE 6-24   Show that the difference changes when the grouping is changed in $(9 - 5) - 3$.

Solution

different groupings $\begin{cases} \rightarrow (9 - 5) - 3 = 4 - 3 = 1 \\ \rightarrow 9 - (5 - 3) = 9 - 2 = 7 \end{cases}$ different answers

Note: Use the Order of Operations first to clear grouping symbols.

EXAMPLE 6-25   Show that the quotient changes when the grouping is changed in $(24 \div 6) \div 2$.

Solution

different groupings $\begin{cases} \rightarrow (24 \div 6) \div 2 = 4 \div 2 = 2 \\ \rightarrow 24 \div (6 \div 2) = 24 \div 3 = 8 \end{cases}$ different answers

D. The distributive properties

- To multiply a sum by a real number, you can first add and then multiply, or you can first multiply and then add.

EXAMPLE 6-26 Compute $2(4 + 3)$ two ways: **(a)** first adding, then multiplying; **(b)** first multiplying, then adding.

Solution

(a) $\qquad\qquad\qquad 2(4 + 3) = 2 \cdot 7 = 14$ ⟵ ⎤ same

(b) $\qquad\qquad\qquad 2(4 + 3) = 2 \cdot 4 + 2 \cdot 3 = 8 + 6 = 14$ ⟵ ⎦ answer

Example 6-26 illustrates the following property:

Distributive Property of Multiplication Over Addition

If a, b, and c are real numbers, then

$$a(b + c) = ab + ac \quad \text{and} \quad (b + c)a = ba + ca$$

- To multiply a difference by a real number, you can first subtract and then multiply, or you can first multiply and then subtract.

EXAMPLE 6-27 Compute $4(5 - 2)$ two ways: **(a)** first subtracting, then multiplying; **(b)** first multiplying, then subtracting.

Solution **(a)** $\qquad\qquad 4(5 - 2) = 4 \cdot 3 = 12$ ⟵ ⎤ same

$\qquad\qquad$ **(b)** $\qquad\qquad 4(5 - 2) = 4 \cdot 5 - 4 \cdot 2 = 20 - 8 = 12$ ⟵ ⎦ answer

Example 6-27 illustrates the following property:

Distributive Property of Multiplication Over Subtraction

If a, b, and c are real numbers, then

$$a(b - c) = ab - ac \quad \text{and} \quad (b - c)a = ba - ca$$

The Distributive Property of Multiplication Over Addition and the Distributive Property of Multiplication Over Subtraction are jointly called the **distributive properties.**

E. Properties involving zero or one

There are several important properties of real numbers that involve zero.

EXAMPLE 6-28 List the important properties of real numbers that involve zero.

Solution If a and b are real numbers, then

$a + 0 = 0 + a = a$ $\qquad\qquad$ *Identity Property for Addition*
$\qquad\qquad\qquad\qquad\qquad\qquad$ Because $a + 0 = a$ for every real number a, 0 is called the
$\qquad\qquad\qquad\qquad\qquad\qquad$ **identity element for addition.**

$a + (-a) = -a + a = 0$ $\qquad$ *Additive Inverse Property*
$\qquad\qquad\qquad\qquad\qquad\qquad$ Because $a + (-a) = 0$ for every real number a, the symbol
$\qquad\qquad\qquad\qquad\qquad\qquad$ $-a$ is called the **opposite** of a or the **additive inverse** of a.

$a - 0 = a + 0 = a$
$0 - a = 0 + (-a) = -a$
$a - b = 0 \quad \text{means} \quad a = b$

$a \cdot 0 = 0 \cdot a = 0$ $\qquad\qquad$ *Zero-Factor Property*
$\qquad\qquad\qquad\qquad\qquad\qquad$ A zero factor always gives a zero *product.*

$ab = 0 \quad \text{means} \quad a = 0 \text{ or } b = 0$ $\qquad$ *Zero-Product Property*
$\qquad\qquad\qquad\qquad\qquad\qquad$ The only way for the product of two real numbers to be zero
$\qquad\qquad\qquad\qquad\qquad\qquad$ is for one or both of the real numbers to be zero.

$$0 \div a = \frac{0}{a} = 0 \ (a \neq 0)$$

Zero-Dividend Property
Zero divided by any nonzero real number equals zero.

$a \div 0$ and $\frac{a}{0}$ are undefined.

Zero-Divisor Property
Dividing by zero is not defined.

There are several important properties of real numbers that involve the number one.

EXAMPLE 6-29 List the important properties of real numbers that involve one.

Solution If a and b are real numbers, then:

$$a \cdot 1 = 1 \cdot a = a$$

Identity Property for Multiplication
Because $a \cdot 1 = a$ for every real number a, 1 is called the **identity element for multiplication.**

$$a \cdot \frac{1}{a} = \frac{1}{a} \cdot a = 1 \ (a \neq 0)$$

Multiplicative Inverse Property
Because $a \cdot \frac{1}{a} = 1$ for every nonzero real number a, $\frac{1}{a} \ (a \neq 0)$ is called the reciprocal of a or the **multiplicative inverse** of a.

$$a \div 1 = \frac{a}{1} = a$$

$$a \div b = \frac{a}{b} = 1 \quad \text{means} \quad a = b$$
$(a \neq 0 \text{ and } b \neq 0)$

Unit-Fraction Property
The only way for the quotient of two real numbers to equal one is for the two real numbers to be the same.

F. Properties involving negative signs

There are several important properties of real numbers that involve negative signs.

EXAMPLE 6-30 List the important properties of real numbers that involve negative signs.

Solution If a and b are real numbers, then:

$$-a = -1 \cdot a = a(-1)$$
Read $-a$ as "the opposite of a."

$$-(-a) = a$$
Read $-(-a)$ as "the opposite of the opposite of a."

$$ab = -a(-b)$$

$$-(ab) = -a(b) = a(-b) = -(-a)(-b)$$

G. Subtraction and division properties

There are two important properties of real numbers that involve subtraction and division.

The subtraction property of real numbers tells us that to subtract two real numbers, we change subtraction to addition while writing the opposite of the subtrahend.

EXAMPLE 6-31 State the subtraction property algebraically.

Solution

Subtraction Property If a and b are real numbers, then $a - b = a + (-b)$.

The division property of real numbers tells us that to divide fractions, we change division to multiplication while writing the reciprocal of the divisor.

EXAMPLE 6-32 State the division property algebraically.

Solution

> ***Division Property***   If a and b are real numbers, then $\dfrac{a}{b} = a \cdot \dfrac{1}{b}$ or $\dfrac{1}{b} \cdot a$.

6-4. Using the Distributive Properties

A. Computing using the distributive properties

To compute with real numbers using the distributive properties, you must multiply each number inside the parentheses separately by the number outside the parentheses.

EXAMPLE 6-33 Evaluate (**a**) $(3 + 1)5$ and (**b**) $3(7 - 2)$ using the distributive properties.

Solution

(**a**) Correct Method

$(3 + 1)5 = 3 \cdot 5 + 1 \cdot 5$

$\qquad = 15 + 5$

$\qquad = 20$ ⟵ correct answer

Wrong Method

$(3 + 1)5 = 3 + 1 \cdot 5$ No! (Multiply both 3 and 1 by 5.)

$\qquad = 3 + 5$

$\qquad = 8$ ⟵ wrong answer

(**b**) Correct Method

$3(7 - 2) = 3 \cdot 7 - 3 \cdot 2$

$\qquad = 21 - 6$

$\qquad = 15$ ⟵ correct answer

Wrong Method

$3(7 - 2) = 3 \cdot 7 - 2$ No! (Multiply both 7 and 2 by 3.)

$\qquad = 21 - 2$

$\qquad = 19$ ⟵ wrong answer

B. Computing without using the distributive properties

Sometimes it is easier to compute with real numbers without using the distributive properties. To compute without using the distributive properties, you use the Order of Operations.

EXAMPLE 6-34 Evaluate (**a**) $(3 + 1)5$ and (**b**) $3(7 - 2)$ without using the distributive properties:

Solution

(**a**) $(3 + 1)5 = 4 \cdot 5$ Use the Order of Operations by first adding inside the parentheses.

$\qquad = 20$ ⟵ same answer as found in Example 6-33 using a distributive property

(**b**) $3(7 - 2) = 3 \cdot 5$ Use the Order of Operations by first subtracting inside the parentheses.

$\qquad = 15$ ⟵ same answer as found in Example 6-33 using a distributive property

6-5. Clearing Parentheses

A. Algebraic expressions

A number or variable—as well as the sum, difference, product, quotient, or square root of numbers or variables—is called an **algebraic expression.**

EXAMPLE 6-35 List several different algebraic expressions.

Solution The following are all algebraic expressions:

(**a**) 3   (**b**) x (**c**) $2y$ (**d**) $\dfrac{abc}{5}$ (**e**) $\dfrac{u}{v}$ (**f**) $2w + 4$ (**g**) $6z - \sqrt{z}$ (**h**) $m^2 - 2mn + n^2$

B. Terms

In an algebraic expression that does not contain grouping symbols, the addition symbols separate the **terms.**

EXAMPLE 6-36 How many terms are there in each of the algebraic expressions listed in Example 6-35? Identify each term.

Solution

(a) One term: 3—a **constant term**

(b) One term: x—a **letter term**

(c) One term: $2y$—a general term

    Note: Terms with only products and/or quotients of numbers and variables are all called **general terms.**

(d) One term: $\dfrac{abc}{5}$—a general term

(e) One term: $\dfrac{u}{v}$—a general term

(f) Two terms: $2w$—a general term

                     4—a constant term

(g) Two terms: $6z$—a general term

                 $-\sqrt{z}$—a **negative term**

    Note: $6z - \sqrt{z} = 6z + (-\sqrt{z})$.

(h) Three terms: $m^2\,(=mm)$—a general term

                   $-2mn$—a negative

                           general term

           $n^2\,(=nn)$—a general term

C. Clearing parentheses using the distributive properties

To identify the terms of algebraic expressions with grouping symbols, like $-2(x + 3)$ or $\dfrac{w - 2}{5}$, you must first **clear grouping symbols.** Often this can be done by using the distributive properties.

Recall that to **clear parentheses** by evaluating an indicated product like $(3 + 1)5$ or $3(7 - 2)$, it is almost always easier to compute without using the distributive properties [see Example 6-34].

However, to clear parentheses in an algebraic expression like $-2(x + 3)$ or $(m - n)3$, you must use the distributive properties because it is otherwise impossible to add different terms like x and 3 or to subtract different terms like m and n.

EXAMPLE 6-37 Clear parentheses in and simplify the following expressions:

    (a) $-2(x + 3)$ **(b)** $(m - n)3$ **(c)** $5(2 + y)$ **(d)** $(-3r - 2s)(-4)$

Solution

(a) $-2(x + 3) = -2 \cdot x + (-2)3$

                 $= -2x + (-6)$

                 $= -2x - 6$

(c) $5(2 + y) = 5 \cdot 2 + 5 \cdot y$

              $= 10 + 5y$

          or $5y + 10$

(b) $(m - n)3 = m \cdot 3 - n \cdot 3$

               $= 3 \cdot m - 3 \cdot n$

               $= 3m - 3n$

(d) $(-3r - 2s)(-4) = -3r(-4) - 2s(-4)$

                   $= -3(-4)r - 2(-4)s$

                   $= +12r - (-8s)$

                   $= 12r + (+8s)$

                   $= 12r + 8s$

Note: The terms of **(a)** $-2(x + 3)$ are $-2x$ and -6 because $-2(x + 3) = -2x - 6$.

          (b) $(m - n)3$ are $3m$ and $-3n$ because $(m - n)3 = 3m - 3n$.

          (c) $5(2 + y)$ are $5y$ and 10 because $5(2 + y) = 5y + 10$.

          (d) $(-3r - 2s)(-4)$ are $12r$ and $8s$ because $(-3r - 2s)(-4) = 12r + 8s$.

When parentheses have just a positive sign (or no sign) in front of them, you can clear parentheses simply by writing the same algebraic expression that is inside of the parentheses.

EXAMPLE 6-38   Clear parentheses in $+(3a + 2)$.

same

Solution $+(3a + 2) = 3a + 2$ because $+(3a + 2) = +1(3a + 2)$

$$= (+1)3a + (+1)2$$

$$= 3a + 2$$

Note: The terms of $+(3a + 2)$ are $3a$ and 2 because $+(3a + 2) = 3a + 2$.

When parentheses have just a negative sign in front of them, you can clear parentheses by writing the *opposite* of each term inside the parentheses.

EXAMPLE 6-39   Clear parentheses in $-(7b - 3)$.

opposite

Solution $-(7b - 3) = -7b + 3$ because $-(7b - 3) = -1(7b - 3)$

opposite

$$= (-1)7b - (-1)3$$

$$= -7b - (-3)$$

$$= -7b + (+3)$$

$$= -7b + 3$$

Note: The terms of $-(7b - 3)$ are $-7b$ and 3 because $-(7b - 3) = -7b + 3$.

6-6. Combining Like Terms

A. The numerical coefficient and the literal part of a term

In a term, the number that multiplies the variable(s) is called the **numerical coefficient.**

EXAMPLE 6-40   Identify the numerical coefficient in (a) $2y$, (b) x, (c) $-w^2$, and (d) $\dfrac{abc}{5}$.

Solution

(a) In $2y$, the numerical coefficient is 2.

(b) In x, the numerical coefficient is 1 because $x = 1x$.

(c) In $-w^2$, the numerical is -1 because $-w^2 = -1w^2$.

(d) In $\dfrac{abc}{5}$, the numerical coefficient is $\dfrac{1}{5}$ because $\dfrac{abc}{5} = \dfrac{1}{5}abc$.

The part of a term that is not the numerical coefficient is called the **literal part** of the term.

EXAMPLE 6-41   Identify the literal part in (a) $2y$, (b) x, (c) $-w^2$, and (d) $\dfrac{abc}{5}$.

Solution

(a) In $2y$, the literal part is the variable y.

(b) In x, the literal part is the variable x.

(c) In $-w^2$, the literal part is the variable w^2.

(d) In $\dfrac{abc}{5}$, the literal part is the product of variables abc.

B. Identifying like terms

Terms that have the same literal part are called **like terms.** Terms that have different literal parts are called **unlike terms.**

EXAMPLE 6-42   Identify the like terms in each of the following:

(a) $2x + 5 - x$ (b) $y - 5y + 3x$ (c) $\dfrac{mn}{2} - 3 + 5mn$ (d) $5w + \dfrac{3}{w}$

(e) $-5y + 3x$ (f) $mn + 2m$ (g) $2x^2 + 3x$

Solution

(a) In $2x + 5 - x$, the like terms are $2x$ and $-x$ because they have the same literal part, x.

(b) In $y - 5y + 3x$, the like terms are y and $-5y$ because they have the same literal part, y.

(c) In $\dfrac{mn}{2} - 3 + 5mn$, the like terms are $\dfrac{mn}{2}$ and $5mn$ because they have the same literal part, mn.

   Note: $\dfrac{mn}{2} = \dfrac{1}{2} \cdot mn$.

(d) In $5w + \dfrac{3}{w}$, there are no like terms because the literal parts of $5w$ and $\dfrac{3}{w}$ are different (w and $\dfrac{1}{w}$, respectively).

(e) In $-5y + 3x$, there are no like terms because the literal parts of $-5y$ and $3x$ are different (y and x, respectively).

(f) In $mn + 2m$, there are no like terms because the literal parts of mn and $2m$ are different (mn and m, respectively).

(g) In $2x^2 + 3x$, there are no like terms because the literal parts of $2x^2$ and $3x$ are different (x^2 and x, respectively).

C. Combining like terms

To **combine like terms** in an algebraic expression, you use the distributive properties to add or subtract the numerical coefficients of like terms and then write the same like literal part on the sum or difference.

EXAMPLE 6-43   Combine like terms in the following expressions:

(a) $2w + 3w$ (b) $8ab - 5ab$ (c) $2x + 5 - x$

(d) $y - 5y + 3x$ (e) $\dfrac{mn}{2} - 3 + 5mn$ (f) $8u - 5v - 6u + v$

Solution

(a) $2w + 3w = (2 + 3)w$ Combine like terms using the distributive property of multiplication over addition: $ac + bc = (a + b)c$

   $= 5w$

(b) $8ab - 5ab = (8 - 5)ab$ Combine like terms using the distributive property of multiplication over subtraction: $ac - bc = (a - b)c$

   $= 3ab$

like terms

(c) $2x + 5 - x = \overbrace{2x - x} + 5$ Group like terms.

$\qquad\qquad = 2x - 1x + 5$ *Think:* $-x = -1x$

$\qquad\qquad = (2 - 1)x + 5$ Combine like terms.

$\qquad\qquad = 1x + 5$

$\qquad\qquad = x + 5$ Simplify.

(d) $y - 5y + 3x = 1y - 5y + 3x$ *Think:* $y = 1y$

$\qquad\qquad = (1 - 5)y + 3x$ Combine like terms.

$\qquad\qquad = -4y + 3x$

$\qquad\qquad = 3x - 4y$ Write terms in alphabetical order.

Note: In $3x - 4y$, $3x$ and $-4y$ cannot be combined because $3x$ and $-4y$ are not like terms.

like terms

(e) $\dfrac{mn}{2} - 3 + 5mn = \overbrace{\dfrac{mn}{2}} + 5mn - 3$ Group like terms.

$\qquad\qquad = \dfrac{1}{2}mn + 5mn - 3$ *Think:* $\dfrac{mn}{2} = \dfrac{1}{2}mn$

$\qquad\qquad = \left(\dfrac{1}{2} + 5\right)mn - 3$ Combine like terms.

$\qquad\qquad = 5\dfrac{1}{2}mn - 3$ or $\dfrac{11}{2}mn - 3$ or $\dfrac{11mn}{2} - 3$

(f) $8u - 5v - 6u + v = (8u - 6u) + (-5v + 1v)$ Group like terms together.

$\qquad\qquad = (8 - 6)u + (-5 + 1)v$ Combine like terms.

$\qquad\qquad = 2u + (-4)v$

$\qquad\qquad = 2u - 4v$ Simplify.

Note: In $2u - 4v$, $2u$ and $-4v$ cannot be combined because $2u$ and $-4v$ are not like terms.

Caution: Only like terms can be combined.

D. Clearing parentheses and combining like terms

To **clear parentheses and combine like terms,** you first combine any like terms inside the parentheses, then clear the parentheses and combine any like terms that remain.

EXAMPLE 6-44 Clear parentheses and combine like terms in the following expressions:

(a) $2(x + 5) - 5x$ **(b)** $-(2y - 3) + (3y - 5)$ **(c)** $3(2a - b + 5a) - 8b$ **(d)** $6m - 2(m + n)$

Solution

(a) $2(x + 5) - 5x = \mathbf{2 \cdot} x + \mathbf{2 \cdot} 5 - 5x$ Clear parentheses.

$\qquad\qquad = 2x + 10 - 5x$ Simplify.

$\qquad\qquad = 2x - 5x + 10$ Group like terms.

$\qquad\qquad = -3x + 10$ Combine like terms.

$\qquad\qquad$ or $10 - 3x$

(b) $-(2y - 3) + (3y - 5) = -2y + 3 + 3y - 5$ Clear parentheses.

$\qquad\qquad = -2y + 3y + 3 - 5$ Group like terms.

$$= 1y + (-2) \qquad \text{Combine like terms.}$$

$$= y - 2 \qquad \text{Simplify.}$$

(c) $3(2a - b + 5a) - 8b = 3(2a + 5a - b) - 8b$ \qquad Combine like terms inside the parentheses.

$$= 3(7a - b) - 8b$$

$$= \mathbf{3} \cdot 7a - \mathbf{3} \cdot b - 8b \qquad \text{Clear parentheses.}$$

$$= 21a - 3b - 8b \qquad \text{Simplify.}$$

$$= 21a - 11b \qquad \text{Combine like terms.}$$

(d) $6m - 2(m + n) = 6m + (-2)(m + n)$ \qquad Rename terms.

$$= 6m + (\mathbf{-2})m + (\mathbf{-2})n \qquad \text{Clear parentheses.}$$

$$= 6m - 2m - 2n \qquad \text{Simplify.}$$

$$= 4m - 2n \qquad \text{Combine like terms.}$$

EXAMPLE 6-45 Show how it is possible to make an error when $6m - 2(m + n)$ is not first renamed as $6m + (-2)(m + n)$.

Solution

Wrong Method

$$6m - 2(m + n) = 6m - \mathbf{2} \cdot m + \mathbf{2} \cdot n \qquad \text{No! Multiply each term inside the parentheses by } -2, \text{ not } 2.$$

$$= 6m - 2m + 2n \qquad \text{Simplify.}$$

$$= (6 - 2)m + 2n \qquad \text{Combine like terms.}$$

$$= 4m + 2n \longleftarrow \text{wrong answer}$$

6-7. Evaluating Expressions

A. Evaluating algebraic expressions in one variable

To **evaluate an algebraic expression in one variable** given a numerical value for that variable, you first substitute the given value for the variable in each term that the variable appears in, then evaluate using the Order of Operations.

EXAMPLE 6-46 Evaluate $2x^2 - 5x - 6$ for **(a)** $x = 3$ and **(b)** $x = -2$.

Solution

(a) $2x^2 - 5x - 6 = 2(\mathbf{3})^2 - 5(\mathbf{3}) - 6$ \qquad Substitute 3 for x.

$$= 2 \cdot 9 - 5(3) - 6 \qquad \text{Evaluate using the Order of Operations}$$
$$\qquad\qquad\qquad\qquad\qquad \text{[see Example 2-16].}$$

$$= 18 - 5(3) - 6$$

$$= 18 - 15 - 6$$

$$= 3 - 6$$

$$= -3$$

(b) $2x^2 - 5x - 6 = 2(\mathbf{-2})^2 - 5(\mathbf{-2}) - 6$ \qquad Substitute -2 for x.

$$= 2 \cdot 4 - 5(-2) - 6 \qquad \text{Evaluate using the Order of Operations.}$$

$$= 8 - 5(-2) - 6$$

$$= 8 + 10 - 6$$

$$= 18 - 6$$

$$= 12$$

Caution: To avoid making an error, always use parentheses when substituting a numerical value for a variable, as in Example 6-46.

B. Evaluating algebraic expressions in two or more variables

To **evaluate an algebraic expression in two or more variables** given a numerical value for each different variable, you first substitute each given value for the associated variable and then evaluate using the Order of Operations.

EXAMPLE 6-47   Evaluate $\dfrac{-b + \sqrt{b^2 - 4ac}}{2a}$ for $a = 2$, $b = -1$, and $c = -3$.

Solution $\dfrac{-b + \sqrt{b^2 - 4ac}}{2a} = \dfrac{-(-1) + \sqrt{(-1)^2 - 4(2)(-3)}}{2(2)}$ Substitute 2 for a, -1 for b, and -3 for c.

$$= \frac{1 + \sqrt{1 - 4(2)(-3)}}{2(2)}$$ Evaluate using the Order of Operations.

$$= \frac{1 + \sqrt{1 + 24}}{2(2)}$$

$$= \frac{1 + \sqrt{25}}{2(2)}$$

$$= \frac{1 + 5}{2(2)}$$

$$= \frac{6}{2(2)}$$

$$= \frac{2(3)}{2(2)}$$

$$= \frac{3}{2} \quad \text{or} \quad 1\tfrac{1}{2} \quad \text{or} \quad 1.5$$

6-8. Evaluating Formulas

A. Parts of an equation

An **equation** is a **mathematical sentence** that contains an equality symbol. Every equation has three parts: a **left member**, an **equality symbol,** and a **right member.**

EXAMPLE 6-48   Identify the three different parts of the following equations:

(a) $m + 3 = 2$ (b) $2n = 6$ (c) $4w - 3 = 1$ (d) $2x + 3y = -5$ (e) $C = \tfrac{5}{9}(F - 32)$

Solution

equality symbol

(a) $m + 3 = 2$

(b) $2n = 6$

(c) left member $4w - 3 = 1$ right member

(d) $2x + 3y = -5$

(e) $C = \tfrac{5}{9}(F - 32)$

B. Parts of a formula

A **formula** is an equation that contains two or more variables and represents a known phenomenon. A formula is *solved for a given variable* when that variable is isolated in one member of the equation. When a formula is solved for a given variable, the algebraic expression in the other member of the equation is called the **solution.**

EXAMPLE 6-49   Identify the variable that each formula is solved for and the solution in each of the following:

(a) Temperature Formula: $C = \frac{5}{9}(F - 32)$ **(b) Distance Formula:** $d = rt$

(c) Volume Formula: $V = lwh$

Solution

(a) The formula $C = \frac{5}{9}(F - 32)$ is solved for C and the solution is $\frac{5}{9}(F - 32)$.

(b) The formula $d = rt$ is solved for d and the solution is rt.

(c) The formula $V = lwh$ is solved for V and the solution is lwh.

C. Evaluating formulas to solve problems

To evaluate a formula that is solved for a specific variable given a measurement value for each of the other variables, you substitute each given value for the associated variable and then evaluate using the Order of Operations.

EXAMPLE 6-50   Water boils at 212 degrees Fahrenheit, F. Find the temperature in degrees Celsius, C, at which water boils by evaluating the temperature formula $C = \frac{5}{9}(F - 32)$.

Solution   The question asks you to evaluate $C = \frac{5}{9}(F - 32)$ for $F = 212$.

$$C = \frac{5}{9}(F - 32) \qquad \text{Write the given formula.}$$

$$= \frac{5}{9}(\mathbf{212} - 32) \qquad \text{Substitute 212 for } F.$$

$$= \frac{5}{9} \cdot 180 \qquad \text{Evaluate using the Order of Operations.}$$

$$= 100 \qquad \textit{Think: } \frac{5}{9} \cdot 180 = \frac{5}{9} \cdot \frac{180}{1} = \frac{5}{9} \cdot \frac{9 \cdot 20}{1} = 100$$

Note: $C = 100$ means the Celsius temperature at which water boils is **100°C.**

D. Evaluating a formula when one of the variables represents a rate

To evaluate a formula when one of the variables represents a rate, you may first need to rename to get a common unit of measure that can be eliminated.

EXAMPLE 6-51   What distance d can a car travel in a time t of 20 minutes at a constant rate r of 45 mph (miles per hour) using the distance formula $d = rt$?

$$\overbrace{\qquad\qquad\qquad\qquad}^{\text{different time units}}$$

Solution   The question asks you to evaluate $d = rt$ for $r = 45$ miles/hour and $t = 20$ minutes.

Caution: Before evaluating $d = rt$, you must have a common time unit.

$$20 \text{ minutes} = \frac{20}{60} \text{hour} = \frac{1}{3} \text{hour} \qquad \text{Rename to get a common time unit.}$$

$$d = rt$$ Write the given formula.

$$= 45\frac{\text{miles}}{\text{hour}} \cdot \frac{1}{3}\text{hour}$$ Substitute 45 mph for r and $\frac{1}{3}$ hour for t.

$$= 45\frac{\text{miles}}{\text{hour}} \cdot \frac{1}{3}\frac{\text{hour}}{1}$$ Eliminate the common time unit.

$$= \left(45 \cdot \frac{1}{3}\right)\text{miles}$$ Multiply a measure by a number.

$$= 15 \text{ miles}$$ Simplify.

Note: In 20 minutes, at a constant rate of 45 mph, a car can travel 15 miles.

SOLVED PROBLEMS

PROBLEM 6-1   Identify which of the following numbers are **(a)** rational numbers and which are **(b)** irrational numbers:

$$0 \qquad 2 \qquad -3 \qquad \tfrac{3}{4} \qquad -\tfrac{5}{3} \qquad 2.3 \qquad -0.25 \qquad 0.\overline{6} \qquad \sqrt{9} \qquad -\sqrt{10} \qquad \pi$$

Solution

(a) Rational numbers:

$$0 \qquad 2 \qquad -3 \qquad \tfrac{3}{4} \qquad -\tfrac{5}{3} \qquad 2.3 \qquad -0.25 \qquad 0.\overline{6} \qquad \sqrt{9} \quad [\textit{note: } \sqrt{9} = 3.]$$

(b) Irrational numbers: $-\sqrt{10} = -3.162277659\ldots$ and $\pi = 3.1415926536\ldots$

PROBLEM 6-2   Rename each fraction in three equivalent ways using sign changes:

(a) $\dfrac{1}{2}$ **(b)** $-\dfrac{-2}{3}$ **(c)** $-\dfrac{5}{-8}$ **(d)** $\dfrac{-1}{-4}$ **(e)** $-\dfrac{3}{5}$ **(f)** $\dfrac{-5}{6}$ **(g)** $\dfrac{2}{-9}$ **(h)** $-\dfrac{-1}{-3}$

Solution   Recall that if exactly two of a fraction's three signs are changed, you always get an equal fraction [see Examples 6-7 and 6-8]:

(a) $\dfrac{1}{2} = -\dfrac{-1}{2} = -\dfrac{1}{-2} = \dfrac{-1}{-2}$ **(b)** $-\dfrac{-2}{3} = -\dfrac{2}{-3} = \dfrac{-2}{-3} = \dfrac{2}{3}$ **(c)** $-\dfrac{5}{-8} = -\dfrac{-5}{8} = \dfrac{-5}{-8} = \dfrac{5}{8}$

(d) $\dfrac{-1}{-4} = \dfrac{1}{4} = -\dfrac{-1}{4} = -\dfrac{1}{-4}$ **(e)** $-\dfrac{3}{5} = \dfrac{-3}{5} = \dfrac{3}{-5} = -\dfrac{-3}{-5}$ **(f)** $\dfrac{-5}{6} = -\dfrac{5}{6} = \dfrac{5}{-6} = -\dfrac{-5}{-6}$

(g) $\dfrac{2}{-9} = -\dfrac{2}{9} = \dfrac{-2}{9} = -\dfrac{-2}{-9}$ **(h)** $-\dfrac{-1}{-3} = -\dfrac{1}{3} = \dfrac{-1}{3} = \dfrac{1}{-3}$

PROBLEM 6-3   Simplify each rational number:

(a) $\dfrac{12}{18}$ **(b)** $-\dfrac{-10}{12}$ **(c)** $-\dfrac{42}{-14}$ **(d)** $\dfrac{-22}{-8}$

(e) $-\dfrac{12}{3}$ **(f)** $\dfrac{-15}{10}$ **(g)** $\dfrac{24}{-36}$ **(h)** $-\dfrac{-12}{-30}$

Solution   Recall the Fundamental Rule for Rational Numbers: If a, b, and c are integers ($b \neq 0$ and $c \neq 0$), then $\dfrac{a \cdot c}{b \cdot c} = \dfrac{a}{b}$ [see Example 6-10]:

(a) $\dfrac{12}{18} = \dfrac{2 \cdot 6}{3 \cdot 6} = \dfrac{2}{3}$ **(b)** $-\dfrac{-10}{12} = \dfrac{10}{12} = \dfrac{2 \cdot 5}{2 \cdot 6} = \dfrac{5}{6}$

(c) $-\dfrac{42}{-14} = \dfrac{42}{14} = \dfrac{3 \cdot \cancel{14}}{1 \cdot \cancel{14}} = \dfrac{3}{1} = 3$

(d) $\dfrac{-22}{-8} = \dfrac{22}{8} = \dfrac{\cancel{2} \cdot 11}{\cancel{2} \cdot 4} = 2\frac{3}{4}$

(e) $-\dfrac{12}{3} = -\dfrac{\cancel{3} \cdot 4}{1 \cdot \cancel{3}} = -\dfrac{4}{1} = -4$

(f) $\dfrac{-15}{10} = -\dfrac{15}{10} = -\dfrac{3 \cdot \cancel{5}}{2 \cdot \cancel{5}} = -\dfrac{3}{2}$ or $-1\frac{1}{2}$

(g) $\dfrac{24}{-36} = -\dfrac{24}{36} = -\dfrac{2 \cdot \cancel{12}}{3 \cdot \cancel{12}} = -\dfrac{2}{3}$

(h) $-\dfrac{-12}{-30} = -\dfrac{12}{30} = -\dfrac{2 \cdot \cancel{6}}{5 \cdot \cancel{6}} = -\dfrac{2}{5}$

PROBLEM 6-4 Add rational numbers:

(a) $3 + \frac{3}{8}$ (b) $0.75 + (-\frac{3}{8})$ (c) $-1\frac{1}{2} + (-1.25)$ (d) $6 + \frac{1}{2} + 0.2 + (-3\frac{1}{4}) + (-2.75) + (-1)$

Solution Recall that to add rational numbers, you use the Addition Rules for Numbers with Like (or Unlike) Signs given in Section 2-4A [see Example 6-11]:

(a) $3 + \dfrac{3}{8} = 3\frac{3}{8}$ or 3.375

(b) $0.75 + \left(-\dfrac{3}{8}\right) = 0.75 + (-0.375) = 0.375$ or $\dfrac{3}{8}$

(c) $-1\frac{1}{2} + (-1.25) = -1\frac{1}{2} + (-1\frac{1}{4}) = -2\frac{3}{4}$ or -2.75

(d) $6 + \dfrac{1}{2} + 0.2 + (-3\frac{1}{4}) + (-2.75) + (-1) = 6 + 0.5 + 0.2 + (-3.25) + (-2.75) + (-1)$

$$= 6.7 + (-7)$$
$$= -0.3 \quad \text{or} \quad -\dfrac{3}{10}$$

PROBLEM 6-5 Subtract rational numbers:

(a) $5 - \frac{3}{4}$ (b) $1.3 - 2.9$ (c) $0.25 - (-\frac{4}{5})$ (d) $-2\frac{1}{2} - (-3.75)$

Solution Recall that to subtract rational numbers, you use the Subtraction Rules for Numbers given in Section 2-4B [see Example 6-12]:

(a) $5 - \dfrac{3}{4} = \dfrac{20}{4} - \dfrac{3}{4} = 4\frac{1}{4}$ or 4.25

(b) $1.3 - 2.9 = -1.6$ or $-1\frac{3}{5}$

(c) $0.25 - \left(-\dfrac{4}{5}\right) = 0.25 + \left(+\dfrac{4}{5}\right) = 0.25 + 0.8$

$$= 1.05 \quad \text{or} \quad 1\frac{1}{20}$$

(d) $-2\frac{1}{2} - (-3.75) = -2\frac{1}{2} + (+3.75)$

$$= -2.5 + 3.75$$
$$= 1.25 \quad \text{or} \quad 1\frac{1}{4}$$

PROBLEM 6-6 Multiply rational numbers:

(a) $\frac{1}{2}(-\frac{3}{4})$ (b) $-0.5(2.6)$ (c) $\frac{1}{4}(0.25)$ (d) $-1.25(-2\frac{3}{4})$

Solution Recall that to multiply rational numbers, you use the Multiplication Rules for Numbers given in Section 2-4C [see Example 6-13]:

(a) $\dfrac{1}{2}\left(-\dfrac{3}{4}\right) = -\dfrac{3}{8}$ or -0.375

(b) $-0.5(2.6) = -1.3$ or $-1\frac{3}{10}$

(c) $\dfrac{1}{4}(0.25) = \dfrac{1}{4} \cdot \dfrac{1}{4} = \dfrac{1}{16}$ or 0.0625

(d) $-1.25(-2\frac{3}{4}) = -1.25(-2.75)$

$$= 3.4375 \quad \text{or} \quad 3\frac{7}{16}$$

PROBLEM 6-7 Divide rational numbers:

(a) $-\frac{2}{3} \div \frac{3}{4}$ (b) $1.2 \div (-0.5)$ (c) $\frac{3}{4} \div 0.25$

(d) $-2.5 \div (-1\frac{1}{2})$ (e) $0 \div 0.5$ (f) $-3.6 \div 0$

Solution Recall that to divide rational numbers, you use the Division Rules for Numbers given in Section 2-4D [see Example 6-14]:

(a) $-\dfrac{2}{3} \div \dfrac{3}{4} = -\dfrac{2}{3} \cdot \dfrac{4}{3} = -\dfrac{8}{9}$ or $-0.\overline{8}$ (b) $1.2 \div (-0.5) = -2.4$ or $-2\frac{2}{5}$

(c) $\dfrac{3}{4} \div 0.25 = \dfrac{3}{4} \div \dfrac{1}{4} = \dfrac{3}{\cancel{4}} \cdot \dfrac{\cancel{4}}{1} = \dfrac{3}{1} = 3$ (d) $-2.5 \div (-1\frac{1}{2}) = -2.5 \div (-1.5) = 1.\overline{6}$ or $1\frac{2}{3}$

(e) $0 \div 0.5 = 0$ (f) $-3.6 \div 0$ is not defined

PROBLEM 6-8 Write the product of (a) 4 times m, (b) -5 times n, and (c) x times y in five ways.

Solution Recall that the traditional "times" symbol $\times$ is never used with variables [see Example 6-15]:

(a) 4 times m $=$ $4 \cdot m =$ $4(m) =$ $(4)m =$ $(4)(m) = 4m$

(b) -5 times $n = -5 \cdot n = -5(n) = (-5)n = (-5)(n) = -5n$ $\rangle$ preferred form

(c) x times y $=$ $x \cdot y =$ $x(y) =$ $(x)y =$ $(x)(y) = xy$

PROBLEM 6-9 Write the quotient of (a) m divided by 8, (b) -3 divided by n, and (c) x divided by y in five ways.

Solution Recall that the traditional division symbol $\div$ and slash division symbol $/$, although acceptable, are seldom used to write an indicated quotient. The preferred form of division notation is $\frac{a}{b}$ [see Example 6-16]:

(a) m divided by 8 $=$ $m \div 8 =$ $m/8 =$ $m \cdot \dfrac{1}{8} = \dfrac{1}{8} \cdot m = \dfrac{m}{8}$

(b) -3 divided by $n = -3 \div n = -3/n = -3 \cdot \dfrac{1}{n} = \dfrac{1}{n}(-3) = \dfrac{-3}{n}$ $\rangle$ preferred form

(c) x divided by y $=$ $x \div y =$ $x/y =$ $x \cdot \dfrac{1}{y} = \dfrac{1}{y} \cdot x = \dfrac{x}{y}$

PROBLEM 6-10 Identify each missing number and identify the property by name when possible:

(a) $3 + ? = 2 + 3$ (b) $?(-5) = -5 \cdot 9$ (c) $(? + 3) + 6 = 8 + (3 + 6)$

(d) $(4 \cdot 6)? = 4(6 \cdot 9)$ (e) $?(5 + 3) = 2 \cdot 5 + 2 \cdot 3$ (f) $(5 - 2)3 = 5 \cdot 3 - 2 \cdot ?$

(g) $? + 0 = 4$ (h) $5 + ? = 0$ (i) $6 - ? = 6$

(j) $0 - ? = -5$ (k) $5 - ? = 0$ (l) $7 \cdot 0 = ?$

(m) $5 \cdot ? = 0$ (n) $0 \div 6 = ?$ (o) $2 \div 0 = ?$

(p) $8 \cdot ? = 1$ (q) $3 \cdot ? = 3$ (r) $7 \div ? = 7$

(s) $? \div 5 = 1$ (t) $-? = -1 \cdot 5$ (u) $-(-2) = ?$

(v) $3 \cdot 4 = -3 \cdot ?$ (w) $-(2 \cdot 5) = ?(-5)$ (x) $3 - 5 = 3 + ?$

Solution

(a) $3 + \mathbf{2} = 2 + 3$ Commutative Property of Addition [see Example 6-18]

(b) $\mathbf{9}(-5) = -5 \cdot 9$ Commutative Property of Multiplication [see Example 6-19]

(c) $(\mathbf{8} + 3) + 6 = 8 + (3 + 6)$ Associative Property of Addition [see Example 6-22]

(d) $(4 \cdot 6)\mathbf{9} = 4(6 \cdot 9)$ Associative Property of Multiplication [see Example 6-23]

(e) $\mathbf{2}(5 + 3) = 2 \cdot 5 + 2 \cdot 3$ Distributive Property of Multiplication Over Addition [see Example 6-26]

(f) $(5 - 2)3 = 5 \cdot 3 - 2 \cdot 3$ Distributive Property of Multiplication over Subtraction [see Example 6-27]

(g) $4 + 0 = 4$ Identity Property for Addition [see Example 6-28]

(h) $5 + (-5) = 0$ Additive Inverse Property [see Example 6-28]

(i) $6 - 0 = 6$ [see Example 6-28]

(j) $0 - 5 = -5$ [see Example 6-28]

(k) $5 - 5 = 0$ [see Example 6-28]

(l) $7 \cdot 0 = 0$ Zero-Factor Property [see Example 6-28]

(m) $5 \cdot 0 = 0$ Zero-Product Property [see Example 6-28]

(n) $0 \div 6 = 0$ Zero-Dividend Property [see Example 6-28]

(o) $2 \div 0$ is **not defined.**   Zero-Divisor Property [see Example 6-28]

(p) $8 \cdot \frac{1}{8} = 1$ Multiplicative Inverse Property [see Example 6-29]

(q) $3 \cdot 1 = 3$ Identity Property for Multiplication [see Example 6-29]

(r) $7 \div 1 = 7$ [see Example 6-29]

(s) $5 \div 5 = 1$ Unit-Fraction Property [see Example 6-29]

(t) $-5 = -1 \cdot 5$ [see Example 6-30]

(u) $-(-2) = 2$ [see Example 6-30]

(v) $3 \cdot 4 = -3(-4)$ [see Example 6-30]

(w) $-(2 \cdot 5) = 2(-5)$ [see Example 6-30]

(x) $3 - 5 = 3 + (-5)$ Subtraction Property [see Example 6-31]

PROBLEM 6-11   Identify each term in the following:

(a) $x + 3$ **(b)** $y - 2$ **(c)** mn **(d)** $\dfrac{w}{2}$

(e) $2a - 3b + 2$ **(f)** $-4h + \dfrac{3}{h} - 2\sqrt{h}$ **(g)** $2(x + 5)$ **(h)** $3(y - 4)$

(i) $+(a^2 - a - 1)$ **(j)** $-(x - y)$ **(k)** $3 + \pi$ **(l)** w

Solution   Recall that in an algebraic expression that does not contain grouping symbols, the addition symbols separate the terms. Also, to identify terms when grouping symbols are present, you must first clear the grouping symbols [see Example 6-36]:

(a) In $x + 3$, the terms are x and 3.

(b) In $y - 2$, the terms are y and -2 because $y - 2 = y + (-2)$.

(c) The product of two or more numbers and variables like mn is a term.

(d) The quotient of numbers and variables like $\dfrac{w}{2}$ is a term.

(e) In $2a - 3b + 2$, the terms are $2a$, $-3b$, and 2.

(f) In $-4h + \dfrac{3}{h} - 2\sqrt{h}$, the terms are $-4h, \dfrac{3}{h}$, and $-2\sqrt{h}$.

(g) In $2(x + 5)$, the terms are $2x$ and 10 because $2(x + 5) = 2 \cdot x + 2 \cdot 5 = 2x + 10$.

(h) In $3(y - 4)$, the terms are $3y$ and -12 because $3(y - 4) = 3 \cdot y - 3 \cdot 4 = 3y - 12$.

(i) In $+(a^2 - a - 1)$, the terms are a^2, $-a$, and -1 because $+(a^2 - a - 1) = a^2 - a - 1$.

(j) In $-(x - y)$, the terms are $-x$ and y because $-(x - y) = -x + y$.

(k) In $3 + \pi$, the number 3 is a constant term, and π is also a number (which *always* equals $3.1416\ldots$) so π is also a constant term.

(l) A variable like w is a letter term.

PROBLEM 6-12   Clear parentheses in the following:

(a) $5(u + 2)$

(b) $(v - 3)(-4)$

(c) $-2(x + y)$

(d) $(2m - 5n)3$

(e) $+(8h - 5k)$

(f) $-(y^2 - 2y + 5)$

Solution   Recall that to clear parentheses in algebraic expressions, you must use the distributive properties because it is otherwise impossible to combine the unlike terms inside the parentheses [see Examples 6-37, 6-38, and 6-39]:

(a) $5(u + 2) = 5 \cdot u + 5 \cdot 2 = 5u + 10$

(b) $(v - 3)(-4) = v(-4) - 3(-4) = -4v + 12$

(c) $-2(x + y) = (-2)x + (-2)y = -2x - 2y$

(d) $(2m - 5n)3 = 2m \cdot 3 - 5n \cdot 3 = 6m - 15n$

(e) $+(8h - 5k) = 8h - 5k$

(f) $-(y^2 - 2y + 5) = -y^2 + 2y - 5$

PROBLEM 6-13   Combine like terms in the following:

(a) $2x + 5x$

(b) $3y - y$

(c) $w - 5w$

(d) $6ab - 5ab$

(e) $3u + 2 - 4u$

(f) $m - 5n + m$

(g) $\dfrac{c}{2} + c - \dfrac{1}{2}$

(h) $6s - 5r + 2s + 3r$

Solution   Recall that to combine like terms in an algebraic expression, you use the distributive properties to add or subtract the numerical coefficients of like terms and then write the same literal part on the sum or difference [see Example 6-43]:

(a) $2x + 5x = (2 + 5)x = 7x$

(b) $3y - y = 3y - 1y = (3 - 1)y = 2y$

(c) $w - 5w = 1w - 5w = (1 - 5)w = -4w$

(d) $6ab - 5ab = (6 - 5)ab = 1ab = ab$

(e) $3u + 2 - 4u = 3u - 4u + 2$
$$= (3 - 4)u + 2$$
$$= -1u + 2$$
$$= -u + 2 \quad \text{or} \quad 2 - u$$

(f) $m - 5n + m = m + m - 5n$
$$= 1m + 1m - 5n$$
$$= (1 + 1)m - 5n = 2m - 5n$$

(g) $\dfrac{c}{2} + c - \dfrac{1}{2} = \dfrac{1}{2}c + 1c - \dfrac{1}{2}$
$$= \left(\dfrac{1}{2} + 1\right)c - \dfrac{1}{2}$$
$$= \dfrac{3}{2}c - \dfrac{1}{2} \quad \text{or} \quad \dfrac{3c - 1}{2}$$

(h) $6s - 5r + 2s + 3r = 6s + 2s + (-5r) + 3r$
$$= (6 + 2)s + (-5 + 3)r$$
$$= 8s + (-2)r$$
$$= 8s - 2r$$

PROBLEM 6-14   Clear parentheses and combine like terms in the following:

(a) $3(x + 2) + 2x$

(b) $-2(3y - 4) - 5$

(c) $-(m - n) + (m + n)$

(d) $2(3a - b + 5a) - 8b$

Solution   Recall that to clear parentheses and combine like terms, you first combine any like terms inside the parentheses, then clear the parentheses and combine like terms [see Examples 6-44 and 6-45]:

(a) $3(x + 2) + 2x = 3 \cdot x + 3 \cdot 2 + 2x = 3x + 6 + 2x = 3x + 2x + 6 = 5x + 6$

(b) $-2(3y - 4) - 5 = (-2)3y - (-2)4 - 5 = -6y + 8 - 5 = -6y + 3$

(c) $-(m - n) + (m + n) = -m + n + m + n = -m + m + n + n = 0 + 2n = 2n$

(d) $2(3a - b + 5a) - 8b = 2(8a - b) - 8b = 2 \cdot 8a - 2 \cdot b - 8b = 16a - 2b - 8b = 16a - 10b$

PROBLEM 6-15   Evaluate the following algebraic expressions:

(a) $5x$ for $x = 2$

(b) $-2y$ for $y = -4$

(c) $w + 3$ for $w = -6$

(d) $u - 2$ for $u = 1$

(e) $2v + 3$ for $v = 8$

(f) $3a - 5$ for $a = -2$

(g) $x^2 + 5x - 6$ for $x = 2$

(h) $x^2 + 5x - 6$ for $x = -2$

(i) $\dfrac{-b - \sqrt{b^2 - 4ac}}{2a}$ for $a = 6, b = 1, c = -12$

Solution   Recall that to evaluate an algebraic expression given a numerical value for each different variable, you first substitute each given value for the associated variable and then evaluate using the Order of Operations [see Examples 6-46 and 6-47]:

(a) $5x = 5(2) = 10$

(b) $-2y = -2(-4) = 8$

(c) $w + 3 = (-6) + 3 = -3$

(d) $u - 2 = (1) - 2 = -1$

(e) $2v + 3 = 2(8) + 3 = 16 + 3 = 19$

(f) $3a - 5 = 3(-2) - 5 = -6 - 5 = -11$

(g) $x^2 + 5x - 6 = (2)^2 + 5(2) - 6 = 4 + 10 - 6 = 8$

(h) $x^2 + 5x - 6 = (-2)^2 + 5(-2) - 6 = 4 - 10 - 6 = -12$

(i) $\dfrac{-b - \sqrt{b^2 - 4ac}}{2a} = \dfrac{-(1) - \sqrt{(1)^2 - 4(6)(-12)}}{2(6)} = \dfrac{-1 - \sqrt{289}}{12} = \dfrac{-1 - 17}{12} = \dfrac{-18}{12} = -\dfrac{3}{2}$

PROBLEM 6-16   Solve each problem by evaluating the given formula:

(a) Water freezes at 0 degrees Celsius, C. Find the temperature in degrees Fahrenheit, F, at which water freezes by evaluating the **temperature formula** $F = \frac{9}{5}C + 32$.

(b) Find the ideal weight w for a man whose height h is 72 inches using the **male ideal weight formula** $w = 5\frac{1}{2}h - 231$.

(c) Find the ideal weight w for a woman whose height h is 60 inches using the **female ideal weight formula** $w = 5\frac{1}{4}h - 216$.

(d) What distance d can a car travel in a time t of 40 minutes at a constant rate r of 60 mph using the **distance formula** $d = rt$.

(e) What is the altitude a in kilometers when the average annual temperature t is 0 degrees Celsius using the **altitude/temperature formula** $a = 0.16(15 - t)$.

(f) What is the amount A at the end of a 9-month time period t if \$500 principal P is invested at a simple interest rate r of 6% per year using the **amount formula** $A = P(1 + rt)$.

Solution   Recall that to evaluate a formula that is solved for a specific variable given a measurement value for each of the other variables, you substitute each given value for the associated variable and then evaluate using the Order of Operations. To evaluate a formula when one of the variables represents a rate, you may first need to rename to get a common unit of measure that can be eliminated [see Examples 6-50 and 6-51]:

(a) The question asks you to evaluate $F = \frac{9}{5}C + 32$ for $C = 0$:

$$F = \tfrac{9}{5}C + 32$$
$$= \tfrac{9}{5}(0) + 32$$
$$= 0 + 32$$
$$= 32 \text{ degrees Fahrenheit}$$

(b) The question asks you to evaluate $w = 5\frac{1}{2}h - 231$ for $h = 72$:

$$w = 5\tfrac{1}{2}h - 231$$
$$= 5\tfrac{1}{2}(72) - 231$$
$$= 396 - 231$$
$$= 165 \text{ pounds}$$

(c) The question asks you to evaluate
$w = 5\frac{1}{4}h - 216$ for $h = 60$:

$$w = 5\tfrac{1}{4}h - 216$$
$$= 5\tfrac{1}{4}(\mathbf{60}) - 216$$
$$= 315 - 216$$
$$= 99 \text{ pounds}$$

(d) The question asks you to evaluate $d = rt$
for $r = 60$ miles/hour and $t = 40$ minutes:
40 minutes $= \frac{40}{60}$ hour $= \frac{2}{3}$ hour

$$d = rt$$
$$= 60\frac{\text{miles}}{\cancel{\text{hour}}} \cdot \frac{2}{3}\frac{\cancel{\text{hour}}}{1}$$
$$= \left(60 \cdot \frac{2}{3}\right) \text{miles}$$
$$= 40 \text{ miles}$$

(e) The question asks you to evaluate
$a = 0.16(15 - t)$ for $t = 0$:

$$a = 0.16(15 - t)$$
$$= 0.16(15 - \mathbf{0})$$
$$= 0.16(15)$$
$$= 2.4 \text{ kilometers}$$

(f) The question asks you to evaluate
$A = P(1 + rt)$ for $P = \$500$, $r = 6\%$,
and $t = 9$ months:
9 months $= \frac{9}{12}$ year $= \frac{3}{4}$ year

$$A = P(1 + rt)$$
$$= \$500\left(1 + \frac{6\%}{\cancel{\text{year}}} \cdot \frac{3}{4}\frac{\cancel{\text{year}}}{1}\right)$$
$$= \$500\left(1 + 0.06 \cdot \frac{3}{4}\right)$$
$$= \$500(1 + 0.045)$$
$$= \$500(1.045)$$
$$= \$522.50$$

Supplementary Exercises

PROBLEM 6-17   Simplify each rational number:

(a) $\dfrac{24}{40}$ (b) $\dfrac{-20}{50}$ (c) $\dfrac{16}{-80}$ (d) $\dfrac{-75}{-90}$ (e) $-\dfrac{-20}{120}$ (f) $-\dfrac{84}{-120}$

(g) $-\dfrac{8}{10}$ (h) $-\dfrac{-3}{-9}$ (i) $\dfrac{3}{3}$ (j) $\dfrac{-12}{12}$ (k) $\dfrac{9}{-3}$ (l) $\dfrac{-18}{-2}$

(m) $-\dfrac{10}{-6}$ (n) $-\dfrac{-14}{4}$ (o) $-\dfrac{48}{24}$ (p) $-\dfrac{-56}{-32}$ (q) $\dfrac{256}{32}$ (r) $\dfrac{-125}{40}$

(s) $\dfrac{15}{-25}$ (t) $\dfrac{-20}{-45}$ (u) $-\dfrac{12}{18}$ (v) $-\dfrac{-9}{27}$ (w) $\dfrac{20}{-1}$ (x) $-\dfrac{-90}{1}$

PROBLEM 6-18   Compute with rational numbers:

(a) $0.5 + 0.9$

(b) $1\frac{1}{5} + (-0.9)$

(c) $-0.8 + 2\frac{3}{5}$

(d) $-3\frac{1}{2} + (-2.1)$

(e) $\frac{1}{2} + \frac{7}{16}$

(f) $\frac{-3}{4} + 0.75$

(g) $-3\frac{1}{2} + 2.5$

(h) $-5\frac{1}{8} + (-7.25)$

(i) $0.2 - 0.6$

(j) $-0.8 - 1\frac{1}{2}$

(k) $-\frac{3}{8} - (-0.25)$

(l) $4\frac{3}{16} - (2.125)$

(m) $0.5(0.2)$

(n) $\frac{1}{2} \cdot \frac{3}{4}$

(o) $-4.8(-\frac{3}{10})$

(p) $-\frac{1}{4}(-0.\overline{3})$

(q) $0.8(-0.7)$

(r) $-\frac{3}{8} \cdot \frac{2}{3}$

(s) $0.7 \div 2\frac{2}{5}$

(t) $\frac{3}{8} \div 0.5$

(u) $-1.44 \div (-4.8)$

(v) $-\frac{1}{12} \div (-0.25)$

(w) $-4.5 \div 1.8$

(x) $\frac{1}{4} \div 1.8$

(y) $-6.125 \div (-1\frac{5}{16})$

PROBLEM 6-19 Write five correct ways to indicate the following expressions:

(a) 7 times u

(b) -3 times v

(c) x times y

(d) w divided by -2

(e) 9 divided by z

(f) a divided by b

PROBLEM 6-20 Identify each missing number and identify the property by name when possible:

(a) $? + 3 = 3 + (-5)$

(b) $2 \cdot ? = -3 \cdot 2$

(c) $(4 + 5) + 2 = 4 + (? + 2)$

(d) $(2 \cdot 3)5 = 2(? \cdot 5)$

(e) $8 + ? = 8$

(f) $? + (-2) = 0$

(g) $? - 0 = 5$

(h) $? - 4 = -4$

(i) $? - 3 = 0$

(j) $? \cdot 5 = 0$

(k) $0 \cdot ? = 0$

(l) $? \div (-2) = 0$

(m) $2 \div ?$ is not defined

(n) $? \cdot 1 = 6$

(o) $? \cdot \frac{2}{3} = 1$

(p) $? \cdot 7 = -7$

(q) $? \div 1 = -6$

(r) $-2 \div ? = 1$

(s) $3(8 + 9) = ? \cdot 8 + ? \cdot 9$

(t) $-(-5) = ?$

(u) $(6 + 2)5 = 6 \cdot ? + 2 \cdot ?$

(v) $-2 \cdot 9 = 2 \cdot ?$

(w) $3 \cdot 4 = -3 \cdot ?$

(x) $2 - 8 = 2 + ?$

PROBLEM 6-21 Identify each term in the following:

(a) $u + 7$ **(b)** $v - 5$ **(c)** $3x^2 - 5x - 4$ **(d)** $2\sqrt{y} - y\sqrt{2} + y - 2$ **(e)** $-(m + n)$

(f) $(a - 6)$ **(g)** $-3(b - 5)$ **(h)** $2(9 - c)$ **(i)** 2 **(j)** p

(k) $-3hk$ **(l)** $\dfrac{xyz}{5}$ **(m)** $\dfrac{x + 3}{x}$ **(n)** $\dfrac{m - n}{n}$ **(o)** $w\sqrt{5^2 - 4^2}$

PROBLEM 6-22 Clear parentheses and/or combine like terms in the following:

(a) $-3(x + 2)$

(b) $(2 - y)8$

(c) $5(m + n)$

(d) $(x - y)(-5)$

(e) $3h^2 - 2h + 5$

(f) $-(5 - 8k)$

(g) $6u + u$

(h) $3v - 7v$

(i) $9p - p$

(j) $3mn - 8mn$

(k) $2x - 5x + x$

(l) $5y - 8 + y - 5$

(m) $\dfrac{w}{4} - \dfrac{3w}{4}$

(n) $5b - 6a + b + 3a - 2b$

(o) $2(h - 2) + 2h$

(p) $m - 4(m - n) - n$

(q) $-(a - b + c) + 2a - 4c$

(r) $3 + 2(v - 5)$

(s) $u - (u + v)$

(t) $2x - 3(4x - 5) + 6$

PROBLEM 6-23 Evaluate the following algebraic expressions:

(a) $3v$ for $v = 8$

(b) $-5u$ for $u = 2$

(c) $x + 5$ for $x = 2$

(d) $y - 7$ for $y = -8$

(e) $5w + 6$ for $w = -1$

(f) $2z - 9$ for $z = 6$

(g) $a^2 + 3a - 9$ for $a = 3$

(h) $a^2 + 3a - 9$ for $a = -3$

(i) $2u - v$ for $u = \dfrac{1}{2}$ and $v = 2$

(j) $\dfrac{-b + \sqrt{b^2 - 4ac}}{2a}$ for $a = 6, b = -13$ and $c = 6$

(k) $\dfrac{-b - \sqrt{b^2 - 4ac}}{2a}$ for $a = 6, b = -13$ and $c = 6$

(l) $-6a - 9b$ for $a = -\dfrac{1}{3}$ and $b = \dfrac{2}{3}$

PROBLEM 6-24   Solve each problem by evaluating each given formula:

(a) Find the shoe-size number n for a man whose feet each measure 12 inches in length l when standing using the **male shoe-size formula** $n = 3l - 25$.

(b) Find the shoe-size number n for a woman whose feet each measure $9\frac{1}{3}$ inches in length l when standing using the **female shoe-size formula** $n = 3l - 22$.

(c) Find the average temperature t in degrees Celsius at an altitude a of 12 kilometers using the **temperature/altitude formula** $t = 15 - 6.25a$.

(d) Find the Fahrenheit temperature F for the answer to problem **(c)** using the **temperature formula** $F = \dfrac{9}{5}C + 32$.

(e) What distance d can a car travel in a time t of 15 minutes at a constant rate r of 15 mph using the **distance formula** $d = rt$?

(f) What is the amount A at the end of an 18-month time period t if \$800 principal P is invested at a simple interest rate r of 8% per year using the **amount formula** $A = P(1 + rt)$?

(g) How much time t will it take to travel a distance d of 300 miles at a constant rate r of 45 mph using the **time formula** $t = d \div r$?

(h) What is the constant rate r that is necessary to travel a distance d of 100 miles in a time t of $1\frac{1}{2}$ hours using the **rate formula** $r = d \div t$?

Answers to Supplementary Exercises

(6-17) **(a)** $\frac{3}{5}$ **(b)** $-\frac{2}{5}$ **(c)** $-\frac{1}{5}$ **(d)** $\frac{5}{6}$ **(e)** $\frac{1}{6}$ **(f)** $\frac{7}{10}$

(g) $-\frac{4}{5}$ **(h)** $-\frac{1}{3}$ **(i)** 1   **(j)** -1 **(k)** -3 **(l)** 9

(m) $\frac{5}{3}$ or $1\frac{2}{3}$ **(n)** $\frac{7}{2}$ or $3\frac{1}{2}$ **(o)** -2 **(p)** $-\frac{7}{4}$ **(q)** 8   **(r)** $-\frac{25}{8}$ or $-3\frac{1}{8}$

(s) $-\frac{3}{5}$ **(t)** $\frac{4}{9}$ **(u)** $-\frac{2}{3}$ **(v)** $\frac{1}{3}$ **(w)** -20 **(x)** 90

(6-18) **(a)** 1.4 or $1\frac{2}{5}$ **(b)** 0.3 or $\frac{3}{10}$ **(c)** 1.8 or $1\frac{4}{5}$ **(d)** -5.6 or $-5\frac{3}{5}$

(e) $\frac{15}{16}$ or 0.9375   **(f)** 0   **(g)** -1 **(h)** -12.375 or $-12\frac{3}{8}$

(i) -0.4 or $-\frac{2}{5}$ **(j)** -2.3 or $-2\frac{3}{10}$ **(k)** -0.125 or $-\frac{1}{8}$ **(l)** 2.0625 or $2\frac{1}{16}$

(m) 0.1 or $\frac{1}{10}$ **(n)** $\frac{3}{8}$ or 0.375   **(o)** 1.44 or $1\frac{11}{25}$ **(p)** $\frac{1}{12}$ or $0.08\overline{3}$

(q) -0.56 or $-\frac{14}{25}$ **(r)** $-\frac{1}{4}$ or -0.25 **(s)** 0.2916 or $\frac{7}{24}$ **(t)** $\frac{3}{4}$ or 0.75

(u) 0.3 or $\frac{3}{10}$ **(v)** $\frac{1}{3}$ or $0.\overline{3}$ **(w)** -2.5 or $-2\frac{1}{2}$ **(x)** $\frac{5}{36}$ or $0.13\overline{8}$

(y) $4\frac{2}{3}$ or $4.\overline{6}$

(6-19) **(a)** 7 times $u = 7 \cdot u = 7(u) = (7)u = (7)(u) = 7u$

(b) -3 times $v = -3 \cdot v = -3(v) = (-3)v = (-3)(v) = -3v$

(c) x times $y = x \cdot y = x(y) = (x)y = (x)(y) = xy$

(d) w divided by $-2 = w \div (-2) = w/(-2) = w \cdot \dfrac{1}{-2} = \dfrac{1}{-2} \cdot w = \dfrac{w}{-2}$

(e) 9 divided by $z = 9 \div z = 9/z = 9 \cdot \dfrac{1}{z} = \dfrac{1}{z} \cdot 9 = \dfrac{9}{z}$

(f) a divided by $b = a \div b = a/b = a \cdot \dfrac{1}{b} = \dfrac{1}{b} \cdot a = \dfrac{a}{b}$

(6-20) **(a)** -5; Commutative Property of Addition

(b) -3; Commutative Property of Multiplication

(c) 5; Associative Property of Addition

(d) 3; Associative Property of Multiplication

(e) 0; Identity Property for Addition

(f) 2; Additive Inverse Property
(g) 5   **(h)** 0
(i) 3   **(j)** 0; Zero-Factor Property
(k) any real number; Zero-Product Property
(l) 0; Zero-Dividend Property
(m) 0; Zero-Divisor Property
(n) 6; Identity Property for Multiplication
(o) $\frac{3}{2}$; Multiplicative Inverse Property   **(p)** -1
(q) -6 **(r)** -2; Unit-Fraction Property
(s) 3; Distributive Property of Multiplication Over Addition
(t) 5
(u) 5; Distributive Property of Multiplication Over Addition
(v) -9
(w) -4 **(x)** -8; Subtraction Property

(6-21) **(a)** $u, 7$ **(b)** $v, -5$ **(c)** $3x^2, -5x, -4$ **(d)** $2\sqrt{y}, -y\sqrt{2}, y, -2$ **(e)** $-m, -n$
(f) $a, -6$ **(g)** $-3b, 15$ **(h)** $18, -2c$ **(i)** 2   **(j)** p
(k) $-3hk$ **(l)** $\frac{xyz}{5}$ **(m)** $1, \frac{3}{x}$ **(n)** $\frac{m}{n}, -1$ **(o)** $3w$

(6-22) **(a)** $-3x - 6$ **(b)** $16 - 8y$ **(c)** $5m + 5n$ **(d)** $5y - 5x$
(e) $3h^2 - 2h + 5$ **(f)** $8k - 5$ **(g)** $7u$ **(h)** $-4v$
(i) $8p$ **(j)** $-5mn$ **(k)** $-2x$ **(l)** $6y - 13$
(m) $-\frac{w}{2}$ or $-\frac{1}{2}w$ **(n)** $4b - 3a$ or $-3a + 4b$ **(o)** $4h - 4$ **(p)** $-3m + 3n$ or $3n - 3m$
(q) $a + b - 5c$ **(r)** $2v - 7$ **(s)** $-v$ **(t)** $-10x + 21$ or $21 - 10x$

(6-23) **(a)** 24   **(b)** -10 **(c)** 7   **(d)** -1 **(e)** 1   **(f)** 3
(g) 9   **(h)** -9 **(i)** -1 **(j)** $\frac{3}{2}$ **(k)** $\frac{2}{3}$ **(l)** -4

(6-24) **(a)** size 11   **(b)** size 6   **(c)** $-60°C$ **(d)** $-76°F$
(e) $3\frac{3}{4}$ miles   **(f)** \$896   **(g)** $6\frac{2}{3}$ hours   **(h)** $66\frac{2}{3}$ mph

7 LINEAR EQUATIONS IN ONE VARIABLE

THIS CHAPTER IS ABOUT

☑ Identifying Linear Equations in One Variable
☑ Solving Linear Equations Using Rules
☑ Solving Linear Equations Containing Like Terms and/or Parentheses
☑ Solving Linear Equations Containing Fractions, Decimals, or Percents
☑ Solving Literal Equations and Formulas for a Given Variable
☑ Translating Words to Symbols
☑ Solving Word Problems Using Linear Equations

7-1. Identifying Linear Equations in One Variable

A. Linear equations in one variable

A **linear equation in one variable** is an equation in one variable that can be written in standard form as $Ax + B = C$, where A, B, and C are real numbers ($A \neq 0$) and x is any variable.

EXAMPLE 7-1 Which of the following are linear equations in one variable?

(a) $y + 5 = -2$ **(b)** $w - 4 = 5$ **(c)** $2z = 6$ **(d)** $\dfrac{u}{5} = -2$ **(e)** $-3v - 6 = -2$

(f) $x\sqrt{2} + 3 = 5$ **(g)** $\dfrac{a}{2} + 0.3 = 1.8$ **(h)** $\dfrac{b}{-3} - 4 = 5$ **(i)** $2m + 3n = 6$ **(j)** $2 + 3 = 5$

(k) $2x^2 + 4 = 3$ **(l)** $2\sqrt{y} - 6 = 4$ **(m)** $\dfrac{2}{w} + 5 = 7$

Solution

(a) $y + 5 = -2$ is a linear equation in one variable: $1y + 5 = -2$
(b) $w - 4 = 5$ is a linear equation in one variable: $1w + (-4) = 5$
(c) $2z = 6$ is a linear equation in one variable: $2z + 0 = 6$
(d) $\dfrac{u}{5} = -2$ is a linear equation in one variable: $\dfrac{1}{5}u + 0 = -2$
(e) $-3v - 6 = -2$ is a linear equation in one variable: $-3v + (-6) = -2$
(f) $x\sqrt{2} + 3 = 5$ is a linear equation in one variable: $(\sqrt{2})x + 3 = 5$
(g) $\dfrac{a}{2} + 0.3 = 1.8$ is a linear equation in one variable: $\dfrac{1}{2}a + 0.3 = 1.8$
(h) $\dfrac{b}{-3} - 4 = 5$ is a linear equation in one variable: $-\dfrac{1}{3}b + (-4) = 5$

(i) $2m + 3n = 6$ is not a linear equation in one variable because it has two variables.

(j) $2 + 3 = 5$ is not a linear equation in one variable because it has no variables.

(k) $2x^2 + 4 = 3$ is not a linear equation in one variable because the variable has an exponent greater than 1.

(l) $2\sqrt{y} - 6 = 4$ is not a linear equation in one variable because the variable is in the radicand.

(m) $\dfrac{2}{w} + 5 = 7$ is not a linear equation in one variable because the variable is in the denominator.

B. The eight special types of linear equations in one variable

Every linear equation in one variable that does not contain like terms or parentheses can be easily categorized as one of eight special types of linear equations in one variable.

EXAMPLE 7-2   List the eight special types of linear equations in one variable.

Solution   If A, B, and C are real numbers ($A \neq 0$) and x is any variable, then

(1) $x + B = C$ **addition equation**

(2) $x - B = C$ **subtraction equation**

(3) $Ax = C$ **multiplication equation**

(4) $\dfrac{x}{A} = C$ **division equation**

(5) $Ax + B = C$ **multiplication-addition equation**

(6) $Ax - B = C$ **multiplication-subtraction equation**

(7) $\dfrac{x}{A} + B = C$ **division-addition equation**

(8) $\dfrac{x}{A} - B = C$ **division-subtraction equation**

Given a linear equation in one variable that does not contain like terms or parentheses, you may need to first write the equation in standard form before it can be categorized as one of the eight special types of linear equations in one variable.

EXAMPLE 7-3   Which special type of linear equation in one variable is given by each of the following equations?

(a) $y + 5 = -2$ **(b)** $w - 4 = 5$ **(c)** $2z = 6$ **(d)** $\dfrac{u}{5} = -2$ **(e)** $2x + 3 = 9$

(f) $-3v - 6 = -2$ **(g)** $\dfrac{a}{2} + 3 = -1$ **(h)** $\dfrac{b}{-3} - 4 = 5$ **(i)** $-c + 2 = -3$

(j) $5 - 2r = 0$ **(k)** $2 = \dfrac{s}{3} + 5$ **(l)** $-1 = -2 - \dfrac{t}{8}$

Solution

(a) $y + 5 = -2$ is an addition equation.

(b) $w - 4 = 5$ is a subtraction equation.

(c) $2z = 6$ is a multiplication equation.

(d) $\dfrac{u}{5} = -2$ is a division equation.

(e) $2x + 3 = 9$ is a multiplication-addition equation.

(f) $-3v - 6 = -2$ is a multiplication-subtraction equation.

(g) $\dfrac{a}{2} + 3 = -1$ is a division-addition equation.

(h) $\dfrac{b}{-3} - 4 = 5$ is a division-subtraction equation.

(i) $-c + 2 = -3$ or $-1c + 2 = -3$ is a multiplication-addition equation.

(j) $5 - 2r = 0$ or $2r - 5 = 0$ is a multiplication-subtraction equation.

(k) $2 = \dfrac{s}{3} + 5$ or $\dfrac{s}{3} + 5 = 2$ is a division-addition equation.

(l) $-1 = -2 - \dfrac{t}{8}$ or $\dfrac{t}{-8} - 2 = -1$ is a division-subtraction equation.

7-2. Solving Linear Equations Using Rules

A. Checking a proposed solution of a linear equation in one variable

Every linear equation in one variable has exactly one solution. The **solution of a linear equation in one variable** is the real number that can replace the variable to make both members of the equation equal.

To **check a proposed solution of a linear equation in one variable,** you first substitute the proposed solution for the variable and then compute using the Order of Operations to get a number sentence. If you get a **true number sentence,** then the proposed solution checks and is the one and only solution of the original linear equation in one variable. If you get a **false number sentence,** then the proposed solution does not check and is not the solution of the original linear equation in one variable.

EXAMPLE 7-4 Which one of the following is a solution of $3x + 2 = 5$: 0 or 1?

Solution Check $x = 0$ in $3x + 2 = 5$.

$3x + 2 = 5$		$\longleftarrow$ given equation
$3(0) + 2$	5	Substitute 0 for x.
$0 + 2$	5	Compute.
2	5	$\longleftarrow$ false ($2 \neq 5$)

$x = 0$ is not a solution because $2 \neq 5$.

Check $x = 1$ in $3x + 2 = 5$.

$3x + 2 = 5$		$\longleftarrow$ given equation
$3(1) + 2$	5	Substitute 1 for x.
$3 + 2$	5	Compute.
5	5	$\longleftarrow$ true ($5 = 5$)

$x = 1$ is a solution because $5 = 5$.

Note: Because every linear equation in one variable has exactly one solution, the one and only solution of $3x + 2 = 5$ is $x = 1$.

B. Solving an addition equation ($x + B = C$)

To **solve an addition equation** in the form $x + B = C$, you subtract using the following Subtraction Rule for Equations.

Subtraction Rule for Equations If a, b, and c are real numbers and $a = b$, then $a - c = b - c$.

Note: The Subtraction Rule for Equations means that if you subtract the same term from both members of an equation, the solution(s) will not change.

To solve an addition equation using the Subtraction Rule for Equations, you subtract the numerical addend from both members of the addition equation to isolate the variable in one member.

EXAMPLE 7-5 Solve $y + 5 = -2$.

$$\text{variable addend}$$
$$\text{numerical addend}$$
$$\text{sum}$$

Solution $y + 5 = -2$ *Think:* In $y + 5 = -2$, the numerical addend is 5.

$y + 5 - 5 = -2 - 5$ Subtract the numerical addend 5 from both members to isolate the variable y in one member.

$y + 0 = -2 - 5$

y is isolated $\longrightarrow y = -2 - 5$

$y = -7 \longleftarrow$ proposed solution

Check: $\underline{y + 5 = -2} \longleftarrow$ original equation

$-7 + 5 \,\big|\, -2$ Substitute the proposed solution -7 for y in the original equation to see if you get a true number sentence.

$-2 \,\big|\, -2 \longleftarrow y = -7$ checks

Note: The one and only solution of $y + 5 = -2$ is $y = -7$.

C. Solving a subtraction equation ($x - B = C$)

To **solve a subtraction equation** in the form $x - B = C$, you add using the Addition Rule for Equations.

Addition Rule for Equations If a, b, and c are real numbers and $a = b$, then $a + c = b + c$.

Note: The Addition Rule for Equations means that if you add the same term to both members of an equation, the solution(s) will not change.

To solve a subtraction equation using the Addition Rule for Equations, you add the subtrahend to both members of the subtraction equation to isolate the variable in one member.

EXAMPLE 7-6 Solve $w - 4 = 5$.

$$\text{minuend}$$
$$\text{subtrahend}$$
$$\text{difference}$$

Solution $w - 4 = 5$ *Think:* In $w - 4 = 5$, 4 is the subtrahend.

$w - 4 + 4 = 5 + 4$ Add the subtrahend 4 to both members to isolate the variable w in one member.

$w + 0 = 5 + 4$

w is isolated $\longrightarrow w = 5 + 4$

$w = 9 \longleftarrow$ proposed solution

Check: $\underline{w - 4 = 5} \longleftarrow$ original equation

$9 - 4 \,\big|\, 5$ Substitute the proposed solution 9 for w.

$5 \,\big|\, 5 \longleftarrow w = 9$ checks

Note: The one and only solution of $w - 4 = 5$ is $w = 9$.

D. Solving a multiplication equation ($Ax = C$)

To **solve a multiplication equation** in the form $Ax = C$, you divide using the Division Rule for Equations.

> ***Division Rule for Equations***   If a, b, and c are real numbers ($c \neq 0$) and $a = b$, then $\dfrac{a}{c} = \dfrac{b}{c}$.

Note: The Division Rule for Equations means that if you divide both members of an equation by the same nonzero term, the solution(s) will not change.

To solve a multiplication equation using the Division Rule for Equations, you divide both members of the multiplication equation by the numerical factor to isolate the variable in one member.

EXAMPLE 7-7   Solve $2z = 6$.

Solution

$$\underset{\text{numerical factor}}{\downarrow}\ \underset{\text{variable factor}}{\downarrow}\ \underset{\text{product}}{\downarrow}$$

$2z = 6$ *Think:* In $2z = 6$, the numerical factor is 2.

$\dfrac{2z}{2} = \dfrac{6}{2}$ Divide both members by the numerical factor 2 to isolate the variable z in one member.

$\dfrac{\cancel{2}z}{\cancel{2}} = \dfrac{6}{2}$

z is isolated $\longrightarrow\ z = \dfrac{6}{2}$

$z = 3$ $\longleftarrow$ proposed solution

Check: $2z = 6$ $\longleftarrow$ original equation

$2(3)\ \vert\ 6$ Substitute the proposed solution 3 for z.

$6\ \vert\ 6$ $\longleftarrow$ $z = 3$ checks

Note: The one and only solution of $2z = 6$ is $z = 3$.

E. Solving a division equation $\left(\dfrac{x}{A} = C\right)$

To **solve a division equation** in the form $\dfrac{x}{A} = C$, you multiply using the Multiplication Rule for Equations.

> ***Multiplication Rule for Equations***   If a, b, and c are real numbers ($c \neq 0$) and $a = b$, then $ac = bc$.

Note: The Multiplication Rule for Equations means that if you multiply both members of an equation by the same nonzero term, the solution(s) will not change.

To solve a division equation using the Multiplication Rule for Equations, you multiply both members of the division equation by the numerical divisor to isolate the variable in one member.

EXAMPLE 7-8 Solve $\dfrac{u}{5} = -2$.

Solution

numerical divisor
variable dividend
quotient

$$\dfrac{u}{5} = -2 \qquad \textit{Think:} \text{ In } \dfrac{u}{5} = -2, \text{ the numerical divisor is 5.}$$

$$\mathbf{5} \cdot \dfrac{u}{5} = \mathbf{5}(-2) \qquad \text{Multiply both members by the numerical divisor 5}$$
$$\text{to isolate the variable } u \text{ in one member.}$$

$$\dfrac{\cancel{5}}{1} \cdot \dfrac{u}{\cancel{5}} = 5(-2)$$

u is isolated $\longrightarrow u = 5(-2)$

$$u = -10 \longleftarrow \text{ proposed solution}$$

Check: $\dfrac{u}{5} = -2 \longleftarrow$ original equation

$$\dfrac{-10}{5} \ \bigg| \ -2 \qquad \text{Substitute the proposed solution } -10 \text{ for } u.$$

$$-2 \ \bigg| \ -2 \longleftarrow u = -10 \text{ checks}$$

Note: The one and only solution of $\dfrac{u}{5} = -2$ is $u = -10$.

The Addition, Subtraction, Multiplication, and Division Rules for Equations are collectively called the **rules for linear equations.**

F. Solving a multiplication-addition equation ($Ax + B = C$)

To **solve a multiplication-addition equation** in the form $Ax + B = C$, you first use the Subtraction Rule for Equations to isolate the **variable term** in one member and then use the Division Rule for Equations to isolate the variable itself.

EXAMPLE 7-9 Solve $2x + 3 = 9$.

Solution

variable term

$$2x + 3 = 9 \qquad \text{Identify the variable term.}$$

$$2x + 3 - 3 = 9 - 3 \qquad \text{Use the Subtraction Rule for Equations to}$$
$$\text{isolate the variable term in one member.}$$

$2x$ is isolated $\longrightarrow 2x = 6$

$$\dfrac{2x}{2} = \dfrac{6}{2} \qquad \text{Use the Division Rule for Equations}$$
$$\text{to isolate the variable } x.$$

x is isolated $\longrightarrow x = 3 \longleftarrow$ proposed solution

Check: $2x + 3 = 9 \longleftarrow$ original equation

$$2(3) + 3 \ \bigg| \ 9 \qquad \text{Substitute the proposed solution 3 for } x.$$

$$6 + 3 \ \bigg| \ 9 \qquad \text{Compute.}$$

$$9 \ \bigg| \ 9 \longleftarrow x = 3 \text{ checks}$$

Note: The one and only solution of $2x + 3 = 9$ is $x = 3$.

G. Solving a multiplication-subtraction equation ($Ax - B = C$)

To **solve a multiplication-subtraction equation** in the form $Ax - B = C$, you first use the Addition Rule for Equations to isolate the variable term in one member and then use the Division Rule for Equations to isolate the variable itself.

EXAMPLE 7-10   Solve $-3v - 6 = -2$.

Solution

variable term
$$\overbrace{-3v} - 6 = -2$$ Identify the variable term.

$$-3v - 6 + 6 = -2 + 6$$ Use the Addition Rule for Equations to isolate the variable term in one member.

$-3v$ is isolated $\longrightarrow -3v = 4$

$$\frac{-3v}{-3} = \frac{4}{-3}$$ Use the Division Rule for Equations to isolate the variable v.

v is isolated $\longrightarrow v = \dfrac{4}{-3}$ or $\dfrac{-4}{3}$ or $-\dfrac{4}{3}$ $\longleftarrow$ proposed solution

Check: $-3v - 6 = -2$ $\longleftarrow$ original equation

$$-3\left(-\frac{4}{3}\right) - 6 \quad\bigg|\quad -2$$ Substitute the proposed solution $-\dfrac{4}{3}$ for v.

$$4 - 6 \quad\bigg|\quad -2$$ Compute.

$$-2 \quad\bigg|\quad -2 \longleftarrow v = -\frac{4}{3} \text{ checks}$$

Note: The one and only solution of $-3v - 6 = -2$ is $v = -\dfrac{4}{3}$.

H. Solving a division-addition equation $\left(\dfrac{x}{A} + B = C\right)$

To **solve a division-addition equation** in the form $\dfrac{x}{A} + B = C$, you first use the Subtraction Rule for Equations to isolate the variable term in one member and then use the Multiplication Rule for Equations to isolate the variable itself.

EXAMPLE 7-11   Solve $\dfrac{a}{2} + 3 = -1$.

Solution

variable term
$$\downarrow$$
$$\frac{a}{2} + 3 = -1$$ Identify the variable term.

$$\frac{a}{2} + 3 - 3 = -1 - 3$$ Use the Subtraction Rule for Equations to isolate the variable term in one member.

$\dfrac{a}{2}$ is isolated $\longrightarrow \dfrac{a}{2} = -4$

$$2 \cdot \frac{a}{2} = 2(-4)$$ Use the Multiplication Rule for Equations to isolate the variable a.

a is isolated $\longrightarrow a = -8$ $\longleftarrow$ proposed solution

Check: $\dfrac{a}{2} + 3 = -1$ ⟵ original equation

$$\dfrac{-8}{2} + 3 \;\Big|\; -1 \qquad \text{Substitute the proposed solution } -8 \text{ for } a.$$

$$-4 + 3 \;\Big|\; -1 \qquad \text{Compute.}$$

$$-1 \;\Big|\; -1 \quad ⟵ \; a = -8 \text{ checks}$$

Note: The one and only solution of $\dfrac{a}{2} + 3 = -1$ is $a = -8$.

I. Solving a division-subtraction equation $\left(\dfrac{x}{A} - B = C\right)$

To **solve a division-subtraction equation** in the form $\dfrac{x}{A} - B = C$, you first use the Addition Rule for Equations to isolate the variable term in one member and then use the Multiplication Rule for Equations to isolate the variable itself.

EXAMPLE 7-12   Solve $\dfrac{b}{-3} - 4 = 5$.

Solution

variable term

$$\dfrac{b}{-3} - 4 = 5 \qquad \text{Identify the variable term.}$$

$$\dfrac{b}{-3} - 4 + 4 = 5 + 4 \qquad \begin{array}{l}\text{Use the Addition Rule for Equations to}\\\text{isolate the variable term in one member.}\end{array}$$

$\dfrac{b}{-3}$ is isolated ⟶ $\dfrac{b}{-3} = 9$

$$-3 \cdot \dfrac{b}{-3} = -3 \cdot 9 \qquad \begin{array}{l}\text{Use the Multiplication Rule for Equations}\\\text{to isolate the variable } b.\end{array}$$

b is isolated ⟶ $b = -27$ ⟵ proposed solution

Check: $\dfrac{b}{-3} - 4 = 5$ ⟵ original equation

$$\dfrac{-27}{-3} - 4 \;\Big|\; 5 \qquad \text{Substitute the proposed solution } -27 \text{ for } b.$$

$$9 - 4 \;\Big|\; 5 \qquad \text{Compute.}$$

$$5 \;\Big|\; 5 \quad ⟵ \; b = -27 \text{ checks}$$

Note: The one and only solution of $\dfrac{b}{-3} - 4 = 5$ is $b = -27$.

J. Solving linear equations in one variable that do not contain like terms or parentheses

Given a linear equation in one variable that does not contain like terms or parentheses and that is not in the form of one of the eight special types of linear equations, you can always rename the given linear equation as one of the special types and then solve using the rules for linear equations.

EXAMPLE 7-13   Solve (a) $-x + 2 = -3$, (b) $5 - 2y = 0$, (c) $2 = \dfrac{m}{3} + 5$, and

(d) $-1 = -3 + \dfrac{n}{-5}$.

Solution

(a) $-x + 2 = -3$ Rename using $-x = -1x$.

$-1x + 2 = -3$ ⟵ multiplication-addition equation

$-1x + 2 - 2 = -3 - 2$ Use the Subtraction Rule for Equations.

$-1x = -5$

$\dfrac{-1x}{-1} = \dfrac{-5}{-1}$ Use the Division Rule for Equations.

$x = 5$ ⟵ solution

(b) $5 - 2y = 0$

$-1(5 - 2y) = -1(0)$ Multiply both members by -1.

$-5 + 2y = 0$

$2y - 5 = 0$ ⟵ multiplication-subtraction equation

$2y - 5 + 5 = 0 + 5$ Use the Addition Rule for Equations.

$2y = 5$

$\dfrac{2y}{2} = \dfrac{5}{2}$ Use the Division Rule for Equations.

$y = \dfrac{5}{2}$ or $2\frac{1}{2}$ or 2.5 ⟵ solution

(c) $2 = \dfrac{m}{3} + 5$ Interchange the left and right members.

$\dfrac{m}{3} + 5 = 2$ ⟵ division-addition equation

$\dfrac{m}{3} + 5 - 5 = 2 - 5$ Use the Subtraction Rule for Equations.

$\dfrac{m}{3} = -3$

$3 \cdot \dfrac{m}{3} = 3(-3)$ Use the Multiplication Rule for Equations.

$m = -9$ ⟵ solution

(d) $-1 = -3 + \dfrac{n}{-5}$

$-3 + \dfrac{n}{-5} = -1$ Interchange the left and right members.

$\dfrac{n}{-5} + (-3) = -1$ Use the commutative property of addition.

$\dfrac{n}{-5} - 3 = -1$ ⟵ division-subtraction equation

$$\frac{n}{-5} - 3 + 3 = -1 + 3 \qquad \text{Use the Addition Rule for Equations.}$$

$$\frac{n}{-5} = 2$$

$$-5 \cdot \frac{n}{-5} = -5(2) \qquad \text{Use the Multiplication Rule for Equations.}$$

$$n = -10 \longleftarrow \text{solution}$$

7-3. Solving Linear Equations Containing Like Terms and/or Parentheses

A. Solving linear equations in one variable containing like terms in only one member

To **solve a linear equation in one variable that contains like terms in only one member,** use the following steps:

(1) Combine like terms.

(2) Rename the equation from Step 1 when necessary to get one of the eight special types of linear equations.

(3) Solve the equation using the rules for linear equations.

EXAMPLE 7-14   Solve $8 = 3x - 4 + x$.

Solution

$$\overset{\text{like terms}}{8 = 3x - 4 + x} \qquad \text{Identify like terms.}$$

$$8 = 4x - 4 \qquad \text{Combine like terms.}$$

$$4x - 4 = 8 \qquad \text{multiplication-subtraction equation}$$

$$4x = 12 \qquad \text{Use the Addition Rule: } 4x - 4 + 4 = 8 + 4$$

$$x = 3 \qquad \text{Use the Division Rule: } \frac{4x}{4} = \frac{12}{4}$$

Check: $8 = 3x - 4 + x \longleftarrow$ original equation

8	$3(3) - 4 + (3)$	Substitute the proposed solution 3 for x.
8	$9 - 4 + 3$	Compute.
8	$5 + 3$	
8	$8 \longleftarrow x = 3$ checks	

Note: The one and only solution of $8 = 3x - 4 + x$ is $x = 3$.

B. Solving linear equations in one variable containing like terms in both members

To **solve a linear equation in one variable that contains like terms in both members,** use the following steps:

(1) Combine like terms in each member when possible.

(2) Collect and combine like terms all in one member.

(3) Rename the equation to get one of the eight special types of linear equations in one variable when necessary.

(4) Solve the equation using the rules for linear equations.

EXAMPLE 7-15 Solve $-y + 8 - 2y = 3y - 6 + y$.

Solution

like terms

$$-y + 8 - 2y = 3y - 6 + y \qquad \text{Identify like terms}$$

$$-3y + 8 = 4y - 6 \qquad \text{Combine like terms in each member.}$$

$$-3y + 3y + 8 = 4y + 3y - 6 \qquad \text{Collect like terms all in one member.}$$

$$8 = 4y + 3y - 6 \longleftarrow \text{like terms are all in one member}$$

$$8 = 7y - 6 \qquad \text{Combine like terms.}$$

$$7y - 6 = 8 \longleftarrow \text{multiplication-subtraction equation}$$

$$7y = 14 \qquad \text{Use the Addition Rule: } 7y - 6 + 6 = 8 + 6$$

$$y = 2 \qquad \text{Use the Division Rule: } \frac{7y}{7} = \frac{14}{7}$$

Check: $\qquad -y + 8 - 2y = 3y - 6 + y \longleftarrow$ original equation

$-(2) + 8 - 2(2)$	$3(2) - 6 + (2)$	Substitute the proposed solution 2 for y.
$-2 + 8 - 4$	$6 - 6 + 2$	Compute.
$6 - 4$	$0 + 2$
2	$2 \longleftarrow y = 2$ checks

Note: The one and only solution of $-y + 8 - 2y = 3y - 6 + y$ is $y = 2$.

C. Solving linear equations in one variable containing parentheses

To **solve a linear equation in one variable containing parentheses,** use the following steps:

(1) Combine like terms inside the parentheses when possible.

(2) Clear parentheses using the distributive properties.

(3) Solve the equation using the rules for linear equations.

EXAMPLE 7-16 Solve $2(7 - 5u - 4) = u - 4(5u - 6 - 3u)$.

Solution

like terms $\qquad$ like terms

$$2(7 - 5u - 4) = u - 4(5u - 6 - 3u) \qquad \text{Identify like terms inside parentheses.}$$

$$2(3 - 5u) = u - 4(2u - 6) \qquad \text{Combine like terms inside parentheses.}$$

$$2(3 - 5u) = u + (-4)(2u - 6)$$

$$2(3) - 2(5u) = u + (-4)(2u) - (-4)(6) \qquad \text{Use the distributive property to clear parentheses.}$$

$$6 - 10u = u - 8u + 24 \longleftarrow \text{parentheses cleared}$$

$$6 - 10u = -7u + 24 \qquad \text{Solve as before [see Example 7-15].}$$

$$6 - 10u + 7u = -7u + 7u + 24$$

$$6 - 3u = 24$$

$$-3u + 6 = 24$$

$$-3u = 18$$

$$u = -6 \longleftarrow \text{proposed solution}$$

Check:

$$\overline{2(7 - 5u - 4) = u - 4(5u - 6 - 3u)} \longleftarrow \text{original equation}$$

$2(7 - 5(-6) - 4)$	$(-6) - 4(5(-6) - 6 - 3(-6))$	Substitute the proposed solution 6 for u.
$2(7 + 30 - 4)$	$-6 - 4(-30 - 6 + 18)$	Compute.
$2(37 - 4)$	$-6 - 4(-36 + 18)$	
$2(33)$	$-6 - 4(-18)$	
66	$-6 + 72$	
66	$66 \longleftarrow u = -6$ checks	

Note: The one and only solution of $2(7 - 5u - 4) = u - 4(5u - 6 - 3u)$ is $u = -6$.

D. Solving equations in one variable that simplify as true number sentences

If an equation in one variable simplifies as a true number sentence, then every real number is a solution of that equation.

EXAMPLE 7-17   Solve $2 - 5w = 4w + 2 - 9w$.

Solution

$$2 - 5w = 4w + 2 - 9w \qquad \text{Identify like terms.}$$
$$2 - 5w = -5w + 2 \qquad \text{Combine like terms.}$$
$$2 - 5w + 5w = -5w + 5w + 2 \qquad \text{Collect like terms.}$$
$$2 = 2 \qquad \textit{Stop!} \text{ The true number sentence } 2 = 2 \text{ means that}$$
every real number is a solution of
$2 - 5w = 4w + 2 - 9w.$

Note: The equation $2 - 5w = 4w + 2 - 9w$ is not a linear equation in one variable because it simplifies as $2 = 2$ or $0w + 2 = 2$ and therefore cannot be written in the form $Ax + B = C$ where $A \neq 0$.

E. Solving equations in one variable that simplify as false number sentences

If an equation in one variable simplifies as a false number sentence then there are no solutions of that equation.

EXAMPLE 7-18   Solve $1 - z = 2 + 3z - 3 - 4z$.

Solution

$$1 - z = 2 + 3z - 3 - 4z \qquad \text{Identify like terms.}$$

$$1 - z = -1 - z \qquad \text{Combine like terms.}$$
$$1 - z + z = -1 - z + z \qquad \text{Collect like terms.}$$
$$1 = -1 \qquad \textit{Stop!} \text{ The false number sentence } 1 = -1 \text{ means}$$
that there is no solution of
$1 - z = 2 + 3z - 3 - 4z.$

Note: The equation $1 - z = 2 + 3z - 3 - 4z$ is not a linear equation in one variable because it simplifies as $1 = -1$ or $0z + 1 = -1$ and therefore cannot be written in the form $Ax + B = C$ where $A \neq 0$.

7-4. Solving Linear Equations Containing Fractions, Decimals, or Percents

A. Solving linear equations in one variable containing fractions

To **solve a linear equation in one variable containing fractions,** you first clear the fractions and then use the rules for linear equations.

To **clear fractions,** you

(1) Find the least common denominator (LCD) of the fractions in the given equation.

(2) Multiply both members of the given equation by the LCD.

(3) Clear parentheses in the equation to clear all fractions.

EXAMPLE 7-19   Solve $\frac{2}{3}x - \frac{1}{2} = \frac{1}{4}$.

Solution   The LCD of $\frac{2}{3}, \frac{1}{2}$, and $\frac{1}{4}$ is 12 [see Section 3-6].

$$12\left(\frac{2}{3}x - \frac{1}{2}\right) = 12\left(\frac{1}{4}\right) \qquad \text{Multiply both members by the LCD 12.}$$

$$12\left(\frac{2}{3}x\right) - 12\left(\frac{1}{2}\right) = 12\left(\frac{1}{4}\right) \qquad \text{Clear parentheses to clear fractions.}$$

$$8x - 6 = 3 \longleftarrow \text{fractions cleared}$$

$$8x = 9 \qquad \text{Solve using the Addition Rule.}$$

$$x = \frac{9}{8} \quad \text{or} \quad 1\tfrac{1}{8} \quad \text{or} \quad 1.125 \longleftarrow \text{proposed solution}$$

Check: $\dfrac{2}{3}x - \dfrac{1}{2} = \dfrac{1}{4} \longleftarrow$ original equation

$$\dfrac{2}{3}\left(\dfrac{9}{8}\right) - \dfrac{1}{2} \;\Big|\; \dfrac{1}{4} \qquad \text{Substitute the proposed solution } \dfrac{9}{8} \text{ for } x.$$

$$\dfrac{3}{4} - \dfrac{1}{2} \;\Big|\; \dfrac{1}{4} \qquad \text{Compute.}$$

$$\dfrac{1}{4} \;\Big|\; \dfrac{1}{4} \longleftarrow x = \dfrac{9}{8} \text{ checks}$$

Note: The one and only solution of $\frac{2}{3}x - \frac{1}{2} = \frac{1}{4}$ is $x = \frac{9}{8}$ or $1\tfrac{1}{8}$ or 1.125.

B. Solving linear equations in one variable containing decimals

To **solve a linear equation in one variable containing decimals,** you can use the rules for linear equations.

EXAMPLE 7-20   Solve $0.5y + 0.2 = 0.3$.

Solution

$$\overset{\text{variable term}}{\underset{\downarrow}{}}$$

$$\overbrace{0.5y} + 0.2 = 0.3$$

$$0.5y + 0.2 - \mathbf{0.2} = 0.3 - \mathbf{0.2} \qquad \text{Use the Subtraction Rule for Equations to isolate the variable term } 0.5y \text{ in one member.}$$

0.5y is isolated $\longrightarrow$ $0.5y = 0.1$

$$\frac{0.5y}{\mathbf{0.5}} = \frac{0.1}{\mathbf{0.5}}$$ 　Use the Division Rule for Equations to isolate the variable y.

y is isolated $\longrightarrow$ $y = 0.2$ or $\dfrac{1}{5}$ $\longleftarrow$ proposed solution

Check: 　$0.5y + 0.2 = 0.3$ $\longleftarrow$ original equation

$0.5(\mathbf{0.2}) + 0.2$	0.3 　Substitute the proposed solution 0.2 for y.
$0.1 + 0.2$	0.3 　Compute.
0.3	0.3 $\longleftarrow$ $y = 0.2$ or $\frac{1}{5}$ checks

Note: The one and only solution of $0.5y + 0.2 = 0.3$ is $y = 0.2$ or $\frac{1}{5}$.

An equation containing decimals can also be solved by first **clearing decimals** and then using the rules for linear equations.

To clear decimals, you

(1) Find the LCD of the fraction form of the decimals from the given equations.

(2) Multiply both members of the given equation by the LCD.

(3) Clear parentheses in the equation to clear all decimals.

EXAMPLE 7-21 　Solve $0.5y + 0.2 = 0.3$ by first clearing decimals.

Solution 　0.5, 0.2, and 0.3 are the same as $\frac{5}{10}$, $\frac{2}{10}$, and $\frac{3}{10}$, respectively. And the LCD of $\frac{5}{10}$, $\frac{2}{10}$, and $\frac{3}{10}$ is 10 [see Section 3-6].

$$10(0.5y + 0.2) = \mathbf{10}(0.3)$$ 　Multiply both members by the LCD 10.

$$\mathbf{10}(0.5y) + \mathbf{10}(0.2) = 10(0.3)$$ 　Clear parentheses to clear decimals.

$$5y + 2 = 3 \longleftarrow \text{decimals cleared}$$

$$5y = 1$$ 　Solve using the rules for linear equations.

$$y = \frac{1}{5} \text{ or } 0.2 \longleftarrow \text{same solution as found in Example 7-20}$$

Note: To solve a linear equation in one variable containing decimals, you can either use the rules for linear equations directly without clearing decimals or you can first clear decimals and then use the rules for linear equations. The best method to use is the method that is easiest for you.

C. Solving linear equations containing percents

To **solve a linear equation in one variable containing percents,** you

(1) **Clear percents** by renaming each percent as either a fraction or a decimal.

(2) Clear fractions or decimals in the equation.

(3) Solve the equation using the rules for linear equations.

EXAMPLE 7-22 　Solve $x + 25\%x = 20$.

Solution

	Renaming the percent as a decimal	Renaming the percent as a fraction
	$x + 25\%x = 20$	$x + 25\%x = 20$
Clear percents:	$x + 0.25x = 20$	$x + \frac{1}{4}x = 20$
Clear decimals or fractions:	$100(x + 0.25x) = 100(20)$	$4(x + \frac{1}{4}x) = 4(20)$
	$100(x) + 100(0.25x) = 100(20)$	$4(x) + 4(\frac{1}{4}x) = 4(20)$
	$100x + 25x = 2000$	$4x + 1x = 80$
Solve as before:	$125x = 2000$	$5x = 80$ Solve as before.
	$x = \frac{2000}{125}$	$x = \frac{80}{5}$
	$x = 16$	$x = 16$

Check:

$$x + 25\%x = 20$$

$16 + 25\%(16)$	20
$16 + 0.25(16)$	20
$16 + 4$	20
	20	20

Check:

$$x + 25\%x = 20$$

$16 + 25\%(16)$	20
$16 + \frac{1}{4}(16)$	20
$16 + 4$	20
	20	20

Note: The one and only solution of $x + 25\%x = 20$ is $x = 16$.

7-5. Solving Literal Equations and Formulas for a Given Variable

A. Literal equations

An equation containing more than one different variable is called a **literal equation.** Every formula is a literal equation because every formula contains two or more different variables.

EXAMPLE 7-23   Which of the following are literal equations?

(a) $P = 2(l + w)$ **(b)** $2x + 3y = 5$ **(c)** $2x + 3x = 5$ **(d)** $2x + 3 = 5$ **(e)** $2 + 3 = 5$

Solution

(a) $P = 2(l + w)$ is a formula (perimeter of a rectangle) and therefore is a literal equation.
(b) $2x + 3y = 5$ is a literal equation because it contains more than one variable.
(c) $2x + 3x = 5$ is not a literal equation because it contains only one variable, x.
(d) $2x + 3 = 5$ is not a literal equation because it contains only one variable, x.
(e) $2 + 3 = 5$ is not a literal equation because it contains no variables.

B. Solving a literal equation for a given variable that is contained in only one term

To **solve a literal equation for a given variable that is contained in only one term,** you first isolate the term containing the given variable in one member of the literal equation and then isolate the given variable itself. When a literal equation is solved for a given variable, the algebraic expression in the other member is called the **solution of the literal equation with respect to the given variable.**

EXAMPLE 7-24   Which variable is the literal equation $y = mx + b$ solved for and what is the solution with respect to that variable?

Solution   The literal equation $y = mx + b$ is solved for y and the solution is $mx + b$.

EXAMPLE 7-25   Solve $2x + 3y = 5$ for x.

Solution

term containing x

$$\overset{\frown}{2x} + 3y = 5$$ Identify the term containing the given variable x.

$$2x + 3y - 3y = 5 - 3y$$ Use the Subtraction Rule for Equations to isolate the term containing the given variable x.

$2x$ is isolated $\longrightarrow$ $2x = 5 - 3y$

$$\frac{2x}{2} = \frac{5 - 3y}{2}$$ Use the Division Rule for Equations to isolate the given variable x.

x is isolated $\longrightarrow$ $x = \dfrac{5 - 3y}{2}$ or $\dfrac{5}{2} - \dfrac{3y}{2}$ or $\dfrac{5}{2} - \dfrac{3}{2}y$ or $-\dfrac{3}{2}y + \dfrac{5}{2}$ $\longleftarrow$ proposed solution

Check: $2x + 3y = 5$ $\longleftarrow$ original equation

$$2\left(\frac{5 - 3y}{2}\right) + 3y \;\Big|\; 5$$ Substitute the proposed solution $\dfrac{5 - 3y}{2}$ for x.

$$5 - 3y + 3y \;\Big|\; 5$$ Simplify.

$$5 + 0 \;\Big|\; 5$$

$$5 \;\Big|\; 5 \longleftarrow \frac{5 - 3y}{2} \text{ checks}$$

C. Solving a literal equation for a given variable that is contained in two or more like terms

To **solve a literal equation for a given variable that is contained in two or more like terms,** you

(1) Collect and combine any like terms.

(2) Isolate the term containing the given variable in one member of the literal equation.

(3) Isolate the variable itself.

EXAMPLE 7-26   Solve $-3m + 7n = 2n + 10$ for n.

Solution

like terms containing n

$$-3m + 7n = 2n + 10$$ Identify the like terms containing the given variable.

$$-3m + 7n - \mathbf{2n} = 2n - \mathbf{2n} + 10$$ Collect like terms.

$$-3m + 5n = 10$$ Combine like terms.

$$-3m + \mathbf{3m} + 5n = 10 + \mathbf{3m}$$ Use the Subtraction Rule for Equations to isolate the term containing the given variable n.

$5n$ is isolated $\longrightarrow$ $5n = 10 + 3m$

$$\frac{5n}{5} = \frac{10 + 3m}{5}$$ Use the Division Rule for Equations to isolate the given variable n.

n is isolated $\longrightarrow$ $n = \dfrac{10 + 3m}{5}$ or $\dfrac{3m + 10}{5}$ or $\dfrac{3m}{5} + 2$ or $\dfrac{3}{5}m + 2$ $\longleftarrow$ solution

(Check as before.)

D. Solving a literal equation for a given variable that is contained in parentheses

To **solve a literal equation for a given variable that is contained in parentheses,** you

(1) Combine any like terms inside the parentheses.

(2) Clear parentheses if the given variable is contained inside the parentheses.

(3) Collect and combine any like terms.

(4) Isolate the term containing the given variable in one member of the literal equation.

(5) Isolate the given variable itself.

EXAMPLE 7-27   Solve $P = 2(l + w)$ for w.

Solution

term containing w

$P = 2(l + w)$ Identify the given variable w as being contained inside the parentheses.

$P = \mathbf{2}(l) + \mathbf{2}(w)$ Use a distributive property to clear parentheses.

$P = 2l + 2w$ ⟵ parentheses cleared

$2w$ is in the left member ⟶ $2l + 2w = P$ Rename to get the term containing the given variable w in the left member.

$2l - 2l + 2w = P - 2l$ Use the Subtraction Rule for Equations to isolate the term containing the given variable w.

$2w$ is isolated ⟶ $2w = P - 2l$

$\dfrac{2w}{2} = \dfrac{P - 2l}{2}$ Use the Division Rule for Equations to isolate the given variable w.

w is isolated ⟶ $w = \dfrac{P - 2l}{2}$ or $\dfrac{P}{2} - l$ or $\dfrac{1}{2}P - l$ ⟵ solution (Check as before.)

Note: The formula $w = \frac{1}{2}P - l$ means that the width w of a rectangle always equals one-half the perimeter P minus the length l.

> *Agreement:* In the remainder of this chapter the words "linear equation" will mean *a linear equation in one variable.*

7-6. Translating Words to Symbols

To help **translate words to symbols,** you can substitute the appropriate symbols given in Table 7-1 for the corresponding **key words.**

TABLE 7-1

Key Words	Math Symbol
the sum of / plus / added to / joined with / increased by / more than / more / and / combined with	+
the difference between / minus / subtracted from / take away / decreased by / less than / less / reduced by / diminished by / exceeds	−

TABLE 7-1 (*continued*)

Key Words	Math Symbol
the product of / times / multiplied by / equal amounts of / of / times as much	·
the quotient of / divides / divided by / ratio / separated into equal amounts of / goes into / over	÷
is the same as / is equal to / equals / is / was / earns are / makes / gives / the result is / leaves / will be	=
twice / double / two times / twice as much as	2·
half / one-half of / one-half times / half as much as	$\frac{1}{2}$·
what number / what part / a number / the number / what amount / what percent / what price	n (any letter can be used)
twice a (the) number / double a (the) number	$2n$
half a (the) number / one-half a (the) number	$\frac{1}{2}n$

EXAMPLE 7-28   Translate each of the following to a linear equation using key words:

(a) Three times a number is 24.

(b) Four less than a number is the same as -1.

(c) Six subtracted from two-thirds of a number gives 2.

(d) Seven times the quantity, x plus 2, is equal to 35.

Solution

(a) *Identify:*   Three times a number is 24.   Identify key words [see Table 7-1].

   Translate: 3   · n $= 24$ Translate to symbols.

$$3n = 24 \longleftarrow \text{equation}$$

(b) *Identify:*   Four less than a number is the same as -1.

   Translate: $n - 4$ $=$ -1 *Think:* "4 less than n" means $n - 4$.

$$n - 4 = -1 \longleftarrow \text{equation}$$

(c) *Identify:*   Six subtracted from two-thirds of a number gives 2.

   Translate: $\frac{2}{3} \cdot n - 6$ $= 2$

$$\frac{2}{3}n - 6 = 2 \longleftarrow \text{equation}$$

(d) *Identify:*   Seven times the quantity, x plus 2, is equal to 35.

   Translate: 7   · $(x + 2)$ $=$ 35

$$7(x + 2) = 35 \longleftarrow \text{equation}$$

7-7. Solving Word Problems Using Linear Equations

To **solve a word problem using a linear equation,** you

(1) *Read* the problem very carefully several times.

(2) *Identify* the unknowns.

(3) *Decide* how to represent the unknowns using one variable.

(4) *Translate* the problem to a linear equation using key words.

(5) *Solve* the linear equation.

(6) *Interpret* the solution of the linear equation with respect to each represented unknown number to find the proposed solutions of the original problem.

(7) *Check* to see if the proposed solutions satisfy all the conditions of the original problem.

EXAMPLE 7-29   Solve the following number problem using a linear equation: Two numbers have a sum of 65. Four times the smaller number is equal to the larger number. Find the two numbers.

Solution

(1) *Read:*   Two numbers have a sum of 65. Four times the smaller number is equal to the larger number. Find the two numbers.

(2) *Identify:*   The unknown numbers are $\begin{cases} \text{the smaller number} \\ \text{the larger number} \end{cases}$.

(3) *Decide:*   Let n = the smaller number
then $65 - n$ = the larger number [see the following *Note*].

(4) *Translate:*   Four times the smaller number is equal to the larger number.   Identify key words.

$$4 \quad \cdot \quad n \quad = \quad 65 - n$$

Translate words to symbols.

$$4n = 65 - n \longleftarrow \text{linear equation}$$

(5) *Solve:* $4n + n = 65 - n + n$ [See Example 7-15].

$$5n = 65$$

$$n = 13$$

(6) *Interpret:* $n = 13$ means the smaller number is 13.

$65 - n = 65 - 13 = 52$ means the larger number is 52.

proposed solutions

(7) *Check:*   Did you find two numbers?   Yes: 13 and 52
Do the two numbers have a sum of 65?   Yes: $13 + 52 = 65$
Is four times the smaller number equal to the larger number?   Yes: $4 \cdot 13 = 52$

Therefore, the two numbers are 13 and 52.

Note: To represent two unknown numbers that have a sum of 65, you can
let n = the smaller number and $65 - n$ = the larger number
or
let n = the larger number and $65 - n$ = the smaller number
because in both cases, $n + (65 - n) = n + 65 - n = n - n + 65 = 0 + 65 = 65$.

EXAMPLE 7-30 Let n = the larger number and $65 - n$ = the smaller number for the problem in Example 7-29.

Solution

Translate: Four times the smaller number is equal to the larger number.

$$4 \cdot (65 - n) = n$$

$$4(65 - n) = n \longleftarrow \text{linear equation}$$

Solve:

$$4(65) - 4(n) = n$$

$$260 - 4n = n$$

$$260 - 4n + 4n = n + 4n$$

$$260 = 5n$$

larger number $\longrightarrow n = 52 \longleftarrow$

smaller number $\longrightarrow 65 - n = 65 - 52 = 13$

same solutions as found in Example 7-29

Among the word problems you may have to solve are problems involving *consecutive integers:*

- **Consecutive integers** are integers that differ by 1.

- **Even consecutive integers** are even integers that differ by 2.

- **Odd consecutive integers** are odd integers that differ by 2.

EXAMPLE 7-31 Write three consecutive integers starting with 5.

Solution

$$5 \longleftarrow \text{first given consecutive integer}$$

$$5 + 1 = 6 \longleftarrow \text{second consecutive integer}$$

$$6 + 1 = 7 \longleftarrow \text{third consecutive integer}$$

EXAMPLE 7-32 Write four consecutive even integers starting with 12.

Solution

$$12 \longleftarrow \text{first given consecutive even integer}$$

$$12 + 2 = 14 \longleftarrow \text{second consecutive even integer}$$

$$14 + 2 = 16 \longleftarrow \text{third consecutive even integer}$$

$$16 + 2 = 18 \longleftarrow \text{fourth consecutive even integer}$$

EXAMPLE 7-33 Write five consecutive odd integers starting with -5.

Solution

$$-5 \longleftarrow \text{first given consecutive odd integer}$$

$$-5 + 2 = -3 \longleftarrow \text{second consecutive odd integer}$$

$$-3 + 2 = -1 \longleftarrow \text{third consecutive odd integer}$$

$$-1 + 2 = 1 \longleftarrow \text{fourth consecutive odd integer}$$

$$1 + 2 = 3 \longleftarrow \text{fifth consecutive odd integer}$$

EXAMPLE 7-34 Solve the following consecutive integer problem using a linear equation: The sum of three consecutive even integers is 78. What are the integers?

Solution

(1) *Read:* The sum of three consecutive even integers is 78. What are the integers?

(2) *Identify:* The unknown numbers are $\begin{cases} \text{the first consecutive even integer} \\ \text{the second consecutive even integer} \\ \text{the third consecutive even integer} \end{cases}$.

(3) *Decide:*   Let n = the first consecutive even integer
then $n + 2$ = the second consecutive even integer
and $n + 4$ = the third consecutive even integer [see the following *Note 1*].

(4) *Translate:*   The sum of three consecutive even integers is 78.

$$n + (n + 2) + (n + 4) \qquad = 78$$

$$n + n + 2 + n + 4 = 78 \longleftarrow \text{linear equation}$$

(5) *Solve:*
$$3n + 6 = 78$$
$$3n = 72$$
$$n = 24$$

(6) *Interpret:* $n = 24$ means the first consecutive even integer is 24.
$n + 2 = 24 + 2 = 26$ means the second consecutive even integer is 26.
$n + 4 = 24 + 4 = 28$ means the third consecutive even integer is 28.

(7) *Check:*   Did you find three consecutive even integers?   Yes: 24, 26, and 28
Do the three consecutive even integers have a sum of 78?   Yes: $24 + 26 + 28 = 78$

Note 1: To represent three consecutive even [or odd] integers, you

let n = the first consecutive even [or odd] integer
then $n + 2$ = the second consecutive even [or odd] integer
and $n + 4$ = the third consecutive even [or odd] integer

because consecutive even [or odd] integers differ by two and $(n + 2) + 2 = n + 4$.

Note 2: To represent three consecutive integers, you

let n = the first consecutive integer
then $n + 1$ = the second consecutive integer
$n + 2$ = the third consecutive integer

because consecutive integers differ by one and $(n + 1) + 1 = n + 2$.

SOLVED PROBLEMS

PROBLEM 7-1   Which of the following are linear equations in one variable?

(a) $2x = 3$ **(b)** $y - 8 = 4$ **(c)** $\dfrac{z}{5} = -8$ **(d)** $\dfrac{w}{2} - 3 = 7$

(e) $m + 3 = 3 + m$ **(f)** $2 + n = -5$ **(g)** $3x + 2y = 6$ **(h)** $0.3x + 0.2 = 0.6$

(i) $3 + 2 = 5$ **(j)** $1 - 2r = 2 + 2r$ **(k)** $\dfrac{a}{2} + 5 = 4$ **(l)** $\dfrac{2}{a} + 5 = 4$

(m) $b\sqrt{5} - 2 = 3$ **(n)** $5\sqrt{b} + 2 = 3$ **(o)** $x^3 = 1$ **(p)** $2m + 3n + 6 = 8 + 3n$

Solution   Recall that a linear equation in one variable is any equation in one variable that can be written in standard form as $Ax + B = C$, where A, B, and C are real numbers ($A \neq 0$) and x is any variable [see Example 7-1]:

standard form
(a) $2x = 3$ or $\overbrace{2x + 0}= 3$ is a linear equation in one variable.

(b) $y - 8 = 4$ or $1y + (-8) = 4$ is a linear equation in one variable.

(c) $\dfrac{z}{5} = -8$ or $\dfrac{1}{5}z + 0 = -8$ is a linear equation in one variable.

(d) $\dfrac{w}{2} - 3 = 7$ or $\dfrac{1}{2}w + (-3) = 7$ is a linear equation in one variable.

(e) $m + 3 = 3 + m$ or $3 = 3$ is not a linear equation in one variable because $3 = 3$ cannot be written as $Ax + B = C$ where $A \neq 0$.

(f) $2 + n = -5$ or $1n + 2 = -5$ is a linear equation in one variable.

(g) $3x + 2y = 6$ is not a linear equation in one variable because there are two different variables.

(h) $0.3x + 0.2 = 0.6$ is a linear equation in one variable in standard form.

(i) $3 + 2 = 5$ is not a linear equation in one variable because there are no variables.

(j) $1 - 2r = 2 + 2r$ or $-4r + 1 = 2$ is a linear equation in one variable.

(k) $\dfrac{a}{2} + 5 = 4$ or $\dfrac{1}{2}a + 5 = 4$ is a linear equation in one variable.

(l) $\dfrac{2}{a} + 5 = 4$ is not a linear equation because there is a variable in the denominator.

(m) $b\sqrt{5} - 2 = 3$ or $(\sqrt{5})b + (-2) = 3$ is a linear equation in one variable.

(n) $5\sqrt{b} + 2 = 3$ is not a linear equation in one variable because there is a variable under the radical symbol.

(o) $x^3 = 1$ is not a linear equation in one variable because the variable has an exponent greater than one.

(p) $2m + 3n + 6 = 8 + 3n$ or $2m + 6 = 8$ is a linear equation in one variable.

PROBLEM 7-2 Determine if the given real number is a solution of the given linear equation in one variable in each of the following:

(a) $\dfrac{3}{2}$; $2x = 3$

(b) -3; $y + 5 = 2$

(c) 6; $\dfrac{w}{-2} = 3$

(d) 0; $z - 4 = -5$

(e) $-\dfrac{9}{2}$; $2m + 3 = -6$

(f) 3; $3n - 5 = 4$

(g) -4; $\dfrac{u}{4} + 3 = 2$

(h) -10; $\dfrac{v}{-2} - 5 = -1$

(i) 9; $2a + 5 = 3a - 4$

(j) 0; $2(b - 3) = -5 - (2 - b)$

(k) -2; $\dfrac{1}{2}x + \dfrac{3}{4} = -\dfrac{1}{4}$

(l) -2.4; $-0.5y - 0.2 = 1$

Solution Recall that to check a proposed solution of a given linear equation in one variable, you first substitute the proposed solution for the variable and then compute using the Order of Operations to get a number sentence. If you get a true number sentence, then the proposed solution checks and it is the one and only solution of the given equation. If you get a false number sentence, then the proposed solution does not check and it is not the solution of the given equation [see Example 7-4]:

(a) $2x = 3$

$2\left(\dfrac{3}{2}\right) \mid 3$

$\quad\quad 3 \mid 3$

true $(3 = 3)$

$\dfrac{3}{2}$ is the solution of $2x = 3$.

(b) $y + 5 = 2$

$-3 + 5 \mid 2$

$\quad\quad 2 \mid 2$

true $(2 = 2)$

-3 is the solution of $y + 5 = 2$.

(c) $\dfrac{w}{-2} = 3$

$\dfrac{6}{-2} \mid 3$

$\quad -3 \mid 3$

false $(-3 \neq 3)$

6 is not the solution of $\dfrac{w}{-2} = 3$.

(d) $z - 4 = -5$

$$\frac{0 - 4 \quad | \quad -5}{}$$
$$-4 \quad | \quad -5$$
false $(-4 \neq -5)$

0 is not the solution of
$z - 4 = -5$.

(e) $2m + 3 = -6$

$$2\left(-\frac{9}{2}\right) + 3 \quad | \quad -6$$
$$-9 + 3 \quad | \quad -6$$
$$-6 \quad | \quad -6$$
true $(-6 = -6)$

$-\dfrac{9}{2}$ is the solution of
$2m + 3 = -6$.

(f) $3n - 5 = 4$

$$3(3) - 5 \quad | \quad 4$$
$$9 - 5 \quad | \quad 4$$
$$4 \quad | \quad 4$$
true $(4 = 4)$

3 is the solution of
$3n - 5 = 4$.

(g) $\dfrac{u}{4} + 3 = 2$

$$\frac{-4}{4} + 3 \quad | \quad 2$$
$$-1 + 3 \quad | \quad 2$$
$$2 \quad | \quad 2$$
true $(2 = 2)$

-4 is the solution of
$\dfrac{u}{4} + 3 = 2$.

(h) $\dfrac{v}{-2} - 5 = -1$

$$\frac{-10}{-2} - 5 \quad | \quad -1$$
$$5 - 5 \quad | \quad -1$$
$$0 \quad | \quad -1$$
false $(0 \neq -1)$

-10 is not the solution of
$\dfrac{v}{-2} - 5 = -1$.

(i) $2a + 5 = 3a - 4$

$$2(9) + 5 \quad | \quad 3(9) - 4$$
$$18 + 5 \quad | \quad 27 - 4$$
$$23 \quad | \quad 23$$
true $(23 = 23)$

9 is the solution of
$2a + 5 = 3a - 4$.

(j) $2(b - 3) = -5 - (2 - b)$

$$\frac{2(0 - 3) \quad | \quad -5 - (2 - 0)}{}$$
$$2(-3) \quad | \quad -5 - 2$$
$$-6 \quad | \quad -7 \quad \longleftarrow \text{ false } (-6 \neq -7)$$

0 is not the solution
of $2(b - 3) = -5 - (2 - b)$.

(k) $\dfrac{1}{2}x + \dfrac{3}{4} = -\dfrac{1}{4}$

$$\frac{1}{2}(-2) + \frac{3}{4} \quad | \quad -\frac{1}{4}$$
$$-1 + \frac{3}{4} \quad | \quad -\frac{1}{4}$$
$$-\frac{1}{4} \quad | \quad -\frac{1}{4} \quad \longleftarrow \text{ true}\left(-\frac{1}{4} = -\frac{1}{4}\right)$$

-2 is the solution of $\dfrac{1}{2}x + \dfrac{3}{4} = -\dfrac{1}{4}$.

(l) $-0.5y - 0.2 = 1$

$$\frac{-0.5(-2.4) - 0.2 \quad | \quad 1}{}$$
$$1.2 - 0.2 \quad | \quad 1$$
$$1 \quad | \quad 1 \quad \longleftarrow \text{ true } (1 = 1)$$

-2.4 is the solution of $-0.5y - 0.2 = 1$.

PROBLEM 7-3 Solve the following equations using the rules for linear equations:

(a) $x + 2 = 5$

(b) $y - 2 = -7$

(c) $3c = 12$

(d) $\dfrac{d}{2} = -4$

(e) $4r + 2 = 14$

(f) $3s - 1 = -10$

(g) $\dfrac{t}{2} + 3 = 5$

(h) $\dfrac{u}{-5} - 3 = 1$

Solution Recall that to solve a linear equation in one variable that does not contain like terms or parentheses, you

(1) Rename when necessary as one of the eight special types of linear equation in one variable.

(2) Use the rules for linear equations to isolate the variable in one member of the linear equation.

(a) $x + 2 = 5$

$x + 2 - 2 = 5 - 2$

$x = 3$ [see Example 7-5.]

(b) $y - 2 = -7$

$y - 2 + 2 = -7 + 2$

$y = -5$ [see Example 7-6.]

(c) $3c = 12$

$$\frac{3c}{3} = \frac{12}{3}$$

$c = 4$ [see Example 7-7.]

(d) $\dfrac{d}{2} = -4$

$$2 \cdot \frac{d}{2} = 2(-4)$$

$d = -8$ [see Example 7-8.]

(e) $4r + 2 = 14$

$4r + 2 - 2 = 14 - 2$

$4r = 12$

$$\frac{4r}{4} = \frac{12}{4}$$

$r = 3$ [see Example 7-9.]

(f) $3s - 1 = -10$

$3s - 1 + 1 = -10 + 1$

$3s = -9$

$$\frac{3s}{3} = \frac{-9}{3}$$

$s = -3$ [see Example 7-10.]

(g) $\dfrac{t}{2} + 3 = 5$

$$\frac{t}{2} + 3 - 3 = 5 - 3$$

$$\frac{t}{2} = 2$$

$$2 \cdot \frac{t}{2} = 2 \cdot 2$$

$t = 4$ [see Example 7-11.]

(h) $\dfrac{u}{-5} - 3 = 1$

$$\frac{u}{-5} - 3 + 3 = 1 + 3$$

$$\frac{u}{-5} = 4$$

$$-5 \cdot \frac{u}{-5} = -5 \cdot 4$$

$u = -20$ [see Example 7-12.]

PROBLEM 7-4 Solve by first combining like terms:

(a) $x - 5 - 6x = -10$

(c) $2 - 5m = 4m + 2 - 9m$

(e) $5(3 - 2u) = -5$

(b) $-w + 8 - 2w = 3w - 6 + w$

(d) $1 - n = 2 + 3n - 3 - 4n$

(f) $3(v + 2) = 2(3 + v - 2v)$

Solution Recall that to solve a linear equation in one variable that contains like terms and/or parentheses, you:

(1) Combine any like terms inside the parentheses.

(2) Clear any parentheses using the distributive properties.

(3) Combine any like terms in each member.

(4) Collect and combine the like terms in one member.

(5) Rename the equation when necessary to get one of the eight special types of linear equations in one variable.

(6) Solve the equation in one variable using the rules for linear equations.

(a) $x - 5 - 6x = -10$

$1x - 6x - 5 = -10$

$-5x - 5 = -10$

$-5x = -5$

$x = 1$

[See Example 7-15.]

(b) $-w + 8 - 2w = 3w - 6 + w$

$-1w - 2w + 8 = 3w + 1w - 6$

$-3w + 8 = 4w - 6$

$-3w - 4w + 8 = 4w - 4w - 6$

$-7w = -14$

$w = 2$

[See Example 7-15.]

(c) $2 - 5m = 4m + 2 - 9m$

$2 - 5m = 4m - 9m + 2$

$2 - 5m = -5m + 2$

$2 - 5m + 5m = -5m + 5m + 2$

$2 = 2 \longleftarrow$ true number sentence

Every real number is a solution of
$2 - 5m = 4m + 2 - 9m$
[See Example 7-17.]

(d) $1 - n = 2 + 3n - 3 - 4n$

$1 - n = 2 - 3 + 3n - 4n$

$1 - n = -1 - n$

$1 - n + n = -1 - n + n$

$1 = -1 \longleftarrow$ false number sentence

There are no solutions of
$1 - n = 2 + 3n - 3 - 4n$
[See Example 7-17.]

(e) $5(3 - 2u) = -5$

$5(3) - 5(2u) = -5$

$15 - 10u = -5$

$-10u + 15 = -5$

$-10u = -20$

$u = 2$

[See Example 7-16.]

(f) $3(v + 2) = 2(3 + v - 2v)$

$3(v + 2) = 2(3 - v)$

$3 \cdot v + 3 \cdot 2 = 2 \cdot 3 - 2 \cdot v$

$3v + 6 = 6 - 2v$

$3v + 2v + 6 = 6 - 2v + 2v$

$5v + 6 = 6$

$5v = 0$

$v = 0$

[See Example 7-16.]

PROBLEM 7-5   Solve by first clearing fractions, decimals, or percents:

(a) $\frac{1}{2}x - \frac{1}{2} = \frac{3}{4}x + \frac{1}{2}$ **(b)** $\frac{y - 2}{6} = \frac{y + 3}{4} - 1$ **(c)** $\frac{1}{2}(2 - w) + 1 = \frac{2}{3}(3w - 1) + \frac{1}{6}$

(d) $0.03m - 1.2 = 0.06$ **(e)** $0.5(2 - n) = 0.3$ **(f)** $u - 25\%u = 1.5$

Solution   Recall that to solve linear equations containing fractions, decimals, or percents, you

(1) Clear all fractions, decimals, and percents.

(2) Combine like terms and/or clear parentheses.

(3) Rename as one of the eight special types of linear equations in one variable.

(4) Use the rules for linear equations. [See Examples 7-19 through 7-22.]

(a) $\frac{1}{2}x - \frac{1}{2} = \frac{3}{4}x + \frac{1}{2}$

The LCD for $\frac{1}{2}$ and $\frac{3}{4}$ is 4.

$$4(\tfrac{1}{2}x - \tfrac{1}{2}) = 4(\tfrac{3}{4}x + \tfrac{1}{2})$$

$$4(\tfrac{1}{2}x) - 4(\tfrac{1}{2}) = 4(\tfrac{3}{4}x) + 4(\tfrac{1}{2})$$

$$2x - 2 = 3x + 2$$

$$2x - 3x - 2 = 3x - 3x + 2$$

$$-1x - 2 = 2$$

$$-1x = 4$$

$$x = -4$$

(b) $\dfrac{y - 2}{6} = \dfrac{y + 3}{4} - 1$

The LCD for $\dfrac{y-2}{6}$ and $\dfrac{y+3}{4}$ is 12.

$$12 \cdot \frac{y - 2}{6} = 12\left(\frac{y + 3}{4} - 1\right)$$

$$12 \cdot \frac{y - 2}{6} = 12 \cdot \frac{y + 3}{4} - 12(1)$$

$$\tfrac{12}{6}(y - 2) = \tfrac{12}{4}(y + 3) - 12$$

$$2(y - 2) = 3(y + 3) - 12$$

$$2y - 4 = 3y + 9 - 12$$

$$2y - 4 = 3y - 3$$

$$2y - 3y - 4 = 3y - 3y - 3$$

$$-1y - 4 = -3$$

$$-1y = 1$$

$$y = -1$$

(c) $\frac{1}{2}(2 - w) + 1 = \frac{2}{3}(3w - 1) + \frac{1}{6}$

The LCD for $\frac{1}{2}$, $\frac{2}{3}$, and $\frac{1}{6}$ is 6.

$$6[\tfrac{1}{2}(2 - w) + 1] = 6[\tfrac{2}{3}(3w - 1) + \tfrac{1}{6}]$$

$$6 \cdot \tfrac{1}{2}(2 - w) + 6(1) = 6 \cdot \tfrac{2}{3}(3w - 1) + 6 \cdot \tfrac{1}{6}$$

$$3(2 - w) + 6 = 4(3w - 1) + 1$$

$$6 - 3w + 6 = 12w - 4 + 1$$

$$-3w + 12 = 12w - 3$$

$$-3w - 12w + 12 = 12w - 12w - 3$$

$$-15w + 12 = -3$$

$$-15w = -15$$

$$w = 1$$

(d) $0.03m - 1.2 = 0.06$

The LCD for $\frac{3}{100}\,(=0.03)$, $\frac{12}{10}\,(=1.2)$, and $\frac{6}{100}\,(=0.06)$ is 100.

$$100(0.03m - 1.2) = 100(0.06)$$

$$100(0.03m) - 100(1.2) = 100(0.06)$$

$$3m - 120 = 6$$

$$3m = 126$$

$$m = 42$$

(e) $0.5(2 - n) = 0.3$

The LCD for $\frac{5}{10}\,(=0.5)$ and $\frac{3}{10}\,(=0.3)$ is 10.

$$10[0.5(2 - n)] = 10(0.3)$$

$$10(0.5)(2 - n) = 10(0.3)$$

$$5(2 - n) = 3$$

$$10 - 5n = 3$$

$$-5n + 10 = 3$$

$$-5n = -7$$

$$n = \tfrac{7}{5} \text{ or } 1.4$$

(f) $u - 25\%u = 1.5$

$$1u - 0.25u = 1.5$$

$$0.75u = 1.5$$

$$u = 2$$

PROBLEM 7-6 Solve the following literal equations and formulas:

(a) Solve for c: $s = c + m$ [retail sales formula]

(b) Solve for x: $x - 4y = -2x + 5$ [literal equation]

(c) Solve for r: $d = rt$ [distance formula]

(d) Solve for l: $n = 3l - 25$ [shoe-size formula]

(e) Solve for a: $P = 2(a + b)$ [perimeter formula]

(f) Solve for b: $A = \frac{1}{2}bh$ [area formula]

Solution Recall that to solve a literal equation for a given variable, you

(1) Clear fractions, decimals, and percents.

(2) Combine like terms and/or clear parentheses.

(3) Isolate the term containing the given variable.

(4) Isolate the given variable itself.

(a)
$$s = c + m$$
$$c + m = s$$
$$c + m - m = s - m$$
$$c = s - m$$
[See Example 7-25.]

(b)
$$x - 4y = -2x + 5$$
$$x + 2x - 4y = -2x + 2x + 5$$
$$3x - 4y = 5$$
$$3x - 4y + 4y = 4y + 5$$
$$3x = 4y + 5$$
$$x = \frac{4y + 5}{3}$$
or $$x = \frac{4y}{3} + \frac{5}{3}$$
or $$x = \frac{4}{3}y + \frac{5}{3}$$
[See Example 7-26.]

(c) $d = rt$
$$rt = d$$
$$\frac{rt}{t} = \frac{d}{t}$$
$$r = \frac{d}{t}$$
[See Example 7-25.]

(d)
$$n = 3l - 25$$
$$3l - 25 = n$$
$$3l - 25 + 25 = n + 25$$
$$3l = n + 25$$
$$l = \frac{n + 25}{3}$$
or $$l = \frac{n}{3} + \frac{25}{3}$$
or $$l = \frac{1}{3}n + \frac{25}{3}$$
[See Example 7-25.]

(e)
$$P = 2(a + b)$$
$$2(a + b) = P$$
$$2a + 2b = P$$
$$2a + 2b - 2b = P - 2b$$
$$2a = P - 2b$$
$$a = \frac{P - 2b}{2}$$
or $$a = \frac{P}{2} - b$$
or $$a = \frac{1}{2}P - b$$
[See Example 7-27.]

(f)
$$A = \frac{1}{2}bh$$
$$\frac{1}{2}bh = A$$
$$2\left(\frac{1}{2}bh\right) = 2(A)$$
$$2 \cdot \frac{1}{2}bh = 2A$$
$$bh = 2A$$
$$\frac{bh}{h} = \frac{2A}{h}$$
$$b = \frac{2A}{h}$$
[See Example 7-25.]

PROBLEM 7-7 Translate the following words to symbols using key words:

(a) The sum of 5 and 2.

(b) The difference between 5 and 2.

(c) The product of 5 and 2.

(d) The quotient of 5 and 2.

(e) Six more than the sum of 5 and 2.   (f) Six less than the difference between 5 and 2.

(g) Six times the product of 5 and 2.   (h) Six divided by the quotient of 5 and 2.

(i) 5 less 2.   (j) 5 subtracted from 2.

(k) The product of 2 and another number, decreased by 5, equals 9.

(l) Five times the quantity, $x + 2$, is equal to 25.

(m) The product of 3 and another number, less twice the same number, is 20.

(n) One-half of a number, increased by 4, is the same as 8.

Solution   Recall that to help translate words to symbols, you can substitute the appropriate symbols given in Table 7-1 for the corresponding key words [see Example 7-28]:

(a) The sum of 5 and 2.

$$5 + 2 \text{ or } 2 + 5$$

(b) The difference between 5 and 2.

larger number $\longrightarrow$ $5 - 2$ $\longleftarrow$ smaller number

(c) The product of 5 and 2

$$5 \cdot 2 \text{ or } 2 \cdot 5$$

(d) The quotient of 5 and 2.

first number given $\longrightarrow$ $5 \div 2$ $\longleftarrow$ second number given

(e) Six more than the sum of 5 and 2

$$6 + (5 + 2)$$

(f) Six less than the difference between 5 and 2

$$(5 - 2) - 6$$

(g) Six times the product of 5 and 2

$$6 \cdot (5 \cdot 2)$$

(h) Six divided by the quotient of 5 and 2

$$6 \div (5 \div 2)$$

(i) 5 less 2

$$5 - 2$$

(j) 5 subtracted from 2

$$2 - 5$$

(k) The product of 2 and another number, decreased by 5, equals 9.

$$2 \cdot n \qquad - \quad 5 \quad = \quad 9, \quad \text{or} \quad 2n - 5 = 9$$

(l) Five times the quantity, $x + 2$, is equal to 25.

$$5 \quad \cdot \quad (x + 2) \quad = \quad 25, \text{ or } 5(x + 2) = 25$$

(m) The product of 3 and another number, less twice the same number, is 20.

$$3 \cdot n \qquad - \qquad 2 \cdot n \qquad = 20, \text{ or } 3n - 2n = 20$$

(n) One-half of a number, increased by 4, is the same as 8.

$$\frac{1}{2} \quad \cdot \quad n \quad + \quad 4 \quad = \quad 8, \quad \text{or} \quad \frac{1}{2}n + 4 = 8, \quad \text{or} \quad \frac{n}{2} + 4 = 8$$

PROBLEM 7-8   Write five of each of the following:

(a) consecutive even integers beginning with 26   (b) consecutive odd integers beginning with -11.

Solution   Recall that consecutive integers differ by one, and consecutive even [or odd] integers differ by two [see Examples 7-31, 7-32, and 7-33]:

(a)

$26 \longleftarrow$ given first consecutive even integer

$26 + 2 = 28 \longleftarrow$ second consecutive even integer

$28 + 2 = 30 \longleftarrow$ third consecutive even integer

$30 + 2 = 32 \longleftarrow$ fourth consecutive even integer

$32 + 2 = 34 \longleftarrow$ fifth consecutive even integer

(b)

$-11 \longleftarrow$ given first consecutive odd integer

$-11 + 2 = -9 \longleftarrow$ second consecutive odd integer

$-9 + 2 = -7 \longleftarrow$ third consecutive odd integer

$-7 + 2 = -5 \longleftarrow$ fourth consecutive odd integer

$-5 + 2 = -3 \longleftarrow$ fifth consecutive odd integer

PROBLEM 7-9   Solve each of the following word problems using a linear equation:

(a) The difference between two numbers is 16. The larger number is 25 more than one-half of the smaller number. What are the numbers?

(b) The sum of three consecutive odd integers is 135. What are the integers?

Solution   Recall that to solve a word problem using a linear equation, you

(1) *Read* the problem very carefully several times.

(2) *Identify* the unknowns.

(3) *Decide* how to represent the unknowns using one variable.

(4) *Translate* the problem to a linear equation using key words.

(5) *Solve* the linear equation.

(6) *Interpret* the solution of the linear equation with respect to each represented unknown number to find the proposed solutions of the original problem.

(7) *Check* to see if the proposed solutions satisfy all of the conditions of the original problem. [See Examples 7-29, 7-30, and 7-34.]

(a) *Identify:*   The unknown numbers are $\begin{cases} \text{the larger number} \\ \text{the smaller number} \end{cases}$.

	Case 1	or	**Case 2**

Decide:

Case 1
Let $\quad n =$ the larger number
then $n - 16 =$ the smaller number
because $n - (n - 16) = n - n + 16$
$\qquad\qquad\qquad\qquad\quad = 16$

Case 2
Let $\quad n =$ the smaller number
then $n + 16 =$ the larger number
because $(n + 16) - n = n + 16 - n$
$\qquad\qquad\qquad\qquad\quad = 16$

Translate: (**Case 1**)   The larger number is 25 more than one-half of the smaller number.

$$n \qquad = 25 \qquad + \qquad \frac{1}{2} \qquad \cdot \qquad (n - 16)$$

$$n = 25 + \frac{1}{2}(n - 16) \longleftarrow \text{linear equation}$$

Solve:

$$2 \cdot n = 2 \cdot 25 + 2 \cdot \frac{1}{2}(n - 16) \qquad \textit{Think: The LCD is 2.}$$

$$2n = 50 + 1(n - 16)$$

$$2n = 50 + n - 16$$

$$2n - n = 50 + n - n - 16$$

$$n = 50 - 16$$

Interpret: $n = 34$
                                                          $\Big\rangle$ solution [Check as before.]
                                   $n - 16 = 18$

Translate: **(Case 2)** The larger number is 25 more than one-half of the smaller number.

$$n + 16 \quad\quad = 25 \quad + \quad\quad \frac{1}{2} \quad\cdot\quad n$$

$$n + 16 = 25 + \frac{1}{2}n \longleftarrow \text{linear equation}$$

Solve: $2(n + 16) = 2 \cdot 25 + 2 \cdot \dfrac{1}{2}n$ *Think:* The LCD is 2.

$$2n + 32 = 50 + n$$

$$2n - n + 32 = 50 + n - n$$

$$n + 32 = 50$$

Interpret: $n = 18$
                                                       $\Big\rangle$ same solutions as found in Case 1
                               $n + 16 = 34$

(b) *Identify:*   The unknown numbers are $\begin{cases} \text{the first consecutive odd integer.} \\ \text{the second consecutive odd integer.} \\ \text{the third consecutive odd integer.} \end{cases}$

Decide: Let $n = $ the first consecutive odd integer
then $n + 2 = $ the second consecutive odd integer
and $n + 4 = $ the third consecutive odd integer because $(n + 2) + 2 = n + 4$.

Translate: The sum of three consecutive odd integers is 135.

$$n + (n + 2) + (n + 4) \quad\quad = 135$$

$$n + n + 2 + n + 4 = 135 \longleftarrow \text{linear equation}$$

Solve: $3n + 6 = 135$

$$3n = 129$$

Interpret: $n = 43$
                               $n + 2 = 45$ $\Big\rangle$ solutions [Check as before.]
                               $n + 4 = 47$

Supplementary Exercises

PROBLEM 7-10   Determine if the given real number is a solution of the given linear equation in one variable:

(a) $-8;\ x + 6 = -2$

(b) $9;\ y - 5 = 4$

(c) $-\dfrac{4}{3};\ -3w = 4$

(d) $-12;\ \dfrac{z}{2} = -5$

(e) $-1;\ -3u + 2 = 5$

(f) $\dfrac{3}{2};\ 2v - 4 = -1$

(g) $-2;\ \dfrac{a}{-2} + 1 = 3$

(h) $15;\ \dfrac{b}{5} - 2 = 1$

(i) $\dfrac{9}{4};\ 2 - 3m = m - 2 + 5m$

(j) $0; -2(5 - n) + 5 = 3(n - 2)$ **(k)** $\frac{1}{3}; \frac{3}{4}h - \frac{1}{2} = \frac{1}{4}$ **(l)** $1; 0.2k + 0.3 = 0.5$

PROBLEM 7-11   Solve the following linear equations:

(a) $x + 5 = -2$ **(b)** $y - 4 = 2$ **(c)** $8 = w + 5$

(d) $7 = z - 2$ **(e)** $3m = 24$ **(f)** $\frac{n}{4} = 5$

(g) $-12 = 6a$ **(h)** $-9 = \frac{b}{-2}$ **(i)** $3n + 7 = -5$

(j) $-6 = 3v + 3$ **(k)** $2h - 10 = 6$ **(l)** $5 = 2k - 5$

(m) $\frac{r}{5} + 6 = 2$ **(n)** $3 = \frac{s}{2} + 5$ **(o)** $\frac{c}{-5} - 3 = -5$

(p) $-5 = \frac{d}{2} - 3$ **(q)** $5p - 12 - 3p = -7$ **(r)** $-q + 4q + 6 = -3q - 4 + q$

(s) $-2(3w - 5) = -8$ **(t)** $3(1 + x) + 2 = 5(x + 1) - 2x$ **(u)** $5y - 2 = 8y + 3(1 - y)$

(v) $\frac{1}{2}z + \frac{5}{6} = -\frac{2}{3}$ **(w)** $\frac{2m + 3}{2} + 1 = \frac{5 - m}{5}$ **(x)** $\frac{3}{4}(n - 4) = 2 - \frac{1}{2}(2n + 3)$

PROBLEM 7-12   Solve the following literal equations and formulas:

(a) Solve for r: $s = r - d$ [retail sales formula]   **(b)** Solve for d: $s = r - d$

(c) Solve for x: $x - y = 5$ [literal equation]   **(d)** Solve for y: $x - y = 5$

(e) Solve for m: $4m - 5n = 2$ [literal equation]   **(f)** Solve for n: $4m - 5n = 2$

(g) Solve for x: $Ax + By = C$ [literal equation]   **(h)** Solve for y: $Ax + By = C$

(i) Solve for b: $y = mx + b$ [slope-intercept formula]   **(j)** Solve for m: $y = mx + b$

(k) Solve for P: $PB = A$ [percent formula]   **(l)** Solve for t: $d = rt$ [distance formula]

(m) Solve for l: $V = lwh$ [volume formula]   **(n)** Solve for r: $I = Prt$ [interest formula]

(o) Solve for h: $A = \frac{1}{2}bh$ [area formula]   **(p)** Solve for B: $V = \frac{1}{3}Bh$ [volume formula]

(q) Solve for P: $A = P(l + rt)$ [amount formula]   **(r)** Solve for r: $A = P(l + rt)$

(s) Solve for l: $n = 3l - 22$ [shoe-size formula]   **(t)** Solve for h: $w = 5\frac{1}{2}(h - 40)$ [weight formula]

(u) Solve for h: $A = \frac{1}{2}h(b_1 + b_2)$ [area formula]   **(v)** Solve for b_1: $A = \frac{1}{2}h(b_1 + b_2)$

(w) Solve for F: $C = \frac{5}{9}(F - 32)$ [temperature formula]   **(x)** Solve for t: $a = 0.16(15 - t)$ [altitude formula]

(y) Solve for v_0: $v = v_0 - 32t$ [velocity formula]   **(z)** Solve for t: $v = v_0 - 32t$

PROBLEM 7-13   Translate each word sentence to a linear equation and then solve the linear equation:

(a) Twelve more than three times a number is thirty.   **(b)** A number added to twenty equals eleven.

(c) Eight less than a number is thirty-six.   **(d)** One-fourth of a number is eighteen.

(e) A number increased by sixteen is the same as nineteen.

(f) One-third of a number increased by twice the same number is thirty-five.

(g) Five times a number diminished by ten leaves seven.

(h) If six is subtracted from four times a number, then the result is three.

(i) The product of a number and nine, less one, is forty-four.

(j) One-half a number added to two times the same number is eighty.

(k) Thirteen more than one-eighth of a number equals two and one-half.

(l) The quotient of a number divided by four is ten.

(m) Six times the sum of two and another number gives one hundred sixty-eight.

(n) A number is multiplied by five and then increased by twenty to get forty-two.

(o) Three-fourths of a number is decreased by six to get twenty-one.

(p) Two-thirds of a number, less thirteen, is equal to two.

(q) A number is 8 more than twice the difference of the same number and 2.

(r) Three-eighths of a number, less one-half the same number, gives negative five.

(s) A number increased by one-third of itself is the same as eight.

(t) One-half of a number equals one-fourth of the same number plus six.

(u) Nine is added to one-third of a number to leave nine.

(v) Three-fifths of a number joined with forty-seven will be eighty.

(w) Six subtracted from six times a number is seventy-two.

(x) Thirty subtracted from eight times a number is equal to five.

PROBLEM 7-14 Solve each problem using a linear equation:

(a) The difference between two numbers is 14. The larger number is equal to 3 times the smaller number, less 4. Find the numbers.

(b) The sum of three numbers is 38. The second number is 2 more than three times the first number. The third number equals one-half the difference between the first two numbers. What are the numbers?

(c) The sum of three consecutive odd integers is 171. Find the integers.

(d) The first consecutive integer is divided by 4 and the second consecutive integer is divided by 2. If the sum of these two quotients is 5, what are the integers?

Answers to Supplementary Exercises

(7-10) **(a)** yes **(b)** yes **(c)** yes **(d)** no **(e)** yes **(f)** yes
(g) no **(h)** yes **(i)** no **(j)** no **(k)** no **(l)** yes

(7-11) **(a)** -7 **(b)** 6 **(c)** 3 **(d)** 9
(e) 8 **(f)** 20 **(g)** -2 **(h)** 18
(i) -4 **(j)** -3 **(k)** 8 **(l)** 5
(m) -20 **(n)** -4 **(o)** 10 **(p)** -4
(q) $\frac{5}{2}$ or $2\frac{1}{2}$ or 2.5 **(r)** -2 **(s)** 3 **(t)** every real number
(u) no solutions **(v)** -3 **(w)** $-\frac{5}{4}$ or $-1\frac{1}{4}$ or -1.25 **(x)** 2

(7-12) **(a)** $r = s + d$ **(b)** $d = r - s$ **(c)** $x = y + 5$ **(d)** $y = x - 5$

(e) $m = \dfrac{5n + 2}{4}$ or $\dfrac{5n}{4} + \dfrac{1}{2}$ or $\dfrac{5}{4}n + \dfrac{1}{2}$

(f) $n = \dfrac{2 - 4m}{-5}$ or $-\dfrac{2}{5} + \dfrac{4m}{5}$ or $\dfrac{4}{5}m - \dfrac{2}{5}$

(g) $x = \dfrac{C - By}{A}$ or $\dfrac{C}{A} - \dfrac{By}{A}$ or $\dfrac{C}{A} - \dfrac{B}{A}y$ or $-\dfrac{B}{A}y + \dfrac{C}{A}$

(h) $y = \dfrac{C - Ax}{B}$ or $\dfrac{C}{B} - \dfrac{Ax}{B}$ or $\dfrac{C}{B} - \dfrac{A}{B}x$ or $-\dfrac{A}{B}x + \dfrac{C}{B}$

(i) $b = y - mx$ **(j)** $m = \dfrac{y - b}{x}$ or $\dfrac{y}{x} - \dfrac{b}{x}$ **(k)** $P = \dfrac{A}{B}$ **(l)** $t = \dfrac{d}{r}$

(m) $l = \dfrac{V}{wh}$ **(n)** $r = \dfrac{I}{Pt}$ **(o)** $h = \dfrac{2A}{b}$ **(p)** $B = \dfrac{3V}{h}$

(q) $P = \dfrac{A}{l + rt}$ **(r)** $r = \dfrac{A - Pl}{Pt}$ or $\dfrac{A}{Pt} - \dfrac{l}{t}$

(s) $l = \dfrac{n + 22}{3}$ or $\dfrac{n}{3} + \dfrac{22}{3}$ or $\dfrac{1}{3}n + \dfrac{22}{3}$

(t) $h = \dfrac{w + 220}{5\frac{1}{2}}$ or $\dfrac{w}{5\frac{1}{2}} + 40$ or $\dfrac{2}{11}w + 40$

(u) $h = \dfrac{2A}{b_1 + b_2}$ **(v)** $b_1 = \dfrac{2A - hb_2}{h}$ or $\dfrac{2A}{h} - b_2$

(w) $F = \dfrac{9C + 160}{5}$ or $\dfrac{9C}{5} + 32$ or $\dfrac{9}{5}C + 32$

(x) $t = \dfrac{2.4 - a}{0.16}$ or $15 - \dfrac{a}{0.16}$ or $15 - 6.25a$

(y) $v_0 = v + 32t$

(z) $t = \dfrac{v - v_0}{-32}$ or $\dfrac{v}{-32} + \dfrac{v_0}{32}$ or $-\dfrac{1}{32}v + \dfrac{1}{32}v_0$ or $\dfrac{1}{32}v_0 - \dfrac{1}{32}v$ or $\dfrac{v_0 - v}{32}$

(7-13) **(a)** $12 + 3n = 30$; $n = 6$ **(b)** $n + 20 = 11$; $n = -9$ **(c)** $n - 8 = 36$; $n = 44$
(d) $\frac{1}{4}n = 18$; $n = 72$ **(e)** $n + 16 = 19$; $n = 3$ **(f)** $\frac{1}{3}n + 2n = 35$; $n = 15$
(g) $5n - 10 = 7$; $n = \frac{17}{5}$ or $3\frac{2}{5}$ **(h)** $4n - 6 = 3$; $n = \frac{9}{4}$ or $2\frac{1}{4}$ **(i)** $9n - 1 = 44$; $n = 5$
(j) $\frac{1}{2}n + 2n = 80$; $n = 32$ **(k)** $13 + \frac{1}{8}n = 2\frac{1}{2}$; $n = -84$ **(l)** $\frac{n}{4} = 10$; $n = 40$
(m) $6(2 + n) = 168$; $n = 26$ **(n)** $5n + 20 = 42$; $n = \frac{22}{5}$ or $4\frac{2}{5}$ **(o)** $\frac{3}{4}n - 6 = 21$; $n = 36$
(p) $\frac{2}{3}n - 13 = 2$; $n = 22\frac{1}{2}$ **(q)** $n = 8 + 2(n - 2)$; $n = -4$ **(r)** $\frac{3}{8}n - \frac{1}{2}n = -5$; $n = 40$
(s) $n + \frac{1}{3}n = 8$; $n = 6$ **(t)** $\frac{1}{2}n = \frac{1}{4}n + 6$; $n = 24$ **(u)** $\frac{1}{3}n + 9 = 9$; $n = 0$
(v) $\frac{3}{5}n + 47 = 80$; $n = 55$ **(w)** $6n - 6 = 72$; $n = 13$ **(x)** $8n - 30 = 5$; $n = \frac{35}{8}$ or $4\frac{3}{8}$

(7-14) **(a)** 23, 9   **(b)** 7, 23, 8   **(c)** 55, 57, 59   **(d)** 6, 7

8 RATIO, RATE, AND PROPORTION

THIS CHAPTER IS ABOUT

☑ Finding the Ratio
☑ Finding the Rate
☑ Writing Proportions
☑ Solving Proportions
☑ Solving Problems Using Proportions

8-1. Finding the Ratio

A. Finding the ratio of like measures

Measures with the same unit of measure are called **like measures.** When two like measures are divided, the quotient will always be a number (with no unit of measure). When two like measures are divided and the number quotient is written in fraction form, the fraction is called a **ratio.** To find the **ratio of two like measures in lowest terms,** you use the first like measure as the numerator of the fraction and the second like measure as the denominator of the fraction. Then you divide the like measures and rename the fraction quotient in lowest terms.

EXAMPLE 8-1 Find the ratio of 20 feet to 8 feet in lowest terms.

Solution The ratio of 20 feet to 8 feet $= \dfrac{20 \text{ feet}}{8 \text{ feet}}$ ⟵ first given measure
⟵ second given measure

$= \dfrac{20 \cancel{\text{ feet}}}{8 \cancel{\text{ feet}}}$ Eliminate the common unit of measure.

$= \dfrac{5}{2}$ ⟵ Simplify $\dfrac{20}{8}$ [see Example 3-24].

Ratios are commonly written in three different forms: **fraction form, word form,** and **colon form.**

EXAMPLE 8-2 Write the following ratio in fraction form, word form, and colon form: 20 feet to 8 feet

Solution The ratio of 20 feet to 8 feet $= \frac{5}{2}$ ⟵ fraction form
$= 5 \text{ to } 2$ ⟵ word form
$= 5:2$ ⟵ colon form

Note: All three forms of the ratio ($\frac{5}{2}$, 5 to 2, and 5:2) are read as "five to two."

EXAMPLE 8-3 What is the meaning of the following sentence? The ratio of 20 feet to 8 feet is $\frac{5}{2}$, or 5 to 2, or 5:2.

Solution   This sentence means there are 5 feet in the first measure (20 feet) for each 2 feet in the second measure (8 feet).

Caution: The ratio of 20 feet to 8 feet is not the same as the ratio of 8 feet to 20 feet.

EXAMPLE 8-4   Find the ratio of 8 feet to 20 feet in lowest terms.

Solution   The ratio of 8 feet to 20 feet $= \dfrac{8 \text{ feet}}{20 \text{ feet}}$ ⟵ first given measure
⟵ second given measure

$$= \dfrac{8 \text{ feet}}{20 \text{ feet}} \quad \text{Eliminate the common unit of measure.}$$

$$= \dfrac{2}{5} \; \longleftarrow \text{ Simplify } \dfrac{8}{20}.$$

Note: The ratio of 20 feet to 8 feet is $\frac{5}{2}$, or 5 to 2, or 5:2.
The ratio of 8 feet to 20 feet is $\frac{2}{5}$, or 2 to 5, or 2:5.

B. Finding the ratio of unlike measures that belong to the same measurement family

Measures can be separated into **measurement families.**

EXAMPLE 8-5   Name the common measurement families.

Solution   The common measurement families are

(a) length measures   (b) capacity measures

(c) weight (mass) measures   (d) area measures

(e) volume measures   (f) time measures

(g) money measures

In the Unites States, measurements are often expressed in **U.S. Customary units.** Other industrial nations use **metric units.** [See Chapter 9 for more details.] To rename unlike U.S. Customary family measures, you can use the unit conversions in Appendix Table 4 on page 348. To rename unlike metric family measures, you can use the unit conversions in Table 4 as well. To rename U.S. Customary measures as metric measures (or vice versa), you can use the unit conversions in Appendix Table 5 on page 349.

To rename unlike time and money measures, you can use the following unit conversions.

Time			Money		
1 century	=	10 decades	1 dollar	=	100 cents
1 decade	=	10 years	1 half-dollar	=	50 cents
1 year	=	12 months	1 quarter	=	25 cents
1 normal year	=	365 days	1 dime	=	10 cents
1 leap year	=	366 days	1 nickel	=	5 cents
1 business year	=	360 days			
1 week	=	7 days			
1 day	=	24 hours			
1 hour	=	60 minutes			
1 minute	=	60 seconds			

To rename time or money measures, you multiply by the correct unit fraction.

EXAMPLE 8-6   Rename $2\frac{1}{2}$ hours as minutes.

Solution The necessary unit conversion is 1 hour = 60 minutes.

$$2\tfrac{1}{2} \text{ hours} = 2\tfrac{1}{2} \text{ hours} \times 1 \qquad \text{Multiply by 1.}$$

Substitute the correct unit conversion.

$$= \frac{2\tfrac{1}{2} \text{ hours}}{1} \times \frac{60 \text{ minutes}}{1 \text{ hour}} \quad \longleftarrow \text{ 1 hour} = 60 \text{ minutes means } 1 = \frac{60 \text{ minutes}}{1 \text{ hour}}$$

$$= \frac{2\tfrac{1}{2} \text{ \sout{hours}}}{1} \times \frac{60 \text{ minutes}}{1 \text{ \sout{hour}}} \qquad \text{Eliminate the common unit of measure.}$$

$$= (2\tfrac{1}{2} \times 60) \text{ minutes} \qquad \text{Multiply.}$$

$$= 150 \text{ minutes}$$

Note: $2\tfrac{1}{2}$ hours can be renamed as 150 minutes.

To find the **ratio of unlike measures that belong to the same measurement family,** you first rename the unlike measures as like measures and then find the ratio of the like measures.

EXAMPLE 8-7 Find the ratio of $2\tfrac{1}{2}$ hours to 30 minutes.

Solution The necessary unit conversion is 1 hour = 60 minutes.

$$2\tfrac{1}{2} \text{ hours} = 150 \text{ minutes} \qquad \text{Rename to get like measures.}$$

$$\frac{2\tfrac{1}{2} \text{ hours}}{30 \text{ minutes}} = \frac{150 \text{ minutes}}{30 \text{ minutes}} \quad \longleftarrow \begin{array}{l} \text{The ratio of } 2\tfrac{1}{2} \text{ hours to 30 minutes} \\ = \text{the ratio of 150 minutes to 30 minutes.} \end{array}$$

$$= \frac{150 \text{ \sout{minutes}}}{30 \text{ \sout{minutes}}} \qquad \text{Eliminate the common unit of measure.}$$

$$= \frac{5}{1} \qquad \text{Simplify } \frac{150}{30}.$$

Note: The ratio of $2\tfrac{1}{2}$ hours to 30 minutes is $\frac{5}{1}$, or 5 to 1, or 5:1. This means that there are 5 minutes in the first given measure ($2\tfrac{1}{2}$ hours) for each 1 minute in the second given measure (30 minutes).

8-2. Finding the Rate

A. Finding the rate of unlike measures that cannot be renamed as like measures

Sometimes unlike measures cannot be renamed as like measures because they come from different measurement families and there is no unit conversion that relates them.

EXAMPLE 8-8 Can 300 miles and 15 gallons be renamed as like measures?

Solution No. 300 miles and 15 gallons cannot be renamed as like measures because they come from different measurement families. 300 miles is a *length* measure and 15 gallons is a *capacity* measure, and there is no unit conversion that relates length measures and capacity measures.

The quotient of two unlike measures that cannot be renamed as like measures is called a **rate.** To find **the rate in lowest terms** of two unlike measures that cannot be renamed as like measures, you use the first measure for the numerator of a fraction and the second measure for the denominator. Then you divide the unlike measures and rename the numerical part of the fractional quotient in lowest terms. The unlike measures in a rate cannot be eliminated.

EXAMPLE 8-9   Find the rate of 300 miles to 8 gallons in lowest terms.

Solution   The rate of 300 miles to 8 gallons $= \dfrac{300 \text{ miles}}{8 \text{ gallons}}$ ←— first given measure
←— second given measure

$$= \frac{300}{8} \frac{\text{miles}}{\text{gallon}} \qquad \text{Divide unlike measures.}$$

$$= \frac{75}{2} \frac{\text{miles}}{\text{gallon}} \qquad \text{Simplify } \frac{300}{8}.$$

Rates are commonly written in three different forms: **fraction form, word form,** and **slash form.**

EXAMPLE 8-10   Write the following rate in fraction form, word form, and slash form: 300 miles to 8 gallons.

Solution   The rate of 300 miles to 8 gallons $= \dfrac{75}{2} \dfrac{\text{miles}}{\text{gallon}}$ ←———————— fraction form

$$= \frac{75}{2} \text{ miles per gallon} \quad \text{←—— word form}$$

$$= \frac{75}{2} \text{ miles/gallon} \quad \text{←——— slash form}$$

Note 1: All three forms of the rate $\left(\dfrac{75}{2} \dfrac{\text{miles}}{\text{gallon}}, \dfrac{75}{2} \text{ miles per gallon, and } \dfrac{75}{2} \text{ miles/gallon} \right)$ are read as "seventy-five-halves miles per gallon."

Note 2: In the rate $\dfrac{75}{2} \dfrac{\text{miles}}{\text{gallon}}$, $\dfrac{75}{2}$ is called the **amount** and $\dfrac{\text{miles}}{\text{gallon}}$ is called the **fractional unit of measure.** In a fractional unit of measure $\left(\dfrac{\text{miles}}{\text{gallon}} \right)$, the numerator is always plural (miles) and the denominator is always singular (gallon).

EXAMPLE 8-11   What is the meaning of the following sentence? The rate of 300 miles to 8 gallons is $\dfrac{75}{2} \dfrac{\text{miles}}{\text{gallon}}$, or $\dfrac{75}{2}$ miles per gallon, or $\dfrac{75}{2}$ miles/gallon.

Solution   This sentence means there are 75 miles in the first measure (300 miles) for each 2 gallons in the second measure (8 gallons).

Note: A car traveling 300 miles on 8 gallons of gasoline would average 75 miles for each 2 gallons of gasoline used, or $\dfrac{75}{2} \dfrac{\text{miles}}{\text{gallon}}$.

Caution: The rate of 300 miles to 8 gallons is not the same as the rate of 8 gallons to 300 miles.

EXAMPLE 8-12   Find the rate of 8 gallons to 300 miles in lowest terms.

Solution   The rate of 8 gallons to 300 miles $= \dfrac{8 \text{ gallons}}{300 \text{ miles}}$ ←—— first given measure
←—— second given measure

$$= \frac{8}{300} \frac{\text{gallons}}{\text{mile}} \quad \begin{array}{l} \text{←—— plural} \\ \text{←—— singular} \end{array}$$

$$= \frac{2}{75} \frac{\text{gallons}}{\text{mile}} \qquad \text{Simplify } \frac{8}{300}.$$

Note 1: The rate of 8 gallons to 300 miles means that there are 2 gallons in the first measure (8 gallons) for each 75 miles in the second measure (300 miles).

Note 2: A car using 8 gallons of gasoline to travel 300 miles would average 2 gallons of gasoline used for each 75 miles traveled, or $\dfrac{2}{75} \dfrac{\text{gallons}}{\text{mile}}$.

Note 3: The rate of 8 gallons to 300 miles is $\dfrac{2}{75} \dfrac{\text{gallons}}{\text{mile}}$, or $\dfrac{2}{75}$ gallons per mile, or $\dfrac{2}{75}$ gallons/mile.

The rate of 300 miles to 8 gallons is $\dfrac{75}{2} \dfrac{\text{miles}}{\text{gallon}}$, or $\dfrac{75}{2}$ miles per gallon, or $\dfrac{75}{2}$ miles/gallon.

B. Finding the unit rate

A rate that has an amount that is a whole number, mixed number, or decimal is called a **unit rate.** Every rate can be renamed as a unit rate by renaming the fractional amount as a whole number, mixed number, or decimal.

EXAMPLE 8-13   Write $\dfrac{75}{2} \dfrac{\text{miles}}{\text{gallon}}$ as a unit rate.

Solution $\dfrac{75}{2} = 75 \div 2$ Rename to get a whole number, mixed number, or decimal quotient.

$$= 37\tfrac{1}{2} \quad \text{or} \quad 37.5$$

$$\frac{75}{2} \frac{\text{miles}}{\text{gallon}} = 37\tfrac{1}{2} \frac{\text{miles}}{\text{gallon}} \quad \text{or} \quad 37.5 \frac{\text{miles}}{\text{gallon}} \longleftarrow \text{unit rate}$$

Note 1: The unit rate is $37\tfrac{1}{2} \dfrac{\text{miles}}{\text{gallon}}$ or $\dfrac{37\tfrac{1}{2}}{1} \dfrac{\text{miles}}{\text{gallon}}$, which means that there are $37\tfrac{1}{2}$ miles in the first measure (300 miles) for each 1 gallon in the second measure (8 gallons).

Note 2: A car traveling 300 miles on 8 gallons of gasoline would average $37\tfrac{1}{2}$ miles for each gallon of gasoline used, or $37\tfrac{1}{2}$ miles per gallon (mpg).

Note 3: Whenever you are asked to find the rate in a problem, your answer should always be the unit rate (unless otherwise stated).

8-3. Writing Proportions

A. Determining if two given ratios are proportional

If two given ratios are equal, they are said to be **proportional. A proportion** is formed by placing an equality symbol (=) between two proportional ratios.

To determine if two given ratios are proportional, you can compare the two fractions using cross multiplication. If the cross products of the fractions are equal, then the ratios are proportional. If the cross products of the fractions are not equal, then the ratios are not proportional.

EXAMPLE 8-14   Are the ratios $\dfrac{20}{8}$ and $\dfrac{5}{2}$ proportional?

Solution   To determine whether the fractions $\dfrac{20}{8}$ and $\dfrac{5}{2}$ are proportional, you can cross multiply [see Example 3-15].

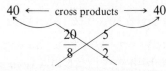

Since the cross products of $\frac{20}{8}$ and $\frac{5}{2}$ are equal $(40 = 40)$, $\frac{20}{8}$ and $\frac{5}{2}$ are proportional ratios.

B. Writing a proportion given two proportional ratios

To write a proportion given two ratios that are proportional (equal), you just place an equality symbol $(=)$ between them.

EXAMPLE 8-15 Write a proportion using the equal ratios $\frac{20}{8}$ and $\frac{5}{2}$.

Solution The proportion for the equal ratios is $\frac{20}{8} = \frac{5}{2}$.

C. Identifying like rates that are directly proportional

Two rates that have identical fractional units of measure are called **like rates.** Two rates that have different fractional units of measure are called **unlike rates.**

EXAMPLE 8-16 Which of the following pairs of rates are like rates?

(a) $\dfrac{8 \text{ people}}{2 \text{ pound}}$ and $\dfrac{12 \text{ people}}{3 \text{ pound}}$ (b) $\dfrac{72 \text{ beats}}{1 \text{ minute}}$ and $\dfrac{18 \text{ beats}}{15 \text{ second}}$

Solution

(a) $\dfrac{8 \text{ people}}{2 \text{ pound}}$ and $\dfrac{12 \text{ people}}{3 \text{ pound}}$ are like rates because they have identical fractional units of measure.

(b) $\dfrac{72 \text{ beats}}{1 \text{ minute}}$ and $\dfrac{18 \text{ beats}}{15 \text{ second}}$ are unlike rates because they have different fractional units of measure.

Two like rates are **directly proportional** if the amount of one like rate is proportional (equal) to the amount of the other like rate.

EXAMPLE 8-17 Show that the two like rates $\dfrac{8 \text{ people}}{2 \text{ pound}}$ and $\dfrac{12 \text{ people}}{3 \text{ pound}}$ are directly proportional.

Solution

(1) *Identify:* To feed (8 people it takes 2 pounds) Circle the two like rates.
of hamburger. At that rate, to
feed (12 people it will take 3 pounds)
of hamburger.

(2) *Understand:* The amounts of like rates are $\dfrac{8}{2}$ and $\dfrac{12}{3}$, respectively.

(3) *Decide:* The amounts $\dfrac{8}{2}$ and $\dfrac{12}{3}$ are proportional because, when you cross-multiply, $3 \times 8 = 24$ and $2 \times 12 = 24$.

(4) *Interpret:* The two like rates $\dfrac{8}{2}$ and $\dfrac{12}{3}$ are directly proportional; this means that $\dfrac{8 \text{ people}}{2 \text{ pound}}$ is proportional (equal) to $\dfrac{12 \text{ people}}{3 \text{ pound}}$.

D. Writing a direct proportion given two like rates that are directly proportional

When two like rates are directly proportional, you can always write a **direct proportion** by using the two like rates to form two ratios and then equating the ratios.

EXAMPLE 8-18   Write a direct proportion given two like rates that are directly proportional.

(1) *Identify:*   The normal human heart rate is   Circle the two like rates.
(72 beats in 60 seconds). At that
rate, the normal human heart
makes (18 beats in 15 seconds).

(2) *Understand:*   The two like rates $\dfrac{72}{60} \dfrac{\text{beats}}{\text{second}}$ and $\dfrac{18}{15} \dfrac{\text{beats}}{\text{second}}$ are directly proportional

because $\dfrac{72}{60}$ and $\dfrac{18}{15}$ are proportional, since their cross products are equal:

$(15 \times 72 = 1080 = 60 \times 18)$.

(3) *Decide:*   When two like rates are directly proportional, you can always write a
direct proportion as follows:

ratios

(4) *Form ratios:*
$$\frac{72 \text{ beats}}{18 \text{ beats}} \longleftarrow \begin{array}{l} 72 \text{ beats in } 60 \text{ seconds} \\ 18 \text{ beats in } 15 \text{ seconds} \end{array} \longrightarrow \frac{60 \text{ seconds}}{15 \text{ seconds}}$$

direct proportion

(5) *Write the proportion:*   beats ratio $\longrightarrow \dfrac{72}{18} = \dfrac{60}{15} \longleftarrow$ seconds ratio

A proportion can always be written in four different ways, using either both original
ratios, or using both reciprocals of the original ratios. Example 8-19 shows how to
form these proportions.

EXAMPLE 8-19   Write the direct proportion $\dfrac{72}{18} = \dfrac{60}{15}$ in four different ways.

Solution

direct proportions

(a) beats ratio $\longrightarrow \dfrac{72}{18} = \dfrac{60}{15} \longleftarrow$ seconds ratio

(b) reciprocal of beats ratio $\longrightarrow \dfrac{18}{72} = \dfrac{15}{60} \longleftarrow$ reciprocal of seconds ratio

(c) seconds ratio $\longrightarrow \dfrac{60}{15} = \dfrac{72}{18} \longleftarrow$ beats ratio

(d) reciprocal of seconds ratio $\longrightarrow \dfrac{15}{60} = \dfrac{18}{72} \longleftarrow$ reciprocal of beats ratio

Note: To write a direct proportion, you use either both original ratios or both reciprocals of the original ratios.

E. Identifying like rates that are indirectly proportional

Two like rates are **indirectly proportional** if the amount of one like rate is proportional
(equal) to the amount of the other like rate after interchanging both numerators or
both denominators.

EXAMPLE 8-20   Show that the two like rates $\dfrac{30}{300} \dfrac{\text{inches}}{\text{rpm}}$ and $\dfrac{15}{600} \dfrac{\text{inches}}{\text{rpm}}$ are indirectly proportional.

(1) *Identify:*   A driving pulley with a diameter of
(30 inches turns at 300 revolutions per minute (rpm)).
The driven pulley with a
diameter of (15 inches turns at 600 rpm).

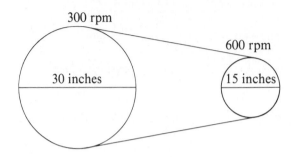

(2) *Understand:*   The amounts of the two like rates are $\dfrac{30}{300}\dfrac{\text{inches}}{\text{rpm}}$ and $\dfrac{15}{600}\dfrac{\text{inches}}{\text{rpm}}$. To make the like rates indirectly proportional, the numerators can be interchanged:

interchange numerators

$$\dfrac{15}{300} \qquad \dfrac{30}{600}$$

(3) *Decide:*   The amounts with the numerators interchanged are $\dfrac{15}{300}$ and $\dfrac{30}{600}$ and are proportional because $600 \times 15 = 300 \times 30 = 9000$.

(4) *Interpret:*   Since $\dfrac{15}{300}$ and $\dfrac{30}{600}$ are proportional, the two like rates $\dfrac{30}{300}\dfrac{\text{inches}}{\text{rpm}}$ and $\dfrac{15}{600}\dfrac{\text{inches}}{\text{rpm}}$ are indirectly proportional.

> *Caution:* Two like rates that are indirectly proportional are never directly proportional too. Likewise, two like rates that are directly proportional are never indirectly proportional too.

EXAMPLE 8-21   Show that the like rates from Example 8-20 are not directly proportional.

Solution   The like rates from Example 8-20, $\dfrac{30}{300}\dfrac{\text{inches}}{\text{rpm}}$ and $\dfrac{15}{600}\dfrac{\text{inches}}{\text{rpm}}$, are not directly proportional since the cross products of $\dfrac{30}{300}$ and $\dfrac{15}{600}$ are not equal ($600 \times 30 = 18,000$ and $300 \times 15 = 4500$).

F. Writing an indirect proportion given two like rates that are indirectly proportional

When two like rates are indirectly proportional, you can always write an **indirect proportion** by using the two like rates to form two ratios and then equating one of the ratios with the *reciprocal* of the other ratio. This is illustrated in Example 8-22.

EXAMPLE 8-22   Write an indirect proportion given two like rates that are indirectly proportional.

(1) *Identify:*   A driving gear turns at
(100 rpm with 48 teeth).
The driven gear turns at
(75 rpm with 64 teeth).

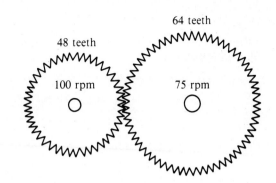

(2) *Understand:* The two like rates $\dfrac{100 \text{ rpm}}{48 \text{ teeth}}$ and $\dfrac{75 \text{ rpm}}{64 \text{ teeth}}$ are indirectly proportional

because $\dfrac{75}{48}$ and $\dfrac{100}{64}$ are proportional ($64 \times 75 = 48 \times 100 = 4800$).

(3) *Decide:* When two like rates are indirectly proportional, you can always write an indirect proportion as follows:

$$\overbrace{\hspace{6cm}}^{\text{ratios}}$$

$$\dfrac{100 \;\text{rpm}}{75 \;\text{rpm}} \longleftarrow \begin{array}{c} 100 \text{ rpm with 48 teeth} \\ 75 \text{ rpm with 64 teeth} \end{array} \longrightarrow \dfrac{48 \;\text{teeth}}{64 \;\text{teeth}}$$

(4) *Form ratios:*

indirect proportion

(5) *Write the proportion:* rpm ratio $\longrightarrow \overbrace{\dfrac{100}{75} = \dfrac{64}{48}}^{} \longleftarrow$ reciprocal of teeth ratio

$$\begin{array}{ccc} 4800 & & 4800 \\ \nwarrow & & \nearrow \\ \dfrac{100}{75} & \times & \dfrac{64}{48} \end{array}$$

(6) *Check:* $\dfrac{100}{75}$ and $\dfrac{64}{48}$ are proportional because $48 \times 100 = 4800 = 75 \times 64$.

Note: To write an indirect proportion, one of the ratios used to form the proportion must be a reciprocal ratio.

Recall: A proportion (direct or indirect) can always be written in four different ways, using various combinations of both original ratios and their reciprocals. Example 8-23 shows the four ways in which an indirect proportion can be written.

EXAMPLE 8-23 Write the indirect proportion $\dfrac{100}{75} = \dfrac{64}{48}$ in four different ways.

Solution

indirect proportions

rpm ratio $\longrightarrow \overbrace{\dfrac{100}{75} = \dfrac{64}{48}}^{} \longleftarrow$ reciprocal of teeth ratio

reciprocal of rpm ratio $\longrightarrow \dfrac{75}{100} = \dfrac{48}{64} \longleftarrow$ teeth ratio

teeth ratio $\longrightarrow \dfrac{48}{64} = \dfrac{75}{100} \longleftarrow$ reciprocal of rpm ratio

reciprocal of teeth ratio $\longrightarrow \dfrac{64}{48} = \dfrac{100}{75} \longleftarrow$ rpm ratio

Caution: An indirect proportion always has exactly one reciprocal ratio.

8-4. Solving Proportions

A. Identifying the terms of a proportion

Each of the two numerators and two denominators that form a proportion is called a **term of the proportion.** Every proportion has exactly four terms.

EXAMPLE 8-24   What are the terms of the proportion $\frac{20}{8} = \frac{5}{2}$?

Solution   The four terms of the proportion $\frac{20}{8} = \frac{5}{2}$ are 20, 5, 8, and 2.

B. Solving a proportion when one of the terms is unknown

The proportion $\frac{n}{8} = \frac{5}{2}$ has one **unknown term** (*n*) and three **known terms** (5, 8, and 2).

To **solve a proportion with one unknown term,** you must find a number that will replace the unknown term and thus form a proportion. This number is called the **solution of the proportion.**

To solve a proportion containing one unknown term, you can use the cross products to write a number sentence and then solve the number sentence.

EXAMPLE 8-25   Solve the proportion $\frac{n}{8} = \frac{5}{2}$.

Solution

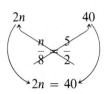

Find the cross products.

$2n = 40$

Think: In a proportion, the cross products are always equal.

$$n = \frac{40}{2}$$

Think: To solve $2n = 40$, divide both members by 2.

$n = 20$ ⟵⸺ proposed solution

Check: $\frac{n}{8} = \frac{5}{2}$ ⟵⸺ original proportion

$\frac{20}{8} \stackrel{?}{=} \frac{5}{2}$

Substitute the proposed solution $n = 20$.

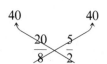

Cross multiply.

$n = 20$ is the correct solution for $\frac{n}{8} = \frac{5}{2}$ because $2 \times 20 = 8 \times 5 = 40$.

Note: The method shown in this example will work when the unknown term is in the numerator or in the denominator.

EXAMPLE 8-26   Solve the proportion $\frac{20}{n} = \frac{5}{2}$.

Solution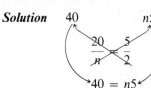

Find the cross products.

$$40 = n5$$

Equate the cross products.

$$n = \frac{40}{5}$$

Divide both members by 5.

$$n = 8 \longleftarrow \text{solution}$$

Check: $\dfrac{20}{8} = \dfrac{5}{2}$

Substitute the solution in the original proportion.

Note: If you cross multiply, you can see that the solution is correct $(2 \times 20 = 40 = 8 \times 5)$.

8-5. Solving Word Problems Using Proportions

A proportion can sometimes be used to solve a word problem. A word problem that can be solved using a proportion is called a **proportion problem.** To **solve a proportion problem,** you write and solve a proportion.

You can use a proportion to solve a word problem if you are given a rate or ratio and a fixed amount for one of the units in that rate or ratio; then you can find the proportional amount of the other unit of measure in the rate or ratio.

A. Solving a proportion problem when a direct proportion is required

When the two units of measure in a given rate (ratio) act in the same way, you solve the proportion problem using a **direct proportion.** Any two units of measure in a rate can be said to act in the same way if, as one unit of measure increases, the other unit of measure increases too. Or, if one unit of measure decreases, the other unit of measure decreases too.

Recall: You write a direct proportion by using two rates to form two ratios and then equating the ratios.

EXAMPLE 8-27 Solve this direct proportion problem:

(1) *Identify:* A certain car can travel 120 miles on 10 gallons of gas. At that rate, how far can the car travel on 25 gallons of gas?

Circle the given rate and the fixed amount of one of the measures in the rate.

(2) *Understand:* The two units of measure in the given rate (miles and gallons) act in the same way. That is, the number of miles increases as the number of gallons increases. (Or if the number of miles decreased, the number of gallons would decrease.)

(3) *Decide:* To solve a proportion problem when the two units of measure in the given rate act in the same way, you write and solve a direct proportion.

$$\overbrace{\qquad\qquad\qquad}^{\text{ratios}}$$

(4) *Form ratios:* $\dfrac{120 \text{ miles}}{n \text{ miles}} \longleftarrow \begin{array}{c} 120 \text{ miles on 10 gallons} \\ n \text{ miles}\quad \text{on 25 gallons} \end{array} \longrightarrow \dfrac{10 \text{ gallons}}{25 \text{ gallons}}$

direct proportion

(5) *Write the proportion:*   miles ratio ⟶ $\dfrac{120}{n} = \dfrac{10}{25}$ ⟵ gallons ratio

(6) *Solve as before:*

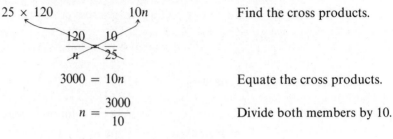

$25 \times 120 \qquad\qquad\qquad 10n$ Find the cross products.

$3000 = 10n$ Equate the cross products.

$n = \dfrac{3000}{10}$ Divide both members by 10.

$n = 300$ ⟵ proposed solution

(7) *Interpret:* $n = 300$ means that at the given rate 25 gallons of gas will allow the car to travel 300 miles.

(8) *Check:*   Are the two like rates, $\dfrac{120}{10} \dfrac{\text{miles}}{\text{gallons}}$ and $\dfrac{300}{25} \dfrac{\text{miles}}{\text{gallons}}$, directly proportional?

Yes: $\dfrac{120}{10}$ and $\dfrac{300}{25}$ are directly proportional because $25 \times 120 = 10 \times 300 = 3000$.

Recall: A proportion can always be written in four different ways.

EXAMPLE 8-28   Write the four direct proportions that can be used to solve the proportion problem just presented in Example 8-27.

Solution

direct proportions

miles ratio ⟶ $\dfrac{120}{n} = \dfrac{10}{25}$ ⟵ gallons ratio

reciprocal of miles ratio ⟶ $\dfrac{n}{120} = \dfrac{25}{10}$ ⟵ reciprocal of gallons ratio

reciprocal of gallons ratio ⟶ $\dfrac{25}{10} = \dfrac{n}{120}$ ⟵ reciprocal of miles ratio

gallons ratio ⟶ $\dfrac{10}{25} = \dfrac{120}{n}$ ⟵ miles ratio

Caution: To write a direct proportion, you either can use both the original ratios from the proportion problem or both of the reciprocals of the original ratios from the proportion problem. *Never* use one original ratio and one reciprocal ratio to write a direct proportion.

B. Solving a proportion problem when an indirect proportion is required

When the two units of measure in the given rate (ratio) act in opposite ways, you solve the proportion problem using an **indirect proportion.** Any two units of measure can be said to act in opposite ways if as one unit of measure increases, the other unit of measure decreases. Or if one unit of measure decreases, the other unit of measure increases.

Recall: You write an indirect proportion by using two like rates to form two ratios and then equating one of the ratios with the reciprocal of the other ratio.

EXAMPLE 8-29 Solve this indirect proportion problem:

(1) *Identify:*   A certain job can be completed by (5 women in 3 hours). At that rate, how long would it take (2 women) to do the same job?

Circle the given rate and the fixed amount of one of the measures involved in the rate.

(2) *Understand:*   The two units of measure in the given rate (women and hours) act in opposite ways. That is, the number of women decreases as the number of hours increases. (Or if the number of women increased the number of hours would decrease.)

(3) *Decide:*   To solve a proportion problem when the two units of measure in the given rate act in opposite ways, you write and solve an indirect proportion.

$$\overbrace{\qquad\qquad\qquad \text{ratios} \qquad\qquad\qquad}$$

(4) *Form ratios:* $\dfrac{5 \text{ women}}{2 \text{ women}}$ ⟵ 5 women take 3 hours ⟶ $\dfrac{3 \text{ hours}}{n \text{ hours}}$
⟵ 2 women take n hours ⟶

indirect proportion

(5) *Write the proportion:*   women ratio ⟶ $\dfrac{5}{2} = \dfrac{n}{3}$ ⟵ reciprocal of hours ratio

(6) *Solve as before:*

15   2n   Find the cross products.

$15 = 2n$ Equate the cross products.

$n = \dfrac{15}{2}$ Divide both members by 2.

$n = 7\frac{1}{2}$ ⟵ solution

(7) *Interpret:* $n = 7\frac{1}{2}$ means that to complete the given job with 2 women would take $7\frac{1}{2}$ hours.

Check: Are the two like rates $\dfrac{5 \text{ women}}{3 \text{ hours}}$ and $\dfrac{2 \text{ women}}{7\frac{1}{2} \text{ hours}}$ indirectly proportional?

Yes: After interchanging the numerators, you can see that $\dfrac{2}{3}$ and $\dfrac{5}{7\frac{1}{2}}$ are proportional because $7\frac{1}{2} \times 2 = 3 \times 5 = 15$.

Does it take 2 women longer to complete the job than it takes 5 women to complete the job? *Yes:* It takes 2 women $7\frac{1}{2}$ hours to complete the job, while it takes 5 women 3 hours to complete the job. The fact that it takes *less* women *more* time to do the job shows the rates to be indirectly proportional: as one unit of measure decreases, the other unit of measure increases. The indirect proportion can also be stated another way: it takes *more* women *less* time to do the job—so as one unit of measure increases, the other unit decreases.

Recall: A proportion can always be written in four different ways.

EXAMPLE 8-30   Write the four different indirect proportions that can be used to solve the proportion problem in Example 8-29.

Solution

indirect proportions

women ratio ⟶ $\dfrac{5}{2} = \dfrac{n}{3}$ ⟵ reciprocal of hour ratio reciprocal of women ratio $\longrightarrow \dfrac{2}{5} = \dfrac{3}{n} \longleftarrow$ hour ratio

hour ratio $\longrightarrow \dfrac{3}{n} = \dfrac{2}{5} \longleftarrow$ reciprocal of women ratio

reciprocal of hour ratio $\longrightarrow \dfrac{n}{3} = \dfrac{5}{2} \longleftarrow$ women ratio

Caution: An indirect proportion always uses exactly one reciprocal of one of the original ratios.

Caution: A direct proportion cannot be used to solve a proportion problem that requires an indirect proportion. Likewise, an indirect proportion cannot be used to solve a direct proportion problem.

EXAMPLE 8-31 Show that the direct proportion $\dfrac{5}{2} = \dfrac{3}{n}$ cannot be used to solve the proportion problem in Example 8-29.

Solution

(1) *Solve as before:* $5n \qquad\qquad 6$ Find the cross products.

$5n = 6$ Equate the cross products.

$n = \dfrac{6}{5}$ Divide both members by 5.

$n = 1\tfrac{1}{5} \longleftarrow$ proposed solution

(2) *Interpret:* $n = 1\tfrac{1}{5}$ would mean that it takes 2 women less time to complete the given job in Example 8-29 than it did for 5 women. This doesn't seem possible. It should take fewer women more time, not less time.

(3) *Conclusion:*   The direct proportion $\dfrac{5}{2} = \dfrac{3}{n}$ cannot be used to solve the indirect proportion problem in Example 8-29.

SOLVED PROBLEMS

PROBLEM 8-1   Find each ratio of each pair of like measures in lowest terms:

(a) 126 meters to 42 meters   (b) 32 wins to 48 wins   (c) $5\tfrac{1}{3}$ days to 24 days   (d) $2.25 to $15

Solution   Recall that to find the ratio of two like measures in lowest terms, you use the first like measure as the numerator of the fraction and the second like measure as the denominator. Then you divide the like measures, and rename the fraction quotient in lowest terms [see Example 8-1]:

(a) 126 meters to 42 meters $= \dfrac{126 \text{ meters}}{42 \text{ meters}}$ (b) 32 wins to 48 wins $= \dfrac{32 \text{ wins}}{48 \text{ wins}}$

$= \dfrac{126}{42} = \dfrac{3}{1}$ $= \dfrac{32}{48} = \dfrac{2}{3}$

(c) $5\frac{1}{3}$ days to 24 days $= \dfrac{5\frac{1}{3}\text{ days}}{24\text{ days}} = \dfrac{5\frac{1}{3}}{24}$

$= 5\frac{1}{3} \div 24 = \dfrac{16}{3} \div \dfrac{24}{1}$

$= \dfrac{16}{3} \times \dfrac{1}{24} = \dfrac{2}{9}$

(d) \$2.25 to \$15 $= \dfrac{\$2.25}{\$15}$

$= \dfrac{2.25}{15} = \dfrac{225}{1500} = \dfrac{3}{20}$

PROBLEM 8-2 Rename each time or money measure as indicated:

(a) 3 decades = ? years   **(b)** 1 normal year = ? hours   **(c)** 615 cents = ? nickels

Solution Recall that to rename unlike time or money measures, you can use the unit conversions on page 270 [see Example 8-6]:

(a) 3 decades $= \dfrac{3\text{ decades}}{1} \times \dfrac{10\text{ years}}{1\text{ decade}} = (3 \times 10)$ years $= 30$ years

(b) 1 normal year $= \dfrac{1\text{ normal year}}{1} \times \dfrac{365\text{ days}}{1\text{ normal year}} \times \dfrac{24\text{ hours}}{1\text{ day}} = (365 \times 24)$ hours $= 8760$ hours

(c) 615 cents $= \dfrac{615\text{ cents}}{1} \times \dfrac{1\text{ nickel}}{5\text{ cents}} = \dfrac{615}{5}$ nickels $= 123$ nickels

PROBLEM 8-3 Find each ratio of two unlike measures that belong to the same measurement family:

(a) 45 minutes to 6 hours   **(b)** 2 dimes to 3 quarters   **(c)** 42 nurses to 126 doctors

Solution Recall that to find the ratio of two unlike measures that belong to the same measurement family, you first rename the unlike measures as like measures and then find the ratio of the like measures [see Example 8-7]:

(a) 6 hours $= \dfrac{6\text{ hours}}{1} \times \dfrac{60\text{ minutes}}{1\text{ hour}} = (6 \times 60)$ minutes $= 360$ minutes

45 minutes to 6 hours $=$ 45 minutes to 360 minutes $= \dfrac{45\text{ minutes}}{360\text{ minutes}} = \dfrac{1}{8}$

(b) 2 dimes $= 20$ cents   3 quarters $= 75$ cents

2 dimes to 3 quarters $=$ 20 cents to 75 cents $= \dfrac{20\text{ cents}}{75\text{ cents}} = \dfrac{4}{15}$

(c) 42 nurses $= 42$ people   126 doctors $= 126$ people

42 nurses to 126 doctors $=$ 42 people to 126 people $= \dfrac{42\text{ people}}{126\text{ people}} = \dfrac{1}{3}$

PROBLEM 8-4 Find each rate in fraction form in lowest terms:

(a) 500 miles to 30 gallons   **(b)** 496 miles to 12 hours   **(c)** \$28 to 8 pounds

Solution Recall that to find the rate in lowest terms of two unlike measures that cannot be renamed as like measures, you use the first given measure for the numerator of a fraction and the second given measure for the denominator. Then you divide the unlike measures, and rename the numerical part of the fractional quotient in lowest terms. Remember that the numerator is always plural and the denominator is always singular in a rate [see Example 8-9]:

(a) 500 miles to 30 gallons $= \dfrac{500\text{ miles}}{30\text{ gallons}} = \dfrac{500}{30}\dfrac{\text{miles}}{\text{gallon}} = \dfrac{50}{3}\dfrac{\text{miles}}{\text{gallon}}$

(b) 496 miles to 12 hours $= \dfrac{496 \text{ miles}}{12 \text{ hours}} = \dfrac{496}{12} \dfrac{\text{miles}}{\text{hour}} = \dfrac{124}{3} \dfrac{\text{miles}}{\text{hour}}$

(c) 28 dollars to 8 pounds $= \dfrac{28 \text{ dollars}}{8 \text{ pounds}} = \dfrac{28}{8} \dfrac{\text{dollars}}{\text{pound}} = \dfrac{7}{2} \dfrac{\text{dollars}}{\text{pound}}$

PROBLEM 8-5   Find the unit rate for each of the following fractional rates:

(a) $\dfrac{50}{3} \dfrac{\text{miles}}{\text{gallon}}$ 　　 **(b)** $\dfrac{124}{3} \dfrac{\text{miles}}{\text{hour}}$ 　　 **(c)** $\dfrac{7}{2} \dfrac{\text{dollars}}{\text{pound}}$ 　　 **(d)** $\dfrac{27}{2} \dfrac{\text{people}}{\text{room}}$ 　　 **(e)** $\dfrac{1}{4} \dfrac{\text{gram}}{\text{cent}}$

Solution   Recall that every rate can be renamed as a unit rate by renaming the fractional amount as a whole number, mixed number, or decimal [see Example 8-13]:

(a) $\dfrac{50}{3} \dfrac{\text{miles}}{\text{gallon}} = 16\frac{2}{3} \dfrac{\text{miles}}{\text{gallon}}$ or $16\frac{2}{3}$ miles/gallon   or $16\frac{2}{3}$ miles per gallon   or $16\frac{2}{3}$ mpg

(b) $\dfrac{124}{3} \dfrac{\text{miles}}{\text{hour}} = 41\frac{1}{3} \dfrac{\text{miles}}{\text{hour}}$ or $41\frac{1}{3}$ miles/hour   or $41\frac{1}{3}$ miles per hour   or $41\frac{1}{3}$ mph

(c) $\dfrac{7}{2} \dfrac{\text{dollars}}{\text{pound}} = 3\frac{1}{2} \dfrac{\text{dollars}}{\text{pound}}$ or 3.5 dollars/pound   or   \$3.50 per pound   or   \$3.50/pound

(d) $\dfrac{27}{2} \dfrac{\text{people}}{\text{room}} = 13\frac{1}{2} \dfrac{\text{people}}{\text{room}}$ or 13.5 people/room   or 13.5 people per room

(e) $\dfrac{1}{4} \dfrac{\text{gram}}{\text{cent}} = 0.25 \dfrac{\text{gram}}{\text{cent}}$ or 0.25 gram/cent   or 0.25 gram per cent   or 0.25 gram/¢

PROBLEM 8-6   Write a proportion for each pair of ratios that are proportional:

(a) $\dfrac{10}{32}$ and $\dfrac{21}{63}$ 　　 **(b)** $\dfrac{150}{60}$ and $\dfrac{45}{18}$ 　　 **(c)** $\dfrac{12}{18}$ and $\dfrac{28}{42}$ 　　 **(d)** $\dfrac{11}{16}$ and $\dfrac{21}{32}$ 　　 **(e)** $\dfrac{2\frac{1}{2}}{3}$ and $\dfrac{\frac{5}{8}}{\frac{3}{4}}$

Solution   Recall that to determine if two given ratios are proportional, you can compare the two fractions using cross multiplication. If the cross products of the fractions are equal, then the ratios are proportional. If the cross products of the fractions are not equal, then the fractions are not proportional. Also recall that to write a proportion given two ratios that are proportional (equal), you just write an equality symbol ($=$) between them [see Examples 8-14 and 8-15]:

(a) $\dfrac{10}{32}$ and $\dfrac{21}{63}$ are not proportional because $63 \times 10 = 630$ and $32 \times 21 = 672$.

(b) $\dfrac{150}{60}$ and $\dfrac{45}{18}$ are proportional because $18 \times 150 = 60 \times 45 = 2700$: $\dfrac{150}{60} = \dfrac{45}{18}$.

(c) $\dfrac{12}{18}$ and $\dfrac{28}{42}$ are proportional because $42 \times 12 = 18 \times 28 = 504$: $\dfrac{12}{18} = \dfrac{28}{42}$.

(d) $\dfrac{11}{16}$ and $\dfrac{21}{32}$ are not proportional because $32 \times 11 = 352$ and $16 \times 21 = 336$.

(e) $\dfrac{2\frac{1}{2}}{3}$ and $\dfrac{\frac{5}{8}}{\frac{3}{4}}$ are proportional because $\frac{3}{4} \times 2\frac{1}{2} = 3 \times \frac{5}{8} = 1\frac{7}{8}$: $\dfrac{2\frac{1}{2}}{3} = \dfrac{\frac{5}{8}}{\frac{3}{4}}$.

PROBLEM 8-7   Write a direct proportion for each pair of like rates that are directly proportional:

(a) $\dfrac{10}{6} \dfrac{\text{miles}}{\text{minute}}$ and $\dfrac{5}{3} \dfrac{\text{miles}}{\text{minute}}$ 　　 **(b)** $\dfrac{35}{45} \dfrac{\text{feet}}{\text{dollar}}$ and $\dfrac{4}{5} \dfrac{\text{feet}}{\text{dollar}}$ 　　 **(c)** $\dfrac{7}{8} \dfrac{\text{days}}{\text{pound}}$ and $\dfrac{42}{48} \dfrac{\text{days}}{\text{pound}}$

Solution   Recall that two like rates are directly proportional if the amount of one like rate is proportional (equal) to the amount of the other like rate. Also recall that when two like rates are directly proportional, you can always write a direct proportion by using the two like rates to form two ratios and then equating the ratios [see Examples 8-17 and 8-18]:

(a) $\dfrac{10}{6}\dfrac{\text{miles}}{\text{minute}}$ and $\dfrac{5}{3}\dfrac{\text{miles}}{\text{minute}}$ are directly proportional because $\dfrac{10}{6} = \dfrac{5}{3}$:

$$\overbrace{\begin{matrix}\dfrac{10\ \cancel{\text{miles}}}{5\ \cancel{\text{miles}}} \longleftarrow \dfrac{10\ \text{miles}/6\ \text{minutes}}{5\ \text{miles}/3\ \text{minutes}} \longrightarrow \dfrac{6\ \cancel{\text{minutes}}}{3\ \cancel{\text{minutes}}}\end{matrix}}^{\text{ratios}} \quad \text{means} \quad \overbrace{\dfrac{10}{5} = \dfrac{6}{3}}^{\text{direct proportion}}$$

(b) $\dfrac{35}{45}\dfrac{\text{feet}}{\text{dollar}}$ and $\dfrac{4}{5}\dfrac{\text{feet}}{\text{dollar}}$ are not directly proportional because $\dfrac{35}{45} \neq \dfrac{4}{5}$:

$$5 \times 35 = 175 \text{ and } 45 \times 4 = 180$$

(c) $\dfrac{7}{8}\dfrac{\text{days}}{\text{pound}}$ and $\dfrac{42}{48}\dfrac{\text{days}}{\text{pound}}$ are directly proportional because $\dfrac{7}{8} = \dfrac{42}{48}$:

$$\dfrac{7\ \cancel{\text{days}}}{42\ \cancel{\text{days}}} \longleftarrow \dfrac{7\ \text{days}/8\ \text{pounds}}{42\ \text{days}/48\ \text{pounds}} \longrightarrow \dfrac{8\ \cancel{\text{pounds}}}{48\ \cancel{\text{pounds}}} \quad \text{means} \quad \dfrac{7}{42} = \dfrac{8}{48}$$

PROBLEM 8-8 Write an indirect proportion for each pair of like rates that are indirectly proportional:

(a) $\dfrac{20}{15}\dfrac{\text{men}}{\text{day}}$ and $\dfrac{12}{25}\dfrac{\text{men}}{\text{day}}$ **(b)** $\dfrac{12}{210}\dfrac{\text{inches}}{\text{rpm}}$ and $\dfrac{14}{180}\dfrac{\text{inches}}{\text{rpm}}$ **(c)** $\dfrac{20}{480}\dfrac{\text{teeth}}{\text{rpm}}$ and $\dfrac{55}{180}\dfrac{\text{teeth}}{\text{rpm}}$

Solution Recall that two like rates are indirectly proportional if the amount of one like rate is proportional (equal) to the amount of the other like rate after interchanging both numerators or both denominators. Also recall that when two like rates are indirectly proportional, you can always write an indirect proportion by using the two like rates to form two ratios and then equating one of the ratios with the reciprocal of the other ratio [see Examples 8-20 and 8-21]:

(a) $\dfrac{20}{15}\dfrac{\text{men}}{\text{day}}$ and $\dfrac{12}{25}\dfrac{\text{men}}{\text{day}}$ are directly proportional because $\overbrace{\dfrac{12}{15} = \dfrac{20}{25}}^{\text{interchange numerators}}$:

$$\overbrace{\begin{matrix}\dfrac{20\ \cancel{\text{men}}}{12\ \cancel{\text{men}}} \longleftarrow \dfrac{20\ \text{men}/15\ \text{days}}{12\ \text{men}/25\ \text{days}} \longrightarrow \dfrac{15\ \cancel{\text{days}}}{25\ \cancel{\text{days}}}\end{matrix}}^{\text{ratio}} \quad \text{means} \quad \overbrace{\dfrac{20}{12} = \dfrac{25}{15}}^{\text{indirect proportion}} \longleftarrow \text{one reciprocal ratio}$$

(b) $\dfrac{12}{210}\dfrac{\text{inches}}{\text{rpm}}$ and $\dfrac{14}{180}\dfrac{\text{inches}}{\text{rpm}}$ are indirectly proportional because $\dfrac{14}{210} = \dfrac{12}{180}$:

$$\dfrac{12\ \cancel{\text{inches}}}{14\ \cancel{\text{inches}}} \longleftarrow \dfrac{12\ \text{inches}/210\ \text{rpm}}{14\ \text{inches}/180\ \text{rpm}} \longrightarrow \dfrac{210\ \cancel{\text{rpm}}}{180\ \cancel{\text{rpm}}} \quad \text{means} \quad \dfrac{12}{14} = \dfrac{180}{210}$$

(c) $\dfrac{20}{480}\dfrac{\text{teeth}}{\text{rpm}}$ and $\dfrac{55}{180}\dfrac{\text{teeth}}{\text{rpm}}$ are not indirectly proportional because $\dfrac{55}{480} \neq \dfrac{20}{180}$:

$$180 \times 55 = 9900 \text{ and } 480 \times 20 = 9600$$

PROBLEM 8-9 Solve each proportion containing one unknown term:

(a) $\dfrac{n}{51} = \dfrac{12}{18}$ **(b)** $\dfrac{3\frac{1}{2}}{a} = \dfrac{36}{24}$ **(c)** $\dfrac{5}{2.5} = \dfrac{7.2}{b}$ **(d)** $\dfrac{2}{7} = \dfrac{c}{35}$

Solution Recall that to solve a proportion containing one unknown term, you can use the cross products of the fractions to write a number sentence and then solve the number sentence [see

Examples 8-25 and 8-26]:

(a) $\dfrac{n}{51} = \dfrac{12}{18}$ means $18 \times n = 51 \times 12$

$$18 \times n = 612$$
$$n = 612 \div 18$$
$$n = 34$$

Check: $18 \times 34 = 51 \times 12 = 612$

(b) $\dfrac{3\frac{1}{2}}{a} = \dfrac{36}{24}$ means $24 \times 3\frac{1}{2} = a \times 36$

$$84 = a \times 36$$
$$a = 84 \div 36$$
$$a = 2\frac{1}{3}$$

Check: $24 \times 3\frac{1}{2} = 2\frac{1}{3} \times 36 = 84$

(c) $\dfrac{5}{2.5} = \dfrac{7.2}{b}$ means $b \times 5 = 2.5 \times 7.2$

$$b \times 5 = 18$$
$$b = 18 \div 5$$
$$b = 3\frac{3}{5} \quad \text{or} \quad 3.6$$

Check: $3.6 \times 5 = 2.5 \times 7.2 = 18$

(d) $\dfrac{2}{7} = \dfrac{c}{35}$ means $35 \times 2 = 7 \times c$

$$70 = 7 \times c$$
$$c = 70 \div 7$$
$$c = 10$$

Check: $35 \times 2 = 7 \times 10 = 70$

PROBLEM 8-10 Solve each proportion problem using a direct proportion:

(a) A photo is 8 inches long and 5 inches wide. If a reduced copy is made so that the width is 3 inches, then what will be the length of the copy?

(b) A women earns \$560 in 15 days. At that rate how much will she earn in 7 days?

(c) An 8-ounce jar of instant coffee contains 12 servings. At that rate, how many servings will a 22-ounce jar of the same coffee contain?

(d) Concrete is made of cement, sand, and rock. The ratio of sand to rock is 2:4 in concrete. How much sand is needed for 125 pounds of rock?

Solution Recall that to solve a proportion problem you write either a direct or an indirect proportion and then solve the proportion. Also recall that you solve a proportion problem using a direct proportion when the two units of measure in the rate (ratio) act in the same way. That is, the number of one amount increases as the number of the other amount increases, or the number of one amount decreases as the number of the other amount decreases [see Example 8-27]:

(a) *Understand:* The problem asks you to compare two ratios. The two units of measure involved in any ratio are always the same (here, inches and inches).

Decide: To solve a proportion problem when the two units of measure are the same, you write and solve a direct proportion.

Form ratios:

$$\begin{array}{ll} \text{length} & \text{width} \\ \dfrac{8 \text{ inches}}{n \text{ inches}} \leftarrow \begin{array}{l} 8 \text{ inches}/5 \text{ inches} \\ n \text{ inches}/3 \text{ inches} \end{array} \rightarrow \dfrac{5 \text{ inches}}{3 \text{ inches}} \end{array}$$

Write the proportion: length ratio $\longrightarrow \dfrac{8}{n} = \dfrac{5}{3} \longleftarrow$ width ratio

Solve as before:

$$3 \times 8 = n \times 5 \qquad \text{Equate the cross products.}$$
$$24 = n \times 5 \qquad \text{Simplify.}$$
$$n = 24 \div 5 \qquad \text{Divide the product by the known factor.}$$
$$n = 4\frac{4}{5} \qquad \text{Check as before.}$$

Interpret: $n = 4\frac{4}{5}$ means a 5 inch by 8 inch photo will reduce to a 3 inch by $4\frac{4}{5}$ inch photo.

(b) *Understand:*   In this problem, you compare two rates. The two units of measure involved in the given rate (dollars and days) act in the same way. That is, the number of dollars increases as the number of days increases. Or, the number of dollars decreases as the number of days decreases.

Decide: To solve a proportion when the two units of measure involved in the given rate act in the same way, you write and solve a direct proportion.

Form ratios: $\dfrac{560 \text{ dollars}}{n \text{ dollars}}$ ⟵ $\begin{array}{c}\$560 \text{ in } 15 \text{ days}\\ \$n \ \text{ in } 7 \text{ days}\end{array}$ ⟶ $\dfrac{15 \text{ days}}{7 \text{ days}}$

Write the proportion: dollars ratio ⟶ $\dfrac{560}{n} = \dfrac{15}{7}$ ⟵ days ratio

Solve as before:

$7 \times 560 = n \times 15$ Equate cross products.

$3920 = n \times 15$ Simplify.

$n = 3920 \div 15$ Divide the product by the known factor.

$n = 261.33\overline{3}$ Divide to the thousandths place.

$n \approx 261.33$ Round to the nearest hundredth.

Interpret: $n \approx 261.33$ means that, at the given rate, the woman will earn \$266.33 (to the nearest cent) in 7 days.

(c) The two units of measure is this rate (ounces and servings) act in the same way: as ounces increase, so do servings. The direct proportion is set up as follows:

$\dfrac{8 \text{ ounces}}{22 \text{ ounces}}$ ⟵ $\begin{array}{c}8 \text{ ounces/12 servings}\\ 22 \text{ ounces/}n \text{ servings}\end{array}$ ⟶ $\dfrac{12 \text{ servings}}{n \text{ servings}}$

ounces ratio ⟶ $\dfrac{8}{22} = \dfrac{12}{n}$ ⟵ serving ratio

$n \times 8 = 22 \times 12$

$n \times 8 = 264$

$n = 264 \div 8$

$n = 33$ ⟵ solution

This means that a 22-ounce jar of the same instant coffee contains 33 servings.

(d) The two units of measure (pounds of sand and pounds of rock) act in the same way, so you would set up the direct proportion as follows:

$\dfrac{2 \text{ pounds sand}}{n \text{ pounds sand}}$ ⟵ $\begin{array}{c}2\text{: } 4\\ n\text{: } 125\end{array}$ ⟶ $\dfrac{4 \text{ pounds rock}}{125 \text{ pounds rock}}$

sand ratio ⟶ $\dfrac{2}{n} = \dfrac{4}{125}$ ⟵ rock ratio

$125 \times 2 = n \times 4$

$250 = n \times 4$

$n = 250 \div 4$

$n = 62\frac{1}{2}$ ⟵ solution

This means that $62\frac{1}{2}$ pounds of sand is needed for 125 pounds of rock.

PROBLEM 8-11   Solve each proportion using an indirect proportion:

(a) It takes 6 men 14 days to do a job. At that rate, how long will it take 8 men to do the same job?

(b) At the rate given in problem (a), what is the fewest number of men needed to do the same job in 9 days?

(c) A driven pulley with a diameter of 10 inches turns at 500 rpm. How many revolutions per minute does the driving pulley turn at if its diameter is 15 inches?

(d) At the rate given in problem (c), what is the diameter of the driving pulley if it turns at 200 rpm?

Solution   Recall that you solve a proportion problem using an indirect proportion when the two units of measure involved in the rate act in opposite ways. That is, the number of one amount increases as the number of the other decreases. Or, the number of one amount decreases as the number of the other amount increases [see Example 8-29]:

(a) *Understand:*   The two units of measure in the rate (men and days) act in opposite ways. That is, the number of men increases as the number of days decreases. Or, the number of men decreases as the number of days increases.

Decide: To solve a proportion problem when the two units of measure act in opposite ways, you write and solve an indirect proportion.

Form ratios:

$$\frac{6 \text{ men}}{8 \text{ men}} \longleftarrow \frac{6 \text{ men/14 days}}{8 \text{ men/}n \text{ days}} \longrightarrow \frac{14 \text{ days}}{n \text{ days}}$$

Write the proportion: men ratio $\longrightarrow \dfrac{6}{8} = \dfrac{n}{14} \longleftarrow$ reciprocal of days ratio

Solve as before:

$14 \times 6 = 8 \times n$ Equate the cross products.

$84 = 8 \times n$ Simplify.

$n = 84 \div 8$ Divide the product by the known factor.

$n = 10\frac{1}{2} \longleftarrow$ solution

Interpret: $n = 10\frac{1}{2}$ means that it will take 8 men $10\frac{1}{2}$ days to do the same job.

(b) Now the unknown term is *men*, not *days*. So you would set up the indirect proportion as follows:

$$\frac{6 \text{ men}}{n \text{ men}} \longleftarrow \frac{6 \text{ men /14 days}}{n \text{ men/ 9 days}} \longrightarrow \frac{14 \text{ days}}{9 \text{ days}}$$

men ratio $\longrightarrow \dfrac{6}{n} = \dfrac{9}{14} \longleftarrow$ reciprocal of days ratio

$14 \times 6 = n \times 9$

$84 = n \times 9$

$n = 84 \div 9$

$n = 9\frac{1}{3} \longleftarrow$ solution

Interpret: $n = 9\frac{1}{3}$ means that it would take more than 9 men, or a minimum of 10 men, to do the same job in 9 days.

(c) The two units of measure in this rate (inches and rpm's) act in opposite ways: as the number of inches increases, the revolutions per minute decrease:

$$\frac{10 \text{ inches}}{15 \text{ inches}} \longleftarrow \frac{10 \text{ inches/500 rpm}}{15 \text{ inches/}n \text{ rpm}} \longrightarrow \frac{500 \text{ rpm}}{n \text{ rpm}}$$

inches ratio $\longrightarrow \dfrac{10}{15} = \dfrac{n}{500} \longleftarrow$ reciprocal of rpm ratio

$$500 \times 10 = 15 \times n$$

$$5000 = 15 \times n$$

$$n = 5000 \div 15$$

$$n = 333\tfrac{1}{3} \longleftarrow \text{ solution}$$

This means that a pulley with a diameter of 15 inches turns at $333\tfrac{1}{3}$ rpm.

(d) Now you must find out the diameter of the driving pulley if it turns at only 200 rpm:

$$\frac{10 \text{ inches}}{n \text{ inches}} \quad \begin{matrix} \longleftarrow \text{ 10 inches/500 rpm} \longrightarrow \\ \longleftarrow \text{ n inches/200 rpm} \longrightarrow \end{matrix} \quad \frac{500 \text{ rpm}}{200 \text{ rpm}}$$

$$\text{inches ratio} \longrightarrow \frac{10}{n} = \frac{200}{500} \longleftarrow \text{ reciprocal of rpm ratio}$$

$$500 \times 10 = n \times 200$$

$$5000 = n \times 200$$

$$n = 5000 \div 200$$

$$n = 25 \longleftarrow \text{ solution}$$

This means that the pulley must be 25 inches in diameter if it turns at 200 rpm.

Supplementary Exercises

PROBLEM 8-12   Find each ratio:

(a) 30 miles to 50 miles   **(b)** 14 hours to 16 hours   **(c)** 36 feet to $4\tfrac{1}{2}$ feet

(d) 48 meters to 0.6 meters   **(e)** $25 to $1.25   **(f)** 1 year to 1 month

(g) 6 days to 2 weeks   **(h)** 2 half-dollars to 3 nickels   **(i)** 2 quarters to 3 dimes

(j) 50 women to 30 men   **(k)** 14 wins to 30 games   **(l)** 3000 students to 120 instructors

(m) 0.5 grams to 500 kilograms   **(n)** 2 feet to 5 yards   **(o)** $1\tfrac{1}{2}$ pounds to 12 ounces

PROBLEM 8-13   Find each rate:

(a) 55 miles to 60 minutes   **(b)** $28 to 8 pounds   **(c)** 210 miles to 4 hours

(d) 280 miles to 16 gallons   **(e)** $1740 to 4 weeks   **(f)** 12 gallons to $4\tfrac{1}{2}$ hours

(g) 2000 calories to 24 hours   **(h)** $31\tfrac{1}{4}$ hours to 5 people   **(i)** 5000 pages to 48 days

(j) 10,000 bricks to 12 walls   **(k)** 260 pounds to 8 cases   **(l)** $19,011 to 12 months

(m) 730.8 miles to 14.5 hours   **(n)** 335 feet to 40 seconds   **(o)** $102 to 8 people

(p) 425 houses to 200 TV sets   **(q)** 160 planes to 3 hours   **(r)** $6\tfrac{1}{4}$ pounds to $2.50

PROBLEM 8-14   Solve each proportion:

(a) $\dfrac{5\frac{1}{2}}{1} = \dfrac{p}{24}$ **(b)** $\dfrac{50}{1} = \dfrac{m}{3}$ **(c)** $\dfrac{400}{20} = \dfrac{250}{n}$ **(d)** $\dfrac{3}{4} = \dfrac{b}{16}$ **(e)** $\dfrac{8}{c} = \dfrac{6}{30}$ **(f)** $\dfrac{9}{12} = \dfrac{18}{d}$

(g) $\dfrac{5}{f} = \dfrac{35}{14}$ **(h)** $\dfrac{g}{9} = \dfrac{4}{12}$ **(i)** $\dfrac{h}{36} = \dfrac{1}{3}$ **(j)** $\dfrac{10}{k} = \dfrac{15}{12}$ **(k)** $\dfrac{p}{3.5} = \dfrac{6}{15}$ **(l)** $\dfrac{4.5}{5} = \dfrac{27}{r}$

(m) $\dfrac{6}{9} = \dfrac{3\frac{1}{3}}{z}$ **(n)** $\dfrac{a}{15} = \dfrac{4}{5}$ **(o)** $\dfrac{900}{54} = \dfrac{b}{3}$ **(p)** $\dfrac{16}{c} = \dfrac{2}{7}$ **(q)** $\dfrac{50}{8} = \dfrac{150}{d}$ **(r)** $\dfrac{f}{32} = \dfrac{3}{4}$

(s) $\dfrac{12}{210} = \dfrac{g}{140}$ (t) $\dfrac{24}{h} = \dfrac{108}{18}$ (u) $\dfrac{6}{7} = \dfrac{54}{k}$ (v) $\dfrac{1.5}{0.3} = \dfrac{m}{8}$ (w) $\dfrac{0.04}{0.12} = \dfrac{0.6}{n}$ (x) $\dfrac{\frac{1}{2}}{10} = \dfrac{2\frac{1}{2}}{v}$

PROBLEM 8-15   Solve each proportion problem:

(a) A map has a scale of $1\frac{1}{4}$ inch to 10 miles. If the map distance between two cities is $3\frac{1}{2}$ inches, then what is the actual distance between the two cities?

(b) Using the scale given in problem (a), if the actual distance between the two cities is 500 miles, what is the map distance between the two cities?

(c) The average 150-pound person should eat 1800 calories each day to maintain that weight. How many calories should be eaten by a 110-pound person to maintain weight?

(d) Using the rate given in problem (c), how much should you weigh to eat 2000 calories a day to maintain weight?

(e) Three men take 5 hours to unload a train car. At that rate, how long will it take 10 men to unload the same train car?

(f) At the rate given in problem (e), how many men will be needed to unload the same train car in a maximum of 2 hours?

(g) Apples are on sale for $1.50 for 3 pounds. At that rate, how much will 5 pounds of apples cost?

(h) At the rate given in problem (g), how many pounds of apples can be purchased for $5?

(i) To feed 12 people, it takes 5 pounds of canned ham. At that rate, how many people can be given a full serving from 3 pounds of canned ham?

(j) At the rate given in problem (i), how many pounds of canned ham will it take to feed 20 people?

(k) A certain car travels 500 miles in 8 hours. At that rate, how far can the car travel in 12 hours?

(l) At the rate given in problem (k), how many hours will it take the car to travel 300 miles?

(m) A driving gear with 36 teeth turns at 90 rpm. How many teeth are in the driven gear if it turns at 60 rpm?

(n) Using the same driving gear in problem (m), how fast is the driven gear turning if it has 20 teeth?

(o) Two windows are to be constructed so that they are proportional. The height and width of one window is 5 feet by 3 feet, respectively. What should be the height of the other window if its width is 5 feet?

(p) Given the dimensions of the windows in problem (o), what should the width of the other window be if its height is 3 feet?

(q) For every $5 invested in a certain savings account, the bank returns a total of $5.30 at the end of one year. How much should be invested to earn $100 in one year?

(r) Using the same bank in problem (q), how much would be earned if $100 were invested?

(s) A driving pulley has a diameter of 12 inches and turns at 400 rpm. How fast does the driven pulley turn if it has a 20-inch diameter?

(t) Using the same driving pulley in problem (s), what is the diameter of the driven pulley if it turns at 600 rpm?

(u) Jose and Rosa are sitting on opposite ends of a seesaw. Jose weighs 160 pounds and is sitting 5 feet from the pivot (balance point). If Rosa weighs 120 pounds, how far must she sit from the pivot to balance the seesaw? (*Hint:* To balance a seesaw, the number of pounds must increase as the distance from the pivot decreases, or the number of pounds must decrease as the distance from the pivot increases.)

(v) Using the information given for Jose in problem (u), how much does Rosa weigh if she balances the seesaw at 6 feet from the pivot?

(w) A crowbar is built so that it provides its own pivot one inch from the end. If the handle is 30 inches long, how much force can a 120-pound woman exert with the crowbar if she applies all her body weight?
[*Hint:* The same principle used for the see-saw in problem **(u)** can be used here.]

(x) A man wants to pry up a 750-pound weight. If he places the pivot one foot from the weight, how long must the total length of the lever be so that his weight of 150 pounds will just balance the weight?
[*Hint:* The same principle used for the see-saw in problem **(u)** can be used here too. Also, the word *total* is important.]

Answers to Supplementary Exercises

(8-12) **(a)** $\dfrac{3}{5}$ **(b)** $\dfrac{7}{8}$ **(c)** $\dfrac{8}{1}$ **(d)** $\dfrac{80}{1}$ **(e)** $\dfrac{20}{1}$ **(f)** $\dfrac{12}{1}$ **(g)** $\dfrac{3}{7}$ **(h)** $\dfrac{20}{3}$

(i) $\dfrac{5}{3}$ **(j)** $\dfrac{5}{3}$ **(k)** $\dfrac{7}{15}$ **(l)** $\dfrac{25}{1}$ **(m)** $\dfrac{1}{1,000,000}$ **(n)** $\dfrac{2}{15}$ **(o)** $\dfrac{2}{1}$

(8-13) (Any one of the three different forms for the rate is acceptable.)

(a) $\frac{11}{12}$ miles/minute   **(b)** \$3.50/pound   **(c)** $52\frac{1}{2}$ mph
(d) $17\frac{1}{2}$ mpg   **(e)** \$435/week   **(f)** $2\frac{2}{3}$ gallons per hour
(g) $83\frac{1}{3}$ calories/hour   **(h)** $6\frac{1}{4}$ hours/person   **(i)** $104\frac{1}{6}$ pages per day
(j) $833\frac{1}{3}$ bricks per wall   **(k)** $32\frac{1}{2}$ pounds/case   **(l)** \$1584.25 per month
(m) 50.4 mph   **(n)** $8\frac{3}{8}$ fps   **(o)** \$12.75/person
(p) $2\frac{1}{8}$ houses per TV set   **(q)** $53\frac{1}{3}$ planes/hour   **(r)** $2\frac{1}{2}$ pounds/dollar

(8-14) **(a)** 132   **(b)** 150   **(c)** $12\frac{1}{2}$ or 12.5   **(d)** 12   **(e)** 40   **(f)** 24   **(g)** 2   **(h)** 3
(i) 12   **(j)** 8   **(k)** 1.4 or $1\frac{2}{5}$ **(l)** 30   **(m)** 5   **(n)** 12   **(o)** 50   **(p)** 56
(q) 24   **(r)** 24   **(s)** 8   **(t)** 4   **(u)** 63   **(v)** 40   **(w)** 1.8   **(x)** 50

(8-15) **(a)** 28 miles   **(b)** $62\frac{1}{2}$ inches   **(c)** 1320 calories   **(d)** $166\frac{2}{3}$ pounds
(e) $1\frac{1}{2}$ hours   **(f)** 8 men   **(g)** \$2.50   **(h)** 10 pounds
(i) 7 people   **(j)** $8\frac{1}{3}$ pounds   **(k)** 750 miles   **(l)** $4\frac{4}{5}$ hours
(m) 54 teeth   **(n)** 162 rpm   **(o)** $8\frac{1}{3}$ feet   **(p)** $1\frac{4}{5}$ feet
(q) \$1666.67   **(r)** \$6   **(s)** 240 rpm   **(t)** 8 inches
(u) $6\frac{2}{3}$ feet   **(v)** $133\frac{1}{3}$ pounds   **(w)** 3600 pounds   **(x)** 6 feet

9 MEASUREMENT

THIS CHAPTER IS ABOUT

☑ Identifying U.S. Customary Measures
☑ Renaming U.S. Customary Measures
☑ Identifying Metric Measures
☑ Renaming Metric Measures
☑ Computing with Measures

9-1. Identifying U.S. Customary Measures

A. Introduction to U.S. Customary measures

Every **measure** has two parts—an **amount** and a **unit of measure.**

EXAMPLE 9-1   Identify the amount and unit of measure in 5 feet.

Solution   In 5 feet, 5 is the amount and
feet is the unit of measure.

There are two different measurement systems used in the world today—the **United States (U.S.) Customary measurement system** and the **metric measurement system.** At the present time, the United States is the only industrial nation in the world that has not completely converted to the metric system of measures.

The basic U.S. Customary measures are **length, capacity** (liquid volume), **weight, area,** and **volume** (space volume).

B. U.S. Customary length measures

The common units of measure for U.S. Customary length measures are the **inch** (in.), **foot** (ft), **yard** (yd), and **mile** (mi).

Note: The only common U.S. Customary unit of measure that ends in a period is the inch (in.).

EXAMPLE 9-2   Describe the distance represented by **(a)** 1 inch, **(b)** 1 foot, **(c)** 1 yard, and **(d)** 1 mile.

Solution

(a) For the average person, the distance between the first and second knuckle on the index finger is about 1 inch (in.).

(b) For the average person, the distance between the elbow and the wrist is about 1 foot (1 ft).

(c) The width of an average house door is about 1 yard (1 yd).

(d) The length of 10 average city blocks is about 1 mile (1 mi).

EXAMPLE 9-3   Which of the following is the approximate height of an average house door?
(a) 7 in.   **(b)** 7 ft   **(c)** 7 yd   **(d)** 7 mi

Solution

(a) Wrong. 7 in. (or 7 inches) is about the length of a new pencil.

(b) Correct! 7 ft (or 7 feet) is about the height of an average door.

(c) Wrong. 7 yd (or 7 yards) is about the height of an average house.

(d) Wrong. 7 mi (or 7 miles) is about the height that a jet airliner cruises above land.

C. U.S. Customary capacity measures

The common units of measure for U.S. Customary capacity measures are the **teaspoon** (tsp), **tablespoon** (tbsp), **fluid ounce** (fl oz), **cup** (c), **pint** (pt), **quart** (qt), and **gallon** (gal).

EXAMPLE 9-4 Describe the amount of liquid represented by (a) 1 teaspoon, (b) 1 tablespoon, (c) 1 fluid ounce, (d) 1 cup, (e) 1 pint, (f) 1 quart, and (g) 1 gallon.

Solution

(a) A household dessert spoon measures about 1 teaspoon (1 tsp) of liquid.

(b) A household serving spoon measures about 1 tablespoon (1 tbsp) of liquid.

(c) Two household tablespoons contain about 1 fluid ounce (1 fl oz) of liquid.

(d) An average coffee mug contains about 1 cup (1 c) of liquid.

(e) A large glass of milk contains about 1 pint (1 pt) of liquid.

(f) An average carton of milk contains about 1 quart (1 qt) of liquid.

(g) An average can of house paint contains about 1 gallon (1 gal) of liquid.

EXAMPLE 9-5 What is the approximate amount of liquid that an average bathtub might contain, (a) 40 tsp, (b) 40 c, or (c) 40 gal?

Solution

(a) Wrong. 40 tsp is about the amount of liquid that an average teacup might contain.

(b) Wrong. 40 c is about the amount of liquid that an average bathroom sink might contain.

(c) Correct! 40 gal is about the amount of liquid that an average bathtub might contain.

D. U.S. Customary weight measures

The common units of measure for U.S. Customary weight measures are the **ounce** (oz), **pound** (lb), and **ton** (T).

Note: In the U.S. Customary system, there is a *long ton* measure and a *short ton* measure. In this text, the word *ton* will always refer to the short ton measure.

EXAMPLE 9-6 Describe the weight represented by (a) 1 ounce, (b) 1 pound, and (c) 1 ton.

Solution

(a) Nine U.S. pennies weigh about 1 ounce (1 oz).

(b) An average can of soda pop weighs about 1 pound (1 lb).

(c) A small compact car weighs about 1 ton (1 T).

EXAMPLE 9-7 What is the approximate weight of an average cat, (a) 5 oz, (b) 5 lb, or (c) 5 T?

Solution

(a) Wrong. 5 oz is about the weight of a hamburger patty.

(b) Correct! 5 lb is about the weight of an average cat.

(c) Wrong. 5 T is about the weight of an African elephant.

E. U.S. Customary area measures

The common units of measure for U.S. Customary area measures are the **square inch** (sq in. or in^2), **square foot** (sq ft or ft^2), **square yard** (sq yd or yd^2), and **acre** (A).

EXAMPLE 9-8 Describe the area represented by (**a**) 1 square inch, (**b**) 1 square foot, (**c**) 1 square yard, and (**d**) 1 acre.

Solution

(**a**) The surface of an average U.S. postage stamp is about 1 square inch (1 in^2).

(**b**) The surface of a $33\frac{1}{3}$ record album is about 1 square foot (1 ft^2).

(**c**) The top (or bottom) half of an average house door surface is about 1 square yard (1 yd^2).

(**d**) The combined area of two ice hockey rinks is a little smaller than 1 acre (1 A).

EXAMPLE 9-9 What is the approximate area of an average kitchen floor, (**a**) 72 in^2, (**b**) 72 ft^2, (**c**) 72 yd^2, or (**d**) 72 A?

Solution

(**a**) Wrong. 72 in^2 (or 72 square inches) is about one-half the surface of a $33\frac{1}{3}$ record album.

(**b**) Correct! 72 ft^2 (or 72 square feet) is about the area of an average kitchen floor.

(**c**) Wrong. 72 yd^2 (or 72 square yards) is larger than the surface of a U.S. basketball court.

(**d**) Wrong. 72 A (or 72 acres) is about the area of a small farm.

F. U.S. Customary volume measures

The common units of measure for U.S. Customary volume measures are the **cubic inch** (cu in. or in^3), **cubic foot** (cu ft or ft^3), and **cubic yard** (cu yd or yd^3).

EXAMPLE 9-10 Describe the volume represented by (**a**) 1 cubic inch, (**b**) 1 cubic foot, and (**c**) 1 cubic yard.

Solution

(**a**) The space occupied by a small pillbox measuring 1 inch on each side is 1 cubic inch (1 in^3).

(**b**) The space occupied by a small portable TV is about 1 cubic foot (1 ft^3).

(**c**) The space occupied by a standard floor model TV is 1 cubic yard (1 yd^3).

EXAMPLE 9-11 What is the approximate space occupied by a standard four-drawer file cabinet? (**a**) 14 in^3, (**b**) 14 ft^3, or (**c**) 14 yd^3?

Solution

(**a**) Wrong. 14 in^3 (or 14 cubic inches) is about the amount of space occupied by a standard soup can.

(**b**) Correct! 14 ft^3 (or 14 cubic feet) is about the amount of space occupied by a standard four-drawer file cabinet.

(**c**) Wrong. 14 yd^3 (or 14 cubic yards) is about the amount of space inside a small kitchen.

9-2. Renaming U.S. Customary Measures

A. Introduction to U.S. Customary unit conversions and unit fractions

A *number sentence* relating two different units of measure in which one of the amounts is 1 is called a **unit conversion**.

EXAMPLE 9-12 Which of the following number sentences are unit conversions?
(**a**) 1 ft = 12 in.   (**b**) 2 c = 1 pt   (**c**) $1\frac{1}{2}$ lb = 24 oz

Solution

(a) The number sentence 1 ft = 12 in. is a unit conversion because it relates two different units of measure (feet and inches) and one of the amounts is 1 (1 ft).

(b) The number sentence 2 c = 1 pt is a unit conversion because it relates two different units of measure (cups and pints) and one of the amounts is 1 (1 pt).

(c) The number sentence $1\frac{1}{2}$ lb = 24 oz is not a unit conversion even though it relates two different units of measure (pounds and ounces) because neither one of the amounts is 1 ($1\frac{1}{2}$ lb and 24 oz).

The following are the common **U.S. Customary unit conversions.**

Length	Capacity	
12 in. = 1 ft	3 tsp = 1 tbsp	2 c = 1 pt
3 ft = 1 yd	2 tbsp = 1 fl oz	2 pt = 1 qt
5280 ft = 1 mi	8 fl oz = 1 c	4 qt = 1 gal

Weight	Area	Volume
16 oz = 1 lb	144 in² = 1 ft²	1728 in³ = 1 ft³
2000 lb = 1 T	9 ft² = 1 yd²	27 ft³ = 1 yd³

A **unit fraction** is formed when one measure in a unit conversion is used for the numerator and the other measure is used for the denominator.

EXAMPLE 9-13 Write two different unit fractions using the unit conversion 1 yd = 3 ft.

Solution

$$\frac{1 \text{ yd}}{3 \text{ ft}} \qquad \frac{3 \text{ ft}}{1 \text{ yd}}$$

Note: Every unit conversion has exactly two different unit fractions associated with it.

Since the numerator and denominator of a unit fraction are equal, every unit fraction is equal to 1.

EXAMPLE 9-14 Write two different number sentences relating the number 1 and the unit fractions associated with the following unit conversion: 1 yd = 3 ft

Solution

$$1 = \frac{1 \text{ yd}}{3 \text{ ft}} \qquad 1 = \frac{3 \text{ ft}}{1 \text{ yd}}$$

Note: Number sentences like those in Example 9-14 are used to rename a given measure in terms of a smaller or larger unit of measure.

B. Renaming a U.S. Customary measure using a unit conversion

To rename a given measure in terms of a different unit of measure when there is a unit conversion relating the two units of measure, you multiply the given measure by the correct unit fraction.

EXAMPLE 9-15 Rename 42 in. as feet.

Solution Rename inches as feet using the unit conversion 12 in. = 1 ft.

42 in. = 42 in. × 1   Multiply by 1.
                             Replace 1 with the correct unit fraction.

$$= 42 \text{ in.} \times \frac{1 \text{ ft}}{12 \text{ in.}} \quad \longleftarrow \quad 12 \text{ in.} = 1 \text{ ft means } 1 = \frac{1 \text{ ft}}{12 \text{ in.}}$$

$$= \frac{42\ \cancel{in.}}{1} \times \frac{1\ ft}{12\ \cancel{in.}} \qquad \text{Eliminate the common unit of measure}$$

(See the following *Note* and *Caution*.)

$$= \frac{42}{12}\ ft \qquad \text{Multiply fractions.}$$

$$= \frac{7}{2}\ ft \qquad \text{Simplify.}$$

$$= 3\tfrac{1}{2}\ ft$$

Note: In this example, the unit fraction $\frac{1\ ft}{12\ in.}$ is used instead of $\frac{12\ in.}{1\ ft}$ so that the common unit of measure (inches) could be eliminated in the next step.

Caution: To eliminate a common unit of measure, one of the common units must be in the numerator and the other common unit must be in the denominator.

EXAMPLE 9-16 Can the common unit of measure be eliminated in (a) $30\ ft \times \frac{3\ ft}{1\ yd}$ or (b) $30\ ft \times \frac{1\ yd}{3\ ft}$?

Solution

(a) The common unit of measure (feet) cannot be eliminated in $\frac{30\ ft}{1} \times \frac{3\ ft}{1\ yd}$ because neither denominator contains feet as a unit of measure.

(b) The common unit of measure (feet) can be eliminated in $\frac{30\ ft}{1} \times \frac{1\ yd}{3\ ft}$ because feet appears in both a numerator and a denominator:

$$\frac{30\ \cancel{ft}}{1} \times \frac{1\ yd}{3\ \cancel{ft}} = \frac{30}{3}\ yd = 10\ yd$$

C. Renaming U.S. Customary measures when more than one unit conversion is necessary

To rename a measure when more than one unit conversion is necessary, you first list each necessary unit conversion in the order needed.

EXAMPLE 9-17 Rename $\tfrac{3}{4}$ gal as cups.

Solution Rename gallons as cups using the following unit conversions:

gallons to quarts	quarts to pints	pints to cups
1 gal = 4 qt	1 qt = 2 pt	1 pt = 2 c

$$\frac{3}{4}\ gal = \frac{3}{4}\ gal \times \underset{\downarrow}{1} \times \underset{\downarrow}{1} \times \underset{\downarrow}{1} \qquad \begin{array}{l}\text{Multiply by 1 for each} \\ \text{necessary unit conversion.}\end{array}$$

$$= \frac{3\ gal}{4} \times \frac{4\ qt}{1\ gal} \times \frac{2\ pt}{1\ qt} \times \frac{2\ c}{1\ pt} \qquad \begin{array}{l}\text{Substitute the correct} \\ \text{unit fractions for 1.}\end{array}$$

$$= \frac{3\ \cancel{gal}}{4} \times \frac{4\ \cancel{qt}}{1\ \cancel{gal}} \times \frac{2\ \cancel{pt}}{1\ \cancel{qt}} \times \frac{2\ c}{1\ \cancel{pt}} \qquad \text{Eliminate all common units of measure.}$$

$$= \frac{3 \times 4 \times 2 \times 2}{4}\ c \qquad \text{Simplify.}$$

$$= \frac{48}{4}\ c$$

$$= 12\ c \longleftarrow \tfrac{3}{4}\ \text{gallon contains 12 cups}$$

Note: In this example, one unit fraction is substituted for each necessary unit conversion in one step so that the common units of measure can all be eliminated in the next step.

D. Renaming U.S. Customary mixed measures

When two or more different units of measure are used to write one single measure, that measure is called a **mixed measure.**

EXAMPLE 9-18 Rename the units of measure 1 T + 800 lb as a mixed measure.

mixed measure

Solution 1 T + 800 lb = $\overbrace{\text{1 T 800 lb}}$ *Think:* 1 T 800 lb means 1 T + 800 lb

To rename a mixed measure, you first write the mixed measure as a sum.

EXAMPLE 9-19 Rename 2 T 800 lb as pounds.

Solution Rename tons as pounds using the unit conversion 2000 lb = 1 T.

$$2 \text{ T } 800 \text{ lb} = 2 \text{ T} + 800 \text{ lb} \qquad \text{Rename as a sum.}$$

$$= (2 \text{ T} \times 1) + 800 \text{ lb} \qquad \text{Multiply 2 T by 1 for the necessary unit conversion.}$$

$$= \left(\frac{2 \text{ T}}{1} \times \frac{2000 \text{ lb}}{1 \text{ T}} \right) + 800 \text{ lb} \qquad \text{Substitute the correct unit fraction.}$$

$$2000 \text{ lb} = 1 \text{ T means } 1 = \frac{2000 \text{ lb}}{1 \text{ T}}$$

$$= (2 \times 2000) \text{ lb} + 800 \text{ lb} \qquad \text{Eliminate the common unit of measure.}$$

$$= 4000 \text{ lb} + 800 \text{ lb} \qquad \text{Multiply.}$$

$$= 4800 \text{ lb} \qquad \text{Add pounds: } 4000 + 800 = 4800$$

Note: 2 T 800 lb = 4800 lb or 2.4 T

9-3. Identifying Metric Measures

A. Introduction to metric measures

The metric measurement system has been used in Europe for about 200 years. Currently, the United States is slowly trying to convert to the metric system. If you take the time and effort to learn the metric system now, you will have a well-planned head start when metric conversion becomes mandatory in the United States.

The following table will help you identify metric units and their abbreviations.

Metric Units of Measure

Place Values	Thousands	Hundreds	Tens	Ones	Tenths	Hundredths	Thousandths
Digit Values	1000	100	10	1	0.1	0.01	0.001
U.S. Money Values	$1000 bill	$100 bill	$10 bill	dollar	dime	cent	mill
Metric Length Units	kilometer km	hectometer hm	dekameter dam	meter m	decimeter dm	centimeter cm	millimeter mm
Metric Capacity Units	kiloliter kL	hectoliter hL	dekaliter daL	liter L	deciliter dL	centiliter cL	milliliter mL
Metric Mass Units	kilogram kg	hectogram hg	dekagram dag	gram g	decigram dg	centigram cg	milligram mg

Note: To rename U.S. Customary measures as metric measures, you can use the unit conversions in Appendix Table 5.

The basic metric measures are **length, capacity** (liquid volume), **mass*** (weight), **area,** and **volume** (space volume).

The basic metric units of measure for length, capacity, and mass (weight) are **meter** (m), **liter** (L), and **gram** (g), respectively.

Note: The symbol *l* (a cursive "el") is sometimes used to represent the word "liter" instead of the symbol L.

To name other metric units of measure, you attach the correct prefix to the correct basic unit.

EXAMPLE 9-20 Identify the meaning and symbol of the following prefixes:

    (a) kilo-   **(b)** hecto-   **(c)** deka-   **(d)** deci-   **(e)** centi-   **(f)** milli-

Solution

	Prefix	Symbol	Meaning
(a)	kilo-	k	1000 (one thousand)
(b)	hecto-	h	100 (one hundred)
(c)	deka-	da	10 (ten)
(d)	deci-	d	$\frac{1}{10}$ or 0.1 (one tenth)
(e)	centi-	c	$\frac{1}{100}$ or 0.01 (one hundredth)
(f)	milli-	m	$\frac{1}{1000}$ or 0.001 (one thousandth)

B. Metric length measures

The common units of measure for metric length measures are the **millimeter** (mm), **centimeter** (cm), **meter** (m), and **kilometer** (km).

EXAMPLE 9-21 Describe the distance represented by **(a)** 1 millimeter, **(b)** 1 centimeter, **(c)** 1 meter, and **(d)** 1 kilometer.

Solution

(a) The thickness of a U.S. dime is about 1 millimeter (1 mm).

(b) For the average person, the thickness at the end of the little finger is about 1 centimeter (1 cm).

(c) The width of an average house door is about 1 meter (1 m).

(d) The length of 6 average city blocks is about 1 kilometer (1 km).

EXAMPLE 9-22 What is the approximate length of a standard paper clip, **(a)** 3 mm, **(b)** 3 cm, **(c)** 3 m, or **(d)** 3 km?

Solution

(a) Wrong. 3 mm (or 3 millimeters) is about the thickness of a standard paper clip.

(b) Correct! 3 cm (or 3 centimeters) is about the length of a standard paper clip.

(c) Wrong. 3 m (or 3 meters) is about the length of a compact car.

(d) Wrong. 3 km (or 3 kilometers) is about the height of a large mountain.

C. Metric capacity measures

The common units of measure for metric capacity measures are the **milliliter** (mL), **liter** (L), and **kiloliter** (kL).

---

* There is a scientific difference between mass and weight. But in everyday situations and problems you can think of mass and weight as being the same thing.

EXAMPLE 9-23 Describe the amount of liquid represented by (**a**) 1 milliliter, (**b**) 1 liter, and (**c**) 1 kiloliter.

Solution

(**a**) It takes about 15 drops of water to make 1 milliliter (1 mL) of liquid.

(**b**) An average carton of milk (quart-size) contains about 1 liter (1 L) of liquid.

(**c**) A shallow children's wading pool contains about 1 kiloliter (1 kL) of liquid.

EXAMPLE 9-24 What is the approximate amount of liquid that an average water bucket might contain, (**a**) 5 mL, (**b**) 5 L, or (**c**) 5 kL?

Solution

(**a**) Wrong. 5 mL (or 5 milliliters) is about the amount of liquid that an average household teaspoon might contain.

(**b**) Correct! 5 L (or 5 liters) is about the amount of liquid that an average water bucket might hold.

(**c**) Wrong. 5 kL (or 5 kiloliters) is about the amount of liquid that a small water tank might contain.

D. Metric mass (weight) measures

The common units of measure for metric mass (weight) measures are the **milligram** (mg), **gram** (g), **kilogram** (kg), and **tonne** (t).

Note: A tonne is sometimes referred to as a **metric ton.** In this text, we'll use the term (and abbreviation) tonne (t).

EXAMPLE 9-25 Describe the mass (weight) represented by (**a**) 1 milligram, (**b**) 1 gram, (**c**) 1 kilogram, or (**d**) 1 tonne.

Solution

(**a**) A long human hair has a mass (weight) of about 1 milligram (1 mg).

(**b**) Two standard paper clips have a mass (weight) of about 1 gram (1 g).

(**c**) A pair of men's shoes have a mass (weight) of about 1 kilogram (1 kg).

(**d**) A large horse has a mass (weight) of about 1 tonne (1 t).

EXAMPLE 9-26 What is the approximate weight of a U.S. nickel (5¢ piece), (**a**) 5 mg, (**b**) 5 g, (**c**) 5 kg, or (**d**) 5 t?

Solution

(**a**) Wrong. 5 mg (or 5 milligrams) is about the mass (weight) of a tiny bird feather.

(**b**) Correct! 5 g (or 5 grams) is about the mass (weight) of a U.S. nickel.

(**c**) Wrong. 5 kg (or 5 kilograms) is about the mass (weight) of an average bowling ball.

(**d**) Wrong. 5 t (or 5 tonnes) is about the mass (weight) of an African elephant.

E. Metric area measures

The common units of measure for metric area measures are the **square centimeter** (cm^2), **square meter** (m^2), and **hectare** (ha).

EXAMPLE 9-27 Describe the area represented by (**a**) 1 square centimeter, (**b**) 1 square meter, and (**c**) 1 hectare.

Solution

(**a**) Any one of the six square surfaces of a standard sugar cube measures about 1 square centimeter ($1\ cm^2$).

(**b**) The top (or bottom) half of a standard house door measures about 1 square meter ($1\ m^2$).

(**c**) The combined area of two U.S. football fields is a little larger than 1 hectare (1 ha).

EXAMPLE 9-28   What is the approximate area of a standard garage door, (**a**) 8 cm², (**b**) 8 m², or (**c**) 8 ha?

Solution   (**a**) Wrong. 8 cm² (or 8 square centimeters) is about the area of a postage stamp.

(**b**) Correct! 8 m² (or 8 square meters) is about the area of a standard garage door.

(**c**) Wrong. 8 ha (or 8 hectares) is about the area of a city park.

F. Metric volume measures

The common units of measure for metric volume measures are the **cubic centimeter** (cm³ or cc) and **cubic meter** (m³).

EXAMPLE 9-29   Describe the volume represented by (**a**) 1 cubic centimeter and (**b**) 1 cubic meter.

Solution

(**a**) The space occupied by a standard sugar cube is about 1 cubic centimeter (1 cm³ or cc).

(**b**) The space occupied by a standard kitchen stove is about 1 cubic meter (1 m³).

EXAMPLE 9-30   What is the approximate space enclosed by an average living room, (**a**) 25 cm³ or (**b**) 25 m³?

Solution

(**a**) Wrong. 25 cm³ (or 25 cubic centimeters) is about the space occupied by a standard finger-ring box.

(**b**) Correct! 25 m³ (or 25 cubic meters) is about the space enclosed by an average living room.

9-4. Renaming Metric Measures

A. Introduction to metric unit conversion and unit fractions

Recall from Section 9-2 that a number sentence relating two different units of measure in which one of the amounts is 1 is called a unit conversion.

The following are the common **metric unit conversions:**

Length	Capacity	Mass (Weight)
10 mm = 1 cm	1000 mL = 1 L	1000 mg = 1 g
100 cm = 1 m	1000 L = 1 kL	1000 g = 1 kg
1000 m = 1 km		1000 kg = 1 t

Area	Volume
100 mm² = 1 cm²	1000 mm³ = 1 cm³
10,000 cm² = 1 m²	1,000,000 cm³ = 1 m³

Also recall that

- a unit fraction is formed when one measure in a unit conversion is used for the numerator and the other measure is used for the denominator
- every unit fraction is equal to 1 (since the denominator is equal to the numerator)
- every unit conversion has two unit fractions associated with it

B. Renaming a metric measure using a unit conversion

Recall from Example 9-15 that to rename a given measure as a different unit of measure when there is a unit conversion relating the two units of measure, you multiply the given measure by the correct unit fraction.

EXAMPLE 9-31 Rename 2.5 m as centimeters.

Solution Rename meters as centimeters using the unit conversion 100 cm = 1 m.

$$2.5 \text{ m} = 2.5 \text{ m} \times 1 \qquad \text{Multiply by 1.}$$

Replace 1 with the correct unit fraction.

$$= 2.5 \text{ m} \times \frac{100 \text{ cm}}{1 \text{ m}} \longleftarrow 100 \text{ cm} = 1 \text{ m means } 1 = \frac{100 \text{ cm}}{1 \text{ m}}$$

$$= \frac{2.5 \cancel{\text{ m}}}{1} \times \frac{100 \text{ cm}}{1 \cancel{\text{ m}}} \qquad \text{Eliminate the common unit of measure.}$$

$$= \left(\frac{2.5 \times 100}{1} \right) \text{cm} \qquad \text{Multiply fractions.}$$

$$= 250 \text{ cm} \qquad \text{Simplify.}$$

Note: In this example the unit fraction $\frac{100 \text{ cm}}{1 \text{ m}}$ is used instead of $\frac{1 \text{ m}}{100 \text{ cm}}$ so that the common unit of measure (meters) can be eliminated in the next step.

C. Renaming metric measures when more than one unit conversion is necessary

To rename a measure when more than one unit conversion is necessary, you first list each necessary unit conversion in the order needed.

EXAMPLE 9-32 Rename 185,000 mL as kiloliters.

Solution Rename milliliters as kiloliters using the following unit conversions:

milliliters to liters	liters to kiloliters
1000 mL = 1 L	1000 L = 1 kL

$$185,000 \text{ mL} = 185,000 \text{ mL} \times \quad 1 \quad \times \quad 1 \qquad \text{Multiply by 1 for each necessary unit conversion.}$$

$$= \frac{185,000 \text{ mL}}{1} \times \frac{1 \text{ L}}{1000 \text{ mL}} \times \frac{1 \text{ kL}}{1000 \text{ L}} \qquad \text{Substitute the correct unit fractions for 1.}$$

$$= \frac{185,000 \cancel{\text{ mL}}}{1} \times \frac{1 \cancel{\text{ L}}}{1000 \cancel{\text{ mL}}} \times \frac{1 \text{ kL}}{1000 \cancel{\text{ L}}} \qquad \text{Eliminate all common units of measure.}$$

$$= \frac{185,000}{1000 \times 1000} \text{ kL} \qquad \text{Simplify.}$$

$$= \frac{185,000}{1,000,000} \text{ kL}$$

$$= \frac{185}{1000} \text{ kL}$$

$$= 0.185 \text{ kL} \longleftarrow 185,000 \text{ mL is contained in } 0.185 \text{ kiloliters}$$

Note: In this Example 9-32, one unit fraction is substituted for each necessary unit conversion in one step so that the common units of measure can all be eliminated in the next step.

To rename metric measures as U.S. Customary measures, you can use the unit conversions in Appendix Table 5.

9-5. Computing with Measures

Recall from Section 8-1 that measures that have the same unit of measure are called **like measures** and measures that have different units of measure are called **unlike measures.**

EXAMPLE 9-33   Which of the following are like measures?

$$\text{(a)} \ \ 5 \text{ m}, 5 \text{ cm} \qquad \text{(b)} \ \ 6 \text{ ft}, 2 \text{ ft}$$

Solution

(a) 5 m and 5 cm are unlike measures because the units of measure (m and cm) are different.

(b) 6 ft and 2 ft are like measures because the unit of measure (ft) is the same.

A. Adding measures

To **add like measures,** you add the amounts and then write the same unit of measure on the sum.

EXAMPLE 9-34   Add: 35.8 g + 26.75 g

Solution   35.8 g + 26.75 g = (35.8 + 26.75) g   Add the amounts.

             = 62.55 g   Write the same unit of measure.

To **add mixed measures,** you first line up corresponding like measures in columns, then add the like measures, and then simplify the sum if possible.

EXAMPLE 9-35   Add: 3 yd 2 ft 8 in. + 5 yd 9 in.

Solution   yards   feet   inches
          ↓   ↓   ↓

           3 yd   2 ft   8 in.   Line up the measures in columns.
          +5 yd   9 in.
          ─────────────────────
           8 yd   2 ft   17 in.   Add like measures.

      8 yd   2 ft   17 in. = 8 yd + 2 ft + 17 in.   Simplify.

                          = 8 yd + 2 ft + 12 in. + 5 in.

                          = 8 yd + 2 ft + 1 ft + 5 in.   *Think:* 12 in. = 1 ft

                          = 8 yd + 3 ft + 5 in.

                          = 8 yd + 1 yd + 5 in.   *Think:* 3 ft = 1 yd

                          = 9 yd + 5 in.

                          = 9 yd 5 in. ⟵— simplest form

Note: 3 yd 2 ft 8 in. + 5 yd 9 in. = 9 yd 5 in., or 329 in.

B. Subtracting measures

To **subtract like measures,** you subtract the amounts and then write the same unit of measure on the difference.

EXAMPLE 9-36   Subtract 500 m − 150 m.

Solution   500 m − 150 m = (500 − 150) m   Subtract the amounts.

             = 350 m   Write the same unit of measure.

To **subtract mixed measures,** you first line up corresponding like measures in columns and then subtract like measures, renaming when necessary.

EXAMPLE 9-37   Subtract 5 gal 4 qt − 6 qt 1 pt.

Solution

           gallons quarts pints

Step 1:   5 gal 4 qt   Line up like measures.
       −   6 qt 1 pt

                  4 qts

Step 2:   5 gal 3 qt 2 pt   Rename quarts as pints to get more pints.
       −   6 qt 1 pt   4 qt = 3 qt + 1 qt
                     1 pt   = 3 qt + 2 pt
                             = 3 qt 2 pt
                    Subtract pints.

         5 gal 3 qt

Step 3:   4 gal 7 qt 2 pt   Rename gallons as quarts to get more quarts.
       −   6 qt 1 pt   5 gal 3 qt = 4 gal + 1 gal + 3 qt
               1 qt 1 pt   = 4 gal + 4 qt + 3 qt
                            = 4 gal 7 qt
                    Subtract quarts.

Step 4:   4 gal 7 qt 2 pt   Subtract gallons.
       −   6 qt 1 pt
       4 gal 1 qt 1 pt ⟵ simplest form

C. Multiplying measures

To **multiply a measure by a number,** you multiply the amount by the number and then write the same unit of measure on the product.

EXAMPLE 9-38   Multiply 5 × 8 L.

Solution   5 × 8 L = (5 × 8) L   Multiply the amount by the number.

              = 40 L   Write the same unit of measure on the product.

To **multiply a mixed measure by a number,** you multiply each individual measure within the mixed measure by the number.

EXAMPLE 9-39   Multiply 6 × 4 lb 8 oz.

Solution   4 lb 8 oz   Write in vertical form.
           ×   6
         24 lb 48 oz   Multiply: 6 × 4 lb = 24 lb
                           6 × 8 oz = 48 oz

        24 lb 48 oz = 24 lb + 48 oz   Simplify.

                 = 24 lb + 3 lb ⟵ 16 oz = 1 lb means 48 oz = 3 lb

                 = 27 lb ⟵ simplest form

Note: 6 × 4 lb 8 oz = 27 lb, or 432 oz.

To **multiply two or more like measures,** you multiply the amounts and multiply the units of measure separately.

Multiplication Rule for Two Like Length Measures
When two like length measures are multiplied, the product will always be an associated area measure (for example, in², ft², m², cm², etc.).

EXAMPLE 9-40 Multiply the following like length measures:

(a) 5 in. × 4 in.   **(b)** 2 ft × 1 ft   **(c)** 6 yd × 3 yd

(d) 1 mm × 8 mm   **(e)** 4 cm × 8 cm   **(f)** 1 m × 1 m

Solution   two like length measures   associated area measure

$$\text{(a)} \quad \overbrace{5 \text{ in.} \times 4 \text{ in.}} = (5 \times 4)(\text{in.} \times \text{in.}) = \overbrace{20 \text{ in}^2}$$

(b) 2 ft × 1 ft $= (2 \times 1)(\text{ft} \times \text{ft})$ $= 2 \text{ ft}^2$

(c) 6 yd × 3 yd $= (6 \times 3)(\text{yd} \times \text{yd})$ $= 18 \text{ yd}^2$

(d) 1 mm × 8 mm $= (1 \times 8)(\text{mm} \times \text{mm}) = 8 \text{ mm}^2$

(e) 4 cm × 8 cm $= (4 \times 8)(\text{cm} \times \text{cm})$ $= 32 \text{ cm}^2$

(f) 1 m × 1 m $= (1 \times 1)(\text{m} \times \text{m})$ $= 1 \text{ m}^2$

Multiplication Rule for Three Length Measures

When three like length measures are multiplied, the product will always be an associated volume measure (for example, in^3, ft^3, m^3, cm^3, etc.).

EXAMPLE 9-41 Multiply the following like length measures:

(a) 1 in. × 4 in. × 1 in.   **(b)** 2 ft × 1 ft × 5 ft

(c) 3 yd × 2 yd × 4 yd   **(d)** 1 cm × 1 cm × 1 cm

(e) 5 m × 4 m × 2 m

Solution   three like length measures   associated volume measure

$$\text{(a)} \quad \overbrace{1 \text{ in.} \times 4 \text{ in.} \times 1 \text{ in.}} = (1 \times 4 \times 1)(\text{in.} \times \text{in.} \times \text{in.}) = \overbrace{4 \text{ in}^3}$$

(b) 2 ft × 1 ft × 5 ft $= (2 \times 1 \times 5)(\text{ft} \times \text{ft} \times \text{ft})$ $= 10 \text{ ft}^3$

(c) 3 yd × 2 yd × 4 yd $= (3 \times 2 \times 4)(\text{yd} \times \text{yd} \times \text{yd})$ $= 24 \text{ yd}^3$

(d) 1 cm × 1 cm × 1 cm $= (1 \times 1 \times 1)(\text{cm} \times \text{cm} \times \text{cm}) = 1 \text{ cm}^3$

(e) 5 m × 4 m × 2 m $= (5 \times 4 \times 2)(\text{m} \times \text{m} \times \text{m})$ $= 40 \text{ m}^3$

When two unlike measures are multiplied, you multiply the amounts and the units of measure separately. For example, in a **work problem,** feet (ft) and pounds (lb) are multiplied to get **foot-pounds** (ft-lb).

EXAMPLE 9-42 Multiply two unlike measures 6 ft × 5 lb.

Solution $6 \text{ ft} \times 5 \text{ lb} = (6 \times 5)(\text{ft} \times \text{lb})$ Multiply the amounts and
                                          units of measure separately.

               $= 30 \text{ ft} \times \text{lb}$ or 30 ft-lb ⟵ simplest form

Note: 30 ft-lb and 30 ft × lb are both read as "thirty foot-pounds."

D. Dividing measures

To **divide a measure by a number,** you divide the amount by the number and then write the same unit of measure on the quotient.

EXAMPLE 9-43 Divide 6 km ÷ 2.

Solution $6 \text{ km} \div 2 = (6 \div 2) \text{ km}$ Divide the amount by the number.

               $= 3 \text{ km}$ Write the same unit of measure on the quotient.

To **divide a mixed measure by a number,** you divide each individual measure within the mixed measure by the number while being careful to leave space in the dividend for any missing units of measure that may be needed.

EXAMPLE 9-44   Divide 9 yd 3 in. ÷ 2.

Solution

yards   feet   inches

$2 \overline{)\ 9\ \text{yd} \qquad\qquad 3\ \text{in.}}$ Write in division box form while leaving space for the missing units of measure: 9 yd 3 in. = 9 yd 0 ft 3 in.

$\begin{array}{r} 4\ \text{yd} \\ 2\overline{)\ 9\ \text{yd} \qquad\qquad 3\ \text{in.}} \\ -8\ \text{yd} \longleftarrow \qquad 2\times 4\ \text{yd} \\ \hline 1\ \text{yd} \longrightarrow 3\ \text{ft} \end{array}$ Divide yards.

Rename yards as feet.

$\begin{array}{r} 4\ \text{yd}\quad 1\ \text{ft}\qquad 7\ \text{in. R1 in.} \\ 2\overline{)\ 9\ \text{yd}\qquad\qquad 3\ \text{in.}} \\ -8\ \text{yd} \\ \hline 1\ \text{yd} \longrightarrow 3\ \text{ft} \\ -2\ \text{ft} \\ \hline 1\ \text{ft} \longrightarrow 12\ \text{in.} \\ 15\ \text{in.} \\ -14\ \text{in.} \longleftarrow 2\times 7\ \text{in.} \\ \hline 1\ \text{in.} \end{array}$ Divide feet.

Rename feet as inches.
Divide inches.

Note: 9 yd 3 in. ÷ 2 = 4 yd 1 ft 7 in. R 1 in.

To **divide two like measures,** you divide the amounts and divide the units of measure separately.

Division Rule for Two Like Measures

When two like measures are divided, the quotient will always be a number, with *no* unit of measure.

EXAMPLE 9-45   Divide the following like measures:

(a) 1 in. ÷ 1 in.   **(b)** 5 lb ÷ 2 lb   **(c)** 18 pt ÷ 10 pt

(d) 1 mm ÷ 2 mm   **(e)** 10 kg ÷ 3 kg   **(f)** 100 L ÷ 1000 L

Solution

two like measures   number

(a) $1\ \text{in.} \div 1\ \text{in.} = (1 \div 1)\dfrac{\cancel{\text{in.}}}{\cancel{\text{in.}}} = 1$

(b) $5\ \text{lb} \div 2\ \text{lb} = (5 \div 2)\dfrac{\cancel{\text{lb}}}{\cancel{\text{lb}}} = \dfrac{5}{2} \text{ or } 2\tfrac{1}{2} \text{ or } 2.5$

(c) $18\ \text{pt} \div 10\ \text{pt} = (18 \div 10)\dfrac{\cancel{\text{pt}}}{\cancel{\text{pt}}} = \dfrac{9}{5} \text{ or } 1\tfrac{4}{5} \text{ or } 1.8$

(d) $1\ \text{mm} \div 2\ \text{mm} = (1 \div 2)\dfrac{\cancel{\text{mm}}}{\cancel{\text{mm}}} = \dfrac{1}{2} \text{ or } 0.5$

(e) $10\ \text{kg} \div 3\ \text{kg} = (10 \div 3)\dfrac{\cancel{\text{kg}}}{\cancel{\text{kg}}} = \dfrac{10}{3} \text{ or } 3\tfrac{1}{3} \text{ or } 3.\overline{3}$

(f) $100\ \text{L} \div 1000\ \text{L} = (100 \div 1000)\dfrac{\cancel{\text{L}}}{\cancel{\text{L}}} = \dfrac{1}{10} \text{ or } 0.1$

To **divide two unlike measures,** you divide the amounts and the units of measure separately.

Division Rule for Two Unlike Measures

When two unlike measures are divided, the quotient is called a **rate.**

EXAMPLE 9-46   Divide the unlike measures 20 lb ÷ 5 ft.

Solution $20 \text{ lb} \div 5 \text{ ft} = (20 \div 5)\dfrac{\text{lb}}{\text{ft}}$ Divide the amounts and units of measure separately.

$$= 4\,\dfrac{\text{lb}}{\text{ft}} \longleftarrow \text{rate}$$

Note: The rate $4\,\dfrac{\text{lb}}{\text{ft}}$ can also be written as 4 lb/ft, or 4 lb per ft, and all three forms are read as "four pounds per foot."

Caution: The unlike measures in a rate cannot be eliminated.

SOLVED PROBLEMS

PROBLEM 9-1   Match each given statement on the left with the most appropriate U.S. Customary measure on the right:

Length

(a) The length of a garage	3 in.
(b) The length of a used pencil	3 ft
(c) The height of a kitchen chair	3 yd
(d) The distance between two towns	3 mi

Capacity

(e) The capacity of a standard can of wall paint	1 tsp
(f) The capacity of a standard can of car oil	1 c
(g) The capacity of an individual soup serving	1 qt
(h) The capacity of 80 drops of water	1 gal

Weight

(i) The weight of a standard bag of sugar	5 oz
(j) The weight of an African elephant	5 lb
(k) The weight of a hamburger patty	5 T

Area

(l) The area of a standard room carpet	90 in²
(m) The floor area of a small house	90 ft²
(n) The area of a farm	90 yd²
(o) The area of the front cover of this text	90 A

Volume

(p) The space in a standard house room ... 800 in³

(q) The space in a large house ... 800 ft³

(r) The space taken up by a portable radio ... 800 yd³

Solution Recall that to help identify U.S. Customary measures, you can use your intuitive idea of 1 foot, 1 quart, 1 pound, and so forth [see Examples 9-2 through 9-11]:

(a) The length of a garage might be 3 yards.

(b) The length of a used pencil might be 3 inches.

(c) The height of a kitchen chair might be 3 feet.

(d) The distance between towns might be 3 miles.

(e) The capacity of a standard can of wall paint is 1 gallon.

(f) The capacity of a standard can of car oil is 1 quart.

(g) The capacity of an individual soup serving is about 1 cup.

(h) The capacity of 80 drops of water is about 1 teaspoon.

(i) The weight of a standard bag of sugar is 5 pounds

(j) The weight of an African elephant might be 5 tons.

(k) The weight of a hamburger patty might be 5 ounces.

(l) The area of a standard room carpet might be 90 square feet.

(m) The floor area of a small house might be 90 square yards.

(n) The area of a farm might be 90 acres.

(o) The area of the front cover of this text is about 90 square inches.

(p) The space in a standard house room might be 800 cubic feet.

(q) The space in a large house might be 800 cubic yards.

(r) The space taken up by a portable radio might be 800 cubic inches.

PROBLEM 9-2 Rename each U.S. Customary measure using the correct unit conversion:

(a) 2 ft = ? in.
(b) 9 ft = ? yd
(c) 2 pt = ? c
(d) 5 pt = ? qt
(e) 2 lb = ? oz

(f) 500 lb = ? T
(g) 2 ft² = ? in²
(h) 27 ft² = ? yd²
(i) 2 ft³ = ? in³
(j) 63 ft³ = ? yd³

Solution Recall that to rename a given measure in terms of a different unit of measure when there is a unit conversion relating the two units of measure, you multiply the given measure by the correct unit fractions [see Example 9-15]:

(a) $2 \text{ ft} = 2 \text{ ft} \times 1 = \dfrac{2 \text{ ft}}{1} \times \dfrac{12 \text{ in.}}{1 \text{ ft}} = (2 \times 12) \text{ in.} = 24 \text{ in.}$

(b) $9 \text{ ft} = 9 \text{ ft} \times 1 = \dfrac{9 \text{ ft}}{1} \times \dfrac{1 \text{ yd}}{3 \text{ ft}} = \dfrac{9}{3} \text{ yd} = 3 \text{ yd}$

(c) $2 \text{ pt} = 2 \text{ pt} \times 1 = \dfrac{2 \text{ pt}}{1} \times \dfrac{2 \text{ c}}{1 \text{ pt}} = (2 \times 2) \text{ c} = 4 \text{ c}$

(d) $5 \text{ pt} = 5 \text{ pt} \times 1 = \dfrac{5 \text{ pt}}{1} \times \dfrac{1 \text{ qt}}{2 \text{ pt}} = \dfrac{5}{2} \text{ qt} = 2\frac{1}{2} \text{ qt}$

(e) $2 \text{ lb} = 2 \text{ lb} \times 1 = \dfrac{2 \text{ lb}}{1} \times \dfrac{16 \text{ oz}}{1 \text{ lb}} = (2 \times 16) \text{ oz} = 32 \text{ oz}$

(f) $500 \text{ lb} = 500 \text{ lb} \times 1 = \dfrac{500 \text{ lb}}{1} \times \dfrac{1 \text{ T}}{2000 \text{ lb}} = \dfrac{500}{2000} \text{ T} = \dfrac{1}{4} \text{ T}$

(g) $2 \text{ ft}^2 = 2 \text{ ft}^2 \times 1 = \dfrac{2 \text{ ft}^2}{1} \times \dfrac{144 \text{ in}^2}{1 \text{ ft}^2} = (2 \times 144) \text{ in}^2 = 288 \text{ in}^2$

(h) $27 \text{ ft}^2 = 27 \text{ ft}^2 \times 1 = \dfrac{27 \text{ ft}^2}{1} \times \dfrac{1 \text{ yd}^2}{9 \text{ ft}^2} = \dfrac{27}{9} \text{ yd}^2 = 3 \text{ yd}^2$

(i) $2 \text{ ft}^3 = 2 \text{ ft}^3 \times 1 = \dfrac{2 \text{ ft}^3}{1} \times \dfrac{1728 \text{ in}^3}{1 \text{ ft}^3} = (2 \times 1728) \text{ in}^3 = 3456 \text{ in}^3$

(j) $63 \text{ ft}^3 = 63 \text{ ft}^3 \times 1 = \dfrac{63 \text{ ft}^3}{1} \times \dfrac{1 \text{ yd}^3}{27 \text{ ft}^3} = \dfrac{63}{27} \text{ yd}^3 = \dfrac{7}{3} \text{ yd}^3 = 2\frac{1}{3} \text{ yd}^3$

PROBLEM 9-3 Rename each U.S. Customary measure when more than one unit conversion is necessary:

(a) 1 mi = ? yd   **(b)** 160 fl oz = ? qt   **(c)** $1 \text{ yd}^2 = ? \text{ in}^2$ **(d)** $72 \text{ in}^2 = ? \text{ yd}^2$

Solution Recall that to rename a given measure when more than one unit conversion is necessary, you first list each necessary unit conversion in the order needed and then multiply by all the correct unit fractions [see Example 9-17]:

(a) 1 mi = 5280 ft and 3 ft = 1 yd are the necessary unit conversions:

$$1 \text{ mi} = \dfrac{1 \text{ mi}}{1} \times \dfrac{5280 \text{ ft}}{1 \text{ mi}} \times \dfrac{1 \text{ yd}}{3 \text{ ft}} = \dfrac{5280}{3} \text{ yd} = 1760 \text{ yd}$$

(b) 8 fl oz = 1 c, 2 c = 1 pt, and 2 pt = 1 qt are the necessary unit conversions:

$$160 \text{ fl oz} = \dfrac{160 \text{ fl oz}}{1} \times \dfrac{1 \text{ c}}{8 \text{ fl oz}} \times \dfrac{1 \text{ pt}}{2 \text{ c}} \times \dfrac{1 \text{ qt}}{2 \text{ pt}} = \dfrac{160}{8 \times 2 \times 2} \text{ qt} = \dfrac{5}{1} \text{ qt} = 5 \text{ qt}$$

(c) $1 \text{ yd}^2 = 9 \text{ ft}^2$ and $1 \text{ ft}^2 = 144 \text{ in}^2$ are the necessary unit conversions:

$$1 \text{ yd}^2 = \dfrac{1 \text{ yd}^2}{1} \times \dfrac{9 \text{ ft}^2}{1 \text{ yd}^2} \times \dfrac{144 \text{ in}^2}{1 \text{ ft}^2} = (9 \times 144) \text{ in}^2 = 1296 \text{ in}^2$$

(d) $144 \text{ in}^2 = 1 \text{ ft}^2$ and $9 \text{ ft}^2 = 1 \text{ yd}^2$ are the necessary unit conversions:

$$72 \text{ in}^2 = \dfrac{72 \text{ in}^2}{1} \times \dfrac{1 \text{ ft}^2}{144 \text{ in}^2} \times \dfrac{1 \text{ yd}^2}{9 \text{ ft}^2} = \dfrac{72}{144 \times 9} \text{ yd}^2 = \dfrac{1}{18} \text{ yd}^2$$

PROBLEM 9-4 Rename each U.S. Customary mixed measure:

(a) 3 yd 2 ft = ? ft   **(b)** 5 qt 1 pt = ? qt   **(c)** $1 \text{ yd}^2 \ 3 \text{ ft}^2 = ? \text{ ft}^2$ **(d)** $2 \text{ yd}^3 \ 18 \text{ ft}^3 = ? \text{ ft}^3$

Solution Recall that to rename a mixed measure, you first write the mixed measure as a sum [see Example 9-19]:

(a) $3 \text{ yd } 2 \text{ ft} = 3 \text{ yd} + 2 \text{ ft} = \dfrac{3 \text{ yd}}{1} \times \dfrac{3 \text{ ft}}{1 \text{ yd}} + 2 \text{ ft} = 9 \text{ ft} + 2 \text{ ft} = 11 \text{ ft}$

(b) $5 \text{ qt } 1 \text{ pt} = 5 \text{ qt} + 1 \text{ pt} = 5 \text{ qt} + \dfrac{1 \text{ pt}}{1} \times \dfrac{1 \text{ qt}}{2 \text{ pt}} = 5 \text{ qt} + \dfrac{1}{2} \text{ qt} = 5\frac{1}{2} \text{ qt}$

(c) $1 \text{ yd}^2 \ 3 \text{ ft}^2 = 1 \text{ yd}^2 + 3 \text{ ft}^2 = \dfrac{1 \text{ yd}^2}{1} \times \dfrac{9 \text{ ft}^2}{1 \text{ yd}^2} + 3 \text{ ft}^2 = 9 \text{ ft}^2 + 3 \text{ ft}^2 = 12 \text{ ft}^2$

(d) $2 \text{ yd}^3 \ 18 \text{ ft}^3 = 2 \text{ yd}^3 + 18 \text{ ft}^3 = \dfrac{2 \text{ yd}^3}{1} \times \dfrac{27 \text{ ft}^3}{1 \text{ yd}^3} + 18 \text{ ft}^3 = 54 \text{ ft}^3 + 18 \text{ ft}^3 = 72 \text{ ft}^3$

PROBLEM 9-5 Match each given item on the left with the most appropriate metric measure on the right:

Length

 (a) The height of a room   3 mm

 (b) The width of a paper clip   3 cm

(c) The distance from home to work   3 m

(d) The length of an eraser   3 km

Capacity

(e) The capacity of a large water bottle   20 mL

(f) The capacity of a large tablespoon   20 L

(g) The capacity of a city water tank   20 kL

Mass

(h) The mass of a high school girl   50 mg

(i) The mass of an alarm clock   50 g

(j) The mass of a train engine   50 kg

(k) The mass of a large bird feather   50 t

Area

(l) The surface area of a double house door   5 cm^2

(m) The surface area of a city block   5 m^2

(n) The surface area of a U.S. nickel   5 ha

Volume

(o) The space taken up by a hand-held calculator   20 cm^3

(p) The space inside a swimming pool   20 m^3

Solution Recall that to help identify metric measures, you can use your intuitive idea of 1 meter, 1 liter, 1 gram, and so forth [see Examples 9-21 through 9-30]:

(a) The height of a room might be 3 m.

(b) The width of a paper clip might be 3 mm.

(c) The distance from home to work might be 3 km.

(d) The length of an eraser might be 3 cm.

(e) The capacity of a large water bottle might be 20 L.

(f) The capacity of a large tablespoon might be 20 mL.

(g) The capacity of a city water tank might be 20 kL.

(h) The mass of a high school girl might be 50 kg.

(i) The mass of an alarm clock might be 50 g.

(j) The mass of a train engine might be 50 t.

(k) The mass of a large bird feather might be 50 mg.

(l) The surface area of a double house door might be 5 m^2.

(m) The surface area of one city block might be 5 ha.

(n) The surface area of a U.S. nickel is about 5 cm^2.

(o) The space taken up by a hand-held calculator might be 20 cm^3.

(p) The space inside a swimming pool might be 20 m^3.

PROBLEM 9-6 Rename each metric measure using the correct unit conversions:

(a) 2 cm = ? mm   (b) 300 cm = ? m   (c) 2.5 km = ? m   (d) 500 mL = ? L

(e) 3 g = ? mg   (f) 250 g = ? kg   (g) 1000 cm^2 = ? m^2 (h) 1.5 cm^3 = ? mm^3

Solution Recall that to rename a given measure in terms of a different unit of measure when there is a unit conversion relating the two units of measure, you multiply the given measure by the correct unit fraction [see Example 9-31]:

(a) $2 \text{ cm} = 2 \text{ cm} \times 1 = \dfrac{2 \cancel{\text{ cm}}}{1} \times \dfrac{10 \text{ mm}}{1 \cancel{\text{ cm}}} = (2 \times 10) \text{ mm} = 20 \text{ mm}$

(b) $300 \text{ cm} = 300 \text{ cm} \times 1 = \dfrac{300 \cancel{\text{ cm}}}{1} \times \dfrac{1 \text{ m}}{100 \cancel{\text{ cm}}} = \dfrac{300}{100} \text{ m} = 3 \text{ m}$

(c) $2.5 \text{ km} = 2.5 \text{ km} \times 1 = \dfrac{2.5 \cancel{\text{ km}}}{1} \times \dfrac{1000 \text{ m}}{1 \cancel{\text{ km}}} = (2.5 \times 1000) \text{ m} = 2500 \text{ m}$

(d) $500 \text{ mL} = 500 \text{ mL} \times 1 = \dfrac{500 \cancel{\text{ mL}}}{1} \times \dfrac{1 \text{ L}}{1000 \cancel{\text{ mL}}} = \dfrac{500}{1000} \text{ L} = \dfrac{1}{2} \text{ L} \quad \text{or} \quad 0.5 \text{ L}$

(e) $3 \text{ g} = 3 \text{ g} \times 1 = \dfrac{3 \cancel{\text{ g}}}{1} \times \dfrac{1000 \text{ mg}}{1 \cancel{\text{ g}}} = (3 \times 1000) \text{ mg} = 3000 \text{ mg}$

(f) $250 \text{ g} = 250 \text{ g} \times 1 = \dfrac{250 \cancel{\text{ g}}}{1} \times \dfrac{1 \text{ kg}}{1000 \cancel{\text{ g}}} = \dfrac{250}{1000} \text{ kg} = \dfrac{1}{4} \text{ kg} \quad \text{or} \quad 0.25 \text{ kg}$

(g) $1000 \text{ cm}^2 = 1000 \text{ cm}^2 \times 1 = \dfrac{1000 \cancel{\text{ cm}^2}}{1} \times \dfrac{1 \text{ m}^2}{10{,}000 \cancel{\text{ cm}^2}} = \dfrac{1000}{10{,}000} \text{ m}^2 = \dfrac{1}{10} \text{ m}^2 \quad \text{or} \quad 0.1 \text{ m}^2$

(h) $1.5 \text{ cm}^3 = 1.5 \text{ cm}^3 \times 1 = \dfrac{1.5 \cancel{\text{ cm}^3}}{1} \times \dfrac{1000 \text{ mm}^3}{1 \cancel{\text{ cm}^3}} = (1.5 \times 1000) \text{ mm}^3 = 1500 \text{ mm}^3$

PROBLEM 9-7   Rename each metric measure when more than one unit conversion is necessary:

(a) $2 \text{ m} = ? \text{ mm}$ (b) $2.5 \text{ kg} = ? \text{ mg}$ (c) $500{,}000 \text{ g} = ? \text{ t}$

Solution   Recall that to rename a measure when more than one unit conversion is necessary, you first list each necessary unit conversion in the order needed and then multiply by all the correct unit fractions [see Example 9-32]:

(a) $1 \text{ m} = 100 \text{ cm}$ and $1 \text{ cm} = 10 \text{ mm}$ are the necessary unit conversions:

$$2 \text{ m} = \frac{2 \cancel{\text{ m}}}{1} \times \frac{100 \cancel{\text{ cm}}}{1 \cancel{\text{ m}}} \times \frac{10 \text{ mm}}{1 \cancel{\text{ cm}}} = (2 \times 100 \times 10) \text{ mm} = 2000 \text{ mm}$$

(b) $1 \text{ kg} = 1000 \text{ g}$ and $1 \text{ g} = 1000 \text{ mg}$ are the necessary unit conversions:

$$2.5 \text{ kg} = \frac{2.5 \cancel{\text{ kg}}}{1} \times \frac{1000 \cancel{\text{ g}}}{1 \cancel{\text{ kg}}} \times \frac{1000 \text{ mg}}{1 \cancel{\text{ g}}} = (2.5 \times 1000 \times 1000) \text{ mg} = 2{,}500{,}000 \text{ mg}$$

(c) $1000 \text{ g} = 1 \text{ kg}$ and $1000 \text{ kg} = 1 \text{ t}$ are the necessary unit conversions:

$$500{,}000 \text{ g} = \frac{500{,}000 \cancel{\text{ g}}}{1} \times \frac{1 \cancel{\text{ kg}}}{1000 \cancel{\text{ g}}} \times \frac{1 \text{ t}}{1000 \cancel{\text{ kg}}} = \frac{500{,}000}{1000 \times 1000} \text{ t} = \frac{1}{2} \text{ t} = 0.5 \text{ t}$$

PROBLEM 9-8   Add like measures:

(a) $5 \text{ m} + 8 \text{ m}$ (b) $2\frac{1}{2} \text{ ft} + 3 \text{ ft}$ (c) $0.3 \text{ g} + 1.8 \text{ g}$ (d) $\frac{1}{2} \text{ c} + \frac{3}{4} \text{ c}$

Solution   Recall that to add like measures, you add the amounts and then write the same unit of measure on the sum [see Example 9-34]:

(a) $5 \text{ m} + 8 \text{ m} = (5 + 8) \text{ m} = 13 \text{ m}$ (b) $2\frac{1}{2} \text{ ft} + 3 \text{ ft} = (2\frac{1}{2} + 3) \text{ ft} = 5\frac{1}{2} \text{ ft}$

(c) $0.3 \text{ g} + 1.8 \text{ g} = (0.3 + 1.8) \text{ g} = 2.1 \text{ g}$ (d) $\frac{1}{2} \text{ c} + \frac{3}{4} \text{ c} = (\frac{1}{2} + \frac{3}{4}) \text{ c} = \frac{5}{4} \text{ c} = 1\frac{1}{4} \text{ c}$

PROBLEM 9-9   Add mixed measures:

(a) $5 \text{ yd } 2 \text{ ft} + 6 \text{ yd } 1 \text{ ft}$ (b) $2 \text{ ft } 8\frac{1}{2} \text{ in.} + 4 \text{ yd } 7\frac{1}{2} \text{ in.}$

(c) $8 \text{ lb } 10 \text{ oz} + 6 \text{ lb } 11 \text{ oz}$ (d) $3 \text{ gal } 2 \text{ qt } 1 \text{ pt} + 3 \text{ qt } 1 \text{ pt}$

Solution Recall that to add mixed measures, you first line up corresponding like measures in columns, then add like measures, and then simplify the sum if possible [see Example 9-35]:

(a)

$$
\begin{array}{rl}
5 \text{ yd} & 2 \text{ ft} \\
+6 \text{ yd} & 1 \text{ ft} \\
\hline
11 \text{ yd} & 3 \text{ ft} = 11 \text{ yd} + 3 \text{ ft} \\
& = 11 \text{ yd} + 1 \text{ yd} \\
& = 12 \text{ yd}
\end{array}
$$

(b)

$$
\begin{array}{rll}
& 2 \text{ ft} & 8\frac{1}{2} \text{ in.} \\
+4 \text{ yd} & & 7\frac{1}{2} \text{ in.} \\
\hline
4 \text{ yd} & 2 \text{ ft} & 16 \text{ in.} = 4 \text{ yd} + 2 \text{ ft} + 16 \text{ in.} \\
& & = 4 \text{ yd} + 2 \text{ ft} + 12 \text{ in.} + 4 \text{ in.} \\
& & = 4 \text{ yd} + 2 \text{ ft} + 1 \text{ ft} + 4 \text{ in.} \\
& & = 4 \text{ yd} + 3 \text{ ft} + 4 \text{ in.} \\
& & = 4 \text{ yd} + 1 \text{ yd} + 4 \text{ in.} \\
& & = 5 \text{ yd} 4 \text{ in.}
\end{array}
$$

(c)

$$
\begin{array}{rl}
8 \text{ lb} & 10 \text{ oz} \\
+6 \text{ lb} & 11 \text{ oz} \\
\hline
14 \text{ lb} & 21 \text{ oz} = 14 \text{ lb} + 21 \text{ oz} \\
& = 14 \text{ lb} + 16 \text{ oz} + 5 \text{ oz} \\
& = 14 \text{ lb} + 1 \text{ lb} + 5 \text{ oz} \\
& = 15 \text{ lb} 5 \text{ oz}
\end{array}
$$

(d)

$$
\begin{array}{rll}
3 \text{ gal} & 2 \text{ qt} & 1 \text{ pt} \\
+ & 3 \text{ qt} & 1 \text{ pt} \\
\hline
3 \text{ gal} & 5 \text{ qt} & 2 \text{ pt} = 3 \text{ gal} + 5 \text{ qt} + 2 \text{ pt} \\
& & = 3 \text{ gal} + 4 \text{ qt} + 1 \text{ qt} + 1 \text{ qt} \\
& & = 3 \text{ gal} + 1 \text{ gal} + 2 \text{ qt} \\
& & = 4 \text{ gal} 2 \text{ qt}
\end{array}
$$

PROBLEM 9-10 Subtract like measures:

(a) 3 L − 1.5 L   (b) 6 lb − $4\frac{1}{2}$ lb   (c) 2.8 km − 0.75 km   (d) $\frac{3}{4}$ in. − $\frac{5}{8}$ in.

Solution Recall that to subtract like measures, you subtract the amounts and then write the same unit of measure on the difference [see Example 9-36]:

(a) 3 L − 1.5 L = (3 − 1.5) L = 1.5 L   (b) 6 lb − $4\frac{1}{2}$ lb = $(6 − 4\frac{1}{2})$ lb = $1\frac{1}{2}$ lb

(c) 2.8 km − 0.75 km = (2.8 − 0.75) km = 2.05 km   (d) $\frac{3}{4}$ in. − $\frac{5}{8}$ in. = $(\frac{3}{4} − \frac{5}{8})$ in. = $\frac{1}{8}$ in.

PROBLEM 9-11 Subtract mixed measures:

(a) 8 lb 9 oz − 3 lb $7\frac{1}{2}$ oz   (b) 2 yd − 2 ft 6 in.

(c) 5 gal 1 pt − 3 qt 1 c   (d) $4\frac{1}{2}$ T 500 lb − 1800 lb

Solution Recall that to subtract mixed measures, you first line up corresponding like measures in columns, then rename when necessary, and then subtract like measures [see Example 9-37]:

(a)

$$
\begin{array}{rl}
8 \text{ lb} & 9 \text{ oz} \\
-3 \text{ lb} & 7\frac{1}{2} \text{ oz} \\
\hline
5 \text{ lb} & 1\frac{1}{2} \text{ oz}
\end{array}
$$

(b)

$$
\begin{array}{rll}
2 \text{ yd} & & \overbrace{= 1 \text{ yd} \quad 3 \text{ ft}}^{2 \text{ yd}} \quad = 1 \text{ yd} \overbrace{\quad 2 \text{ ft} \quad 12 \text{ in.}}^{3 \text{ ft}} \\
- \quad 2 \text{ ft} \ 6 \text{ in.} & = & \quad 2 \text{ ft} \ 6 \text{ in.} = \qquad 2 \text{ ft} \quad 6 \text{ in.} \\
\hline
& & 1 \text{ yd} \quad 0 \text{ ft} \quad 6 \text{ in.} = 1 \text{ yd } 6 \text{ in.}
\end{array}
$$

(c)

$$
\begin{array}{rll}
5 \text{ gal} & 1 \text{ pt} & = \overbrace{4 \text{ gal} \quad 4 \text{ qt}}^{5 \text{ gal}} \quad \overbrace{0 \text{ pt} \quad 2 \text{ c}}^{1 \text{ pt}} \\
- \quad 3 \text{ qt} & 1 \text{ c} & = \qquad 3 \text{ qt} \qquad\quad 1 \text{ c} \\
\hline
& & 4 \text{ gal} \quad 1 \text{ qt} \quad 0 \text{ pt} \quad 1 \text{ c} = 4 \text{ gal } 1 \text{ qt } 1 \text{ c}
\end{array}
$$

(d)

$$
\begin{array}{rl}
4\frac{1}{2} \text{ T} & 500 \text{ lb} = \overbrace{3\frac{1}{2} \text{ T} \quad 2500 \text{ lb}}^{4\frac{1}{2}\text{ T } 500 \text{ lb}} \longleftarrow \quad 4\frac{1}{2} \text{ T} + 500 \text{ lb} = 3\frac{1}{2} \text{ T} + 1 \text{ T} + 500 \text{ lb} \\
- \qquad\quad 1800 \text{ lb} = \qquad\quad 1800 \text{ lb} \qquad\qquad\qquad\qquad = 3\frac{1}{2} \text{ T} + 2000 \text{ lb} + 500 \text{ lb} \\
\hline
\qquad 3\frac{1}{2} \text{ T} \quad\ 700 \text{ lb}
\end{array}
$$

PROBLEM 9-12   Multiply a measure by a number:

(a) 3×5 cm
(b) 5.6 mL $\times 0.25$
(c) $8 \times 4\frac{1}{2}$ ft
(d) $\frac{3}{4}$ lb $\times \frac{2}{3}$

Solution   Recall that to multiply a measure by a number, you multiply the amounts and then write the same unit of measure on the product [see Example 9-38]:

(a) 3×5 cm $= (3 \times 5)$ cm $= 15$ cm
(b) 5.6 mL $\times 0.25 = (5.6 \times 0.25)$ mL $= 1.4$ mL

(c) $8 \times 4\frac{1}{2}$ ft $= (8 \times 4\frac{1}{2})$ ft $= 36$ ft
(d) $\frac{3}{4}$ lb $\times \frac{2}{3} = (\frac{3}{4} \times \frac{2}{3})$ lb $= \frac{1}{2}$ lb

PROBLEM 9-13   Multiply a mixed measure by a number:

(a) 2×5 mi 30 yd
(b) 5 T 400 lb $\times 5$
(c) $\frac{1}{2} \times 2$ gal 3 qt 1 pt
(d) 5 yd 6 ft 10 in. $\times 2\frac{1}{2}$

Solution   To multiply a mixed measure by a number, you multiply each individual measure within the mixed measure by the number [see Example 9-39]:

(a)
$$
\begin{array}{rr}
5\text{ mi} & 30\text{ yd} \\
\times & 2 \\
\hline
10\text{ mi} & 60\text{ yd}
\end{array}
$$

(b)
$$
\begin{array}{rr}
5\text{ T} & 400\text{ lb} \\
\times & 5 \\
\hline
25\text{ T} & 2000\text{ lb}
\end{array} = 25\text{ T} + 2000\text{ lb}
$$
$$
= 25\text{ T} + 1\text{ T}
$$
$$
= 26\text{ T}
$$

(c)
$$
\begin{array}{rrr}
2\text{ gal} & 3\text{ qt} & 1\text{ pt} \\
\times & & \frac{1}{2} \\
\hline
1\text{ gal} & 1\frac{1}{2}\text{ qt} & \frac{1}{2}\text{ pt}
\end{array}
$$

(d)
$$
\begin{array}{rrr}
5\text{ yd} & 6\text{ ft} & 10\text{ in.} \\
\times & & 2\frac{1}{2} \\
\hline
12\frac{1}{2}\text{ yd} & 15\text{ ft} & 25\text{ in.}
\end{array} = 12\frac{1}{2}\text{ yd} + 15\text{ ft} + 25\text{ in.}
$$
$$
= 12\frac{1}{2}\text{ yd} + \overbrace{5\text{ yd}} + \overbrace{2\text{ ft} + 1\text{ in.}}
$$
$$
= 17\frac{1}{2}\text{ yd } 2\text{ ft } 1\text{ in.}
$$

PROBLEM 9-14   Multiply two like length measures:

(a) 3 in. $\times$ 5 in.
(b) $1\frac{1}{2}$ ft $\times$ 4 ft
(c) $\frac{1}{2}$ yd $\times \frac{3}{4}$ yd
(d) 2 mm $\times$ 3 mm
(e) 5 cm $\times$ 0.5 cm
(f) 2.3 m $\times$ 0.8 m

Solution   Recall that when two like length measures are multiplied, the product will always be the associated area measure [see Example 9-40]:

(a) 3 in. $\times$ 5 in. $= (3 \times 5)$ in$^2 = 15$ in^2
(b) $1\frac{1}{2}$ ft $\times$ 4 ft $= (1\frac{1}{2} \times 4)$ ft$^2 = 6$ ft^2

(c) $\frac{1}{2}$ yd $\times \frac{3}{4}$ yd $= (\frac{1}{2} \times \frac{3}{4})$ yd$^2 = \frac{3}{8}$ yd^2
(d) 2 mm $\times$ 3 mm $= (2 \times 3)$ mm$^2 = 6$ mm^2

(e) 5 cm $\times$ 0.5 cm $= (5 \times 0.5)$ cm$^2 = 2.5$ cm^2
(f) 2.3 m $\times$ 0.8 m $= (2.3 \times 0.8)$ m$^2 = 1.84$ m^2

PROBLEM 9-15   Multiply three like length measures:

(a) 2 in. $\times$ 3 in. $\times$ 5 in.   **(b)** $\frac{1}{2}$ yd $\times$ 3 yd $\times 2\frac{1}{2}$ yd   **(c)** $\frac{1}{2}$ ft $\times \frac{3}{4}$ ft $\times \frac{5}{8}$ ft   **(d)** 5 cm $\times$ 2 cm $\times$ 4 cm

Solution   Recall that when three like length measures are multiplied, the product will always be an associated volume measure [see Example 9-41]:

(a) 2 in. $\times$ 3 in. $\times$ 5 in. $= (2 \times 3 \times 5)$ in$^3 = 30$ in^3

(b) $\frac{1}{2}$ yd $\times$ 3 yd $\times 2\frac{1}{2}$ yd $= (\frac{1}{2} \times 3 \times 2\frac{1}{2})$ yd$^3 = \frac{15}{4}$ yd$^3 = 3\frac{3}{4}$ yd^3

(c) $\frac{1}{2}$ ft $\times \frac{3}{4}$ ft $\times \frac{5}{8}$ ft $= (\frac{1}{2} \times \frac{3}{4} \times \frac{5}{8})$ ft$^3 = \frac{15}{64}$ ft^3

(d) 5 cm $\times$ 2 cm $\times$ 4 cm $= (5 \times 2 \times 4)$ cm$^3 = 40$ cm^3

PROBLEM 9-16   Divide a measure by a number:

(a) 8 ft $\div 2$
(b) $7\frac{1}{2}$ lb $\div 3$
(c) 18 m $\div 1.5$
(d) 0.75 L $\div 0.25$

Solution   Recall that to divide a measure by a number, you divide the amounts and then write the same unit of measure on the quotient [see Example 9-43]:

(a) $8 \text{ ft} \div 2 = (8 \div 2) \text{ ft} = 4 \text{ ft}$

(b) $7\frac{1}{2} \text{ lb} \div 3 = (7\frac{1}{2} \div 3) \text{ lb} = 2\frac{1}{2} \text{ lb}$

(c) $18 \text{ m} \div 1.5 = (18 \div 1.5) \text{ m} = 12 \text{ m}$

(d) $0.75 \text{ L} \div 0.25 = (0.75 \div 0.25) \text{ L} = 3 \text{ L}$

PROBLEM 9-17   Divide a mixed measure by a number:

(a) 16 lb 8 oz ÷ 4   **(b)** 3 T 500 lb ÷ 5   **(c)** 8 yd 5 in. ÷ 3   **(d)** 9 gal 3 qt 1 c ÷ 2

Solution   Recall that to divide a mixed measure by a number, you divide each individual measure within the mixed measure by the number while being careful to leave space in the dividend for any missing units of measure that may be needed [see Example 9-44]:

```
                4 lb    2 oz
(a)  4 ) 16 lb   8 oz
         −16 lb
            0    8 oz
                −8 oz
                  0
```

```
                        1300 lb
(b)  5 ) 3 T      500 lb
         ↳→6000 lb
           6500 lb
          −6500 lb
              0
```

```
            2 yd    2 ft    1 in.  R2 in.
(c)  3 )  8 yd            5 in.
         −6 yd
          2 yd    6 ft
                 −6 ft
                   0     5 in.
                        −3 in.
                         2 in.
```

```
            4 gal    3 qt    1 pt      R1 c
(d)  2 )  9 gal     3 qt             1 c ↱
         −8 gal
          1 gal     4 qt
                    7 qt
                   −6 qt
                    1 qt    2 pt
                           −2 pt
                             0
```

PROBLEM 9-18   Divide like measures:

(a) $8 \text{ ft} \div 2 \text{ ft}$ **(b)** $\frac{3}{4} \text{ c} \div \frac{1}{2} \text{ c}$ **(c)** $10 \text{ m} \div 5 \text{ m}$ **(d)** $0.25 \text{ mL} \div 0.4 \text{ mL}$

Solution   Recall that when two like measures are divided, the quotient will always be a number (no unit of measure) [see Example 9-45]:

(a) $8 \text{ ft} \div 2 \text{ ft} = (8 \div 2)\dfrac{\cancel{\text{ft}}}{\cancel{\text{ft}}} = 4$

(b) $\dfrac{3}{4} \text{ c} \div \dfrac{1}{2} \text{ c} = \left(\dfrac{3}{4} \div \dfrac{1}{2}\right)\dfrac{\cancel{\text{c}}}{\cancel{\text{c}}} = 1\frac{1}{2}$

(c) $10 \text{ m} \div 5 \text{ m} = (10 \div 5)\dfrac{\cancel{\text{m}}}{\cancel{\text{m}}} = 2$

(d) $0.25 \text{ mL} \div 0.4 \text{ mL} = (0.25 \div 0.4)\dfrac{\cancel{\text{mL}}}{\cancel{\text{mL}}} = 0.625$

PROBLEM 9-19   Divide unlike measures:

(a) $8 \text{ gal} \div 2 \text{ ft}$ **(b)** $6\frac{1}{2} \text{ oz} \div 2\frac{1}{2} \text{ in.}$ **(c)** $12 \text{ kg} \div 0.5 \text{ m}$ **(d)** $1.25 \text{ g} \div 2 \text{ mL}$

Solution   Recall that when unlike measures are divided, the quotient will always be a rate [see Example 9-46]:

(a) $8 \text{ gal} \div 2 \text{ ft} = (8 \div 2)\dfrac{\text{gal}}{\text{ft}} = 4\,\dfrac{\text{gal}}{\text{ft}}$

(b) $6\frac{1}{2} \text{ oz} \div 2\frac{1}{2} \text{ in.} = (6.5 \div 2.5)\dfrac{\text{oz}}{\text{in.}} = 2.6\,\dfrac{\text{oz}}{\text{in.}}$

(c) $12 \text{ kg} \div 0.5 \text{ m} = (12 \div 0.5)\dfrac{\text{kg}}{\text{m}} = 24\,\dfrac{\text{kg}}{\text{m}}$

(d) $1.25 \text{ g} \div 2 \text{ mL} = (1.25 \div 2)\dfrac{\text{g}}{\text{mL}} = 0.625\,\dfrac{\text{g}}{\text{mL}}$

Supplementary Exercises

PROBLEM 9-20   Match each given statement on the left with the most appropriate U.S. Customary measure on the right:

(a) The weight of a portable typewriter   2 ft^3

(b) The length of a jumbo paper clip   7 ft

(c) The capacity of a small carton of cream   2 ft^2

(d) The weight of a golf ball   1 c

(e) The height of a basketball player   2 oz

(f) The capacity of a small milk glass   2 in.

(g) The top area of a coffee table   8 lb

(h) The capacity of a fish tank   1 pt

PROBLEM 9-21   Rename U.S. Customary measures:

(a) 30 in. = ? ft   (b) $3\frac{1}{2}$ gal = ? pt   (c) 2 T 800 lb = ? lb   (d) 8 yd = ? ft

(e) $1\frac{1}{2}$ mi = ? yd   (f) 2 yd $1\frac{1}{2}$ ft = ? yd   (g) 3 pt = ? c   (h) $2\frac{1}{2}$ qt = ? c

(i) $\frac{1}{2}$ gal 1 qt = ? qt   (j) 24 oz = ? lb   (k) 3 lb 6 oz = ? lb   (l) 36 in. = ? ft

(m) 2 yd = ? ft   (n) 6600 ft = ? mi   (o) 1980 yd = ? mi   (p) 60 in. = ? yd

(q) 5 yd 2 ft = ? ft   (r) 7 c = ? pt   (s) 9 pt = ? qt   (t) 11 qt = ? gal

(u) 20 c = ? qt   (v) 3 qt 1 pt = ? qt   (w) 48 oz = ? lb   (x) 6000 lb = ? T

PROBLEM 9-22   Match each given statement on the left with the most appropriate metric measure on the right:

(a) The length of a standard paper clip   5 kL

(b) The capacity of a car's gas tank   2 m^3

(c) The mass of a thumbtack   400 mg

(d) The distance from Los Angeles to New York   75 L

(e) The capacity of a farm water tank   2 m^2

(f) The mass of a 10-speed bicycle   3 cm

(g) The area of a standard house door   4500 km

(h) The capacity of a bathtub   13 kg

PROBLEM 9-23   Rename metric measures:

(a) 325 cm = ? m   (b) 0.5 kL = ? mL   (c) 750 mg = ? g   (d) 250 mm = ? cm

(e) 2.5 m = ? cm   (f) 1.5 L = ? mL   (g) 3250 mL = ? L   (h) 2.5 kg = ? g

(i) 0.2 kg = ? mg   (j) 800 mm = ? m   (k) 900 cm = ? m   (l) 2000 m = ? km

(m) 1 m = ? mm   (n) 7000 mL = ? L   (o) 8500 mL = ? L   (p) 5250 L = ? kL

(q) 1 kL = ? mL   (r) 900 mg = ? g   (s) 215 g = ? kg   (t) 4000 kg = ? t

(u) 1 kg = ? mg   (v) 0.3 g = ? mg   (w) 0.8 m = ? cm   (x) 0.05 km = ? cm

PROBLEM 9-24   Add measures:

(a) 6 m + 2 m   (b) 5 yd 2 ft 10 in. + 3 yd 8 in.   (c) 3 qt 1 pt 1 c + 1 pt 1 c (**d**) 46 t + 25 t

(**e**) 8 mm + 7 mm

(**f**) 11 L + 35 L

(**g**) 1.5 t + 2.6 t

(**h**) 0.6 km + 0.85 km

(**i**) $5\frac{1}{2}$ mi + $6\frac{1}{4}$ mi

(**j**) $1\frac{1}{3}$ c + $3\frac{3}{4}$ c

(**k**) 5 yd 1 ft + 8 yd 1 ft

(**l**) 3 ft $6\frac{1}{2}$ in. + 2 ft $5\frac{1}{4}$ in.

(**m**) 8 lb 10 oz + 3 lb 9 oz

(**n**) 4 yd $7\frac{1}{4}$ in. + $9\frac{3}{8}$ in.

(**o**) 4 gal 3 qt 1 pt + 2 qt 1 pt

PROBLEM 9-25 Subtract measures:

(**a**) 9 mL − 5 mL

(**b**) 72 m − 15 m

(**c**) 2.5 g − 0.8 g

(**d**) 8.5 kL − 3.9 kL

(**e**) $3\frac{1}{3}$ T − $2\frac{1}{2}$ T

(**f**) $5\frac{1}{8}$ pt − $1\frac{1}{4}$ pt

(**g**) 8 lb 10 oz − 4 lb 7 oz

(**h**) 3 gal $2\frac{1}{2}$ qt − 1 gal $1\frac{3}{4}$ qt

(**i**) 6 yd 1 ft − 2 yd 2 ft

(**j**) 5 mi 25 yd − 750 yd

(**k**) 5 yd 2 ft − 5 ft 6 in.

(**l**) 5 gal 1 pt − 3 qt 1 c

(**m**) 6 m − 2 m

(**n**) 4 gal 2 qt − 3 qt 1 pt

(**o**) 5 yd 6 in. − 2 ft 8 in.

PROBLEM 9-26 Multiply with measures:

(**a**) 7 × 8 mg

(**b**) 85 kg × 4

(**c**) 0.6 × 1.5 m

(**d**) 2.8 L × 1.5

(**e**) 3 × 6 mi 5 yd

(**f**) 5 T 200 lb × 8

(**g**) $2\frac{1}{2}$ × 6 gal 3 qt 1 pt

(**h**) 6 yd 5 ft 10 in. × $3\frac{1}{2}$

(**i**) 7 in. × 8 in.

(**j**) $2\frac{1}{2}$ ft × 6 ft

(**k**) $\frac{1}{2}$ yd × $3\frac{1}{2}$ yd

(**l**) $\frac{3}{4}$ mi × $\frac{2}{3}$ mi

(**m**) 5 mm × 9 mm

(**n**) 0.5 cm × 8 cm

(**o**) 1.5 m × 3.2 m

(**p**) 0.1 km × 0.1 km

(**q**) 2 in. × 5 in. × 8 in.

(**r**) $2\frac{1}{2}$ ft × 2 ft × $\frac{1}{4}$ ft

(**s**) $\frac{1}{2}$ yd × $\frac{1}{2}$ yd × $\frac{1}{2}$ yd

(**t**) 0.5 mm × 2 mm × 1 mm

(**u**) 2 cm × 1.25 cm × 0.25 cm

(**v**) 0.1 m × 0.1 m × 0.1 m

(**w**) 4 × 8 lb 5 oz

(**x**) 6 × 2 m

PROBLEM 9-27 Divide measures:

(**a**) 6 m ÷ 2

(**b**) 7 yd 5 in. ÷ 4

(**c**) 1 gal 1 pt ÷ 3

(**d**) 2.4 m ÷ 1.5

(**e**) 15 g ÷ 3

(**f**) 81 t ÷ 6

(**g**) 930 m ÷ 0.6

(**h**) 8.52 mm ÷ 0.04

(**i**) $\frac{1}{4}$ ft ÷ $\frac{1}{2}$

(**j**) $\frac{5}{6}$ gal ÷ $\frac{2}{3}$

(**k**) $2\frac{2}{3}$ mi ÷ $1\frac{1}{3}$

(**l**) $1\frac{7}{8}$ in. ÷ $2\frac{1}{4}$

(**m**) 180 lb 15 oz ÷ 5

(**n**) 21 gal 3 qt ÷ 3

(**o**) 7 T 14 oz ÷ $3\frac{1}{2}$

(**p**) 25 yd 11 in. ÷ $2\frac{1}{2}$

(**q**) 12 in. ÷ 3 in.

(**r**) 20 ft ÷ $\frac{1}{2}$ ft

(**s**) $10\frac{1}{2}$ yd ÷ 3 yd

(**t**) $\frac{1}{2}$ mi ÷ $\frac{3}{4}$ mi

(**u**) 0.5 m ÷ 1.5 m

(**v**) 10.8 cm ÷ 2 cm

(**w**) 18 lb ÷ 2 ft

(**x**) 36 g ÷ 0.5 cm

Answers to Supplementary Exercises

(**9-20**) (**a**) 8 lb (**b**) 2 in. (**c**) 1 pt (**d**) 2 oz (**e**) 7 ft (**f**) 1 c (**g**) 2 ft² (**h**) 2 ft³

(**9-21**) (**a**) $2\frac{1}{2}$ ft (**b**) 28 pt (**c**) 4800 lb (**d**) 24 ft (**e**) 2640 yd (**f**) $2\frac{1}{2}$ yd
(**g**) 6 c (**h**) 10 c (**i**) 3 qt (**j**) $1\frac{1}{2}$ lb (**k**) $3\frac{3}{8}$ (**l**) 3 ft
(**m**) 6 ft (**n**) $1\frac{1}{4}$ mi (**o**) $1\frac{1}{8}$ mi (**p**) $1\frac{2}{3}$ yd (**q**) 17 ft (**r**) $3\frac{1}{2}$ pt
(**s**) $4\frac{1}{2}$ qt (**t**) $2\frac{3}{4}$ gal (**u**) 5 qt (**v**) $3\frac{1}{2}$ qt (**w**) 3 lb (**x**) 3 T

(**9-22**) (**a**) 3 cm (**b**) 75 L (**c**) 400 mg (**d**) 4500 km
(**e**) 5 kL (**f**) 13 kg (**g**) 2 m² (**h**) 2 m³

(9-23) (a) 3.25 m   (b) 500,000 mL   (c) 0.75 g   (d) 25 cm   (e) 250 cm   (f) 1500 mL
       (g) 3.25 L   (h) 2500 g   (i) 200,000 mg   (j) 0.8 m   (k) 9 m   (l) 2 km
       (m) 1000 mm   (n) 7 L   (o) 8.5 L   (p) 5.25 kL   (q) 1,000,000 mL   (r) 0.9 g
       (s) 0.215 kg   (t) 4 t   (u) 1,000,000 mg   (v) 300 mg   (w) 80 cm   (x) 5000 cm

(9-24) (a) 8 m   (b) 9 yd 6 in.   (c) 4 qt 1 pt   (d) 71 t   (e) 15 mm
       (f) 46 L   (g) 4.1 t   (h) 1.45 km   (i) $11\frac{3}{4}$ mi   (j) $5\frac{1}{12}$ c
       (k) 13 yd 2 ft   (l) 5 ft $11\frac{3}{4}$ in.   (m) 12 lb 3 oz   (n) 4 yd 1 ft $4\frac{5}{8}$ in.   (o) 5 gal 2 qt

(9-25) (a) 4 mL   (b) 57 m   (c) 1.7 g   (d) 4.6 kL   (e) $\frac{5}{6}$ T
       (f) $3\frac{7}{8}$ pt   (g) 4 lb 3 oz   (h) 2 gal $\frac{3}{4}$ qt   (i) 3 yd 2 ft   (j) 4 mi 1035 yd
       (k) 3 yd 2 ft 6 in.   (l) 4 gal 1 qt 1 c   (m) 4 m   (n) 3 gal 2 qt 1 pt   (o) 4 yd 10 in.

(9-26) (a) 56 mg   (b) 340 kg   (c) 0.9 m   (d) 4.2 L
       (e) 18 mi 15 yd   (f) 40 T 1600 lb   (g) 17 gal $\frac{1}{2}$ qt $\frac{1}{2}$ pt   (h) 27 yd 2 ft 5 in.
       (i) 56 in^2 (j) 15 ft^2 (k) $1\frac{3}{4}$ yd^2 (l) $\frac{1}{2}$ mi^2
       (m) 45 mm^2 (n) 4 cm^2 (o) 4.8 m^2 (p) 0.01 km^2
       (q) 80 in^3 (r) $1\frac{1}{4}$ ft^3 (s) $\frac{1}{8}$ yd^3 (t) 1 mm^3
       (u) 0.625 cm^3 (v) 0.001 m^3 (w) 33 lb 4 oz   (x) 12 m

(9-27) (a) 3 m   (b) 1 yd 2 ft 4 in. R 1 in.   (c) 3 pt   (d) 1.6 m   (e) 5 g   (f) 13.5 t
       (g) 1550 m   (h) 213 mm   (i) $\frac{1}{2}$ ft   (j) $1\frac{1}{4}$ gal   (k) 2 mi   (l) $\frac{5}{6}$ in.
       (m) 36 lb 3 oz   (n) 7 gal 1 qt   (o) 2 T 4 oz   (p) 10 yd 4 in. R 1 in.   (q) 4   (r) 40
       (s) $3\frac{1}{2}$ (t) $\frac{2}{3}$ (u) $\frac{1}{3}$ or $0.\overline{3}$ (v) 5.4   (w) $9\,\frac{\text{lb}}{\text{ft}}$ (x) $72\,\frac{\text{g}}{\text{cm}}$

10 GEOMETRY

THIS CHAPTER IS ABOUT

☑ Identifying Geometric Figures
☑ Finding the Perimeter
☑ Finding the Circumference
☑ Finding Areas
☑ Finding Volumes
☑ Solving Geometry Problems Using Linear Equations

10-1. Identifying Geometric Figures

A. Points, lines, and line segments

A **point** is an exact location in space. A **line** is an *infinite* (uncountable) collection of points that has no beginning and no ending. A **line segment** is a complete piece of a line that includes two **endpoints.**

EXAMPLE 10-1 Draw a representation of (**a**) a point, (**b**) a line, and (**c**) a line segment.

Solution

(**a**) • point (A simple dot can be used to represent a point.)

(**b**) ⟷ line (The arrowheads show that the line continues on forever in both directions.)

(**c**) —— line segment

B. Two-dimensional figures

A **two-dimensional figure** is formed by lines that are **straight, curved,** and/or **broken.** A two-dimensional figure in which you cannot get from the **inside** to the **outside** without crossing over a point *on* the figure itself is called a **closed figure.** A two-dimensional figure that is not a closed figure is called an **open figure.** An open figure does not have an inside or an outside.

EXAMPLE 10-2 Identify each of the following as an open or closed figure:

(**a**)   (**b**) 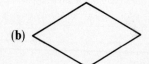   (**c**)

Solution

(**a**)   This is an open figure because it has no inside and no outside.

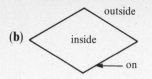

(b)   This is a closed figure because you cannot get from the inside to the outside without crossing over a point on the figure.

(c)   This is an open figure.

A closed figure with three or more **sides** that are all straight line segments is called a **polygon.**

EXAMPLE 10-3   Draw a polygon.

Solution

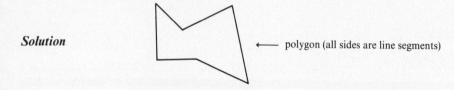

← polygon (all sides are line segments)

Some polygons have special names.

A polygon with three sides is called a **triangle.**

- A triangle with a right angle (square corner) is called a **right triangle.**
- A triangle with three equal sides is called an **equilateral triangle.**
- A triangle with only two equal sides is called an **isosceles triangle.**

EXAMPLE 10-4   Draw a triangle.

Solution

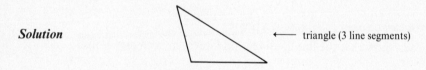

← triangle (3 line segments)

EXAMPLE 10-5   Draw a right triangle.

Solution

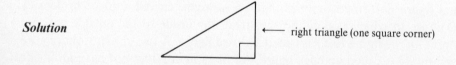

← right triangle (one square corner)

EXAMPLE 10-6   Draw an equilateral triangle.

Solution

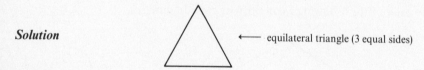

← equilateral triangle (3 equal sides)

EXAMPLE 10-7   Draw an isosceles triangle.

Solution

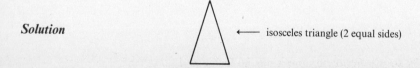

← isosceles triangle (2 equal sides)

A polygon with four sides is called a **quadrilateral.**

- A quadrilateral with four right angles is called a **rectangle.**
- A rectangle with four equal sides is called a **square.**
- A quadrilateral with two pairs of **parallel sides** (always the same distance apart, like railroad tracks) is called a **parallelogram.**
- A quadrilateral with one pair of parallel sides is called a **trapezoid.**

EXAMPLE 10-8 Draw a quadrilateral.

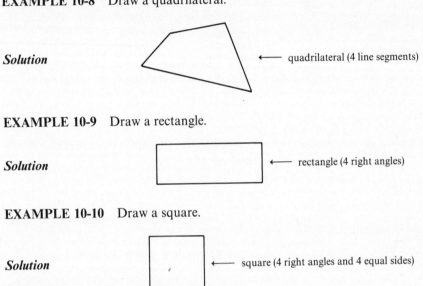

Solution ← quadrilateral (4 line segments)

EXAMPLE 10-9 Draw a rectangle.

Solution ← rectangle (4 right angles)

EXAMPLE 10-10 Draw a square.

Solution ← square (4 right angles and 4 equal sides)

EXAMPLE 10-11 Draw a parallelogram.

Solution ← parallelogram (2 pairs of parallel sides)

EXAMPLE 10-12 Draw a trapezoid.

Solution ← trapezoid (1 pair of parallel sides)

A polygon with five sides is called a **pentagon.**

- A polygon with five equal sides and five equal angles is called a **regular pentagon.**

Note: "*Regular*" means each side and each angle of the polygon are equal.

EXAMPLE 10-13 Draw a regular pentagon.

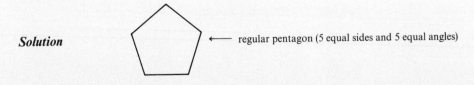

Solution ← regular pentagon (5 equal sides and 5 equal angles)

A polygon with six sides is called a **hexagon.**

- A polygon with six equal sides and six equal angles is called a **regular hexagon.**

EXAMPLE 10-14   Draw a regular hexagon.

Solution

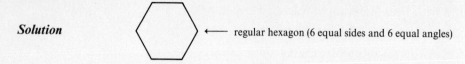

regular hexagon (6 equal sides and 6 equal angles)

A polygon with eight sides is called an **octagon.**

• A polygon with eight equal sides and eight equal angles is called a **regular octagon.**

EXAMPLE 10-15   Draw a regular octagon.

Solution

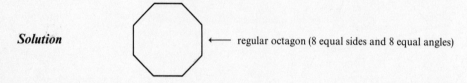

regular octagon (8 equal sides and 8 equal angles)

A **circle** is a closed figure in which every point is an equal distance from a fixed point inside the figure. The fixed point is called the **center** of the circle.

EXAMPLE 10-16   Draw a circle.

Solution

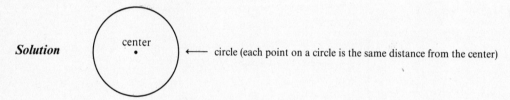

circle (each point on a circle is the same distance from the center)

C. Three-dimensional figures

A **three-dimensional figure** has length, width, and height. A three-dimensional figure shaped like a standard cardboard box is called a **rectangular prism.**

EXAMPLE 10-17   Draw a rectangular prism.

Solution

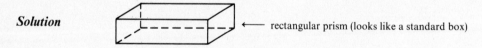

rectangular prism (looks like a standard box)

A rectangular prism with all square sides is called a **cube.**

EXAMPLE 10-18   Draw a cube.

Solution

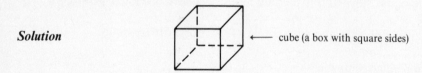

cube (a box with square sides)

A three-dimensional figure shaped like a standard soup can is called a **cylinder.**

EXAMPLE 10-19   Draw a cylinder.

Solution

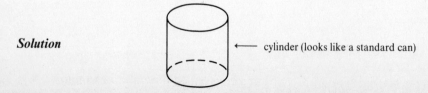

cylinder (looks like a standard can)

A three-dimensional figure shaped like a round ball is called a **sphere.**

EXAMPLE 10-20 Draw a sphere.

Solution ←— sphere (looks like a round ball)

One-half of a sphere is called a **hemisphere.**

EXAMPLE 10-21 Draw a hemisphere.

Solution ←— hemisphere (half of a sphere)

10-2. Finding the Perimeter

A. Finding the perimeter using addition

The total distance around a polygon is called the **perimeter** P. To find the perimeter of any polygon, you add the lengths of all the sides together.

EXAMPLE 10-22 Find the perimeter of the polygon shown in Figure 10-1.

Figure 10-1:

4 ft
3 ft
1 ft
2 ft
2 ft
6 ft

Solution The figure in Figure 10-1 is a polygon with sides of 2 ft, 3 ft, 4 ft, 1 ft, 2 ft, and 6 ft. Add the lengths of all the sides together:

$$P = 2 \text{ ft} + 3 \text{ ft} + 4 \text{ ft} + 1 \text{ ft} + 2 \text{ ft} + 6 \text{ ft}$$

$$= 18 \text{ ft} \longleftarrow \text{total distance around Figure 10-1}$$

B. Finding the perimeter using a formula

You can also find the perimeter of a polygon using a **perimeter formula.** Most of the more common regular polygons have specific perimeter formulas.

EXAMPLE 10-23 Write some common perimeter formulas.

Solution

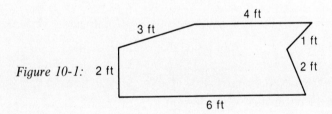

Common Perimeter Formulas		
Equilateral Triangle	$P = 3 \times \overset{s}{\overbrace{\text{side}}}$	$P = 3s$
Rectangle	$P = 2 \times \overset{l}{\overbrace{\text{length}}} + 2 \times \overset{w}{\overbrace{\text{width}}}$	$P = 2l + 2w$ or $P = 2(l + w)$

| **Square** | $P = 4 \times \text{side}$ | $P = 4s$ |

| **Parallelogram** | $P = 2 \times \overbrace{\text{one side}}^{a} + 2 \times \overbrace{\text{adjoining side}}^{b}$ | $P = 2a + 2b$
 or
 $P = 2(a + b)$ |

Regular Pentagon	$P = 5 \times \text{side}$	$P = 5s$
Regular Hexagon	$P = 6 \times \text{side}$	$P = 6s$
Regular Octagon	$P = 8 \times \text{side}$	$P = 8s$

To find the perimeter of a given figure using its perimeter formula, you write the correct perimeter formula, then substitute the known measures for the corresponding letters, and compute the measures.

EXAMPLE 10-24   Find the perimeter of the polygon shown in Figure 10-2 using the appropriate perimeter formula:

Figure 10-2:

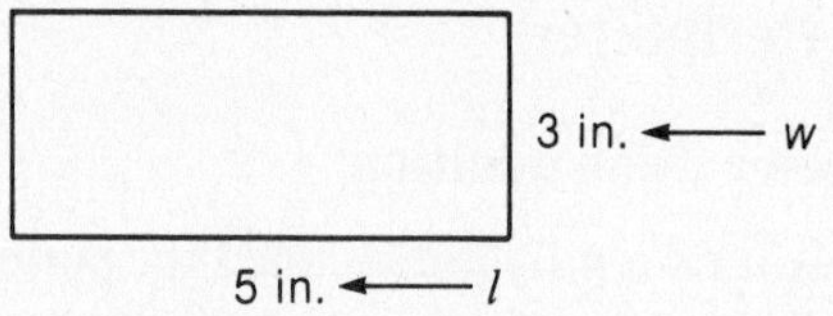

3 in. ⟵ w

5 in. ⟵ l

Solution   The figure in Figure 10-2 is a rectangle with a length l of 5 inches and a width w of 3 inches.

$P = 2(l + w)$ ⟵ perimeter formula for a rectangle

$\quad = 2(5 \text{ in.} + 3 \text{ in.})$ Substitute 5 in. for l (length) and 3 in. for w (width).

$\quad = 2(8 \text{ in.})$ Add inside the parentheses first.

$\quad = 16 \text{ in.}$ Then multiply.

Note 1:   Whenever parentheses are present in a problem, you always compute inside the parentheses first.

Note 2:   If you used addition to find the perimeter of Figure 10-2, you would see that the total distance around Figure 10-2 is 16 inches (5 in. + 3 in. + 5 in. + 3 in. = 16 in.).

EXAMPLE 10-25   Show that the perimeter of Figure 10-2 is 16 inches using the optional perimeter formula for a rectangle ($P = 2l + 2w$):

Solution $P = 2l + 2w$ ⟵ optional perimeter formula for a rectangle

$\quad = 2 \times 5 \text{ in.} + 2 \times 3 \text{ in.}$ Substitute 5 in. for l (length) and 3 in. for w (width).

$\quad = 10 \text{ in.} + 6 \text{ in.}$ Multiply first.

$\quad = 16 \text{ in.}$ Then add.

10-3. Finding the Circumference

A. Renaming the radius or diameter

Any straight line segment that connects a point on a circle with its center is called a **radius** r. (The plural of *radius* is *radii*.)

EXAMPLE 10-26   Draw a circle and one of its radii.

Solution   Every circle has an infinite number of radii, so any line segment drawn from the center to a point on the circle is a radius.

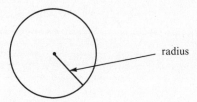

Any straight line segment that connects two points on a circle and also passes through the circle's center is called a **diameter** *d*.

EXAMPLE 10-27   Draw a circle and one of its diameters.

Solution

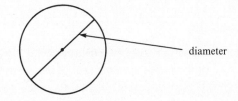

The length of a diameter for a given circle is always twice the length of a radius.

EXAMPLE 10-28   Write a formula relating the length of a diameter for a circle given the length of a radius.

Solution   diameter = 2 × radius, or **$d = 2r$** ⟵ **diameter/radius formula**

To find the length of a diameter of a circle given the length of a radius of the same circle, you can use the diameter/radius formula $d = 2r$.

EXAMPLE 10-29   Find the length of a diameter for the circle shown in Figure 10-3.

Figure 10-3: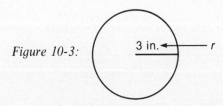

Solution $d = 2r$ Write the diameter/radius formula.

            $= 2 \times 3$ in.   Substitute 3 in. for *r* (radius).

            $= 6$ in.   Multiply a measure by a number [see Example 9-38].

Note: The length of any diameter for Figure 10-3 is 6 inches.

The length of a radius for a given circle is always one-half the length of a diameter.

EXAMPLE 10-30   Write a formula relating the length of a radius for a circle given the length of a diameter.

Solution   radius = one-half diameter or $r = \frac{1}{2}d$, or $r = d \div 2$ ⟵ **radius/diameter formula**

To find the length of a radius of a circle given the length of a diameter for the same circle, you can use the radius/diameter formula $r = d \div 2$.

EXAMPLE 10-31   Find the length of a radius for the circle and diameter shown in Figure 10-4.

Figure 10-4:

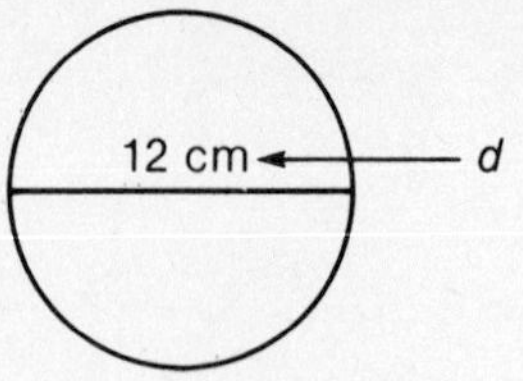

12 cm ◄———— *d*

Solution $r = d \div 2$ Write the radius/diameter formula.

$\qquad\quad = 12 \text{ cm} \div 2$ Substitute 12 cm for *d* (diameter).

$\qquad\quad = 6 \text{ cm}$ Divide a measure by a number [see Example 9-43].

Note: The length of any radius for Figure 10-4 is 6 cm.

B. The circumference of a circle

The distance around a circle is called its **circumference** *C*. The ratio of circumference to diameter for any given circle is always **pi** (π). The number π is a nonrepeating, nonterminating decimal that is approximately equal to $3\frac{1}{7}$, $\frac{22}{7}$, or 3.14. (One or the other of these forms of π is sometimes preferred when using certain measures.)

EXAMPLE 10-32   Write the approximate value of π that is usually used for (**a**) U.S. Customary measures and (**b**) metric measures.

Solution

(**a**) When U.S. Customary measures are given, $\pi \approx 3\frac{1}{7}$ or $\frac{22}{7}$ is usually used.

(**b**) When metric measures are given, $\pi \approx 3.14$ is usually used.

To find the circumference *C* of a circle given the length of a radius or a diameter, you can use the correct circumference formula.

EXAMPLE 10-33   Write the common circumference formulas.

Solution   Circle with a given radius length: Circumference = 2 × π × radius, or $C = 2\pi r$
Circle with a given diameter length: Circumference = π × diameter,   or $C = \pi d$

Note: If you memorize only $C = \pi d$, you can always derive $C = 2\pi r$ by using the diameter/radius formula $d = 2r$.

EXAMPLE 10-34   Find $C = 2\pi r$ using $C = \pi d$ and $d = 2r$.

Solution $C = \pi d$ ◄——— memorized formula

$\qquad\qquad\qquad\qquad C = \pi(2r)$ Substitute 2r for *d* because $d = 2r$.

$\qquad\qquad\qquad\qquad C = 2\pi r$ ◄——— derived formula

To find the **approximate circumference** of a circle with a given length of a diameter, you first write the circumference formula $C = \pi d$, then substitute the given diameter length for *d* and either $3\frac{1}{7}$ or 3.14 for π, and then multiply.

EXAMPLE 10-35   Find the approximate circumference of the circle in Figure 10-5.

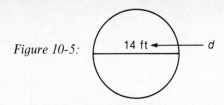

Figure 10-5:

Solution The circle in Figure 10-5 has a diameter d of 14 feet.

$C = \pi d \longleftarrow$ circumference formula for a circle with a given diameter length

$\approx 3\frac{1}{7} \times 14$ ft   Substitute 14 ft for d and $3\frac{1}{7}$ for π because 14 feet is a U.S. Customary measure.

$= \dfrac{22}{\cancel{7}} \times \dfrac{2 \times \cancel{7}}{1}$ ft   Multiply.

$= 44$ ft $\longleftarrow$ approximate circumference of Figure 10-5.

Note: The circumference of the circle in Figure 10-5 is only approximately equal to 44 feet ($C \approx 44$ ft) because π is only approximately equal to $3\frac{1}{7}$ ($\pi \approx 3\frac{1}{7}$).

To find the approximate circumference of a circle given the length of a radius, you first write the circumference formula $C = 2\pi r$, then substitute the given radius length for r and either $3\frac{1}{7}$ or 3.14 for π, and then multiply.

EXAMPLE 10-36 Find the approximate circumference of the circle in Figure 10-6.

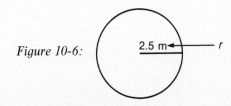

Figure 10-6:

Solution The circle in Figure 10-6 has a radius r of 2.5 meters.

$C = 2\pi r \longleftarrow$ circumference formula for a circle with a given radius length

$\approx 2 \times 3.14 \times 2.5$ m   Substitute 2.5 m for r and 3.14 for π because 2.5 m is a metric measure.

$= 6.28 \times 2.5$ m   Multiply.

$= 15.7$ m $\longleftarrow$ approximate circumference of Figure 10-6.

10-4. Finding Areas

A. Common area units

When the surface of a closed figure is measured by finding how many squares of a given size are needed to completely cover the surface, you are finding the **area A** of that surface in **square units of measure.**

Recall: The common square units of measure are

     (a) square inches (in²)   **(b)** square feet (ft²)   **(c)** square yards (yd²)
     (d) square centimeters (cm²)   **(e)** square meters (m²)

B. Finding the areas of common polygons

The **area formula for a rectangle** is Area = length × width, or $A = lw.$

EXAMPLE 10-37   Find the area of the rectangle in Figure 10-7.

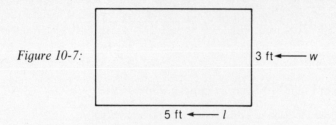

Figure 10-7:

5 ft ◄——— *l*

3 ft ◄——— *w*

Solution   The rectangle in Figure 10-7 has a length *l* of 5 feet and a width *w* of 3 feet.

$A = lw$ ◄——— area formula for a rectangle

$= 5 \text{ ft} \times 3 \text{ ft}$ Substitute 5 ft for *l* (length) and 3 ft for *w* (width).

$= (5 \times 3) \text{ ft}^2$ Multiply two like length measures [see Example 9-40].

$= 15 \text{ ft}^2$ ◄——— area of Figure 10-7

Note: It would take 15 squares each measuring 1 foot on a side to completely cover the surface of the rectangle in Figure 10-7.

EXAMPLE 10-38   Show by counting that it takes 15 squares, each measuring 1 foot on a side, to completely cover the surface of the rectangle in Figure 10-7.

Solution

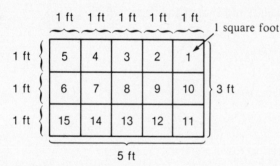

Subdivide Figure 10-7 into squares of 1 square foot each; then count the number of square feet.

15 square feet in all

Note: The total number of square units it takes to completely cover a surface is the area of that surface.

The **area formula for a square** is Area = side × side or $A = s \times s$, or $\boldsymbol{A = s^2}$.

EXAMPLE 10-39   Find the area of the square in Figure 10-8.

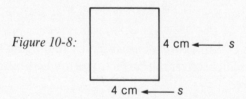

Figure 10-8:

4 cm ◄——— *s*

4 cm ◄——— *s*

Solution   The square in Figure 10-8 has sides *s* with a length of 4 centimeters.

$A = s^2$ ◄——— area formula for a square

$= (4 \text{ cm})^2$ Substitute 4 cm for *s* (side).

$= 4 \text{ cm} \times 4 \text{ cm}$ Rename as a product of repeated factors.

$= (4 \times 4) \text{ cm}^2$ Multiply two like length measures.

$= 16 \text{ cm}^2$ ◄——— area of Figure 10-8

Note: It takes 16 squares, each measuring 1 centimeter on a side, to completely cover the surface of the square in Figure 10-8.

The **area formula for a parallelogram** is Area = base × height, or $A = bh$.

EXAMPLE 10-40   Find the area of the parallelogram in Figure 10-9.

Figure 10-9:
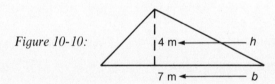

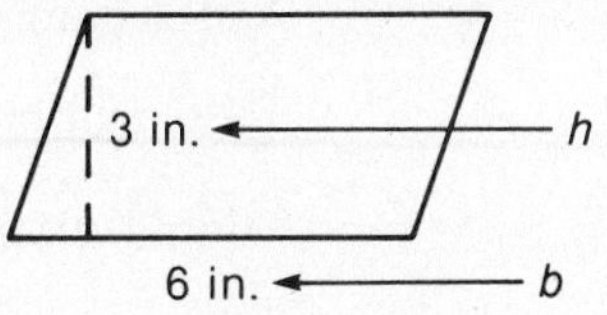

Solution   The parallelogram in Figure 10-9 has a base b of 6 inches and a height h of 3 inches.

$A = bh$ ⟵ area formula for a parallelogram

$= 6$ in. $\times 3$ in.   Substitute 6 in. for b (base) and 3 in. for h (height).

$= (6 \times 3)$ in^2 Multiply two like length measures.

$= 18$ in^2 ⟵ area of Figure 10-9

Note: It takes the equivalent of 18 squares each measuring 1 inch on a side to completely cover the surface of the parallelogram in Figure 10-9.

The **area formula for a triangle** is Area = $\frac{1}{2}$ × base × height, or $A = \frac{1}{2}bh$.

EXAMPLE 10-41   Find the area of the triangle in Figure 10-10.

Figure 10-10:

Solution   The triangle in Figure 10-10 has a base b of 7 meters and a height h of 4 meters.

$A = \frac{1}{2}bh$ ⟵ area formula for a triangle

$= \frac{1}{2} \times 7$ m $\times 4$ m   Substitute 7 m for b (base) and 4 m for h (height).

$= \frac{1}{2} \times (7 \times 4)$ m^2 Multiply two like length measures.

$= \frac{1}{2} \times 28$ m^2 Multiply by $\frac{1}{2}$.

$= 14$ m^2 ⟵ area of Figure 10-10

Note: It takes the equivalent of 14 squares, each measuring 1 meter on a side, to completely cover the surface of the triangle in Figure 10-10.

The **area formula for a trapezoid** is $A = \frac{1}{2}$ × (one base + other base) × height, or $A = \frac{1}{2}(b_1 + b_2)h$.

EXAMPLE 10-42   Find the area of the trapezoid in Figure 10-11:

Figure 10-11:
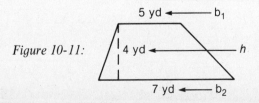

Solution The trapezoid in Figure 10-11 has bases b_1 and b_2 of 5 yards and 7 yards and height h of 4 yards.

$A = \frac{1}{2}(b_1 + b_2)h$ ⟵ area formula for a trapezoid

$\quad = \frac{1}{2}(5 \text{ yd} + 7 \text{ yd}) 4 \text{ yd}$ — Substitute 5 yd for b_1 (one base), 7 yd for b_2 (other base), and 4 yd for h (height).

$\quad = \frac{1}{2} \times 12 \text{ yd} \times 4 \text{ yd}$ — Add in parentheses first.

$\quad = \frac{1}{2} \times (12 \times 4) \text{ yd}^2$ — Then multiply two like length measures.

$\quad = \frac{1}{2} \times 48 \text{ yd}^2$ — Multiply by $\frac{1}{2}$.

$\quad = 24 \text{ yd}^2$ ⟵ area of Figure 10-11

Note: It takes the equivalent of 24 squares, each measuring 1 yard on a side, to completely cover the surface of the trapezoid in Figure 10-11.

C. Finding the area of a circle

The **area formula for a circle** is Area $= \pi \times$ radius $\times$ radius, or $A = \pi rr$ or $A = \pi r^2$.

EXAMPLE 10-43 Find the area of the circle in Figure 10-12.

Figure 10-12:
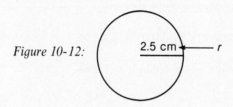
2.5 cm ⟵ r

Solution The circle in Figure 10-12 has a radius r of 2.5 centimeters.

$A = \pi r^2$ ⟵ area formula for a circle

$\quad \approx 3.14 (2.5 \text{ cm})^2$ — Substitute 2.5 cm for r (radius) and 3.14 for π because 2.5 cm is a metric measure.

$\quad = 3.14 \times 2.5 \text{ cm} \times 2.5 \text{ cm}$ — Rename as a product of repeated factors.

$\quad = 3.14 \times (2.5 \times 2.5) \text{ cm}^2$ — Multiply two like length measures.

$\quad = 3.14 \times 6.25 \text{ cm}^2$ — Multiply by 3.14.

$\quad = 19.625 \text{ cm}^2$ ⟵ approximate area of Figure 10-12

Note: It takes the equivalent of about 19.625 ($19\frac{5}{8}$) squares, each measuring 1 meter on a side, to completely cover the surface of the circle in Figure 10-12.

To find the area of a circle with a given diameter, you first use $r = d \div 2$ to find the length of a radius and then compute $A = \pi r^2$.

EXAMPLE 10-44 Find the area of the circle in Figure 10-13.

Figure 10-13:
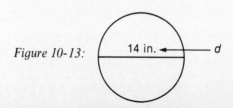
14 in. ⟵ d

Solution The circle in Figure 10-13 is a circle with diameter d 14 inches.

$r = d \div 2$ ⟵ radius/diameter formula

$= 14$ in. $\div 2$ Substitute 14 in. for d (diameter).

$= (14 \div 2)$ in.   Divide a measure by a number.

$= 7$ in. ⟵ length of a radius r for the circle in Figure 10-13

$A = \pi r^2$ ⟵ area formula for a circle

$\approx 3\frac{1}{7}\,(7\text{ in.})^2$ Substitute 7 in. for r (radius) and $3\frac{1}{7}$ for π because 7 in. is a U.S. Customary measure.

$= 3\frac{1}{7} \times 7$ in. $\times 7$ in.   Rename as a product of repeated factors.

$= \dfrac{22}{\cancel{7}} \times \dfrac{\cancel{7}\text{ in.}}{1} \times \dfrac{7\text{ in.}}{1}$ Multiply.

$= 22 \times 7$ in.^{2}

$= 154$ in^2 ⟵ approximate area of Figure 10-13

Note: It takes the equivalent of about 154 squares, each measuring 1 inch on a side, to completely cover the surface of the circle in Figure 10-13.

10-5. Finding Volumes

A. Common volume units

When the space occupied by a three-dimensional object is measured by finding how many cubes of a given size are needed to completely fill that space, you are finding the **volume** V of that object in **cubic units of measure.**

Recall: The common cubic units of measure are

(a) cubic inches (in^3)   **(b)** cubic feet (ft^3)   **(c)** cubic yards (yd^3)

(d) cubic centimeters (cm^3)   **(e)** cubic meters (m^3)

B. The volume of three-dimensional figures with flat surfaces

The **volume formula for a rectangular prism** is Volume = length × width × height, or $V = lwh.$

EXAMPLE 10-45 Find the volume of the rectangular prism in Figure 10-14.

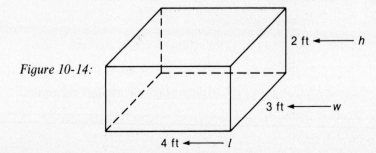

Figure 10-14:

2 ft ⟵ h

3 ft ⟵ w

4 ft ⟵ l

Solution The rectangular prism in Figure 10-14 has a length l of 4 feet, a width w of 3 feet, and a height h of 2 feet.

$V = lwh$ ⟵ volume formula for a rectangular prism

$\quad = 4 \text{ ft} \times 3 \text{ ft} \times 2 \text{ ft}$ Substitute 4 ft for *l* (length), 3 ft for *w* (width), and 2 ft for *h* (height).

$\quad = (4 \times 3) \text{ ft}^2 \times 2 \text{ ft}$ Multiply three like length measures [see Example 9-41].

$\quad = 12 \text{ ft}^2 \times 2 \text{ ft}$

$\quad = (12 \times 2) \text{ ft}^3$

$\quad = 24 \text{ ft}^3$ ⟵ area of Figure 10-14

Note: It takes 24 cubes, each measuring 1 foot on an edge, to completely fill the space occupied by the rectangular prism in Figure 10-14.

EXAMPLE 10-46   Show by counting that it takes 24 cubes, each measuring 1 foot on an edge, to completely fill the space occupied by the rectangular prism in Figure 10-14.

Solution

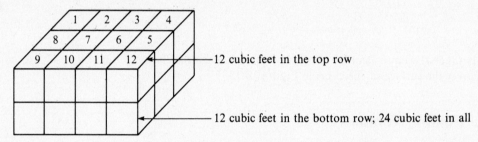

12 cubic feet in the top row

12 cubic feet in the bottom row; 24 cubic feet in all

Note: The total number of cubic units it takes to completely fill the space occupied by an object is the volume of that object.

The **volume formula for a cube** is Volume = edge × edge × edge, or $V = (e)(e)(e)$, or $V = e^3$.

EXAMPLE 10-47   Find the volume of the cube in Figure 10-15.

Figure 10-15:

3 cm ⟵ *e*

3 cm ⟵ *e*

3 cm ⟵ *e*

Solution   The cube in Figure 10-15 has an edge (*e*) of 3 centimeters.

$\quad V = e^3$ ⟵ volume formula for a cube

$\quad\quad = (3 \text{ cm})^3$ Substitute 3 cm for *e* (edge).

$\quad\quad = 3 \text{ cm} \times 3 \text{ cm} \times 3 \text{ cm}$ Rename as a product of repeated factors.

$\quad\quad = (3 \times 3) \text{ cm}^2 \times 3 \text{ cm}$ Multiply three like length measures.

$\quad\quad = 9 \text{ cm}^2 \times 3 \text{ cm}$

$\quad\quad = (9 \times 3) \text{ cm}^3$

$\quad\quad = 27 \text{ cm}^3$ ⟵ volume of Figure 10-15

Note: It takes 27 cubes, each measuring 1 centimeter on an edge, to completely fill the space occupied by the cube in Figure 10-15.

C. The volume of three-dimensional figures with curved surfaces

The **volume formula for a cylinder** is Volume = π × radius × radius × height, or $V = \pi rrh$ or $V = \pi r^2 h$.

EXAMPLE 10-48 Find the volume of the cylinder in Figure 10-16.

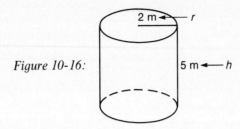

Figure 10-16:

Solution The cylinder in Figure 10-16 has a radius r of 2 meters and a height h of 5 meters.

$V = \pi r^2 h$ ⟵ volume formula for a cylinder	
$\approx 3.14(2 \text{ m})^2\, 5 \text{ m}$	Substitute 2 m for r (radius), 5 m for h (height), and 3.14 for π since meters is a metric measure.
$= 3.14 \times 2 \text{ m} \times 2 \text{ m} \times 5 \text{ m}$	Rename as a product of repeated factors.
$= 3.14 \times (2 \times 2 \times 5) \text{ m}^3$	Multiply three like length measures.
$= 3.14 \times 20 \text{ m}^3$	Multiply by 3.14.
$= 62.8 \text{ m}^3$ ⟵ approximate volume of Figure 10-16.	

Note: It takes the equivalent of about 62.8 cubes each measuring 1 meter on an edge to completely fill the space occupied by the cylinder in Figure 10-16.

The **volume formula for a sphere** is Volume = $\frac{4}{3} \times \pi$ × radius × radius × radius, or $V = \frac{4}{3}\pi rrr$ or $V = \frac{4}{3}\pi r^3$.

EXAMPLE 10-49 Find the volume of the sphere in Figure 10-17.

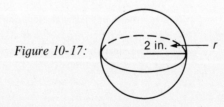

Figure 10-17:

Solution The sphere in Figure 10-17 has a radius r of 2 inches.

$V = \frac{4}{3}\pi r^3$ ⟵ volume formula for a sphere	
$\approx \frac{4}{3} \times 3\frac{1}{7}(2 \text{ in.})^3$	Substitute 2 in. for r (radius) and $3\frac{1}{7}$ for π because inches is a U.S. Customary measure.
$= \frac{4}{3} \times 3\frac{1}{7} \times 2 \text{ in.} \times 2 \text{ in.} \times 2 \text{ in.}$	Rename as a product of repeated factors.
$= \frac{4}{3} \times \frac{22}{7} \times (2 \times 2 \times 2) \text{ in}^3$	Multiply three like length measures.
$= \frac{88}{21} \times 8 \text{ in}^3$	Multiply $\frac{4}{3}$ and $\frac{22}{7}$.
$= \frac{704}{21} \text{ in}^3$	Multiply by $\frac{88}{21}$.
$= 33\frac{11}{21} \text{ in}^3$ ⟵ approximate volume of Figure 10-17	

Note: It takes the equivalent of about $33\frac{11}{21}$ cubes, each measuring 1 inch on an edge, to completely fill the space occupied by the sphere in Figure 10-17.

10-6. Solving Geometry Problems Using Linear Equations

To **solve a geometry problem using a linear equation,** you

(1) *Read* the problem very carefully several times.

(2) *Draw a picture* to help visualize the problem.

(3) *Identify* the unknown measures.

(4) *Decide* how to represent the unknown measures using one variable.

(5) *Translate* the problem to a linear equation using the correct geometry formula (see Appendix Table 6).

(6) *Solve* the linear equation.

(7) *Interpret* the solution of the linear equation with respect to each represented unknown measure to find the proposed solutions of the original problem.

(8) *Check* to see if the proposed solutions satisfy all the conditions of the original problem.

EXAMPLE 10-50   Solve the following geometry problem using a linear equation.

Solution

(1) *Read:* The perimeter of a rectangle is 64 feet. The length of the rectangle is 10 feet longer than the width. Find the area of the rectangle.

(2) *Draw a picture:*

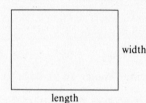

width

length

(3) *Identify:* The unknown measures are $\begin{cases} \text{the length of the rectangle} \\ \text{the width of the rectangle} \end{cases}$.

(4) *Decide:* Let l = the length of the rectangle
then $l - 10$ = the width of the rectangle

(5) *Translate:* $P = 2(l + w)$ ⟵ perimeter formula for a rectangle from Appendix Table 6

$64 = 2[l + (l - 10)]$ ⟵ linear equation   *Think:* $w = l - 10$

(6) *Solve:* $64 = 2(l + l - 10)$

$64 = 2(2l - 10)$

$64 = 2(2l) - 2(10)$

$64 = 4l - 20$

$84 = 4l$

$21 = l$

(7) *Interpret:* $l = 21$ means the length of the rectangle is 21 feet.
$l - 10 = 21 - 10 = 11$ means the width of the rectangle is 11 feet.
$l = 21$ ft and $w = 11$ ft means the area of the rectangle is

$A = lw = (21 \text{ ft})(11 \text{ ft}) = 231 \text{ ft}^2$ ⟵ proposed solution

(8) *Check:*   Did you find the length and the width?   Yes: $l = 21$ feet and $w = 11$ feet
Is the length of the rectangle 10 feet more than the width?   Yes: $21 - 10 = 11$ [feet]
Is the perimeter of the rectangle 64 feet?   Yes: $2(21 + 11) = 2(32) = 64$ [feet]

The area of the rectangle is 231 square feet, or 231 ft².

Note: To represent the unknown length of a rectangle that is 10 units longer than the width, you can let l = the length and $l - 10$ = the width

or

let w = the width and $w + 10$ = the length
because l is 10 more than $l - 10$ and $w + 10$ is 10 more than w.

EXAMPLE 10-51 Let w = the width of the rectangle and $w + 10$ = the length of the rectangle for the problem in Example 10-50.

Solution

Translate: $\qquad P = 2(l + w)$

$\qquad\qquad 64 = 2[(w + 10) + w] \longleftarrow$ linear equation $\qquad$ *Think:* $l = w + 10$

Solve: $\qquad 64 = 2(w + 10 + w)$

$\qquad\qquad 64 = 2(2w + 10)$

$\qquad\qquad 64 = 2(2w) + 2(10)$

$\qquad\qquad 64 = 4w + 20$

$\qquad\qquad 44 = 4w$

$\qquad\qquad w = 11$ [feet] $\longleftarrow$ width

$\qquad w + 10 = 11 + 10 = 21$ [feet] $\longleftarrow$ length

area of rectangle $\longrightarrow A = lw = (21 \text{ ft})(11 \text{ ft}) = 231 \text{ ft}^2 \longleftarrow$ same solution as found in Example 10-50

SOLVED PROBLEMS

PROBLEM 10-1 Identify each geometric figure:

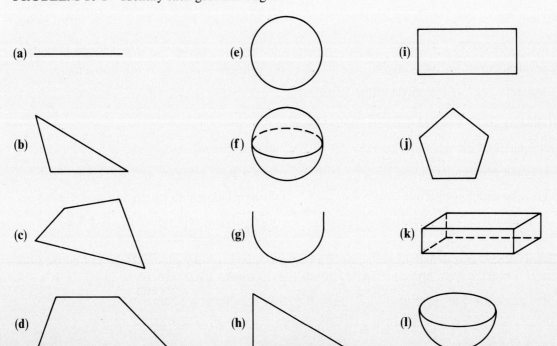

(a)

(b)

(c)

(d)

(e)

(f)

(g)

(h)

(i)

(j)

(k)

(l)

(m)

(r)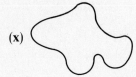

(w)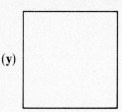

(n)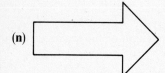

(s)

(x)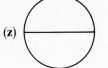

(o)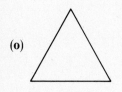

(t)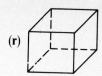

(y)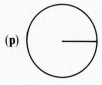

(p)

(u)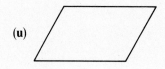

(z)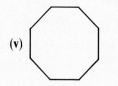

(q)

(v) (octagon figure)

Solution [see Examples 10-1 through 10-21]:

(a) line segment   (b) triangle   (c) quadrilateral   (d) trapezoid

(e) circle   (f) sphere   (g) open figure   (h) right triangle

(i) rectangle   (j) regular pentagon   (k) rectangular prism   (l) hemisphere

(m) point   (n) polygon   (o) equilateral triangle   (p) circle with a radius

(q) regular hexagon   (r) cube   (s) line   (t) isosceles triangle

(u) parallelogram   (v) octagon   (w) cylinder   (x) closed figure

(y) square   (z) circle with a diameter

PROBLEM 10-2 Find the perimeter for the following:

(a) an equilateral triangle with side 5 in.

(b) a square with side 3 cm

(c) a regular pentagon with side 8 ft

(d) a regular hexagon with side 2.5 m

(e) a regular octagon with side $2\frac{1}{2}$ yd

(f) a rectangle with length 8 in. and width 5 in.

(g) a parallelogram with sides 7 cm and 6 cm

(h) a polygon with sides 2 ft, 3 ft, 8 ft, $6\frac{1}{2}$ ft, 10 ft, and $12\frac{3}{4}$ ft.

Solution Recall that to find the perimeter of a polygon, you can either add the lengths of all the sides together or use the appropriate perimeter formula [see Examples 10-22 through 10-25]:

(a) $P = 3s = 3 \times 5$ in. $= 15$ in.

(b) $P = 4s = 4 \times 3$ cm $= 12$ cm

(c) $P = 5s = 5 \times 8$ ft $= 40$ ft

(d) $P = 6s = 6 \times 2.5$ m $= 15$ m

(e) $P = 8s = 8 \times 2\frac{1}{2}$ yd $= 20$ yd

(f) $P = 2(l + w) = 2(8$ in. $+ 5$ in.$) = 2 \times 13$ in. $= 26$ in.

(g) $P = 2(a + b) = 2(7 \text{ cm} + 6 \text{ cm}) = 2 \times 13 \text{ cm} = 26 \text{ cm}$

(h) $P = 2 \text{ ft} + 3 \text{ ft} + 8 \text{ ft} + 6\frac{1}{2} \text{ ft} + 10 \text{ ft} + 12\frac{3}{4} \text{ ft} = 42\frac{1}{4} \text{ ft}$

PROBLEM 10-3   Find the length of a diameter of a circle with the following radius:

(a) $r = 5 \text{ in.}$ **(b)** $r = 7 \text{ cm}$ **(c)** $r = 2\frac{1}{4} \text{ ft}$ **(d)** $r = 1.3 \text{ m}$

Solution   Recall that to find the length of a diameter of a circle given the length of a radius for the same circle, you can use the diameter/radius formula $d = 2r$ [see Examples 10-28 and 10-29]:

(a) $d = 2r = 2 \times 5 \text{ in.} = 10 \text{ in.}$ **(b)** $d = 2r = 2 \times 7 \text{ cm} = 14 \text{ cm}$

(c) $d = 2r = 2 \times 2\frac{1}{4} \text{ ft} = 4\frac{1}{2} \text{ ft}$ **(d)** $d = 2r = 2 \times 1.3 \text{ m} = 2.6 \text{ m}$

PROBLEM 10-4   Find the length of a radius of a circle with the following diameter:

(a) $d = 5 \text{ in.}$ **(b)** $d = 7 \text{ cm}$ **(c)** $d = 2\frac{1}{4} \text{ ft}$ **(d)** $d = 1.3 \text{ m}$

Solution   Recall that to find the length of a radius of a circle given the length of a diameter for the same circle, you can use the radius/diameter formula $r = d \div 2$ [see Examples 10-30 and 10-31]:

(a) $r = d \div 2 = 5 \text{ in.} \div 2 = 2\frac{1}{2} \text{ in.}$ **(b)** $r = d \div 2 = 7 \text{ cm} \div 2 = 3.5 \text{ cm}$

(c) $r = \frac{1}{2}d = \frac{1}{2} \times 2\frac{1}{4} \text{ ft} = 1\frac{1}{8} \text{ ft}$ **(d)** $r = d \div 2 = 1.3 \text{ m} \div 2 = 0.65 \text{ m}$

PROBLEM 10-5   Find the approximate circumference for a circle with

(a) $r = 3\frac{1}{2} \text{ ft}$ **(b)** $r = 2 \text{ m}$ **(c)** $d = 7 \text{ yd}$ **(d)** $d = 1.5 \text{ cm}$

Solution   Recall that to find the circumference of a circle given the length of a radius or a diameter, you use the circumference formula $C = 2\pi r$ or $C = \pi d$ using $\pi \approx 3\frac{1}{7}$ for U.S. customary measures and $\pi \approx 3.14$ for metric measures [see Example 10-36]:

(a) $C = 2\pi r \approx 2 \times 3\frac{1}{7} \times 3\frac{1}{2} \text{ ft} = \frac{\cancel{2}}{1} \times \frac{22}{\cancel{7}} \times \frac{\cancel{7}}{\cancel{2}} \text{ ft} = 22 \text{ ft}$

(b) $C = 2\pi r \approx 2 \times 3.14 \times 2 \text{ m} = 6.28 \times 2 \text{ m} = 12.56 \text{ m}$

(c) $C = \pi d \approx 3\frac{1}{7} \times 7 \text{ yd} = \frac{22}{\cancel{7}} \times \frac{\cancel{7}}{1} \text{ yd} = 22 \text{ yd}$

(d) $C = \pi d \approx 3.14 \times 1.5 \text{ cm} = 4.71 \text{ cm}$

PROBLEM 10-6   Find the area of the following:

(a) rectangle with length 9 m and width 5 m   **(b)** square with side 3 in.

(c) parallelogram with base 6 ft and height 4 ft   **(d)** triangle with base 2 cm and height 8 cm

(e) trapezoid with bases 2 in. and 3 in. and height 4 in.   **(f)** circle with radius $2\frac{1}{3}$ yd

Solution   Recall that to find the area of a square, rectangle, parallelogram, triangle, or circle with a given radius or diameter, you can use the appropriate area formula [see Examples 10-37 and 10-39 through 10-44]:

(a) $A = lw = 9 \text{ m} \times 5 \text{ m} = (9 \times 5) \text{ m}^2 = 45 \text{ m}^2$

(b) $A = s^2 = (3 \text{ in.})^2 = 3 \text{ in.} \times 3 \text{ in.} = (3 \times 3) \text{ in}^2 = 9 \text{ in}^2$

(c) $A = bh = 6 \text{ ft} \times 4 \text{ ft} = (6 \times 4) \text{ ft}^2 = 24 \text{ ft}^2$

(d) $A = \frac{1}{2}bh = \frac{1}{2} \times 2 \text{ cm} \times 8 \text{ cm} = (\frac{1}{\cancel{2}} \times \cancel{2} \times 8) \text{ cm}^2 = 8 \text{ cm}^2$

(e) $A = \frac{1}{2}(b_1 + b_2)h = \frac{1}{2}(2 \text{ in.} + 3 \text{ in.})\, 4 \text{ in.} = \frac{1}{2} \times 5 \text{ in.} \times 4 \text{ in.} = (\frac{1}{2} \times 5 \times 4) \text{ in}^2 = 10 \text{ in}^2$

(f) $A = \pi r^2 \approx 3\frac{1}{7}(2\frac{1}{3} \text{ yd})^2 = 3\frac{1}{7} \times 2\frac{1}{3} \text{ yd} \times 2\frac{1}{3} \text{ yd} = \frac{22}{7} \times \frac{7}{3} \text{ yd} \times \frac{7}{3} \text{ yd} = (\frac{22}{\cancel{7}} \times \frac{7}{3} \times \frac{7}{3}) \text{ yd}^2 = 17\frac{1}{9} \text{ yd}^2$

PROBLEM 10-7   Find the volume of the following:

(a) rectangular prism with length 3 in., width 5 in., and height 2 in.

(b) cube with edge 4 cm

(c) cylinder with radius $1\frac{3}{4}$ ft and height 6 ft

(d) sphere with radius 3 m

Solution   Recall that to find the volume of a rectangular prism, cube, cylinder, or sphere, you can use the appropriate volume formula [see Examples 10-45, 10-47, 10-48, and 10-49, respectively]:

(a) $V = lwh = 3 \text{ in.} \times 5 \text{ in.} \times 2 \text{ in.} = (3 \times 5 \times 2)\text{ in}^3 = 30 \text{ in}^3$

(b) $V = e^3 = (4 \text{ cm})^3 = 4 \text{ cm} \times 4 \text{ cm} \times 4 \text{ cm} = (4 \times 4 \times 4)\text{ cm}^3 = 64 \text{ cm}^3$

(c) $V = \pi r^2 h \approx 3\frac{1}{7}\left(1\frac{3}{4}\text{ ft}\right)^2 6 \text{ ft} = \frac{22}{7} \times \frac{7}{4}\text{ ft} \times \frac{7}{4}\text{ ft} \times 6 \text{ ft} = \frac{22}{7} \times \left(\frac{7}{4} \times \frac{7}{4} \times 6\right)\text{ ft}^3 = 57\frac{3}{4}\text{ ft}^3$

(d) $V = \frac{4}{3}\pi r^3 \approx \frac{4}{3} \times 3.14\,(3 \text{ m})^3$

$\qquad = \frac{4}{3} \times 3.14 \times 3 \text{ m} \times 3 \text{ m} \times 3 \text{ m}$

$\qquad = \frac{4}{3} \times 3.14 \times (3 \times 3 \times 3)\text{ m}^3$

$\qquad = 113.04 \text{ m}^3$

PROBLEM 10-8   Solve each geometry problem using a linear equation:

(a) The perimeter of a triangle is 41 meters. The longest side is 3 times the shortest side. The third side is 8 meters shorter than the longest side. How long is each side?

(b) The area of a rectangle is 192 in^2 [square inches]. Find the perimeter of the rectangle if its length is 16 inches.

Solution   Recall that to solve a geometry problem using a linear equation, you

(1) *Read* the problem very carefully several times.

(2) *Draw a picture* to help visualize the problem.

(3) *Identify* the unknown measures.

(4) *Decide* how to represent the unknown measures using one variable.

(5) *Translate* the problem to a linear equation using the correct geometry formula from Appendix Table 6.

(6) *Solve* the linear equation.

(7) *Interpret* the solution of the linear equation with respect to each represented unknown measure to find the proposed solutions of the original problem.

(8) *Check* to see if the proposed solutions satisfy all the conditions of the original problem. [See Examples 10-50 and 10-51.]

(a) *Identify:*   The unknown measures are $\begin{cases} \text{the shortest side} \\ \text{the longest side} \\ \text{the third side} \end{cases}$.

   Draw a picture:

   shortest side   third side
   longest side

   Decide: Let $s =$ the shortest side
   then $3s =$ the longest side because $3s$ is 3 times s
   and $3s - 8 =$ the third side because $3s - 8$ is 8 less than $3s$.

   Translate: $P = a + b + c \longleftarrow$ perimeter formula for a triangle from Appendix Table 6
   $\qquad\qquad\quad \downarrow \quad\; \downarrow \quad\; \downarrow$
   $\qquad\quad 41 = s + 3s + (3s - 8)$

$$41 = s + 3s + 3s - 8 \longleftarrow \text{linear equation}$$

Solve:
$$41 = 7s - 8$$
$$49 = 7s$$

Interpret:
$$s = 7 \text{ [shortest side]}$$
$$3s = 21 \text{ [longest side]} \longrightarrow \text{solutions [Check as before.]}$$
$$3s - 8 = 13 \text{ [third side]}$$

(b) *Identify:* The unknown measures are $\left\{ \begin{array}{l} \text{the width} \\ \text{the perimeter} \end{array} \right\}$.

Draw a picture:

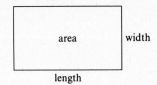

Decide:
Let $l = $ the length
and $w = $ the width
then $2(l + w) = $ the perimeter

Translate:
$$A = lw \longleftarrow \text{area formula for a rectangle}$$
$$192 = (16)w$$
$$16w = 192 \longleftarrow \text{linear equation}$$

Solve:
$$w = 12 \text{ [inches]}$$

Interpret:
$$2(l + w) = 56 \text{ [inches]} \longleftarrow \text{solution [Check as before.]}$$

Supplementary Exercises

PROBLEM 10-9   Find the approximate perimeter for the following:

(a) an equilateral triangle with side 8 cm

(b) a square with side 6 in.

(c) a regular pentagon with side 3 ft

(d) a regular hexagon with side 5.2 m

(e) a regular octagon with side $1\frac{1}{2}$ yd

(f) a rectangle with length 5 cm and width 6 cm

(g) a parallelogram with sides 2.5 cm and 3.5 cm

(h) a polygon with sides 2.3 m, 1.8 m, 6.4 m, 0.25 m, 3.18 m, and 0.08 m.

PROBLEM 10-10   Find the circumference for a circle given the following dimensions:

(a) $r = 3\frac{1}{2}$ in.   **(b)** $r = 4$ cm   **(c)** $r = 14$ yd   **(d)** $r = 5$ mm   **(e)** $r = 7$ in.

(f) $r = 1.5$ m   **(g)** $r = 8$ cm   **(h)** $r = 1\frac{2}{5}$ ft   **(i)** $r = 1.4$ m   **(j)** $r = 2\frac{4}{5}$ in.

(k) $r = 2.8$ mm   **(l)** $r = 1\frac{1}{2}$ ft   **(m)** $r = 3.25$ cm   **(n)** $d = 2.8$ m   **(o)** $d = 7$ ft

(p) $d = 21$ yd   **(q)** $d = 3$ cm   **(r)** $d = 28$ in.   **(s)** $d = 1.3$ cm   **(t)** $d = 4\frac{2}{3}$ ft

(u) $d = 4.7$ m   **(v)** $d = 10\frac{1}{2}$ in.   **(w)** $d = 10.5$ cm   **(x)** $d = 2\frac{2}{3}$ yd   **(y)** $d = 5.625$ mm

PROBLEM 10-11   Find the area of these shapes:

(a) a rectangle with length 4 ft and width 2 ft

(b) a square with side 5 cm

(c) a parallelogram with base 3.4 mm and height 5.6 mm

(d) a triangle with base 1 ft 6 in. and height 6 in.

(e) a trapezoid with bases 8 m and 1.5 m and height 3.2 m

(f) a circle with radius $3\frac{1}{2}$ yd

(g) a circle with diameter 2.5 m

PROBLEM 10-12   Find the volume of the following:

(a) a rectangular prism with length 6 in., width 8 in., and height 4 in.

(b) a cube with edge 1.5 mm

(c) a cylinder with radius 7 cm and height 10 cm

(d) a cylinder with diameter 2 ft and height 7 ft

(e) a sphere with radius 1.5 m

(f) a sphere with diameter 7 yd

PROBLEM 10-13   Solve each geometry problem using a linear equation:

(a) The sum of three angles of a triangle is 180 degrees [180°]. One angle of the triangle is 4 times the second angle. The third angle of the triangle is 30° more than the second angle. Find the measures of all the angles.

(b) Two angles of a triangle are equal. The third angle is equal to the sum of the other two angles. How many degrees are in each angle?

(c) The perimeter of a rectangle is 110 feet. The width is 5 feet shorter than twice the length. Find the area of the rectangle.

(d) The area of a rectangle is 240 square meters [240 m²]. The width of the rectangle is 15 m. Find the perimeter of the rectangle.

Answers to Supplementary Exercises

(10-9) (a) 24 cm   (b) 24 in.   (c) 15 ft   (d) 31.2 m   (e) 12 yd   (f) 22 cm   (g) 12 cm   (h) 14.01 m

(10-10) (a) 22 in.   (b) 25.12 cm   (c) 88 yd   (d) 31.4 mm   (e) 44 in.
(f) 9.42 m   (g) 50.24 cm   (h) $8\frac{4}{5}$ ft   (i) 8.792 m   (j) $17\frac{3}{5}$ in.
(k) 17.584 mm   (l) $9\frac{3}{7}$ ft   (m) 20.41 cm   (n) 8.792 m   (o) 22 ft
(p) 66 yd   (q) 9.42 cm   (r) 88 in.   (s) 4.082 cm   (t) $14\frac{2}{3}$ ft
(u) 14.758 m   (v) 33 in.   (w) 32.97 cm   (x) $8\frac{8}{21}$ yd   (y) 17.6625 mm

(10-11) (a) 8 ft²   (b) 25 cm²   (c) 19.04 mm²   (d) 54 in² or $\frac{3}{8}$ ft²   (e) 15.2 m²
(f) $38\frac{1}{2}$ yd²   (g) 4.90625 m²

(10-12) (a) 192 in³   (b) 3.375 mm³   (c) 1538.6 cm³   (d) 22 ft³   (e) 14.13 m³   (f) $179\frac{2}{3}$ yd³

(10-13) (a) 100°, 25°, 55°   (b) 45°, 45°, 90°   (c) 700 ft²   (d) 62 m

FINAL EXAM

Chapters 6-10

Part 1: Skills and Concepts (60 questions)

1. The reciprocal of $\frac{3}{4}$ is

   (a) $\frac{3}{4}$ (b) $-\frac{3}{4}$ (c) $\frac{4}{3}$ (d) all of these   (e) none of these

2. Which of the following are rational numbers?

   (a) integers   (b) whole numbers   (c) natural numbers   (d) all of these   (e) none of these

3. Which of the following are not rational numbers?

   (a) $\frac{2}{3}$ (b) $\sqrt{4}$ (c) $-0.\overline{3}$ (d) all of these   (e) none of these

4. $\frac{3}{4}$ is equal to

   (a) $\frac{-3}{4}$ (b) $\frac{-3}{-4}$ (c) $-\frac{-3}{-4}$ (d) all of these   (e) none of these

5. $-\dfrac{3}{4}$ is equal to

   (a) $\frac{-3}{4}$ (b) $\frac{3}{-4}$ (c) $-\frac{-3}{-4}$ (d) all of these   (e) none of these

6. $-\dfrac{-12}{-18}$ simplified is

   (a) $-\frac{-2}{3}$ (b) $-\frac{2}{3}$ (c) $\frac{2}{3}$ (d) all of these   (e) none of these

7. The sum of $\frac{1}{2} + 0.75$ is

   (a) 1.25   (b) $1\frac{1}{4}$ (c) $\frac{5}{4}$ (d) all of these   (e) none of these

8. The difference of $2.5 - 3\frac{3}{4}$ is

   (a) 1.25   (b) 6.25   (c) -6.25 (d) all of these   (e) none of these

9. The product of $0.5(-\frac{1}{2})(-4)(0)(-\frac{3}{4})$ is

   (a) 0   (b) $-\frac{3}{4}$ (c) $\frac{3}{4}$ (d) all of these   (e) none of these

10. The quotient of $-\frac{1}{2} \div (-0.125)$ is

    (a) -4 (b) 4   (c) 0.0625   (d) all of these   (e) none of these

11. $\dfrac{-1 - \sqrt{1^2 - 4(6)(-12)}}{2(6)}$ evaluated is

    (a) $\frac{4}{3}$ (b) $-\frac{3}{2}$ (c) $-\frac{17}{12}$ (d) all of these   (e) none of these

12. "2 times x" can be written as

    (a) $2 \cdot x$ (b) $2(x)$ (c) $2x$ (d) all of these   (e) none of these

13. "2 divided by x" can be written as

(a) $x \div 2$ (b) $x \cdot \frac{1}{2}$ (c) $\frac{x}{2}$ (d) all of these   (e) none of these

14. $2(3 \cdot 4) = (3 \cdot 4)2$ is an example of

(a) a commutative property   (b) an associative property

(c) a distributive property   (d) all of these   (e) none of these

15. $2(3 \cdot 4) = (2 \cdot 3)4$ is an example of

(a) a commutative property   (b) an associative property

(c) a distributive property   (d) all of these   (e) none of these

16. $2(3 - 4) = 2 \cdot 3 - 2 \cdot 4$ is an example of

(a) a commutative property   (b) an associative property

(c) a distributive property   (d) all of these   (e) none of these

17. $2(3 - 4) = 2(3 + (-4))$ is an example of

(a) a commutative property   (b) an associative property

(c) a distributive property   (d) all of these   (e) none of these

18. $-2(x - 3)$ equals

(a) $-2x + 6$ (b) $6 - 2x$ (c) $6 + (-2x)$ (d) all of these   (e) none of these

19. $2x + 3y + 5 - 3x - 2y - 4$ equals

(a) $-x + y + 1$ (b) $-x + y - 1$ (c) $x + y + 1$ (d) all of these   (e) none of these

20. $4m - (5m - 10n - m)$ equals

(a) $-2m - 10n$ (b) $-10n$ (c) $10n$ (d) all of these   (e) none of these

21. For $x = -2; 3x^2 + 5x - 6$ equals

(a) 16   (b) -4 (c) -28 (d) -22 (e) none of these

22. If $a = 3, b = 5,$ and $c = -2,$ then $\dfrac{-b + \sqrt{b^2 - 4ac}}{2a}$ equals

(a) $\frac{1}{3}$ (b) 2   (c) 1   (d) $-\frac{2}{3}$ (e) none of these

23. Which of the following are linear equations in one variable?

(a) $3x - 5 = 2x + 4$ (b) $2x + 3 = 2x - 1$ (c) $2x + 3y = 5$ (d) all of these

(e) none of these

24. -3 is a solution of

(a) $2x = 6$ (b) $\dfrac{x}{-3} = -1$ (c) $2x + 6 = 12$ (d) all of these   (e) none of these

25. The solution of $x + 5 = -2$ is

(a) 3   (b) -7 (c) -3 (d) 7   (e) none of these

26. The solution of $y - 3 = 8$ is

(a) 11   (b) 5   (c) -5 (d) -11 (e) none of these

27. The solution of $2w = -8$ is

(a) 4   (b) -16 (c) -4 (d) -10 (e) none of these

28. The solution of $\dfrac{z}{-2} = -4$ is

(a) -8 (b) 2   (c) -2 (d) 8   (e) none of these

29. The solution of $2m + 3 = 1$ is

(a) -1 (b) 2   (c) -4 (d) 8   (e) none of these

30. The solution of $\dfrac{n}{5} - 3 = -8$ is

(a) -1 (b) -25 (c) -55 (d) $-\frac{11}{5}$ (e) none of these

31. The solution of $3x - 5 = 8x + 5$ is

(a) 0   (b) 2   (c) $\frac{11}{10}$ (d) -2 (e) none of these

32. The solution of $-2(x - 3) = -4$ is

(a) 5   (b) -1 (c) 1   (d) -5 (e) none of these

33. The solution of $2(x - 3) = 5x - (3 + x)$ is

(a) $-\frac{4}{3}$ (b) -3 (c) -1 (d) $-\frac{3}{4}$ (e) none of these

34. The solution of $\frac{1}{2}y + \frac{3}{4} = \frac{1}{8}$ is

(a) $\frac{5}{4}$ (b) $-\frac{5}{4}$ (c) $\frac{5}{2}$ (d) $-\frac{5}{2}$ (e) none of these

35. The solution of $0.2w - 0.3 = 0.1$ is

(a) 0.2   (b) -1 (c) 20   (d) 2   (e) none of these

36. The solution of $x + 33\frac{1}{3}\% x = 16$ is

(a) 12   (b) 0   (c) 1   (d) 24   (e) none of these

37. The solution of $2x + 3y = 5$ for y is

(a) $\dfrac{5 - 3y}{2}$ (b) $2x - 5$ (c) $\dfrac{5 - 2x}{3}$ (d) $\dfrac{2x - 5}{3}$ (e) none of these

38. The solution of $A = 2(l + w)$ for w is

(a) $\dfrac{A}{2l}$ (b) $\dfrac{A}{2} + l$ (c) $l - \dfrac{A}{2}$ (d) $\frac{1}{2}A - l$ (e) none of these

39. The ratio in lowest terms of 30 seconds to 2 minutes is

(a) $\dfrac{4}{1}$ (b) $\dfrac{30}{120}$ (c) $\dfrac{1}{4}$ (d) $\dfrac{120}{30}$ (e) none of these

40. The unit rate of 200 miles on 12 gallons is

(a) $\dfrac{50}{3}$ (b) $\dfrac{50}{3} \dfrac{\text{miles}}{\text{gallon}}$ (c) $16\frac{2}{3}$ (d) $16\frac{2}{3} \dfrac{\text{miles}}{\text{gallon}}$ (e) none of these

41. Which of the following fraction pairs are directly proportional?

(a) $\frac{1}{2}, \frac{50}{100}$　　　(b) $\frac{30}{3}, \frac{1}{90}$　　　(c) $\frac{12}{18}, \frac{20}{24}$　　　(d) all of these　　　(e) none of these

42. Which of the following fraction pairs are indirectly proportional?

(a) $\frac{1}{2}, \frac{50}{100}$　　　(b) $\frac{30}{3}, \frac{1}{10}$　　　(c) $\frac{12}{18}, \frac{20}{24}$　　　(d) all of these　　　(e) none of these

43. The solution of $\dfrac{n}{12} = \dfrac{5}{8}$ is

(a) $7\frac{1}{2}$　　　(b) $3\frac{1}{3}$　　　(c) 19.2　　　(d) all of these　　　(e) none of these

44. $\dfrac{n}{12} = \dfrac{5}{8}$ is equivalent to

(a) $\dfrac{12}{n} = \dfrac{8}{5}$　　　(b) $\dfrac{5}{8} = \dfrac{n}{12}$　　　(c) $\dfrac{8}{5} = \dfrac{12}{n}$　　　(d) all of these　　　(e) none of these

45. The height of a tall milk glass might be

(a) 6 in.　　　(b) 6 ft　　　(c) 6 yd　　　(d) 6 mi　　　(e) none of these

46. The amount of milk that a tall milk glass might contain is

(a) 1 c　　　(b) 1 pt　　　(c) 1 qt　　　(d) 1 gal　　　(e) none of these

47. The weight of a tall milk glass full of milk might be

(a) 1 oz　　　(b) 1 lb　　　(c) 1 T　　　(d) all of these　　　(e) none of these

48. 54 inches renamed as an equal amount of yards is

(a) $4\frac{1}{2}$ yd　　　(b) 18 yd　　　(c) 6 yd　　　(d) $1\frac{1}{2}$ yd　　　(e) none of these

49. The height of a tall milk glass might be

(a) 15 mm　　　(b) 15 cm　　　(c) 15 m　　　(d) 15 km　　　(e) none of these

50. The amount of milk that a tall milk glass might contain is

(a) 470 mL　　　(b) 470 L　　　(c) 470 kL　　　(d) all of these　　　(e) none of these

51. The mass of a tall milk glass full of milk might be

(a) 470 mg　　　(b) 470 g　　　(c) 470 kg　　　(d) 470 t　　　(e) none of these

52. 1.5 km renamed as an equal amount of centimeters is

(a) 150 cm　　　(b) 1500 cm　　　(c) 15,000 cm　　　(d) 150,000 cm　　　(e) none of these

53. The difference of 2 yd 1 ft − 16 in. is

(a) 20 in.　　　(b) 5 ft 6 in.　　　(c) 5 ft 8 in.　　　(d) 65 in.　　　(e) none of these

54. The quotient of 2 yd 1 ft ÷ 16 in. is

(a) $5\frac{1}{4}$　　　(b) $2\frac{1}{4}$　　　(c) $\frac{3}{16}$　　　(d) 3　　　(e) none of these

55. The perimeter of a rectangle with length 4 m and width 3 m is

(a) 12 m　　　(b) 14 m　　　(c) 14 m^2　　　(d) all of these　　　(e) none of these

56. The approximate circumference of a circle with diameter 14 in. is

  (a) 44 in.   (b) 88 in.   (c) 154 in.   (d) all of these   (e) none of these

57. The area of a triangle with base 9 in. and height 6 in. is

  (a) 15 in^2 (b) 27 in.   (c) 27 in^2 (d) all of these   (e) none of these

58. The approximate area of a circle with radius 2 m is

  (a) 6.28 m^2 (b) 12.56 m   (c) 12.56 m^2 (d) all of these   (e) none of these

59. The volume of a rectangle prism with length 3 ft, width 4 ft, and height 6 in. is

  (a) 72 ft^3 (b) 13 ft^3 (c) 6 ft^3 (d) all of these   (e) none of these

60. The volume of a cylinder with radius 7 in. and height 5 in. is

  (a) 154 in^3 (b) 770 in^3 (c) 110 in^3 (d) all of these   (e) none of these

Part 2: Problem Solving (12 questions)

61. One number is 4 more than 3 times another number. The difference between the two numbers is 18. What are the numbers?

62. The sum of three consecutive odd integers is 105. Find the integers.

63. A 120-mile trip took $7\frac{1}{2}$ gallons of gas. At that rate, how much gas will be needed for a 300-mile trip?

64. At the rate given in problem 63, how many miles can be traveled on a full 25-gallon gas tank?

65. At $4\frac{1}{2}$ mph, it takes 5 minutes to walk around a certain track. How long will it take to walk around the track at 6 mph?

66. At the rate given in problem 65, how fast will you need to run to get around the track in 2 minutes?

67. The length (l) of a certain man's foot is 12 inches when standing normally. Find the man's shoe-size number (n) by evaluating the formula $n = 3l - 25$.

68. The height (h) of a certain woman is 64 inches (5 ft 4 in.). Find the woman's *ideal weight* (w) by evaluating the formula $w = 5\frac{1}{4}h - 216$.

69. What distance (d) can a car travel in a time (t) of 15 minutes at a constant rate (r) of 50 mph using the distance formula $d = rt$?

70. How much time (t) will it take a car to travel a distance (d) of 200 miles at a constant rate (r) of 60 mph using the distance formula $d = rt$?

71. The area of a rectangle is 104 m^2. If the width of the rectangle is 8 m, what is the perimeter?

72. The perimeter of a rectangle is 582 ft. The length of the rectangle is twice the width. What is the area of the rectangle?

Final Exam Answers

Part 1

1. c	**2.** d	**3.** e	**4.** b	**5.** d
6. b	**7.** d	**8.** e	**9.** a	**10.** b
11. b	**12.** d	**13.** e	**14.** a	**15.** b
16. c	**17.** e	**18.** d	**19.** a	**20.** c

21. b	**22.** a	**23.** a	**24.** e	**25.** b
26. a	**27.** c	**28.** d	**29.** a	**30.** b
31. d	**32.** a	**33.** e	**34.** b	**35.** d
36. a	**37.** c	**38.** d	**39.** c	**40.** d
41. a	**42.** b	**43.** a	**44.** d	**45.** a
46. b	**47.** b	**48.** d	**49.** b	**50.** a
51. b	**52.** d	**53.** c	**54.** a	**55.** b
56. a	**57.** c	**58.** c	**59.** c	**60.** b

Part 2

61. 7 and 25

62. 33, 35, and 37

63. $18\frac{3}{4}$ gal

64. 400 mi

65. $3\frac{3}{4}$ min (or 3 min 45 s)

66. $11\frac{1}{4}$ mph

67. size 11

68. 120 lb

69. $12\frac{1}{2}$ mi

70. $3\frac{1}{3}$ hr (or 3 hr 20 min)

71. 42 m

72. 18,818 ft^2

APPENDIX TABLE 1:
Numbers and Properties

NUMBERS

Natural Numbers or *Counting Numbers:* 1, 2, 3, . . .

Whole Numbers: 0, 1, 2, 3, . . .

Integers: . . . , $-3, -2, -1, 0, 1, 2, 3, . . .$

Rational Numbers: All numbers that can be written as $\frac{a}{b}$ where a and b are integers and $b \neq 0$.

Irrational Numbers: All decimal numbers that neither terminate nor repeat.

Real Numbers: All rational and irrational numbers.

PROPERTIES

Commutative Properties: $a + b = b + a$ $\qquad\qquad ab = ba$

Associative Properties: $(a + b) + c = a + (b + c)$ $\qquad (ab)c = a(bc)$

Distributive Properties: $a(b + c) = ab + ac$ $\qquad a(b - c) = ab - ac$

Identity Properties: $a + 0 = 0 + a = a$ $\qquad a(1) = 1(a) = a$

Zero Property: $a(0) = 0(a) = 0$

Zero-Product Property: $ab = 0$ means $a = 0$ or $b = 0$

Subtraction Properties: $a - b = a + (-b)$ $\qquad a - 0 = a + 0 = a$

$\qquad\qquad 0 - a = 0 + (-a) = -a$ $\qquad a - b = 0$ means $a = b$

Division Properties: $\dfrac{a}{b} = \dfrac{1}{b} \cdot a = a \cdot \dfrac{1}{b}$ $\qquad 0 \div a = \dfrac{0}{a} = 0 \quad (a \neq 0)$

$\qquad\qquad a \div 0$ or $\dfrac{a}{0}$ is not defined $\qquad a \div 1 = \dfrac{a}{1} = a$

$\qquad\qquad \dfrac{a}{b} = 1$ means $a = b \quad (a \neq 0$ and $b \neq 0)$

Inverse Properties: $a + (-a) = -a + a = 0$ $\qquad a \cdot \dfrac{1}{a} = \dfrac{1}{a} \cdot a = 1 \ (a \neq 0)$

Negati e Properties: $-(-a) = a$ $\qquad\qquad -1(a) = a(-1) = -a$

$\qquad\qquad a(-b) = (-a)b = -(ab) = -ab$ $\qquad (-a)(-b) = ab$

$\qquad\qquad \dfrac{a}{b} = \dfrac{-a}{-b} = -\dfrac{-a}{b} = -\dfrac{a}{-b}$ $\qquad \dfrac{-a}{b} = \dfrac{a}{-b} = -\dfrac{a}{b} = -\dfrac{-a}{-b}$

APPENDIX TABLE 2:
Squares and Square Roots
[for whole numbers 1–100]

Number N	Square N^2	Square Root $\sqrt{N}$	Number N	Square N^2	Square Root $\sqrt{N}$	Number N	Square N^2	Square Root $\sqrt{N}$
0	0	0	35	1225	5.916	70	4900	8.367
1	1	1	36	1296	6	71	5041	8.426
2	4	1.414	37	1369	6.083	72	5184	8.485
3	9	1.732	38	1444	6.164	73	5329	8.544
4	16	2	39	1521	6.245	74	5476	8.602
5	25	2.236	40	1600	6.325	75	5625	8.660
6	36	2.449	41	1681	6.403	76	5776	8.718
7	49	2.646	42	1764	6.481	77	5929	8.775
8	64	2.828	43	1849	6.557	78	6084	8.832
9	81	3	44	1936	6.633	79	6241	8.888
10	100	3.162	45	2025	6.708	80	6400	8.944
11	121	3.317	46	2116	6.782	81	6561	9
12	144	3.464	47	2209	6.856	82	6724	9.055
13	169	3.606	48	2304	6.928	83	6889	9.110
14	196	3.742	49	2401	7	84	7056	9.165
15	225	3.873	50	2500	7.071	85	7225	9.220
16	256	4	51	2601	7.141	86	7396	9.274
17	289	4.123	52	2704	7.211	87	7569	9.327
18	324	4.243	53	2809	7.280	88	7744	9.381
19	361	4.359	54	2916	7.348	89	7921	9.434
20	400	4.472	55	3025	7.416	90	8100	9.487
21	441	4.583	56	3136	7.483	91	8281	9.539
22	484	4.690	57	3249	7.550	92	8464	9.592
23	529	4.796	58	3364	7.616	93	8649	9.644
24	576	4.899	59	3481	7.681	94	8836	9.695
25	625	5	60	3600	7.746	95	9025	9.747
26	676	5.099	61	3721	7.810	96	9216	9.798
27	729	5.196	62	3844	7.874	97	9409	9.849
28	784	5.292	63	3969	7.937	98	9604	9.899
29	841	5.385	64	4096	8	99	9801	9.950
30	900	5.477	65	4225	8.062	100	10,000	10
31	961	5.568	66	4356	8.124			
32	1024	5.657	67	4489	8.185			
33	1089	5.745	68	4624	8.246			
34	1156	5.831	69	4761	8.307			

APPENDIX TABLE 3:
Product of Primes
[for whole numbers 1–200]

N	Product of Primes	N	Product of Primes	N	Product of Primes	N	Product of Primes	N	Product of Primes
1		41	p	81	3^4	121	11^2	161	$7 \cdot 23$
2	p^*	42	$2 \cdot 3 \cdot 7$	82	$2 \cdot 41$	122	$2 \cdot 61$	162	$2 \cdot 3^4$
3	p	43	p	83	p	123	$3 \cdot 41$	163	p
4	2^2	44	$2^2 \cdot 11$	84	$2^2 \cdot 3 \cdot 7$	124	$2^2 \cdot 31$	164	$2^2 \cdot 41$
5	p	45	$3^2 \cdot 5$	85	$5 \cdot 17$	125	5^3	165	$3 \cdot 5 \cdot 11$
6	$2 \cdot 3$	46	$2 \cdot 23$	86	$2 \cdot 43$	126	$2 \cdot 3^2 \cdot 7$	166	$2 \cdot 83$
7	p	47	p	87	$3 \cdot 29$	127	p	167	p
8	2^3	48	$2^4 \cdot 3$	88	$2^3 \cdot 11$	128	2^7	168	$2^3 \cdot 3 \cdot 7$
9	3^2	49	7^2	89	p	129	$3 \cdot 43$	169	13^2
10	$2 \cdot 5$	50	$2 \cdot 5^2$	90	$2 \cdot 3^2 \cdot 5$	130	$2 \cdot 5 \cdot 13$	170	$2 \cdot 5 \cdot 17$
11	p	51	$3 \cdot 17$	91	$7 \cdot 13$	131	p	171	$3^2 \cdot 19$
12	$2^2 \cdot 3$	52	$2^2 \cdot 13$	92	$2^2 \cdot 23$	132	$2^2 \cdot 3 \cdot 11$	172	$2^2 \cdot 43$
13	p	53	p	93	$3 \cdot 31$	133	$7 \cdot 19$	173	p
14	$2 \cdot 7$	54	$2 \cdot 3^3$	94	$2 \cdot 47$	134	$2 \cdot 67$	174	$2 \cdot 3 \cdot 29$
15	$3 \cdot 5$	55	$5 \cdot 11$	95	$5 \cdot 19$	135	$3^3 \cdot 5$	175	$5^2 \cdot 7$
16	2^4	56	$2^3 \cdot 7$	96	$2^5 \cdot 3$	136	$2^3 \cdot 17$	176	$2^4 \cdot 11$
17	p	57	$3 \cdot 19$	97	p	137	p	177	$3 \cdot 59$
18	$2 \cdot 3^2$	58	$2 \cdot 29$	98	$2 \cdot 7^2$	138	$2 \cdot 3 \cdot 23$	178	$2 \cdot 89$
19	p	59	p	99	$3^2 \cdot 11$	139	p	179	p
20	$2^2 \cdot 5$	60	$2^2 \cdot 3 \cdot 5$	100	$2^2 \cdot 5^2$	140	$2^2 \cdot 5 \cdot 7$	180	$2^2 \cdot 3^2 \cdot 5$
21	$3 \cdot 7$	61	p	101	p	141	$3 \cdot 47$	181	p
22	$2 \cdot 11$	62	$2 \cdot 31$	102	$2 \cdot 3 \cdot 17$	142	$2 \cdot 71$	182	$2 \cdot 7 \cdot 13$
23	p	63	$3^2 \cdot 7$	103	p	143	$11 \cdot 13$	183	$3 \cdot 61$
24	$2^3 \cdot 3$	64	2^6	104	$2^3 \cdot 13$	144	$2^4 \cdot 3^2$	184	$2^3 \cdot 23$
25	5^2	65	$5 \cdot 13$	105	$3 \cdot 5 \cdot 7$	145	$5 \cdot 29$	185	$5 \cdot 37$
26	$2 \cdot 13$	66	$2 \cdot 3 \cdot 11$	106	$2 \cdot 53$	146	$2 \cdot 73$	186	$2 \cdot 3 \cdot 31$
27	3^3	67	p	107	p	147	$3 \cdot 7^2$	187	$11 \cdot 17$
28	$2^2 \cdot 7$	68	$2^2 \cdot 17$	108	$2^2 \cdot 3^3$	148	$2^2 \cdot 37$	188	$2^2 \cdot 47$
29	p	69	$3 \cdot 23$	109	p	149	p	189	$3^3 \cdot 7$
30	$2 \cdot 3 \cdot 5$	70	$2 \cdot 5 \cdot 7$	110	$2 \cdot 5 \cdot 11$	150	$2 \cdot 3 \cdot 5^2$	190	$2 \cdot 5 \cdot 19$
31	p	71	p	111	$3 \cdot 37$	151	p	191	p
32	2^5	72	$2^3 \cdot 3^2$	112	$2^4 \cdot 7$	152	$2^3 \cdot 19$	192	$2^6 \cdot 3$
33	$3 \cdot 11$	73	p	113	p	153	$3^2 \cdot 17$	193	p
34	$2 \cdot 17$	74	$2 \cdot 37$	114	$2 \cdot 3 \cdot 19$	154	$2 \cdot 7 \cdot 11$	194	$2 \cdot 97$
35	$5 \cdot 7$	75	$3 \cdot 5^2$	115	$5 \cdot 23$	155	$5 \cdot 31$	195	$3 \cdot 5 \cdot 13$
36	$2^2 \cdot 3^2$	76	$2^2 \cdot 19$	116	$2^2 \cdot 29$	156	$2^2 \cdot 3 \cdot 13$	196	$2^2 \cdot 7^2$
37	p	77	$7 \cdot 11$	117	$3^2 \cdot 13$	157	p	197	p
38	$2 \cdot 19$	78	$2 \cdot 3 \cdot 13$	118	$2 \cdot 59$	158	$2 \cdot 79$	198	$2 \cdot 3^2 \cdot 11$
39	$3 \cdot 13$	79	p	119	$7 \cdot 17$	159	$3 \cdot 53$	199	p
40	$2^3 \cdot 5$	80	$2^4 \cdot 5$	120	$2^3 \cdot 3 \cdot 5$	160	$2^5 \cdot 5$	200	$2^3 \cdot 5^2$

* p means prime.

APPENDIX TABLE 4:
Systems of Measure

U.S. Customary System of Measures	Metric System of Measures

Length

1 ft	= 12 in.	in.: inch(es)
1 yd	= 3 ft	ft: foot (feet)
1 mi	= 1760 yd	yd: yard(s)
1 mi	= 5280 ft	mi: mile(s)

Length

1 cm	= 10 mm	mm: millimeter(s)
1 m	= 100 cm	cm: centimeter(s)
1 km	= 1000 m	m: meter(s)
		km: kilometer(s)

Capacity

1 tsp	= 80 gtt	gtt: drop(s)
1 tbsp	= 3 tsp	tsp: teaspoon(s)
1 fl oz	= 2 tbsp	tbsp: tablespoon(s)
1 c	= 8 fl oz	fl oz: fluid ounce(s)
1 pt	= 2 c	c: cup(s)
1 qt	= 2 pt	pt: pint(s)
1 gal	= 4 qt	qt: quart(s)
		gal: gallon(s)

Capacity

1 L	= 1000 mL	mL: milliliter(s)
1 kL	= 1000 L	L: liter(s)
		kL: kiloliter(s)

Weight

1 lb	= 16 oz	oz: ounce(s)
1 T	= 2000 lb	lb: pound(s)
		T: ton(s) or short ton(s)

Mass

1 g	= 1000 mg	mg: milligram(s)
1 kg	= 1000 g	g: gram(s)
1 t	= 1000 kg	kg: kilogram(s)
		t: tonne(s) or metric ton(s)

Area

1 ft^2	$= 144 \text{ in}^2$	in^2: square inch(es)
1 yd^2	$= 9 \text{ ft}^2$	ft^2: square foot (feet)
1 A	$= 4840 \text{ yd}^2$	yd^2: square yard(s)
1 mi^2	$= 640 \text{ A}$	A: acre(s)
		mi^2: square mile(s)

Area

1 cm^2	$= 100 \text{ mm}^2$	mm^2: square millimeter(s)
1 m^2	$= 10{,}000 \text{ cm}^2$	cm^2: square centimeter(s)
1 ha	$= 10{,}000 \text{ m}^2$	m^2: square meter(s)
1 km^2	$= 100 \text{ ha}$	ha: hectare(s)
		km^2: square kilometer(s)

Volume

1 ft^3	$= 1728 \text{ in}^3$	in^3: cubic inch(es)
1 yd^2	$= 27 \text{ ft}^3$	ft^3: cubic foot (feet)
		yd^3: cubic yard(s)

Volume

1 cm^3	$= 1000 \text{ mm}^3$	mm^3: cubic millimeter(s)
1 m^3	$= 1{,}000{,}000 \text{ cm}^3$ (cc)	cm^3 (cc): cubic centimeter(s)
		m^3: cubic meter(s)

Temperature

Water boils at 212°F.   °F: degrees Fahrenheit
The normal human body temperature is
98.6°F. Water freezes at 32°F.

Temperature

Water boils at 100°C.   °C: degrees Celsius
The normal human body temperature is 37°C.
Water freezes at 0°C.

Time

1 min	= 60 sec	sec: second(s)	1 hr	= 60 min	min: minute(s)
1 da	= 24 hr	hr: hour(s)	1 wk	= 7 da	wk: week(s)
1 yr	= 12 mo	mo: month(s)	1 yr	≈ 365 da	da: day(s)
					yr: year(s)

APPENDIX TABLE 5:
Conversion Factors
[U.S./Metric]

U.S. Customary/Metric		Paper-and-Pencil Conversion Factors:	Calculator Conversion Factors:
From	To	Multiply By	Multiply By
Length			
inches (in.)	millimeters (mm)	25	**25.4**
inches	centimeters (cm)	2.5	**2.54**
feet (ft)	meters (m)	0.3	**0.3048**
yards (yd)	meters	0.9	**0.9144**
miles (mi)	kilometers (km)	1.6	1.609
Capacity			
drops (gtt)	milliliters (mL)	16	16.23
teaspoons (tsp)	milliliters	5	4.929
tablespoons (tbsp)	milliliters	15	14.79
fluid ounces (fl oz)	milliliters	30	29.57
cups (c)	liters (L)	0.24	0.2366
pints (pt)	liters	0.47	0.4732
quarts (qt)	liters	0.95	0.9464
gallons (gal)	liters	3.8	3.785
Weight (Mass)			
ounces (oz)	grams (g)	28	28.35
pounds (lb)	kilograms (kg)	0.45	0.4536
tons (T)	tonnes (t)	0.9	0.9072
Area			
square inches (in^2)	square centimeters (cm^2)	6.5	6.452
square feet (ft^2)	square meters (m^2)	0.09	0.09290
square yards (yd^2)	square meters	0.8	0.8361
square miles (mi^2)	square kilometers (km^2)	2.6	2.590
acres (A)	hectares (ha)	0.4	0.4047
Volume			
cubic inches (in^3)	cubic centimeters (cm^3 or cc)	16	16.39
cubic feet (ft^3)	cubic meters (m^3)	0.03	0.02832
cubic yards (yd^3)	cubic meters	0.8	0.7646
Temperature			
degrees Fahrenheit (°F)	degrees Celsius (°C)	$\frac{5}{9}$ (after subtracting 32)	**0.5556** (after subtracting 32)

Note: All conversion factors in bold type are exact. All others are rounded.

APPENDIX TABLE 5 [Continued]:
Conversion Factors
[Metric/U.S.]

| | Metric/U.S. Customary | | Paper-and-Pencil Conversion Factors: | Calculator Conversion Factors: |
	From	*To*	*Multiply By*	*Multiply By*
Length	millimeters (mm)	inches (in.)	0.04	0.03937
	centimeters (cm)	inches	0.4	**0.3937**
	meters (m)	feet (ft)	3.3	**3.280**
	meters	yards (yd)	1.1	**1.094**
	kilometers (km)	miles (mi)	0.6	0.6214
Capacity	milliliters (mL)	drops (gtt)	0.06	0.06161
	milliliters	teaspoons (tsp)	0.2	0.2029
	milliliters	tablespoons (tbsp)	0.07	0.06763
	milliliters	fluid ounces (fl oz)	0.03	0.03381
	liters (L)	cups (c)	4.2	4.227
	liters	pints (pt)	2.1	2.113
	liters	quarts (qt)	1.1	1.057
	liters	gallons (gal)	0.26	0.2642
Mass (Weight)	grams (g)	ounces (oz)	0.035	0.03527
	kilograms (kg)	pounds (lb)	2.2	2.205
	tonnes (t)	tons (T)	1.1	1.102
Area	square centimeters (cm^2)	square inches (in^2)	0.16	0.1550
	square meters (m^2)	square feet (ft^2)	11	10.76
	square meters	square yards (yd^2)	1.2	1.196
	square kilometers (km^2)	square miles (mi^2)	0.4	0.3861
	hectares (ha)	acres (A)	2.5	2.471
Volume	cubic centimeters (cm^3)	cubic inches (in^3)	0.06	0.06102
	cubic meters (m^3)	cubic feet (ft^3)	35	35.31
	cubic meters	cubic yards (yd^3)	1.3	1.308
Temperature	degrees Celsius (°C)	degrees Fahrenheit (°F)	$\frac{9}{5}$ (then add 32)	**1.8** (then add 32)

Note: All conversion factors in bold type are exact. All others are rounded.

APPENDIX TABLE 6:
Geometry Formulas

	Figure	Perimeter (P)	Area (A)
Square		$P = 4s$	$A = s^2$
Rectangle		$P = 2(l + w)$	$A = lw$
Parallelogram		$P = 2(a + b)$	$A = bh$
Triangle		$P = a + b + c$	$A = \frac{1}{2}bh$
Circle		**Circumference (C)**	**Area (A)**
		$C = \pi d = 2\pi r$	$A = \pi r^2$

	Figure	Volume (V)	Surface Area (SA)
Cube		$V = e^3$	$SA = 6e^2$
Rectangular Prism (box)		$V = lwh$	$SA = 2(lw + lh + wh)$
Cylinder		$V = \pi r^2 h$	$SA = 2\pi r(r + h)$
Sphere		$V = \frac{4}{3}\pi r^3$	$SA = 4\pi r^2$

INDEX

Page numbers in bold type indicate solved problems.